International Association of Fire Chiefs

Fire Officer
Principles and Practice
THIRD EDITION

Michael Ward, MGA, FIFireE
Senior Consultant
Fitch and Associates
Platte City, Missouri

JONES & BARTLETT
LEARNING

Jones & Bartlett Learning
World Headquarters
5 Wall Street
Burlington, MA 01803
978-443-5000
info@jblearning.com
www.jblearning.com

National Fire Protection Association
1 Batterymarch Park
Quincy, MA 02169-7471
www.NFPA.org

International Association of Fire Chiefs
4025 Fair Ridge Drive
Fairfax, VA 22033
www.IAFC.org

Jones & Bartlett Learning books and products are available through most bookstores and online booksellers. To contact Jones & Bartlett Learning directly, call 800-832-0034, fax 978-443-8000, or visit our website, www.jblearning.com.

Substantial discounts on bulk quantities of Jones & Bartlett Learning publications are available to corporations, professional associations, and other qualified organizations. For details and specific discount information, contact the special sales department at Jones & Bartlett Learning via the above contact information or send an email to specialsales@jblearning.com.

Copyright © 2015 by Jones & Bartlett Learning, LLC, an Ascend Learning Company and National Fire Protection Association®.

All rights reserved. No part of the material protected by this copyright may be reproduced or utilized in any form, electronic or mechanical, including photocopying, recording, or by any information storage and retrieval system, without written permission from the copyright owner.

The content, statements, views, and opinions herein are the sole expression of the respective authors and not that of Jones & Bartlett Learning, LLC. Reference herein to any specific commercial product, process, or service by trade name, trademark, manufacturer, or otherwise does not constitute or imply its endorsement or recommendation by Jones & Bartlett Learning, LLC and such reference shall not be used for advertising or product endorsement purposes. All trademarks displayed are the trademarks of the parties noted herein. *Fire Officer: Principles and Practice, Third Edition* is an independent publication and has not been authorized, sponsored, or otherwise approved by the owners of the trademarks or service marks referenced in this product.

There may be images in this book that feature models; these models do not necessarily endorse, represent, or participate in the activities represented in the images. Any screenshots in this product are for educational and instructive purposes only. Any individuals and scenarios featured in the case studies throughout this product may be real or fictitious, but are used for instructional purposes only.

The procedures and protocols in this book are based on the most current recommendations. The International Association of Fire Chiefs (IAFC), National Fire Protection Association (NFPA), and the publisher, however, make no guarantee as to, and assume no responsibility for, the correctness, sufficiency, or completeness of such information or recommendations. Other or additional safety measures may be required under particular circumstances.

Notice: The individuals described in "You are the Fire Officer" and "Fire Officer in Action" throughout the text are fictitious.

Production Credits
Chief Executive Officer: Ty Field
President: James Homer
Chief Product Officer: Eduardo Moura
Executive Publisher: Kimberly Brophy
Vice President of Sales, Public Safety Group: Matthew Maniscalco
Director of Sales, Public Safety Group: Patricia Einstein
Executive Acquisitions Editor: Bill Larkin
Associate Editor: Amanda Brandt
Production Editor: Cindie Bryan
Senior Marketing Manager: Brian Rooney
VP, Manufacturing and Inventory Control: Therese Connell
Composition: Aptara®, Inc.
Cover Design: Kristin E. Parker
Director of Photo Research and Permissions: Amy Wrynn
Photo Research and Permissions Associate: Ashley Dos Santos
Cover Image: © Glen E. Ellman
Printing and Binding: Courier Companies
Cover Printing: Courier Companies

Library of Congress Cataloging-in-Publication Data
Ward, Michael J., FIFireE.
 Fire officer: principles and practice/Michael J. Ward.—Third edition.
     pages cm
 "This book provides information to meet the standards of the National Fire Protection Association (NFPA) 1021, Standard for Fire Officer Professional Qualifications, at the Fire Officer I and Fire Officer II levels"—Introduction.
 Includes bibliographical references and index.
 ISBN 978-1-284-02667-2 (paperback)—ISBN 1-284-02667-1 (paperback)
 1. Fire departments—Management—Vocational guidance. 2. Fire extinction—Vocational guidance. 3. Fire chiefs—Certification.
 I. National Fire Protection Association. II. Title.
 TH9119.F568 2015
 363.37068–dc23
                            2013029392

6048

Printed in the United States of America
17 16 15 14 13   10 9 8 7 6 5 4 3 2 1

Brief Contents

- **CHAPTER 1** Introduction to the Fire Officer ... 2
- **CHAPTER 2** Preparing for Promotion ... 22
- **CHAPTER 3** Fire Fighters and the Fire Officer ... 40
- **CHAPTER 4** Fire Officer Communications ... 60
- **CHAPTER 5** Safety and Risk Management ... 84
- **CHAPTER 6** Understanding People: Management Concepts ... 110
- **CHAPTER 7** Leading the Fire Company ... 126
- **CHAPTER 8** Training and Coaching ... 140
- **CHAPTER 9** Evaluation and Discipline ... 162
- **CHAPTER 10** Organized Labor and the Fire Officer ... 182
- **CHAPTER 11** Working in the Community ... 202
- **CHAPTER 12** Handling Problems, Conflicts, and Mistakes ... 222
- **CHAPTER 13** Preincident Planning and Code Enforcement ... 240
- **CHAPTER 14** Budgeting ... 270
- **CHAPTER 15** Managing Incidents ... 290
- **CHAPTER 16** Rules of Engagement ... 316
- **CHAPTER 17** Fire Attack ... 338
- **CHAPTER 18** Fire Cause Determination ... 362
- **CHAPTER 19** Crew Resource Management and Leading Change ... 384
- Appendix A: An Extract from: NFPA 1021, *Standard for Fire Officer Professional Qualifications*, 2014 Edition ... 400
- Appendix B: ProBoard Assessment Methodology Matrices for NFPA 1021, 2014 Edition ... 404
- Appendix C: Principles of Fire and Emergency Service Administration (FESHE) Correlation Guide ... 410
- Glossary ... 411
- Index ... 419

Contents

CHAPTER 1   Introduction to the Fire Officer 2

Introduction 4
Requirements of the Fire Officer I 4
Roles and Responsibilities
    of the Fire Officer I 5
Fire Service in the United States 5
    History of the Fire Service 7
    Fire Equipment 8
    Communications 8
    Building Codes 9
    Paying for Fire Service 10
    Training and Education 10
Fire Department Organization 10
    Source of Authority 10
    Chain of Command 11
    Basic Principles of Organization 11
    Other Views of Organization 12
The Functions of Management 13
Rules and Regulations, Policies,
    and Standard Operating
    Procedures 14
    Ethics 14
Requirements for Fire Officer II 16
Additional Roles and
    Responsibilities for Fire Officer II ... 16
Working with Other Organizations 16
Challenges for the Captain 16
    Supervision and Motivation 16
    Increase in Nonfire Incidents 17
    Deterioration of the Built
        Environment 17
    Related Duties 17
    Cultural Diversity 17

CHAPTER 2   Preparing for Promotion 22

Introduction 24
The Origin of Promotional
    Examinations 24
Sizing up Promotion Opportunities ... 25
    Postexamination Promotional
        Considerations 25
When Fire Officers Are Voted In 26

Preparing a Promotional Examination ... 26
    Charting the Required Knowledge,
        Skills, and Abilities 26
Promotional Examination
    Components 29
    Multiple-Choice Written
        Examination 29
    Assessment Centers 30
    Emergency Incident Simulations ... 31
    Interpersonal Interaction 31
    Writing or Speaking Exercise 32
    Technical Skills Demonstration 32
Preparing for a Promotional
    Examination 33
    Building a Personal Study Journal ... 35
    Preparing for Role Playing 35

CHAPTER 3   Fire Fighters and the Fire Officer ... 40

Introduction 42
The Fire Officer's Tasks 42
    The Beginning of Shift Report 42
    Notifications 43
    Decision Making and Problem
        Solving 44
    Example of a Typical Fire Station
        Workday 44
    Example of a Typical Volunteer
        Duty Night 45
The Transition from Fire Fighter
    to Fire Officer 45
Fire Officer as Supervisor–
    Commander–Trainer 46
    Supervisor 46
    Commander 47
    Trainer 47
The Fire Officer's Supervisor 49
Integrity and Ethical Behavior 50
    Integrity 50
    Ethical Behavior 51
Workplace Diversity 51
    The Fire Officer's Role in
        Workplace Diversity 53
    The Fire Station as a Business
        Work Location 55

CHAPTER 4 Fire Officer Communications 60

- Introduction 62
- The Communication Process 62
  - The Communication Cycle 62
- Effective Communication Skill Basics 63
  - Active Listening 63
  - Stay Focused 65
  - Ensure Accuracy 65
  - Keep Your Supervisor Informed 65
  - The Grapevine 65
  - Overcoming Environmental Noise . . . 65
- Emergency Communications 66
  - "Unit Calling, Repeat . . ." 67
  - Initial On-Scene Radio Report 67
  - Using the Communications Order Model 67
  - Radio Reports 67
- Reporting 68
- Types of Reports .68
  - Verbal Reports 68
  - Written Reports 68
  - Using Information Technology 72
- Written Communications 72
  - Informal Communications 72
  - Formal Communications 72
- Writing a Report .75
- Presenting a Report 75
- Preparing a News Release 76
- Social Media 76
  - Establishing a Fire Department Social Media Policy 76
  - Helmet Cams and On-the-Job Digital Images 77
  - Engaging the Community Through Social Media 77

CHAPTER 5 Safety and Risk Management 84

- Introduction 86
- Fire Fighter Death and Injury Trends 86
  - Everyone Goes Home® 86
  - National Fire Fighter Near-Miss Reporting System 87
  - Reducing Deaths from Sudden Cardiac Arrest 87
  - Reducing Deaths from Motor Vehicle Collisions 88
  - Reducing Deaths from Fire Suppression Operations 89
- Incident Safety Officer 92
  - Incident Safety Officer and Incident Management 93
  - Qualifications to Operate as an Incident Safety Officer 93
  - Assistant Incident Safety Officers at Large or Complex Incidents 94
  - Incident Scene Rehabilitation 94
- Creating and Maintaining a Safe Work Environment 95
  - Safety Policies and Procedures 95
  - Emergency Incident Injury Prevention 95
  - Fire Station Safety 96
- Infection Control 97
  - Infectious Disease Exposure 97
- Accident Investigation 99
  - Accident Investigation and Documentation 99
- Postincident Analysis 99
- Analyzing Death and Injury Data 101
  - Sudden Cardiac Arrest 101
  - Struck by or Contact with an Object 102
  - Caught or Trapped 102
- Analyzing Near-Miss Reports 103
  - HFACS Level 1: Unsafe Acts 103
  - HFACS Level 2: Preconditions to Unsafe Acts 103
  - HFACS Level 3: Unsafe Supervision 103
  - HFACS Level 4: Organizational Influences 103
- Data Analysis 104
- Mitigating Hazards 105

CHAPTER 6 Understanding People: Management Concepts 110

Introduction 112
Managing People 112
Scientific Management 113
   Frederick Winslow Taylor 113
   Multiphase Fire Fighter Safety
     and Deployment Study 113
Humanistic Management............ 114
   McGregor: Theory X and
     Theory Y 114
   Maslow: Hierarchy of Need 115
   Blake and Mouton's Managerial
     Grid 117
Human Resources Management...... 118
Utilizing Human Resources 120
   Mission Statement 120
   Getting Assignments Completed ... 120
Human Resources for Managers...... 121
The Four Borders of Managing
   Human Resources 122

CHAPTER 7 Leading the Fire Company 126

Introduction 128
The Fire Officer as a Follower 128
Leadership Styles.................. 129
   Autocratic 129
   Democratic 129
   Laissez-Faire 129
Power............................ 129
Leadership in Routine Situations 130
Emergency Scene Leadership 130
   Methods of Assigning Tasks 132
   Critical Situations............... 132
   The Dispatch Center 133
   Other Responding Units 134
Leadership Challenges.............. 134
   Fire Station as Municipal
     Work Location Versus
     Fire Fighter Home............ 134
   Leadership in the Volunteer
     Fire Service 135

Motivation....................... 135
   Reinforcement Theory 135
   Motivation–Hygiene Theory....... 136
   Goal-Setting Theory............. 136
   Equity Theory 136
   Expectancy Theory 136

CHAPTER 8 Training and Coaching 140

Introduction 142
Overview of Training................ 142
Fire Officer Training
   Responsibilities................ 143
   Review of the Four-Step Method ... 143
   Ensure Proficiency of Existing
     Skill Sets 146
   Mentoring 146
   Provide New or Revised
     Skill Sets 146
   Ensure Competence and
     Confidence 147
When New Member Training
   Is On-the-Job 147
   Skills That Must Be Learned
     Immediately 147
   Skills Necessary for Staying
     Alive 148
Live Fire Training................... 148
   Student Prerequisites 148
   Fire Officer Preparation
     Responsibilities 150
   Prohibited Live Fire Training
     Activities 151
Developing a Specific Training
   Program 151
   Assess Needs 151
   Establish Objectives............ 151
   Develop the Training Program 152
   Deliver the Training............. 152
   Evaluate the Impact 152
Professional Development 152
   Training Versus Education 152
   Cultural Change Issues
     in Education 153
   Academic Accreditation 153

Fire Fighter Certification
   Programs 154
Accreditation of Certification
   Programs 154
**Building a Professional
   Development Plan 155**
Supervising Fire Officer
   Preparation 155
Managing Fire Officer 156
CPSE Designations 156

CHAPTER 9 Evaluation and Discipline 162

Introduction 164
Evaluation 164
Starting the Evaluation Process
   with a New Fire Fighter 165
Providing Feedback After
   an Incident or Activity 165
Discipline 166
Positive Discipline: Reinforcing
   Positive Performance 167
**Documentation and Record
   Keeping 171**
Formal Evaluation and Discipline 171
Annual Evaluations 171
Conducting the Annual
   Evaluation 171
Evaluation Errors 173
Negative Discipline: Correcting
   Unacceptable Behavior 174
**Employee Assistance
   Program 177**

CHAPTER 10 Organized Labor and the Fire Officer 182

Introduction 184
**The International Association
   of Fire Fighters 185**
**Establishing a Strong Supervisor/
   Employee Relationship 185**
**Positive Labor–Management
   Relations 186**
**The Fire Officer's Role as a
   Supervisor 186**
Grievance Procedure 186
**Legislative Framework for
   Collective Bargaining 190**
Norris-LaGuardia Act of 1932 191
Wagner-Connery Act of 1935 191
Taft-Hartley Labor Act of 1947 191
Landrum-Griffin Act of 1959 191
Collective Bargaining for Federal
   Employees 192
State Labor Laws 192
**Organizing Fire Fighters into
   Labor Unions 192**
Labor Actions in the Fire Service 193
Striking for Better Working
   Conditions: 1918–1921 194
Striking to Preserve Wages and
   Staffing: 1931–1933 194
Staffing, Wages, and Contracts:
   1973–1980 194
Negative Impacts of Strikes 194
The Fair Labor Standards Act
   and Fire-Based EMS 194
**The Growth of IAFF as a Political
   Influence 195**
Labor–Management Alliances 195
Fire Fighter Safety and
   Deployment Study 195
EMS Systems Performance
   Measurement 196
The IAFC/IAFF
   Labor–Management Initiative . . . 196
Fire Service Joint
   Labor–Management
   Wellness–Fitness Task Force 197

CHAPTER 11 Working in the Community 202

Introduction 204
Understanding the Community 204
Risk Reduction 206
Responding to Public
   Inquiries 206

Public Education..................207
   National and Regional Public
      Education Programs..........209
   Locally Developed Programs......212
Media Relations...................213
   Fire Department Public
      Information Officer............213
   Press Releases...................215
   The Fire Officer as
      Spokesperson..................215
Social Media Outreach.............217
   Social Media Challenges..........217

CHAPTER 12 Handling Problems, Conflicts, and Mistakes..................222

Introduction......................224
Complaints, Conflicts,
   and Mistakes...................224
General Decision-Making
   Procedures....................225
   Define the Problem..............225
   Generate Alternative
      Solutions.....................226
   Select a Solution.................227
   Implement the Solution..........227
   Evaluate the Results.............228
Managing Conflict................230
   Personnel Conflicts and
      Grievances...................230
   Conflict Resolution Model........230
   Investigate....................232
   Take Action...................232
   Follow Up....................232
   Emotions and Sensitivity.........232
Policy Recommendations and
   Implementation................234
   Recommending Policies and
      Policy Changes................234
   Implementing Policies...........234
Citizen Complaints................235
Customer Service versus Customer
   Satisfaction....................235
   Complainant Expectations........236

CHAPTER 13 Preincident Planning and Code Enforcement..........240

Introduction......................242
The Fire Officer's Role in
   Community Fire Safety..........242
Preincident Planning...............243
   A Systematic Approach..........244
   Putting the Data to Use..........247
Understanding Fire Codes..........247
   Building Code versus Fire Code....247
   State Fire Codes.................248
   Local Fire Codes................248
   Model Codes...................248
   Retroactive Code Requirements....248
Understanding Built-in Fire
   Protection Systems..............249
   Water-Based Fire Protection
      Systems.....................249
   Special Extinguishing Systems.....250
   Fire Alarm and Detection Systems..251
Understanding Fire Code
   Compliance Inspections.........252
   Fire Company Inspections........252
Classifying by Building or
   Occupancy....................252
   Construction Type..............252
   Occupancy and Use Group.......254
   NFPA 704 Marking System.......256
Preparing for an Inspection.........258
   Reviewing the Fire Code.........258
   Review Prior Inspection Reports,
      Fire History, and Preincident
      Plans........................258
   Coordinate Activity with the Fire
      Prevention Division............258
   Arrange a Visit.................258
   Assemble Tools and References....258
Conducting the Inspection..........259
   General Overview..............259
   Meet with the Representative.....259
   Inspecting from the Outside In,
      Bottom to Top................259
   Exit Interview..................260

Writing the Inspection/Correction Report.................. 260
General Inspection Requirements.... 260
   Access and Egress 260
   Exit Signs and Emergency Lighting.................... 261
   Portable Fire Extinguishers 261
   Built-in Fire Protection Systems.... 261
   Electrical 261
   Special Hazards 261
   Hazard Identification Signs 261
Selected Use Group–Specific Concerns......................261
   Public Assembly................ 261
   Business...................... 262
   Educational 262
   Factory Industrial 262
   Hazardous 263
   Health Care 263
   Mercantile 263
   Residential.................... 263
   Special Properties............... 263
   Detention..................... 263
   Storage....................... 263
   Mixed 264
Emergency Management and Business Continuity Plans 264
   Risk Assessment................ 264
   Incident Prevention 264
   Mitigation 264
   Resource Management and Logistics 264
   Publishing the Plan 265
   Training, Exercises, and Evaluation................... 265

CHAPTER 14 Budgeting 270

Introduction 272
The Budget Cycle 272
   Base and Supplemental Budgets.... 272
Revenue Sources 274
   Local Government Sources........ 274
   Volunteer Fire Departments....... 274
Lower Revenue Means Fewer Resources...................... 275
   Lower Revenue Options......... 275
The Purchasing Process............. 276
   Petty Cash 277
   Purchase Orders................ 277
   Requisitions................... 279
   The Bidding Process............. 279
Revenue Sources 280
   Grants 280
   Nontraditional Revenue Sources ... 280
Expenditures......................281
   Personnel Expenditures 281
   Operating Expenditures.......... 281
   Capital Expenditures 282
Bond Referendums and Capital Projects........................ 282
Navigating the Budgetary Process ... 283
   Developing a Budget Proposal 283
   Overview of Fire Ground Location Program 283
   FGL Annual Personnel and Operating Expenditures 283
   FGL Capital Budget 284
   Ask for Everything You Need...... 284
   Cost Recovery and Reduction 284

CHAPTER 15 Managing Incidents 290

Introduction 292
The Origin of Incident Management 292
   FIRESCOPE and Fire Ground Commander 292
   Developing One System.......... 293
National Incident Management System 293
Postincident Review................ 294
   Preparing Information for an Incident Review.............. 294
   Conducting a Critique 295
   Documentation and Follow-up 296
The Fire Officer's Role in Incident Management 296
   Levels of Command 297

Strategic-Level Incident
    Management 297
    Responsibilities of Command...... 297
    Establishing Command 298
    Command Options 298
    Functions of Command 299
    Transfer of Command 300
    Fire Fighter Accountability........ 300
After the Transfer of Command:
    Building the Incident
    Management System 302
    Command Staff 302
    General Staff Functions 302
    Location Designators 304
National Response Framework 304
Tactical-Level Incident
    Management 306
    Divisions, Groups, and Units...... 306
    Branches 307
Fire Officer Greater Alarm
    Responsibilities................. 307
    Staging...................... 308
Task-Level Incident Management 308
    Task Forces and Strike Teams...... 308
    Greater Alarm Infrastructure 309

CHAPTER 16 Rules of Engagement 316

Introduction 318
How the Rules Came to Be 318
Understanding the Scope of the
    Problem 319
    Survivability Profiling 319
    Fire Fighter Survivability inside
        Structure Fires................ 319
Rule 1. Size up Your Tactical Area
    of Operation................... 320
Rule 2. Determine the Occupant
    Survival Profile................. 320
    Today's Smoke Is More Toxic 321
Rule 3. Do Not Risk Your Life for
    Lives or Property That Cannot
    Be Saved 321
    Lives That Could Not Be Saved 321
Rule 4. Extend Limited Risk to
    Protect Savable Property 322
Rule 5. Extend Vigilance and
    Measured Risk to Protect and
    Rescue Savable Lives 323
    Deteriorating Conditions 323
Rule 6. Go in Together, Stay
    Together, Come out Together 324
Rule 7. Maintain Continuous
    Awareness of Your Air Supply,
    Situation, Location, and Fire
    Conditions 326
    Air Maintenance as a Situational
        Awareness Tool 327
Rule 8. Constantly Monitor
    Fire-Ground Communications
    for Critical Radio Reports 327
Rule 9. You Are Required to Report
    Unsafe Practices or Conditions
    That Can Harm You. Stop,
    Evaluate, and Decide............. 328
    Learning from the Aviation
        Industry 328
    Raise the Red Flag 329
Rule 10. You Are Required to
    Abandon Your Position and
    Retreat Before Deteriorating
    Conditions Can Harm You. 330
    Melted Helmets and Heat-Crazed
        Face Pieces 330
Rule 11. Declare a Mayday As Soon
    As You Think You Are in Danger.... 330
    Mayday versus Emergency
        Traffic 331

CHAPTER 17 Fire Attack 338

Introduction 340
Structure Fire Research 340
    New Fire Behavior Graph.......... 340
    Modern versus Legacy
        Single-Family Dwellings 340
    Flow Path..................... 342
Supervising a Single Company 342
    Closeness of Supervision 343

Situational Leadership 344
Standardized Actions 344
Command Staff Assignments 344
Sizing up the Incident 344
Prearrival Information 345
On-Scene Observations 345
Lloyd Layman's Five-Step Size-up
 Process 345
National Fire Academy Size-up
 Process 346
Risk–Benefit Analysis 347
Developing an Incident
 Action Plan 348
Incident Priorities 348
Tactical Priorities 349
Tactical Safety Considerations 349
Scene Safety 350
Rapid Intervention Crews 350
Personnel Accountability Report . . . 352
Supervising Multiple Companies 352
Determining Task Assignments 353
Assigning Resources 353
General Structure Fire
 Considerations 354
Single-Family Dwellings 354
Low-Rise Multiple-Family
 Dwellings 354
High-Rise Considerations 356

CHAPTER 18 Fire Cause Determination 362

Introduction 364
Common Causes of Fires 364
Requesting an Investigator 365
Fire Growth and Development 365
Disabled Built-in Fire
 Protection 366
Delayed Notification or Difficulty
 in Getting to the Fire 366
Tampered or Altered Equipment . . . 366
Legal Considerations 366
Searches 366
Securing the Scene 367
Evidence 367
Protecting Evidence 367

The Nature of Fire Investigation 368
Finding the Point of Origin 368
Fire Patterns 369
Determining the Cause of
 the Fire 369
Source and Form of Heat
 Ignition 370
Material First Ignited 370
Ignition Factor or Cause 370
Fire Analysis 370
Conducting Interviews 370
Vehicle Fire Cause
 Determination 372
Wildland Fire Cause
 Determination 373
Fire Cause Classifications 374
Accidental Fire Causes 374
Natural Fire Causes 374
Incendiary Fire Causes 375
Undetermined Fire Causes 375
Indicators of Incendiary Fires 375
Accelerants and Trailers 375
Multiple Points of Origin 377
Arson 377
Arson Motives 377
Documentation and Reports 378
Preliminary Investigation
 Documentation 378
Investigation Report 379
Legal Proceedings 379
After the Fire Officials Are Gone 380

CHAPTER 19 Crew Resource Management and Leading Change 384

Introduction 386
Origins of Crew Resource
 Management 386
Researching and Validating CRM
 Concepts 386
Human Error 387
Active Failures and Latent
 Conditions 387
Error Management Model 388

The CRM Model. 388	Appendix A: An Extract from: NFPA 1021, *Standard for Fire Officer Professional Qualifications*, 2014 Edition. 400
Communication Skills 388	
Teamwork 389	
Task Allocation 391	Appendix B: ProBoard Assessment Methodology Matrices for NFPA 1021, 2014 Edition. 404
Critical Decision Making 392	
Situational Awareness. 392	
Recommending Change. 395	Appendix C: Principles of Fire and Emergency Service Administration (FESHE) Correlation Guide 410
Implementing Change 395	
	Glossary 411
	Index. 419

Instructor, Student and Technology Resources

■ Instructor's ToolKit

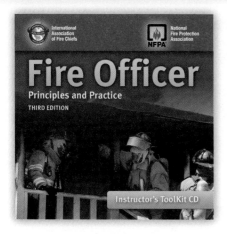

Preparing for class is easy with the resources on this CD. Instructors can choose resources for the Fire Officer I, Fire Officer II, or a combination of the two. The CD includes the following resources:
- PowerPoint® Presentations
- Lesson Plans
- Student Workbook Answers
- Image and Table Bank

■ Instructor's Test Bank

The Instructor's Test Bank contains over 600 multiple-choice questions and allows instructors to create tailor-made classroom tests and quizzes quickly and easily by selecting, editing, organizing, and printing a test along with an answer key, including page references to the text. Each test question includes the NFPA 1021 reference.

Student Workbook

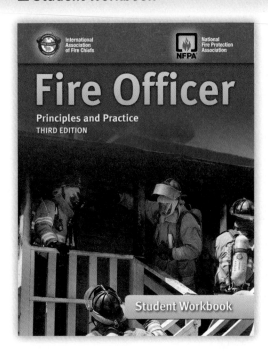

This resource is designed to encourage critical thinking and aid comprehension of the course materials. The student workbook uses the following activities to enhance mastery of the material.
- Case studies with corresponding questions
- In-basket exercises
- Matching, fill-in-the-blank, short-answer, and multiple-choice questions

Navigate TestPrep: Fire Officer

Navigate TestPrep: Fire Officer is a dynamic program designed to prepare students to sit for Fire Officer I and II certification examinations by including the same type of questions they will likely see on the actual examination.

It provides a series of self-study modules, organized by chapter and level, offering practice examinations and simulated certification examinations using multiple-choice questions. All questions are page referenced to *Fire Officer: Principles and Practice, Third Edition* for remediation to help students hone their knowledge of the subject matter.

Students can begin the task of studying for Fire Officer I and II certification examinations by concentrating on those subject areas where they need the most help. Upon completion, students will feel confident and prepared to complete the final step in the certification process—passing the examination.

Digital Curriculum Solution Packages

Digital Curriculum Solution Packages allow educators to offer their students cutting-edge digital resources based on high-quality technical content. The innovative, multifaceted online tools facilitate absolute understanding.

Gold-standard content joins sound instructional design in a user-friendly online interface to give students a truly interactive, engaging learning experience through the following components:

- Navigate Course Manager. This easy-to-use and fully hosted online learning platform seamlessly combines authoritative textbook content with interactive tools, assessments, and robust reporting tools that enable instructors to track real-time student progress and engagement.
- Navigate eFolio. eFolio is an exciting new alternative for instructors and students looking for more interactive learning opportunities than a textbook provides. This enhanced eBook combines authoritative content with interactive learning activities.
- Navigate TestPrep. This dynamic tool is designed to prepare students to sit for local, regional, or state examinations and simulated certification examinations.
- Interactive Lectures. Take the lectures out of the classroom! These interactive lectures provide anytime, anywhere access for students to an innovative educational environment.

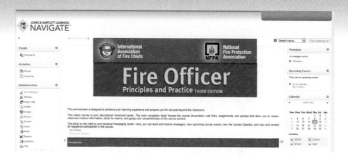

Available packages include:

Advantage Package, Print Edition
- Printed Textbook
- Printed Student Workbook
- Navigate Course Manager

Advantage Package, Digital Edition
- Navigate eFolio
- Navigate Course Manager

Preferred Package
- Printed Textbook
- Printed Student Workbook
- Navigate eFolio
- Navigate Course Manager
- Navigate TestPrep

Premier Package
- Printed Textbook
- Printed Student Workbook
- Navigate eFolio
- Navigate Course Manager
- Navigate TestPrep
- Interactive Lectures

Acknowledgments

Jones & Bartlett Learning, the National Fire Protection Association, and the International Association of Fire Chiefs would like to thank the current editors, contributors, and reviewers of *Fire Officer: Principles and Practice, Third Edition*.

Editorial Board

Shawn Kelley
International Association of Fire Chiefs
Fairfax, Virginia

Steven F. Sawyer
Senior Fire Service Specialist
National Fire Protection Association
Quincy, Massachusetts

Author

Michael Ward, MGA, FIFireE
Senior Consultant
Fitch and Associates
Platte City, Missouri

Contributors

Dave Belcher
Lieutenant, Violet Township Fire Department
President, Ohio Society of Fire Service Instructors
Pickerington, Ohio

Jason W. Burrow
Firefighter Medic III—Training Officer
Hanover Fire EMS
Hanover, Virginia

Aaron R. Byington, MA, NRP
Captain, EMS Training
Layton City Fire Department
Layton City, Utah

Sean DeCrane
Battalion Chief/Director of Training
Cleveland Division of Fire
Cleveland, Ohio

Billy E. Floyd, Jr.
Division Chief of Fire Training
North Myrtle Beach Department of Public Safety
North Myrtle Beach, South Carolina

Gregg Gerner
Fire Chief
GM Wentzville Assembly Center
Wentzville, Missouri

Doug Goodings
Manager of Academic Standards and Evaluations
Office of the Fire Marshal/Ontario Fire College
Gravenhurst, Ontario, Canada

Jim Grady III
Chief, Frankfort Fire Protection District
Illinois Fire Chiefs Association
Illinois Fire Service Institute
Frankfort, Illinois

Bill Guindon
Director
Maine Fire Service Institute
Brunswick, Maine

Robert L. Havens
Captain/Deputy EMC
Port Arthur Fire Department
Port Arthur, Texas

Jennifer Henson, MS
Director of Fire Protection Technology and Firefighter Certification
Catawba Valley Community College
Hickory, North Carolina

Joseph Knitter, EFO
Fire Chief
South Milwaukee Fire Department
Milwaukee Area Technical College
South Milwaukee, Wisconsin

Jim Lovell
Captain, Fire Training Division
Santa Fe County Fire Department
Santa Fe, New Mexico

Stephen K. Lovette
Captain
High Point Fire Department, Training Division
High Point, North Carolina

Stephen E. McClure
Operations Chief, Charleston, West Virginia Fire Department (ret.)
Director, Jackson County EMS
Ripley, West Virginia

Ed Mezulis
Battalion Chief, Sedona Fire District
Adjunct Faculty, Yavapai College
Clarkdale, Arizona

Gary P. Morris
Fire Chief (ret.)
Seattle, Washington

Christopher M. Riley
Captain
Portsmouth Fire, Rescue, and Emergency Services
Portsmouth, Virginia

Reviewers

Michael J. Barakey, MPA, EFO, CFO, MIFireE
District Chief
Virginia Beach Fire Department
Virginia Beach, Virginia

Brent G. Batla, MS, EFO, CFO, MIFireE
Battalion Chief
City of Burleson
Burleson, Texas

Dave Belcher, OFE
Lieutenant, Violet Township Fire Department
President, Ohio Society of Fire Service Instructors
Pickerington, Ohio

Joseph C. Bennett
West Virginia Department of Education RESA 7
Clarksburg, West Virginia

Matthew Brixey
Captain
Toledo Fire and Rescue Department
Toledo, Ohio

Jason W. Burrow
Firefighter Medic III—Training Officer
Hanover Fire EMS
Hanover, Virginia

Aaron R. Byington, MA, NRP
Captain, EMS Training
Layton City Fire Department
Layton City, Utah

Douglas K. Cline
Assistant Chief
Horry County Fire Rescue
Conway, South Carolina

Derrick S. Clouston
Fire and Rescue Training, Field Service Supervisor
North Carolina Office of State Fire Marshal
Kaplan University
Raleigh, North Carolina

Frank Dalton
Course Coordinator, Fire Officer Training Program
Ohio Fire Chiefs Association
Columbus, Ohio

Sean DeCrane
Battalion Chief/Director of Training
Cleveland Division of Fire
Cleveland, Ohio

Billy E. Floyd, Jr.
Division Chief of Fire Training
North Myrtle Beach Department of Public Safety
North Myrtle Beach, South Carolina

Joseph D. Fowler
Instructor
Moultrie Technical College, Fire Science Division
Tifton, Georgia

Dave Goldsmith
Fire Chief
Highview Fire Protection District
Louisville, Kentucky

Doug Goodings
Manager of Academic Standards and Evaluations
Office of the Fire Marshal/Ontario Fire College
Gravenhurst, Ontario, Canada

Bryan Goustos
Captain, Training Division
Brampton Fire and Emergency Services
Brampton, Ontario, Canada

J. Robert Griffin
Instructor
Asheville Buncombe Technical College
Asheville, North Carolina

Bill Guindon
Director
Maine Fire Service Institute
Brunswick, Maine

Ian Harmon AS, NREMT-P
Firefighter/Paramedic
Lexington, South Carolina

Jeffry J. Harran
Battalion Chief
Lake Havasu City Fire Department
Lake Havasu City, Arizona

Robert L. Havens
Captain/Deputy EMC
Port Arthur Fire Department
Port Arthur, Texas

Jennifer Henson, MS
Director of Fire Protection Technology and Firefighter Certification
Catawba Valley Community College
Hickory, North Carolina

Edward S. Janke, MS, EFO, FBINA
Director of Public Safety, Village of Howard
Adjunct Instructor, Northeast Wisconsin Technical College
Green Bay, Wisconsin

Paul Johnson
Lieutenant
Downers Grove Fire Department
Downers Grove, Illinois

William Justiz
Coordinator, Emergency Medical Technology and Emergency Management Programs
Triton College
River Grove, Illinois

Bill Kinsey
Division Chief
Bullhead City Fire Department
Bullhead City, Arizona

Joseph Knitter, EFO
Fire Chief
South Milwaukee Fire Department
Milwaukee Area Technical College
South Milwaukee, Wisconsin

Kenneth C. Krebbs, Jr.
Director, Fire Science and Emergency Management Applications
Yavapai College
Prescott Valley, Arizona

Curtis Kyer
Fire/EMS Assistant Coordinator, Belmont College
Lieutenant, Cumberland Trail Fire District #4
St. Clairsville, Ohio

Joshua Livermore
Captain
Bullhead City Fire Department
Bullhead City, Arizona

Jim Lovell
Captain, Fire Training Division
Santa Fe County Fire Department
Santa Fe, New Mexico

Stephen K. Lovette
Captain
High Point Fire Department, Training Division
High Point, North Carolina

Daniel Manning, MS
Professor, Anna Maria College, Colorado State University–Global Campus, Pima Community College
Interim Chief of Training/Fire Captain, Fort Huachuca Fire Department
Fort Huachuca, Arizona

Stephen E. McClure
Operations Chief, Charleston, West Virginia Fire Department (ret.)
Director, Jackson County EMS
Ripley, West Virginia

Dean Meenach, RN, BSN, CEN, CCRN, CPEN, EMT-P
Director of EMS Education
Mineral Area College
Park Hills, Missouri

Ed Mezulis
Battalion Chief, Sedona Fire District
Adjunct Faculty, Yavapai College
Clarkdale, Arizona

Irvin "Butch" Miller, MPA, SAPM
Division Chief (ret.), Reno Fire Department
Program Director/Professor, Texas A&M University–San Antonio
San Antonio, Texas

James P. Moore
Fire Rescue Chief, Crystal Lake Fire Rescue
Crystal Lake, Illinois
Field Staff Instructor, University of Illinois Fire Service Institute
Champaign, Illinois

Gary Patrick
Fire Chief
Silverhill Volunteer Fire Department
Silverhill, Alabama

Gary D. Peaton
Fire Academy Instructor
South Carolina Fire Academy
Columbia, South Carolina

James D. Reynolds
Emergency Response Services Instructor
Capterton Center for Applied Technology/RESA V
Parkersburg, West Virginia

Christopher M. Riley
Captain
Portsmouth Fire, Rescue, and Emergency Services
Portsmouth, Virginia

Gerald F. Robinson, AAS, BS, CFO
Deputy Superintendent
Ohio Fire Academy
Reynoldsburg, Ohio

Barbara R. Russo
Program Director/Assistant Professor
Fayetteville State University
Fayetteville, North Carolina

Patrick San Julian
Lieutenant
Clarksburg Fire Department
Clarksburg, West Virginia

Stephen Scionti
Training Captain
Tucson Fire Department
Tucson, Arizona

John R. Smoot
Fire Chief
Teays Valley Fire Department
Scott Depot, West Virginia

Adon W. Snyder, Jr.
Fire Chief
Clinton Fire Department
Clinton, North Carolina

Rich Solomon
Captain/EMS Coordinator
Sable Altura Fire Rescue
Arapahoe County, Colorado

Norman Staines
Director, Emergency Services
Caldwell Community College & Technical Institute
Hudson, North Carolina

Jim Stedman
Battalion Chief (ret.), Chicago Fire Department
Chicago, Illinois
Program Director, Fire Science Administration, Penn Foster College
Scranton, Pennsylvania

Gavin Summers
Captain
Burnaby Fire Department
Burnaby, British Columbia, Canada

Robert Swiger
Assistant Chief
Clinton Fire Department
Clinton, North Carolina

Mark van der Feyst
Woodstock Fire Department
Woodstock, Ontario, Canada

Peter Van Dorpe
District Chief, Training Division, Chicago Fire Department (ret.)
Assistant Chief, Algonquin–Lake in the Hills Fire Protection District
Lake in the Hills, Illinois

Chris Watson
Battalion Chief, Austin Fire Department
Austin Community College
Austin, Texas

■ Photographic Contributors

We would like to extend a huge "thank you" to Glen E. Ellman, the photographer for this project. Glen is a commercial photographer and fire fighter based in Fort Worth, Texas. His expertise and professionalism are unmatched.

Thank you to the following organizations that opened up their facilities for this photo shoot:

Fairfax County Fire and Rescue
Fairfax, Virginia
Richard Bowers, Fire Chief
Manuel Barrero, Deputy Chief
Donald Brasfield, Captain II
Rebecca Kelly, Captain II
William S. Moreland, Captain I
John Macinyak, Lieutenant
Gary W. Vozzola, Lieutenant
Gregory B. Barnett, Master Technician
Joel Kozersteen, Master Technician
Rodney Snapp, Master Technician
Scott Zugner, Master Technician
Alfred Doughty, Technician
Ivam Holmes, Jr., Technician
Sarah Joliat, Technician
Jason Menifee, Technician
Kevin Ngo, Technician
Christian Waelder, Technician
Namasté Bosse, Fire Fighter
Brian McNew, Fire Fighter
Joseph C. Morris, Fire Fighter

Fort Worth Fire Department
Fort Worth, Texas

Portland Fire Department
Portland, Maine

Introduction to the Fire Officer

Fire Officer I

Knowledge Objectives

After studying this chapter, you will be able to:
- Identify the requirements of a Fire Officer I (NFPA 4.1) (NFPA 4.1.1). (pp 4–5)
- Describe the roles and responsibilities of the Fire Officer I (NFPA 4.1.1). (p 5)
- Describe the fire service in the United States (NFPA 4.1.1). (pp 5–10)
- Describe fire department organization (NFPA 4.1.1). (pp 10–13)
- Discuss the functions of management (NFPA 4.1.1). (pp 13–14)
- Discuss the roles of rules and regulations, policies, and standard operating procedures (NFPA 4.1.1). (pp 14, 16)

Skills Objectives

There are no Fire Officer I skills objectives for this chapter.

Fire Officer II

Knowledge Objectives

After studying this chapter, you will be able to:
- Identify the requirements for Fire Officer II (NFPA 5.1) (NFPA 5.1.1). (p 16)
- Describe the roles and responsibilities of the Fire Officer II (NFPA 5.1.1). (p 16)
- Describe working with other organizations (NFPA 5.1.2). (p 16)
- Describe additional challenges for the fire captain (NFPA 5.1.1) (NFPA 5.1.2). (pp 16–17)

Skills Objectives

There are no Fire Officer II skills objectives for this chapter.

CHAPTER 1

Additional NFPA Standards

- NFPA 1001, *Standard for Fire Fighter Professional Qualifications*
- NFPA 1041, *Standard for Fire Service Instructor Professional Qualifications*

Principles of Fire and Emergency Service Administration (FESHE) Course Outcomes

1. Acknowledge career development opportunities and strategies for success. (p 10)
3. Identify and explain the concepts of span and control, effective delegation, and division of labor. (pp 11–13)
5. Explain the history of management and supervision methods and procedures. (pp 7–10)
6. Discuss the various levels of leadership, roles, and responsibilities within the organization. (pp 11–14, 16)
8. Identify the importance of ethics as it relates to fire and emergency services. (pp 14, 16)

You Are the Fire Officer

It is 45 minutes before "A" shift will start its day at Municipal City Fire and Rescue Department Fire Station 100. The newly promoted A shift fire officers are already in the fire station kitchen planning their first day. Captain Jean Davis has been with the department for 15 years, spending 8 years as a truck company lieutenant and a training officer. Lieutenant Taylor Williams has been with the department for 7 years; most of that time was spent on engine companies.

Fire Station 100 runs an engine company, a paramedic ambulance, and the only quint in the department. Captain Davis will command Engine 100 and Lieutenant Williams will command Quint 100. Battalion Chief Frank Johnson runs the battalion and is coming by at 10 A.M. to meet with the new officers.

1. What are the responsibilities of a supervising fire officer?
2. How does a new supervising officer work within the department's organizational structure?
3. How can you convert the theories and concepts memorized for a promotional exam into effective unit officer activities?

Introduction

This text provides information to meet the criteria outlined in National Fire Protection Association (NFPA) 1021, *Standard for Fire Officer Professional Qualifications*, at the Fire Officer I and Fire Officer II levels. The professional qualification standards for fire officers are documented in NFPA 1021.

The NFPA 1021 standard defines four levels of fire officer. The technical committee that developed the current edition of NFPA 1021 performed a task analysis to validate the existence of four distinct fire officer levels and the specific requirements that should apply at each level.

The Fire Officer I level is the first step in a progressive sequence and is generally associated with an officer supervising a single fire company or apparatus. A Fire Officer I could also be assigned to supervise a small administrative or technical group. The next step, Fire Officer II, generally refers to the senior non–chief officer level in a larger fire department. An officer at this level could be the overall supervisor of a multiple-unit fire station. A Fire Officer II could also be in charge of a larger group performing a specialized service or a significant administrative section within the fire department.

Fire Officer III and IV generally refer to chief officer positions. An individual who is qualified at the Fire Officer III level might work as a battalion or district chief in a large department and possibly as a deputy or assistant chief in a smaller organization. Personnel at the Fire Officer IV level tend to be fire chiefs or hold senior positions in charge of a major component of the fire department.

An officer is responsible for being a leader and supervisor to a crew of fire fighters, managing a budget for the station, understanding the response district, knowing departmental operational procedures, and being able to manage an incident. The officer must also understand fire prevention methods, fire and building codes and applicable ordinances (laws enacted by a local government or municipality), and the department's records management system. At each higher level, there are increasing requirements for knowledge and management skills.

The foundation of company officer practice came from World War II combat experience. These elements were codified when the NFPA established the Professional Qualification standards, adopting NFPA 1021 for fire officers in 1976. The International Association of Fire Chiefs (IAFC) expanded company officer development in 2003, providing an *Officer Development Handbook* to encourage company officers to acquire the appropriate levels of training, experience, self-development, and education throughout their professional journey to prepare for the CPSE (Center for Public Safety Excellence) Chief Fire Officer (CFO) designation, the pinnacle of professional development.

Fire Officer I

Requirements of the Fire Officer I

The Fire Officer I classification is bestowed upon an individual who supervises a single fire suppression unit or a small administrative group within a fire department. At this level, emphasis is placed on accomplishing the department's goals and objectives by working through subordinates to achieve desired results. The Fire Officer I must be able to prioritize multiple demands on the time of the company or work group members and to delegate tasks to subordinates. These demands may be related to emergency operations, nonemergency tasks, or administrative functions.

The Fire Officer I performs administrative duties and supervisory functions that are related to a small group of fire department members. Typical administrative duties include record keeping, managing projects, preparing budget requests, initiating and completing station maintenance requisitions, and conducting preliminary accident investigations. Supervisory duties include making work assignments and ensuring that health and safety procedures are followed. Nonemergency duties could include developing preincident plans, providing company-level training, delivering public education programs, and responding to community inquiries.

Emergency duties include supervising a group of fire fighters who are performing company-level tasks, functioning as the initial arriving officer at an emergency scene, performing a size-up, establishing the Incident Management System, developing and implementing an incident action plan, deploying resources, and maintaining personnel accountability. Once the emergency incident has been mitigated, the Fire Officer I is expected to conduct a preliminary investigation to determine the origin and cause, secure the scene to preserve evidence, and conduct a postincident analysis. Fire Officer I candidates are also required to meet all of the requirements of Fire Fighter II as defined in NFPA 1001, *Standard for Fire Fighter Professional Qualifications*, and of Fire Instructor I as defined in NFPA 1041, *Standard for Fire Service Instructor Professional Qualifications*.

The IAFC uses another term to distinguish the different company officers. The IAFC calls the Fire Officer I level a Supervising Fire Officer within its *Officer Development Handbook*. In this text, the Fire Officer I will be the lieutenant.

Roles and Responsibilities of the Fire Officer I

The roles and responsibilities of a fire officer differ from those of a fire fighter. Understanding the new roles is essential for the new fire officer to succeed. A Fire Officer I has the following roles and responsibilities:

- Supervises and directs the activities of a single unit
- Instructs members of the company regarding operating procedures, including duty assignments and giving special instructions when fighting fires
- Responds to alarms for fires, vehicle extrications, hazardous materials incidents, emergency medical incidents, and other emergencies as required
- Assumes command of emergency scenes, per the Incident Command System; analyzes situations; and determines proper procedures until being relieved by a higher-ranking officer
- Administers emergency medical first aid and cardiopulmonary resuscitation (CPR), and attends to victims until primary medical personnel arrive
- Oversees routine and preventive maintenance and makes periodic inspections of the assigned apparatus
- Receives direction and instruction from the fire captain and battalion chief regarding station operations, grounds and building maintenance, and overall fire scene action
- Provides training to crew members regarding the apparatus operations, including leading practical training exercises; participates in departmental in-service training and drills
- Evaluates employee performance and conducts performance reviews
- Reads, studies, interprets, and applies departmental procedures, technical manuals, building plans, and so on
- Completes and maintains manual or computer records and prepares necessary reports on incidents, accidents, and personnel training
- Performs preincident planning activities, including touring and studying businesses for physical layout, possible hazards, location of water sources, exposure problems, potential life loss, and other factors
- Determines a preliminary origin and cause of a fire
- Participates, prepares, and delivers various public education programs regarding fire prevention and safety and conducts tours of the fire station as required
- Assists in fire safety inspections of public and private buildings or property
- Participates in and oversees the periodic inspection and testing of equipment, such as hoses, ladders, and engines
- Works directly in firefighting activities; utilizes tools, equipment, portable extinguishers, hoses, ladders, and other items as necessary
- Takes appropriate action on the maintenance needs of equipment, buildings, and grounds
- Supervises and performs maintenance and cleaning work on fire equipment, buildings, and grounds

The transition to fire officer is a big step in a fire fighter's career. It involves not only increased responsibility, but also a different role from that of a fire fighter. The officer is a part of management and is responsible for the conduct of others. The officer has to apply policies, procedures, and rules to subordinates and to different situations. Doing so successfully means being consistent and fair and not playing favorites. These changes often involve difficult adjustments for the new officer. As an officer, you will be required to take actions that might not make you happy or popular, but they are your responsibility. It is like the role of a parent—often difficult, but ultimately rewarding.

Fire Service in the United States

The U.S. fire service originated as communities of citizens who responded when a fire broke out. Fighting fires was considered a civic duty in early America, and no compensation was provided for such activities. Citizens volunteered their time to answer the call of public service, and each fire fighter had a regular occupation that provided a living. Over time, however, the fire service evolved into many different methods of providing personnel when the alarm sounds.

Many fully volunteer departments are composed of members who are notified when an alarm occurs. These men and women drop whatever they are doing to respond to the

Near-Miss REPORT

Report Number: 12-0000270

Event Description: While fighting a fire on the second floor in a single-family residence, one fire fighter had his leg fall through the floor and became temporarily trapped. His captain immediately issued a mayday. Command acknowledged the mayday, deployed the rapid intervention team (RIT), and requested additional resources and notifications. Interior crews were able to remove the trapped fire fighter by utilizing hand tools already inside the structure. The trapped fire fighter and captain went to rehab at the on-scene ambulance and the fire fighter was evaluated. No injuries occurred to the fire fighter, the captain, or anyone else operating interior. Personnel accountability reports were completed per guideline for normal firefighting operations and mayday situations. Two crews (one engine and one truck company) responded from another fire and had not completed shift change. Accountability was critical.

The company involved with the mayday was the first company on the scene, with the second company arriving seconds behind the first. The first company issued a brief incident report, established command, and made an interior attack. They reported heavy smoke from the "C" side of the structure. The second engine company "bypassed" the RIT role due to the time of day and the vehicles in the driveway. This decision was communicated to the responding battalion chief (BC) and communications. At this time, the first-due BC was still assigned to the previous fire and the BC who responded to the second incident was from the other side of the city. The BC arrived on scene, assumed command, and assigned the engine companies to the interior division. Reports were given by the interior crews. The truck company (two personnel) arrived on scene at approximately the same time as the BC, and its members were assigned to secure utilities. The third engine arrived on scene and was assigned as the safety officer and RIT.

All department engine companies were paramedic units and minimum staffing was four. Normal response to a structure fire is three engines, one truck, one battalion chief, and one ambulance (provided by two contract companies within the city for transport). Normally, the second-arriving engine automatically assumes safety and RIT duties based on department SOGs.

Lesson Learned: This was the first mayday declared at an incident since our inception of the two-in/two-out rule and mayday standard operating guideline (SOG) approximately 15 years ago. The "success" of not having an injured fire fighter(s) from this incident was a direct result of being aggressive in mayday training, reviewing NIOSH reports, and reviewing the *Rules of Engagement for Firefighter Survival #11* prior to this incident, which saved a life. The #11 document was in that month's training plan for the department. The department took the time to review with all companies on duty the importance of this training and reminded all fire fighters to review this information on a regular basis.

Accountability is huge when dealing with a mayday. The department uses a passport system for its accountability, but this system is only as strong as the fire fighters using it and the officers enforcing it. It was critical during the near-miss incident given that fire fighters from another shift responded. BC vehicles also carry shift rosters. The incident commander (IC) utilized both the passports and rosters to work accountability.

Courtesy of the National Fire Fighter Near-Miss Reporting Systems (firefighternearmiss.com)

---

emergency. Some volunteer departments have a sufficient number of personnel and volume of calls that members are scheduled to be on standby or present at the fire station for specific shifts, according to a duty roster. In both forms of fully volunteer departments, the personnel are still unpaid for their services.

Some departments have moved away from the purely volunteer method of staffing due to an increasing number of alarms and a decreasing availability of volunteers. Some departments provide an incentive for fire fighters by paying them for each response to an alarm. These departments are termed "paid on call" or use part-time paid personnel.

Recruitment and retention are a challenge in volunteer fire departments that respond to several calls each day, particularly in areas where few individuals are available to serve as volunteers. The demands often exceed the amount of time volunteers are able to commit to the fire department, even if compensation is provided. A combination department uses full-time career personnel along with volunteer or paid-on-call personnel. This system usually provides faster response times because some personnel are on duty at the stations, ready to respond immediately. Frequently, the full-time staff consists of a minimum number of fire fighters, allowing for apparatus to respond and handle routine emergencies, such as requests for medical assistance, motor vehicle collisions, and incipient fires. The volunteer or paid-on-call staff are dispatched as a backup force when an incident exceeds the capabilities of the full-time personnel, such as a working structure fire.

A career department is staffed by full-time, paid personnel whose regular job is working for the fire department. These departments are typically found where the level of risk and call volumes require personnel to be on duty at the station at all times.

Although four forms of staffing fire department organizations are commonly used, most discussions divide fire fighters into two categories: career and volunteer. NFPA provides a statistical snapshot with its *The United States Fire Service Fact Sheet*. According to the 2012 snapshot, there are 1.1 million fire fighters in the United States. Of this total, approximately 31 percent are full-time, career fire fighters and 69 percent are volunteers, a group that includes both part-time and paid-on-call fire fighters. Three out of four career fire fighters work in communities with populations of 25,000 or more. Most volunteer fire fighters work in fire departments that protect small, rural communities with populations of 2500 or less.

There are 30,145 fire departments in the United States. Combination departments include varying proportions of career and volunteer members and can be mostly career or mostly volunteer. Two-thirds of fire department responses in 2011 were for medical emergencies; only 5 percent of the responses involved actual fires.

Private industry and nongovernmental organizations may operate their own fire brigades or plant emergency response teams to protect factories, processing plants, large private facilities, and isolated communities. Although many of these groups are established to handle incipient fire situations, some are organized along the lines of a municipal fire department, including fire officers.

■ History of the Fire Service

Since prehistoric times, controlled fire has been a source of comfort and warmth, but uncontrolled fire has brought death and destruction. Accounts from the Roman Empire describe community efforts to suppress uncontrolled fires. In 24 B.C., the Roman emperor Augustus Caesar created what was probably the first fire department. Called the Familia Publica, it was composed of approximately 600 slaves who were stationed around the city to watch for and fight fires. Because the members of the Familia Publica were slaves, they had little interest in preserving the homes of their masters and little desire to take risks, so fires continued to be a problem. By about A.D. 60, under the emperor Nero, the Corps of Vigiles had been established as the fire protectors. This group of 7000 free men was responsible for firefighting, fire prevention, and building inspections. The Corps adopted the formal rank structure of the Roman military, an organizational model that is still used by most fire departments.

The first documented fire in North America occurred in Jamestown, Virginia, in 1607. The fire started in the community blockhouse and almost burned down the entire settlement. At that time, most structures were built entirely of combustible materials, such as straw and wood. Local ordinances soon required the use of less flammable building materials and mandated that fires be "banked," or covered over, throughout the night. In 1630, Boston, Massachusetts, established the first fire regulations in North America when it banned wood chimneys and thatched roofs. In the Dutch colony of New York in 1647, Governor Peter Stuyvesant not only banned wood chimneys and thatched roofs but also required that chimneys be swept out regularly. Fire wardens imposed fines on those residents who did not obey the regulations; the money collected was used to pay for firefighting equipment.

Colonial fire fighters had only buckets, ladders, and fire hooks (tools used to pull down burning structures). Homeowners were required to keep buckets filled with water outside their doors and to bring them to the scene of the fire. Some towns also required that ladders be available so fire fighters could access the roof to extinguish small fires. If all else failed, the fire hook was used to pull down a burning building and prevent the fire from spreading to nearby structures. The "hook-and-ladder truck" evolved from this early equipment.

The first organized volunteer fire company was established in Philadelphia. The Union Fire Company was formed in 1735, under the leadership of Benjamin Franklin. Franklin recognized the many dangers of fire and continually sought ways to prevent it. For example, he developed the lightning rod to help draw lightning strikes—a common cause of fires—away from homes. Another early volunteer fire fighter, George Washington, imported one of the first fire engines from England, which he donated to the Alexandria (Virginia) Fire Department in 1765.

In 1871, two major fires significantly affected the development of both the fire service and fire codes. At the time, the city of Chicago was a boom town with 60,000 buildings—40,000 of which were constructed wholly of wood, with roofs of tar and felt or wooden shingles. With the lax construction regulations and no rain for 3 weeks, the city was extremely dry and vulnerable to fire on October 8, 1871, when a fire started in a barn on the west side of the city. The fire department was already exhausted from fighting a four-block fire earlier in the day. Errors in judging the situation and signaling the alarm resulted in a delayed department response. The Great Chicago Fire burned through the city for 3 days **FIGURE 1-1**. When it was over, more than 2000 acres and 17,000 homes had been destroyed, the city had suffered more than $200 million in damage, 300 people were dead, and 90,000 were homeless.

At the same time, another fire was raging 262 miles north of Chicago in Peshtigo, Wisconsin. Throughout the summer, the north woods of Wisconsin had experienced drought-like conditions. Logging operations had left pine branches carpeting the forest floor. A flash forest fire created a "tornado of fire" more than 1000 feet high and 5 miles wide. Ultimately more than 2400 square miles of forest land burned, several small communities were destroyed, and more than 2200 people lost their lives. The Peshtigo firestorm even jumped the 60-mile-wide Green Bay to destroy several hundred square miles of land and additional settlements on Wisconsin's northeast peninsula. Although not as widely publicized as the Chicago fire, Peshtigo is the deadliest fire in United States history.

These events would forever change the U.S. fire service and, along with it, the role of the fire officer. Communities

FIGURE 1-1 The Great Chicago Fire of 1871 caused the deaths of 300 people and led to changes in firefighting operations in the United States.

began to enact strict building and fire codes. The development of water pumping systems, advances in firefighting equipment, and improvements in communications and alarm systems all helped ensure that such tragedies would not recur.

■ Fire Equipment

Buckets of water gave way to hand-powered pumpers in 1720 when Richard Newsham developed the first such pumper in London, England. Dozens of individuals powered this type of pump, making it possible to propel a steady stream of water from a safe distance. By 1829, more powerful, steam-powered pumpers had been developed and began to replace the hand-powered pumpers. Many volunteer fire fighters felt threatened by the steam engines and fought against their use. Steam engines were heavy machines that were pulled to the fire by a trained team of horses. They required constant attention, which limited their use to larger cities that could meet the costs of maintaining the horses and the steamers.

The advent of the internal combustion engine in the early 1900s enabled most communities to have machine-powered pumpers. Today, both staffed and unstaffed firehouses have fire engines ready to respond at any hour of the day or night. Although they require regular maintenance, current equipment

Safety Zone

NFPA Fire Statistics—2011
In 2011 in the United States there were:
- 484,500 residential fires
- 2640 civilian fire deaths
- 15,635 civilian fire injuries
- 26,500 intentionally set fires
- 686,000 outside fires

Every 23 seconds, a fire department responds to a fire somewhere in the United States.

Data source: *Fire Loss in the United States During 2011.*

Fire Marks

An Ounce of Prevention
Along with the first U.S. fire company, Benjamin Franklin organized the first fire insurance company in the United States and coined the phrase "An ounce of prevention is worth a pound of cure," an apt motto for fire safety. Early insurance companies marked the homes of their policy holders with a plaque, or **fire mark**, that showed the name or logo of the insurance company. The insurance company paid fire departments to respond to those buildings displaying their fire mark. Sadly, others were left to burn. Some communities in the United States continue to fund their fire departments through annual subscription fees; however, the image of a fire truck being present at the scene of a structure fire, yet not extinguishing the fire because the owner has not paid for a subscription, appears to be a contradiction of organizational values.

does not require the constant attention that was demanded by horses and steam engines. Modern fire apparatus carry water, a pumping mechanism, hoses, equipment, and personnel. One new fire suppression vehicle can often outperform several older vehicles.

The progress in fire protection equipment extends beyond vehicles. Without an adequate water supply, after all, the fire apparatus would be helpless. The advent of municipal water systems provided large quantities of water to extinguish major fires.

The Romans developed the first municipal water systems, just as they had developed the first fire companies. It was not until the 1800s, however, that water distribution systems were developed to support fire suppression efforts. George Smith, a fire fighter in New York City, developed the first fire hydrants in 1817. He realized that using a valve to control access to the water in the city's pipes would enable fire fighters to tap into the system when there was a fire. These valves, or fireplugs, were used with both above-ground and below-ground piping systems.

■ Communications

Because small fires are more easily controlled, the sooner a fire department is notified of an incident, the more likely it will be able to extinguish the fire and minimize losses. During the Colonial period, a fire warden or night watchman patrolled neighborhoods and sounded the alarm if a fire was discovered. Some towns, including Charleston, South Carolina, built a series of fire towers where wardens would watch for fires. In many towns, ringing the community fire bell or church bells alerted citizens to a fire.

The introduction of public call boxes in Washington, D.C., during the 1850s was a major advance. Call boxes located around the city enabled citizens to send a coded telegraph signal to the fire department dispatch center. The fire department could determine the location of the fire alarm box by the number of bells in the coded signal. When fire fighters arrived at the alarm box, the caller could direct them to the exact location of the fire. Similar systems are still operating in some places today, but most have been replaced by more immediate and effective communications systems. Cellular telephones enable citizens to report an emergency from almost anywhere, anytime. The introduction of computer-aided dispatch and automatic vehicle locators has improved response times because the closest available fire units can be quickly sent to the emergency.

Communications are also vital for a fire officer to coordinate the firefighting efforts effectively. During the firefight, officers must be able to communicate with fire fighters or summon additional resources. Improvements in communication systems are tied to the history of the fire service. Today's two-way radios enable fire units and individual fire fighters to remain in contact with one another at all times. Before electronic amplification and two-way radios became available, the chief officer would shout commands through his trumpet FIGURE 1-2. The chief's trumpet, or bugle, eventually became a symbol of authority. Although chief officers no longer use trumpets for communicating, the use of trumpets to symbolize the rank of chief also signifies the chief's need to communicate clearly.

■ Building Codes

Fires have served as an impetus for communities to establish building codes. Although the first building codes, developed in ancient Egypt, focused on preventing building collapse, building codes were quickly recognized as an effective means of preventing, limiting, and containing fires.

FIGURE 1-2 The chief's trumpet, used for amplification before electronic devices, eventually became a symbol of authority.

> ### Safety Zone
>
> **Canadian Fire Statistics**
> In Canada, fire loss statistics are compiled in each province by Human Resources Development Canada. The results are forwarded to the Canadian Council of Fire Marshals and Fire Commissioners, which produces an annual fire loss report for Canada. In 2007, the latest year for published results, the statistics revealed:
> - An estimated 69 percent of the 42,753 fires occurred in residences.
> - There were 224 civilian fire deaths.
>
> Data source: *Fire Losses in Canada: Year 2007 and Selected Years.*

Colonial communities had few building codes. The first settlers had a difficult time erecting even primitive shelters, which often were constructed of wood with straw-thatched roofs. The fireplaces used for cooking and heating may have had chimneys constructed of smaller logs. The all-wood construction and open fires meant that fires were a constant threat. As communities developed, they enacted codes restricting the hours during which open fires were permitted and the materials that could be used for roofs and chimneys. In 1678, Boston required that "tyle" or slate be used for all roofs. After British troops burned Washington, D.C., in 1814, codes prohibiting the building of wooden houses were adopted. Building codes also began to require the construction of a fire-resistive wall, or firewall, of brick or mortar between two buildings.

Building codes not only govern construction materials but also frequently require built-in fire prevention and safety measures. Required fire detection equipment notifies both building occupants and the fire department of a fire's presence. Built-in fire suppression or sprinkler systems help contain a fire to a small area and prevent small fires from growing into major fires. Fire escapes, stairways, doors that unlock when the alarm sounds, and doors that open outward enable occupants to escape a burning building safely. Without modern building code requirements, high-rise buildings and large shopping centers could not be built safely.

Fire officers are often on the front line of ensuring that the fire codes are obeyed. They must also understand built-in fire protection systems and recognize how they affect firefighting operations.

> ### Fire Marks
>
> **Chief's Trumpet**
> The historic symbol of the chief officer's trumpet is still used on a chief's badge. This series of crossed trumpets is one of the cherished traditions of the fire service.

Code Development

Building codes and fire codes have evolved over many years. The first codes were locally developed and were often influenced by local preferences and customs. The insurance industry played a major role in the development of the first model codes, which were offered to local jurisdictions as a proposed minimum standard. Local communities could make the code stricter as needed.

Today, model codes and standards are written by national organizations, such as the NFPA. Volunteer committees of citizens and representatives of businesses, insurance companies, and government agencies explore and develop proposals that are debated and reviewed by various groups. The final result, called the consensus document, is then presented to the public. Many states and municipalities adopt selected NFPA codes and standards into their administrative laws.

■ Paying for Fire Service

Many early volunteer fire departments were funded by donations or subscriptions, and many volunteer departments still rely on this source of revenue to purchase equipment and pay operating expenses. The first fire wardens were employed by communities and paid from community funds.

Fire insurance companies were established in England soon after the Great Fire of London in 1666, to help residents and business owners cope with the financial loss from fires. The companies would collect fees (premiums) from homeowners and businesses and pledged to repay the owner for any losses due to fire.

Because the insurance companies could save money if the fire was put out before much damage was done, they often agreed to pay a fire company for trying to extinguish a fire. As mentioned earlier, houses that had insurance were designated with a fire mark **FIGURE 1-3**. Most fire companies were loosely governed, and more than one company might show up to fight the same fire. If two fire companies arrived at a fire, however, a dispute might arise over which company would collect the money. This type of conflict hastened many municipalities' decisions to begin assuming the responsibility for providing fire protection. Today, local tax revenues pay for most career fire departments and support many volunteer organizations. Regardless of the form of the organization, good fire officers are vital to efficient and effective operations.

■ Training and Education

The first fire fighters needed just muscular strength and endurance to pass buckets or operate a hand pumper. As equipment became more complex, the importance of formalized training and good judgment increased.

Modern-day fire fighters operate sophisticated technical equipment, including large vehicles, radios, thermal imagers, and self-contained breathing apparatus. These tools, as well as better fire-detection devices, increase the safety and effectiveness of fire services personnel. Nevertheless, the most important resources on the fire scene remain the knowledgeable, well-trained, physically capable fire fighters who have the ability and the determination to attack a fire.

Fire Department Organization

As soon as firefighting exceeded one person and one bucket, there was a need for an organization to focus individual efforts and provide structure for the endeavor. The model adopted reflects the unique characteristics of the community and the conditions that resulted in the organization of a fire department.

Organizations will evolve as the department grows, the leadership changes, and the community determines which services and activities are needed by the fire department. This section examines the formal conditions and practices found in most departments.

■ Source of Authority

Governments—whether municipal, county, state, provincial, or national—are charged with protecting the welfare of the public against common threats. Fire is one such peril; an uncontrolled fire threatens everyone in the community. Citizens accept certain restrictions on their behavior and pay taxes to protect themselves and support the common good. People charged with protecting the public are given certain authority to enable them to perform effectively. For example, fire departments can legally enter a locked home without permission to extinguish a fire and protect the public. Extinguishing a fire is considered to be an important measure to protect the community.

In most areas, the fire service draws its authority from the governing entity responsible for protecting the public from fire—whether it is a town, a city, a county, a township, or a special fire district. The head of the fire department, usually the fire chief, is accountable to the leader of the governing body, such as the city council, the county commission, the mayor, or the city manager. Because of the relationship between a fire department and a local government, fire fighters should consider

FIGURE 1-3 A fire mark indicated the homeowner had insurance that would pay the fire company for extinguishing the blaze.

themselves civil servants, working for the tax-paying citizens who fund the fire department.

Federal and state governments also grant authority to fire departments and operate their own fire departments and fire protection agencies to protect federal or state properties, particularly for military installations and wildland areas. Some private corporations have government contracts to provide fire protection services or offer subscription services to private property owners.

Most urban and suburban fire departments are organized by a municipal or county government. Typically, these agencies fall under the organizational umbrella as a department, just like the police department, the public works department, and the human resources department.

Another form of organization that is usually similar to that of a municipal or county department is a fire protection district. A fire protection district is a special political subdivision that can be established by a state or a county, with the single purpose of providing fire protection within a defined geographic area. Its operation is overseen by a fire district board that is usually elected by the voters in the district. A fire district operates very much like a school district and has the ability to set a tax rate, collect taxes, and issue bonds.

In some states, a volunteer fire department can be established by charter and is independent of any local government body. The fire department may be a private association rather than a governmental entity. This type of department is not funded directly through taxes, although it may receive a grant or contract with the local government to provide services. Funding can also be obtained from fund-raising events, donations, subscriptions, and fees for service.

■ Chain of Command

The organizational structure of a fire department consists of a chain of command. The ranks may vary in different departments, but the basic concept is generally the same. The chain of command creates a structure for managing the department as well as for directing fire-ground operations. Fire fighters usually report to a supervising officer who is responsible for a single fire company (e.g., an engine company) on a single shift. Some fire departments have only one officer rank, whereas others have two or more officer ranks (sergeants, lieutenants, and captains).

When there are two levels of officers in a fire department, the senior officer has more authority, functioning as a managing fire officer. A managing fire officer, or captain, could be directly responsible for supervising a fire company on one shift and also responsible for coordinating all of the company's activities with other shifts. A managing fire officer could also be in charge of all of the companies on one shift in a multiunit fire station.

Supervising and managing fire officers report directly to an administrative fire officer. In a large organization, there are often several levels of administrative fire officers, usually called chiefs. Battalion chiefs, or district chiefs, are responsible for managing the activities of several fire companies within a defined geographic area, usually in more than one fire station. A battalion chief is usually the officer in charge of a single-alarm working fire.

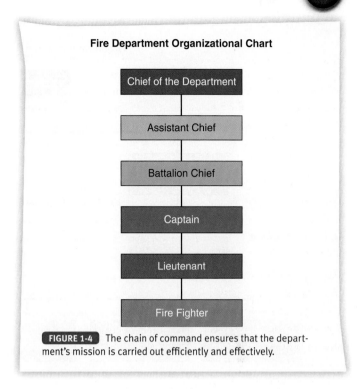

FIGURE 1-4 The chain of command ensures that the department's mission is carried out efficiently and effectively.

Above battalion chiefs in the fire department hierarchy are division chiefs, deputy chiefs, and/or assistant chiefs. Officers at these levels are usually in charge of major functional areas, such as training, emergency operations, support services, and fire prevention within the department. They can also have responsibility for relatively large geographic areas, including several battalions or districts. These officers report directly to the chief of the department.

The fire chief (or chief of the department) is the executive fire officer who has overall responsibility for the administration and operations of the department. The fire chief can delegate responsibilities to other members of the department but is still responsible for ensuring that these activities are properly carried out.

The chain of command is used to implement department rules, policies, and procedures. This organizational structure enables a fire department to determine the most efficient and effective way to fulfill its mission and to communicate this information to all members of the department 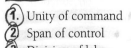. Using the chain of command ensures that a given task is carried out in a uniform manner.

■ Basic Principles of Organization

The fire department uses a paramilitary style of leadership. Most fire departments are structured on the basis of four management principles:

1. Unity of command
2. Span of control
3. Division of labor
4. Discipline

Unity of Command

Unity of command is the management concept that each fire fighter answers to only one supervisor, and each supervisor

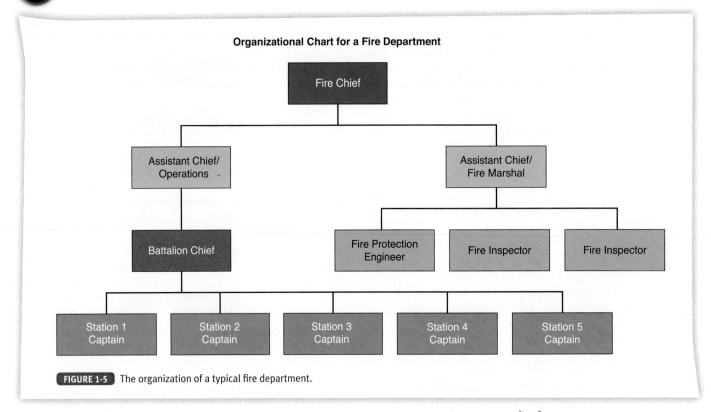

FIGURE 1-5 The organization of a typical fire department.

answers to only one boss **FIGURE 1-5**. In this way, the chain of command ensures that everyone is answerable to the fire chief and establishes a direct route of responsibility from the chief to the fire fighter.

At the fire ground, all functions are assigned according to incident priorities. A fire fighter with more than one supervising officer during an emergency could be overwhelmed with various conflicting assignments. The incident priorities may not get accomplished in a timely and efficient manner. Unity of command is designed to prevent such problems.

Span of Control

Span of control refers to the maximum number of personnel or activities that can be effectively controlled by one individual (usually three to seven). Most experts believe that span of control should extend to no more than five people, but this number can change, depending on the assignment or the task to be completed. A fire officer must recognize his or her own span of control to be effective.

Division of Labor

Division of labor is a way of organizing an incident by breaking down the overall strategy into smaller tasks. Some fire departments are divided into units based on function. For example, the functions of engine companies are to establish water supplies and flow water; truck companies perform forcible entry and rescue functions. Each of these functions can be divided into multiple assignments, which can then be assigned to individual fire fighters. With division of labor, the specific assignment of a task to an individual makes that person responsible for completing the task and prevents duplication of job assignments.

Discipline

Discipline is the set of guidelines that a department establishes for fire fighters. Discipline encompasses behavioral requirements, such as always following orders from superior officers and performing up to expectations. Standard operating procedures, suggested operating guidelines, policies, and procedures are all forms of discipline because they outline how things are to be done, and usually how far a person can go without requesting further guidance. Firefighting demands strong discipline if all personnel are to operate safely and effectively. Discipline can be either positive, when it defines appropriate action, or corrective when it responds to inappropriate actions or behaviors.

■ **Other Views of Organization**

There are several different ways to look at the organization of a fire department. Today's community expectations are shaping new roles and responsibilities unknown to the Civil War generals who organized individual fire companies into the first municipal fire departments. Technology and wartime experiences reinforce the value of a well-prepared fire fighter in imposing order on any crisis that endangers lives and property.

Function

Fire departments can be organized along functional lines. For example, the training division is responsible for leading and coordinating department-wide training activities. Likewise, engine companies and truck companies have certain defined functional responsibilities at a fire. Hazardous material squads have different functional responsibilities that support the overall mission of the fire department.

Geography

Each fire department is responsible for a specific geographic area. Fire stations are located throughout a community to ensure a rapid response time to every area, and each station is responsible for its own specific geographic area. This arrangement also enables the fire department to distribute and use specialized equipment efficiently throughout the community.

Staffing

A fire department must have sufficient trained personnel available to respond to a fire at any hour of the day, every day of the year. Staffing issues affect all fire departments—career departments, combination departments, and volunteer departments alike.

In volunteer departments, it is important to ensure that enough responders are available at all times, but particularly during the day. In the past, when many people worked in or near the communities where they lived, volunteer response time was not an issue. People often now have longer commutes to their places of employment and work longer hours, however, so the number of people available to respond during the weekday can be severely limited.

Members of volunteer departments are also challenged with ever-increasing time demands to obtain additional emergency service certifications, licenses, and continuing education. Although some of this training can be accomplished while on duty at the fire station, other activities will require additional time at an academy, specialized training sessions, or a testing facility.

In combination departments, the challenge is to make sure the appropriate mix of qualifications, certifications, and capabilities is present to deliver prompt and safe response. Within the mix of volunteer and career members, there needs to be a credentialed fire company commander, a qualified apparatus operator, and structural or wildland fire fighters. In many communities the career employee will have multiple certifications, such as apparatus operator and Paramedic, and is expected to fill any position that is not covered by a volunteer member.

In career departments, the challenge is to ensure that all of the assigned positions are covered by appropriately qualified personnel. Although each career position has a home fire station and fire company, it is common for fire fighters to be assigned to a different fire company or a different fire station. The company officer needs to make work assignments based on individual qualifications, experience, and capability.

In all three types of departments, the company officer functions as the staffing coordinator and gatekeeper, implementing the department's rules, regulations, and procedures to ensure proper fire company staffing.

The Functions of Management

Fire officers are managers, and like all other managers, they have specific functions that they must perform, regardless of the size or type of organization. The four functions of management were originally identified by Henri Fayol and published in the *Bulletin de la Société de l'Industrie Minérale* in 1916; Constance Storrs provided the English translation in the 1949 textbook *General and Industrial Management*. These four functions are as follows:

- Planning
- Organizing
- Leading
- Controlling

Planning means developing a scheme, program, or method that is worked out beforehand to accomplish an objective. The fire officer is responsible for developing a work plan for a fire company or an administrative work group. The fire officer develops plans to achieve departmental, work unit, and individual objectives. Short-range planning covers developing a plan that extends up to a year. Medium-range planning covers planning that is 1–3 years in advance. Long-range planning covers events longer than 3 years in advance.

Planning includes establishing goals and objectives and then developing a way to meet and evaluate those goals and objectives. This task could be as simple as planning the daily activities for the fire company or as complex as developing an annual budget. Planning can also include emergency activities, such as developing strategies and tactics and an incident action plan.

Organizing means putting resources together into an orderly, functional, structured whole. The fire officer takes the available people, equipment, structure, and time and develops them into an orderly, functional, and structural unit to implement the plan and deliver the expected services. Fire officers decide which companies will perform certain duties when they arrive on an emergency scene. They also decide which station duties each member of the crew will perform.

Leading means guiding or directing in a course of action. The act of leadership is a complex process of influencing others to accomplish a task. When most people think of managing, they envision the leading function of management. Leading is the human side of managing. It includes motivating, training, guiding, and directing employees.

Controlling means restraining, regulating, governing, counteracting, or overpowering. Fire officers implement the controlling function when they consider the impact on the budget before making purchases, when they conduct employee performance appraisals, and when they ensure compliance with departmental policies.

Fire officers use the functions of management to get work accomplished by and through others. The four functions constitute a continuous cycle; they are never truly "finished." Although fire officers at all levels use all four functions, each level may use them to different degrees.

Assessment Center Tips

Providing a Perfect Response to a Situation
Throughout this text, hints are provided on how to complete assessment center exercises successfully in the promotional exam process. The goal of a promotional exam is to see how well a candidate can perform the typical and extraordinary tasks appropriate for a managing or supervising fire officer. In each assessment center, the candidate must demonstrate the most appropriate response as determined by the department.

> **Getting It Done**
>
> **Digging Through the Management Toolbox**
> Suggestions and tips are provided throughout this text on how to apply the theories, knowledge, and skills to complete a task. A fire officer student starts filling a management toolbox with theories, concepts, and practices. A new fire officer should consider senior fire officers to be valuable resources when developing his or her own toolbox. Consider which theory, concept, or practice an admired and experienced fire officer would use in the same situation. If possible, ask senior fire officers for suggestions or feedback when considering which management tool to use to resolve an issue.

Rules and Regulations, Policies, and Standard Operating Procedures

Fire officers must thoroughly know the department's regulations, policies, and standard operating procedures. This knowledge is essential to ensure a safe and harmonious working environment. Although a fire fighter is required to follow all regulations, policies, and procedures, the fire officer must not only follow these directives, but also ensure compliance with them by subordinates. The transition to fire officer is often a difficult one because of the new role the officer is now required to fill. For officers to enforce the organizational rules, they must understand the differences among rules and regulations, policies, and standard operating procedures.

Rules and regulations are developed by various government or government-authorized organizations to implement a law that has been passed by a government body. Rules may also be established by the local jurisdiction that sets the conditions of employment or internally within a fire department. For example, an organization may have a rule stating that employees with more than 15 years of service will receive 10 shifts of vacation. A fire department may have a rule that requires all members to wear their seat belts when riding in vehicles. Rules and regulations do not leave any room for latitude or discretion.

Policies are developed to provide definite guidelines for present and future actions. Fire department policies outline what is expected in stated conditions. Policies often require personnel to make judgments and to determine the best course of action within the stated policy. Policies governing parts of a fire department's operations may be enacted by other government agencies, as in the case of personnel policies that cover all employees of a city or county. An example of a policy is one stating that the fire officer shall ensure that station sidewalks are maintained to provide safety from slips and falls during snow and ice accumulation. Because it gives the officer latitude in determining how to ensure the safety of pedestrians, this directive is a policy.

Standard operating procedures (SOPs) are written organizational directives that establish or prescribe specific operational or administrative methods to be followed routinely for the performance of designated operations or actions. SOPs are developed within the fire department, are approved by the chief of the department, and ensure that all members of the department approach a situation or perform a given task in the same manner. SOPs provide a uniform way to deal with emergency situations, enabling different stations or companies to work together smoothly, even if they have never worked together before. These procedures are vital because they enable everyone in the department to function properly and know what is expected for each task. Fire officers must learn and frequently review departmental SOPs. An example of an SOP is a statement of the step-by-step process (procedure) to be used whenever vertical ventilation is required.

Some fire departments prefer the term *suggested (or standard) operating guidelines* (SOGs) instead of SOPs because conditions often require the fire officer to use personal judgment in determining the most appropriate action for a given situation. The term SOG suggests that a specific step-by-step procedure should be used, but it allows the officer to deviate from this procedure if the conditions warrant doing so. The distinction between an SOP and an SOG is very subjective.

■ Ethics

Inappropriate behavior by individuals in power or holding the public trust is a frequent target of attention for the media, and the fire service is not exempt from such scrutiny. Most of the time fire officers make ethical decisions, choosing to do the "right" thing; however, sometimes officers make unethical choices (which may then influence fire fighters to make unethical choices). When they engage in such behavior, officers' poor choices often appear in the newspaper and have very negative consequences for both the individual and the fire service organization.

Ethical choices are based on a value system. The officer must consider each situation, often subconsciously, and make a decision based on his or her values. If the organization has clear values that are part of a strong organizational culture, the officer uses the organization's value system. If the values are not clear, the individual substitutes his or her own value system.

The key to improving ethical choices is to have clear organizational values. This can be accomplished by:

- Having a code of ethics that is well known throughout the organization
- Selecting employees who share the values of the organization
- Ensuring that top management exhibits ethical behavior
- Having clear job goals

> **Company Tips**
>
> **Fire Fighters Determine Fire Officer Success**
> Throughout this text, practical advice and information are provided on teamwork and communication. Even the best and brightest officers will fail if the company does not support the supervisor. New fire officers must dedicate considerable effort to develop the members of the unit into a well-prepared team. Fire fighters want to do the best job they can; the fire officer provides the opportunities to meet that need.

VOICES OF EXPERIENCE

In 1994, early in my career, I, like many young fire fighters, had aspirations of someday pursuing a promotion as a company officer. I had always felt that I was good at my job and was an easy employee to manage. I was a hard worker who reported to work early and easily found tasks to keep myself occupied in the down time around the station. Basically, I felt that I did everything I could to be the kind of fire fighter I myself would want to manage as a company officer.

I believed that I was on the path toward advancement, so I continued what I was doing and waited for that "someday" to come. Years went by, and I watched with discouragement as my peers (some with less seniority) were easily promoted to various positions within the department, all while I continued to wait for my time to come. I soon came to realize that my "seniority"-based path to promotion was not working.

I sought advice from my company officer. After receiving some brutally honest constructive criticism, I learned that while I was waiting for my turn at promotion, others were taking educational classes through both formal and informal opportunities to further their knowledge, increase their skills, and fine-tune their abilities. My co-workers recognized opportunities and began practicing for the positions they wanted to be promoted to, and I did not.

> *While I was waiting for my turn at promotion, others were taking classes to further their knowledge, increase their skills, and fine-tune their abilities.*

I was initially upset about the criticism but soon realized that I was the only one to blame. I had the same opportunities that my co-workers did; however, I had failed to recognize and act on them. This wake-up call changed my perception and gave me a clearer understanding of what I needed to do to better prepare for promotion. I took this lesson to heart and began recognizing and taking advantage of multiple opportunities. I started practicing to become a company officer and prepared myself for this promotion by furthering my education, increasing my skills, and fine-tuning my abilities. This desire and motivation paid off, and I was ultimately promoted to the rank of captain.

As an aspiring company officer, it is imperative that you continually seek opportunities to improve your knowledge, skills, and abilities. Self-improvement also consists of looking to others for advice and being willing to accept constructive criticism. This is a small, but important part of the preparation process. A famous quote from Henry Hartman that I often remind myself of is this: "Success always comes when preparation meets opportunity." Preparation for promotion takes a lot of effort, but when your preparation meets opportunity, the possibilities are endless. Congratulations for seizing this opportunity, and I wish you luck and good fortune in your journey.

Aaron R. Byington, MA, NRP
Captain
Layton City Fire Department
Layton City, Utah

- Having performance appraisals that reward ethical behavior
- Implementing an ethics training program

Even at the company level, these values can be implemented to help prevent undesirable ethical choices. One way to help judge a decision is to ask yourself three questions:

- What would my parents and friends say if they knew?
- Would I mind if the paper ran it as a headline story?
- How does it make me feel about myself?

Asking these questions can help prevent an event that could devastate the department's and the fire officer's reputation for years to come.

Fire Officer II

Requirements for Fire Officer II

The requirements for Fire Officer II begin with meeting all of the requirements for Fire Officer I as defined in NFPA 1021. As is true for the Fire Officer I, the duties of the Fire Officer II can be divided into administrative, nonemergency, and emergency activities.

Administrative duties include evaluating a subordinate's job performance, correcting unacceptable performance, and completing formal performance appraisals on each member. Other duties include developing a project or divisional budget, including the related activities of purchasing, soliciting, and rewarding bids, and preparing news releases and other reports to supervisors.

Nonemergency duties include conducting inspections to identify hazards and address fire code violations; reviewing accident, injury, and exposure reports to identify unsafe work environments or behaviors; and taking approved action to prevent reoccurrence of an accident, injury, or exposure. Other duties could include developing a preincident plan for a large complex or property; developing policies and procedures appropriate for this level of supervision; analyzing reports and data to identify problems, trends, or conditions that require corrective action; and then developing and implementing the required actions.

Emergency duties include supervising a multiunit emergency operation using the Incident Command System (ICS) and developing an operational plan to deploy resources to mitigate the incident safely. The ICS defines the roles and responsibilities to be assumed by personnel and the operating procedures to be used in the management and direction of emergency operations. The Fire Officer II is also expected to determine the area of origin and preliminary cause of a fire and to develop and perform a postincident analysis of a multicompany operation.

The IAFC identifies the Fire Officer II level as a Managing Fire Officer. The goal of the IAFC *Officer Development Handbook* is to encourage company officers to acquire the appropriate levels of training, experience, self-development, and education throughout their professional journey to prepare for the Chief Fire Officer designation as the pinnacle of professional development. In this text, the Fire Officer II will be the captain.

Additional Roles and Responsibilities for Fire Officer II

A Fire Officer II has the same roles and responsibilities as a Fire Officer I, along with the following additional items:

- Supervises and directs the activities of a multiunit station
- Completes employee performance appraisals
- Creates a professional development plan for members of the organization
- Leads water rescue, hazardous materials, or other special teams as assigned
- Ensures the safe and proper use of equipment, clothing, and protective gear and enforces departmental policies
- Participates in the formulation or evaluation of departmental or agency policies as assigned, implements new or revised policies, and encourages team efforts of fire personnel
- Participates in the formulation of the departmental budget and makes purchases within it
- Develops emergency incident operational plans requiring multiunit operations
- Prepares written reports so major causes for local service demand are identified for various planning areas within the service area of the organization

This text covers only the roles and responsibilities of Fire Officers I and II according to NFPA 1021. Fire Officer III, IAFC's Administrative Fire Officer, and Fire Officer IV, IAFC's Executive Fire Officer, have more training and responsibilities.

Working with Other Organizations

Fire departments are part of the structure of the community. To fulfill its mission, a fire department must often interact with other organizations. A motor vehicle crash provides a good example of this need. In an area with a centralized 911 call center, the fire department, a separate emergency medical services provider, law enforcement officials, and tow-truck operators might all be dispatched to the same incident. At the scene, all of these personnel must work together to solve the problem. Fire officers frequently have to request assistance from and then interact with other agencies.

Challenges for the Captain

The managing fire officer is more engaged in working with other organizations and groups. The slow recovery from the 2009 recession has created unprecedented challenges that will require changes in the structure, task, and mission of the fire department.

■ Supervision and Motivation

The paramilitary structure of the fire department was established in the 1860s, when cities were replacing independent volunteer companies with municipal fire departments based on the Civil War military deployment model. A rigid command and control process remains essential when operating at emergency scenes. Away from emergencies, however, departments are using the concepts of employee empowerment, decentralized decision making, and delegation to fully engage fire fighters in the required tasks to prepare and maintain readiness for a wide range of community needs. This "all-hazards" approach is especially important when considering nonfire incidents, a crumbling built environment, homeland security, cultural diversity, and ethics.

■ Increase in Nonfire Incidents

Battling a structure fire is a common perception of a "typical day at work" for fire fighters, and both the Fire Fighter I and II levels focus on this task. The NFPA, however, notes that firefighting is actually one of the activities least frequently performed by fire companies. It accounts for only 5 percent of the response workload based on data submitted to the National Fire Incident Reporting System (NFIRS). Of the runs requiring fire suppression tasks, almost half are for structure fires.

Emergency medical services (EMS) calls are now the most frequent activity undertaken by the fire service, accounting for a minimum of 66 percent of fire company responses. In the last decade, the increase in EMS workload exceeded growth on a demographic basis, with some cities noticing the number of EMS calls increasing even while the population was declining. In some fire departments, EMS calls represent more than 80 percent of a fire company's response workload.

Activated fire protection system alarms are the second most common reason for fire service response. The majority of these activations are due to faulty alarm systems, good intentions, or false calls. The company officer must work to remain vigilant in events that result in no service in most of the responses, with an occasional incipient fire or an inferno occurring in fewer than 1 out of 100 activated fire alarm responses.

Investigating an odor, a hazardous condition, or other service call is the third most common reason for fire department response. Fire fighters encounter many opportunities to be creative problem solvers and deliver outstanding customer service during these events.

■ Deterioration of the Built Environment

Although the number of structure fires continues to decline, the rate at which fire fighters are being killed or seriously injured while operating in burning structures continues to climb. Flashover and structural collapse are the primary causes of death within a burning structure. Decades of deferred maintenance and repair all too often make many structures unstable and dangerous to operate in.

Century-old buildings, while robust, may have many worn-out or rusted-out building components. Fire escapes may be pulling out of the walls and structural components crumbling. Many modern-day renovations involve the substitution of lightweight wood elements, some without the benefit of a fire code inspection after changes are made to the occupancy or use of the building.

■ Related Duties

Fire departments have been recognized as the hometown first responder to disasters, catastrophes, terrorist acts, and any other threat to the local community. Community expectations change, however, and the fire department is providing a wider variety of services, often as a part of a multiagency effort.

Conversely, other public safety agencies are getting into areas that were once the sole domain of the fire department. Hundreds of police departments have created hazardous materials response teams, training police officers to the hazardous materials technician level and procuring equipment that may exceed the fire department resources.

To manage this kind of change effectively, a fire officer must understand the roles played by other agencies and recognize how they interact with the fire department. Federal, state, and local response plans identify which organizations are responsible for each area of the incident.

For a local incident, the local fire service is often given primary responsibility for search, rescue, and fire extinguishment. Law enforcement is given responsibility for criminal investigation and scene security. Emergency management is responsible for evacuation notification. The American Red Cross may be responsible for establishing evacuation shelters.

As incidents grow, the Federal Bureau of Investigation (FBI) may take a lead role in an investigation. The Federal Emergency Management Agency (FEMA) may become the primary agency responsible for coordination of the incident, which could include the use of urban search and rescue (USAR) teams. Fire officers must understand their role within the local, state, and federal response plans.

Fire officers must also be aware that some individuals and organizations wish to create chaos and harm emergency service workers. They research fire department activities to exploit weaknesses and identify opportunities. The fire officer must be vigilant for threats to fire fighters and the department.

■ Cultural Diversity

Fifty years ago, the fire service was made up of virtually all white males. This composition began to change in the 1960s. The initial integration of women and minorities focused on assimilation—that is, the fire service desired to make those who were different fit into the mold of the traditional fire service. When the fire service began including a few women and minorities, the new employees either assimilated to the existing culture or they left.

It is now more commonplace for fire departments to have a blend of men, women, Caucasians, Hispanics, African Americans, Asian Americans, Native Americans, and others. Each individual brings his or her own strengths and unique perspectives to the fire service. This integration of diverse employees is far from complete, however, and many departments continue to struggle with bridging these relationships. The fire officer of the future must look beyond the physical attributes of individuals and match each individual's strengths with the organization's needs for the organization, the individual, and the officer to be successful.

You Are the Fire Officer Conclusion

At 10 A.M., Battalion Chief Johnson arrives and has Captain Davis and Lieutenant Williams assemble in the shift captain's office. Closing the door, the chief welcomes them to the battalion and shares his expectations: "Make sure the crews are properly trained in all the tools and equipment. I expect them, and you, to stay physically and mentally prepared to work at any emergency incident. Keep the apparatus, tools, and station in great shape. Train every day. Get intimately familiar with the target hazards."

"Do not surprise me. If there is an injury, property damage accident, or incident, I need to hear about it from you and not from headquarters, or the *News at Noon*. We will work these issues out together."

Pointing to the *Everyone Goes Home* poster from the National Fallen Firefighters Foundation, Chief Johnson concludes his orientation with two directives:

1. Everyone has their seat belt on before the rig starts to respond.
2. The rigs come to a complete stop at every stop sign and red light intersection.

Wrap-Up

Chief Concepts

- The professional qualification standards for fire officers are documented in NFPA 1021, *Standard for Fire Officer Professional Qualifications*.
- At the Fire Officer I level, the emphasis is placed on accomplishing the department's goals and objectives by working through subordinates to achieve the desired results.
- The officer is a part of management and is responsible for the conduct of others. The officer has to apply policies, procedures, and rules to subordinates and to different situations.
- The U.S. fire service originated as communities of citizens who responded when a fire broke out.
- In 24 B.C., the Roman emperor Augustus Caesar created what was probably the first fire department, called the Familia Publica.
- Richard Newsham developed the first hand-powered water pumper in 1720 in London, England. By 1829, more powerful, steam-powered pumpers had been developed and began to replace the hand-powered pumpers.
- During the Colonial period, a fire warden or night watchman patrolled neighborhoods and sounded the alarm if a fire was discovered. Some towns built a series of fire towers where wardens would watch for fires.
- Although the first building codes, developed in ancient Egypt, focused on preventing building collapse, building codes were quickly recognized as an effective means of preventing, limiting, and containing fires.
- Many early volunteer fire departments were funded by donations or subscriptions, and many volunteer departments continue to rely on this source of revenue to purchase equipment and pay operating expenses.
- The first fire fighters needed just muscular strength and endurance to pass buckets or operate a hand pumper. As equipment became more complex, the importance of formalized training and good judgment increased.
- Fire service organizations will evolve as the department grows, the leadership changes, and the community determines which services and activities are needed from the fire department.
- Governments—whether municipal, county, state, provincial, or national—are charged with protecting the welfare of the public against common threats. Fire is one such peril; an uncontrolled fire threatens everyone in the community.
- The chain of command creates a structure for managing the fire department as well as for directing fire-ground operations.

Wrap-Up, continued

- The fire department uses a paramilitary style of leadership. Most fire departments are structured on the basis of four management principles:
  - Unity of command
  - Span of control
  - Division of labor
  - Discipline
- There are several different ways to look at the organization of a fire department—for example, in terms of function, geography, or staffing.
- The four functions of managing are planning, organizing, leading, and controlling.
- Fire officers must thoroughly know the department's regulations, policies, and standard operating procedures. This knowledge is essential to ensure a safe and harmonious working environment.
- Ethical choices are based on a value system. The officer has to consider each situation, often subconsciously, and make a decision based on his or her values.
- The administrative duties of the Fire Officer II include evaluating a subordinate's job performance, correcting unacceptable performance, and completing formal performance appraisals on each member.
- Other duties of a Fire Officer II include developing a project or divisional budget, including the related activities of purchasing, soliciting, and rewarding bids, and preparing news releases and other reports to supervisors.
- The fire department is part of the structure of its community. To fulfill its mission, the fire department must often interact with other organizations. The managing fire officer is more engaged in working with other organizations and groups.
- A rigid command and control process remains essential when operating at emergency scenes. Away from emergencies, however, departments are using the concepts of employee empowerment, decentralized decision making, and delegation.
- Firefighting is one of the activities least frequently performed by fire companies; it accounts for only 5 percent of the response workload.
- Flashover and structural collapse are the primary causes of death for fire fighters within a burning structure.
- As community expectations change, fire departments are providing a wider variety of services, often as a part of a multiagency effort.
- The fire officer of the future must look beyond the physical attributes of individuals and match each individual's strengths with the organization's needs for the organization, the individual, and the officer to be successful.

Hot Terms

<u>Assistant or division chief</u> A midlevel chief who often has a functional area of responsibility, such as training, and answers directly to the fire chief.

<u>Battalion chief</u> Usually the first level of fire chief; also called a district chief. These chiefs are often in charge of running calls and supervising multiple stations or districts within a city. A battalion chief is usually the officer in charge of a single-alarm working fire.

<u>Chain of command</u> The superior–subordinate authority relationship that starts at the top of the organization hierarchy and extends to the lowest levels.

<u>Chief's trumpet</u> An obsolete amplification device that enabled a chief officer to give orders to fire fighters during an emergency; a precursor to the bullhorn and portable radio.

<u>Consensus document</u> A code or standard developed through agreement between people representing different organizations and interests. NFPA codes and standards are consensus documents.

<u>Controlling</u> Restraining, regulating, governing, counteracting, or overpowering.

<u>Decision making</u> The process of identifying problems and opportunities and resolving them.

<u>Discipline</u> A moral, mental, and physical state in which all ranks respond to the will of the leader. Also, the guidelines that a department sets for fire fighters to work within.

<u>Division of labor</u> The production process in which each worker repeats one step over and over, achieving greater efficiencies in the use of time and knowledge; also, the formal assignment of authority and responsibility to job holders.

<u>Fire chief</u> The highest-ranking officer in charge of a fire department; the individual assigned the responsibility for management and control of all matters and concerns pertaining to the fire service organization.

<u>Fire mark</u> Historically, an identifying symbol on a building to let fire fighters know that the building was insured by a company that would pay them for extinguishing the fire.

<u>Incident Command System (ICS)</u> A system that defines the roles and responsibilities to be assumed by personnel and the operating procedures to be used in the management and direction of emergency operations; also referred to as an Incident Management System (IMS).

<u>Leadership</u> A complex process by which a person influences others to accomplish a mission, task, or objective and directs the organization in a way that makes it more cohesive and coherent.

<u>Leading</u> Guiding or directing in a course of action.

<u>Managing Fire Officer</u> The description from the IAFC *Officer Development Handbook* for the tasks and expectations for a Fire Officer II. In this role, the company officer is encouraged to acquire the appropriate levels of training, experience, self-development, and education to prepare for the Chief Fire Officer designation.

Wrap-Up, continued

<u>Organizing</u> Putting resources together into an orderly, functional, structured whole.

<u>Planning</u> Developing a scheme, program, or method that is worked out beforehand to accomplish an objective.

<u>Policies</u> Formal statements that provide guidelines for present and future actions. They often require personnel to make judgments.

<u>Rules and regulations</u> Directives developed by various government or government-authorized organizations to implement a law that has been passed by a government body.

<u>Span of control</u> The maximum number of personnel or activities that can be effectively controlled by one individual (usually three to seven).

<u>Standard operating procedures (SOPs)</u> Written organizational directives that establish or prescribe specific operational or administrative methods to be followed routinely for the performance of designated operations or actions.

<u>Supervising Fire Officer</u> The description from the IAFC *Officer Development Handbook* for the tasks and expectations for a Fire Officer I. In this role, the company officer is encouraged to acquire the appropriate levels of training, experience, self-development, and education to prepare for the Chief Fire Officer designation.

<u>Unity of command</u> The management concept that a subordinate should have only one direct supervisor, and that a decision can be traced back through subordinates to the manager who originated it.

References and Additional Resources

NFPA reprinted material is not the complete and official position of the NFPA on the referenced subject, which is represented only by the standard in its entirety.

Benoit, J., and K. B. Perkins. (2001). *Leading Career and Volunteer Firefighters: Searching for Buried Treasure*. Halifax, NS, Canada: Henson College, Dalhousie University.

Bosanko, E. (1990). *Triumph and Tradition: Firefighting in Prince George's County, Maryland 1887–1990*. Baltimore, MD: John D. Lucas Publishing.

Bugbee, P. (1971). *Man Against Fire: The Story of the National Fire Protection Association, 1896–1971*. Boston, MA: National Fire Protection Association.

Cole, D. (2002). *The Incident Command System: A 25-Year Evaluation by California Practitioners*. Emmitsburg, MD: National Fire Academy, Executive Fire Officer Program.

Fire Analysis & Research Division. (2012). *The United States Fire Service Fact Sheet*. Quincy, MA: National Fire Protection Association.

Flood, J. (2010). *The Fires: How a Computer Formula, Big Ideas and the Best of Intentions Burned Down New York City—and Determined the Future of Cities*. New York, NY: Riverhead Books.

Gess, D., and W. Lutz. (2002). *Firestorm at Peshtigo: A Town, Its People, and the Deadliest Fire in American History*. New York, NY: Henry Holtz and Company.

Greenberg, A. S. (1998). *Cause for Alarm: The Volunteer Fire Department in the Nineteenth-Century City*. Princeton, NJ: Princeton University Press.

Griffins, J. S. (2012). *Fire Department of New York: An Operational Reference*. 9th ed. Los Alamos, NM: James S. Griffin.

Halberstam, D. (2002). *Firehouse*. New York, NY: Hyperion.

Hashagen, P. (1995). *A Distant Fire: History of FDNY Heroes*. Dover, NH: dmc associates.

Hashagen, P. (2000). "New York City Fire Department History." In: *Fire Department City of New York*, J. Kimmerly, ed. Paducah, KY: Turner Publishing Company, 17–230.

Hensler, B. (2011). *Crucible of Fire: Nineteenth-Century Urban Fires and the Making of the Modern Fire Service*. Washington, DC: Potomac Books.

International Association of Fire Chiefs. (2010). *Officer Development Handbook: In Pursuit of the Planned, Progressive, Life-long Process Of Education, Learning, Self-Development, and Expertise*. 2nd ed. Fairfax, VA: International Association of Fire Chiefs.

Karter, M. J. Jr. (2012). *Fire Loss in the United States During 2011*. Quincy, MA: National Fire Protection Association.

Karter, M. J. Jr., and G. P. Stein. (2012). *U.S. Fire Department Profile Through 2011*. Quincy, MA: National Fire Protection Association.

LoSasso, C. (2012). *Mentoring Volunteer Officers Pre- and Post-Promotion*. Emmitsburg, MD: National Fire Academy, Executive Fire Officer Program.

Matejka, M. G. (2002). *Fiery Struggle: Illinois Fire Fighters Build a Union, 1901–1985*. Chicago, IL: Labor History Society.

McAniff, E. P., and J. J. Cunningham. (1974). *Leadership in the Fire Service*. Bayside, NY: McAniff Associates.

McCarl, R. (1985). *The District of Columbia Fire Fighter's Project: A Case Study in Occupational Folklife*. Washington, DC: Smithsonian Institution Press.

Myers, B. (2004). *Mentoring: Preparing Company Officers*. Emmitsburg, MD: National Fire Academy, Executive Fire Officer Program.

Page, J. O. (1973). *Effective Company Command*. Alhambra, CA: Borden Publishing.

Perkins, K. B., and J. Benoit. (1996). *The Future of Volunteer Fire and Rescue Services: Taming the Dragons of Change*. Stillwater, OK: Fire Protection Publications.

Robinson, G. J. (2011). *Lasting Leadership: Preparing for the Transition to Company Officer*. Emmitsburg, MD: National Fire Academy, Executive Fire Officer Program.

Wijayasinghe, M. (2011). *Fire Losses in Canada: Year 2007 and Selected Years*. Calgary, Alberta, Canada: Office of the Fire Commissioner.

FIRE OFFICER in action

When the captain or lieutenant is not on duty, a senior fire fighter is assigned to be the acting fire officer. On emergency incidents, the fire fighter is expected to function fully as a supervising fire officer. The department expects acting officers to function in any of the incident management roles assumed by a lieutenant.

Back at the station, the acting officer functions as a substitute teacher. The only administrative or supervisory tasks accomplished are to handle issues that affect immediate readiness and resources.

Lieutenant Williams plans to spend some one-on-one time with the acting officer. Williams wants to be sure that all the long-term readiness and administrative issues are identified and handled.

1. What is unity of command?
   A. The concept that fire fighters answer to only one supervisor
   B. The number of people one fire officer can supervise effectively
   C. A method of directing an incident
   D. A set of guidelines established for fire fighters to follow

2. _____ means "putting resources together into an orderly, functional, structured whole."
   A. Standard operating procedures
   B. Span of control
   C. Organizing
   D. Controlling

3. The location and year of the deadliest fire in United States history are _____.
   A. Boston, 1919
   B. San Francisco, 1906
   C. Baltimore, 1904
   D. Peshtigo, 1871

4. "Selecting employees who share the values of the organization" describes _____.
   A. cultural diversity
   B. ethics
   C. policies
   D. planning

Fire Captain *Activity*

After the group meeting, Chief Johnson asks Captain Davis to stay for a second meeting. The chief explains that one of the chief's responsibilities is to prepare captains to fill in as a chief. Part of that preparation entails issuing assignments or tasks that increase the captain's skill set and experiences. The chief recommends that the captain keep a pristine shirt and Class A uniform at the station for those last-minute occasions.

NFPA Fire Officer II Job Performance Requirement 5.1.2
Intergovernmental and interagency cooperation.

Application of 5.1.2
1. Using policies and procedures from a fire department familiar to you, develop an action plan for handling "active shooter" events at a middle or high school that includes all appropriate agencies.
2. Using policies and procedures from a fire department familiar to you, describe the initial roles and responsibilities for handling "chemical suicide" events.
3. Using policies and procedures from a fire department familiar to you, develop an interagency agreement for evacuating a community in a flooding incident.

Preparing for Promotion

Fire Officer I

Knowledge Objectives

After studying this chapter, you will be able to:

- Discuss the origin of civil service promotional examinations. (p 24)
- Discuss promotional processes that can be used by fire departments. (pp 25–26)
- Describe how a promotional examination is prepared. (pp 26–29)
- Identify the elements of a promotional examination. (pp 29–32)
- Identify the components of an assessment center. (pp 30–31)
- List techniques for studying for a promotional examination. (pp 33, 35–36)

Skills Objectives

There are no Fire Officer I skills objectives for this chapter.

Fire Officer II

Knowledge Objectives

After studying this chapter, you will be able to:

- Discuss the origin of civil service promotional examinations. (p 24)
- Discuss promotional processes that can be used by fire departments. (pp 25–26)
- Describe how a promotional examination is prepared. (pp 26–29)
- Identify the elements of a promotional examination. (pp 29–32)
- Identify the components of an assessment center. (pp 30–31)
- List techniques for studying for a promotional examination. (pp 33, 35–36)

Skills Objectives

There are no Fire Officer II skills objectives for this chapter.

CHAPTER 2

Principles of Fire and Emergency Service Administration (FESHE) Course Outcomes

1. Acknowledge career development opportunities and strategies for success. (pp 26, 29–33, 35–36)

You Are the Fire Officer

Battalion Chief Johnson walked into Captain Davis's office carrying a box of three-ring binders and textbooks. "As the chair of the lieutenant promotional committee, I have appointed you to serve as one of the three test-item writers. I need you to write questions covering incident management, building construction, and engine operations."

Municipal City has a two-part promotional process for lieutenants. Part 1 is a 100-question multiple-choice exam, with content coming from 15 references. Candidates who score 70 percent or higher on the exam will proceed to Part 2, which takes place at the assessment center. This segment of the promotional process includes four components: an in-basket exercise, an emergency incident simulation, an interpersonal interaction, and a presentation to an interview board.

Captain Davis posts the lieutenant promotion announcement. The written exam will be administered 8 months from today. The announcement generates a discussion with Fire Fighter Kinders at the Station 100 kitchen about preparing for the exam, as this is the first lieutenant exam for which he is eligible. Apparatus Operator Anders says that others started studying months ago, so Kinders should anticipate a poor result on the first attempt. "Promotional exams are like the Olympics," he says. "They involve years of preparation for minutes of performance."

Captain Davis disagrees, noting that Lieutenant Williams placed third on the lieutenant exam after only 4 months of preparation. Davis encourages Kinders to order the reference books today. Studying for a few hours every day, on and off duty, will pay off, he says.

Apparatus Operator Rollo says it is a waste of time to study obsessively for the test. She guesses that the department will need to make six promotions a year to cover anticipated retirements. The promotion list is good for 4 years, so you need to place in the top 25 of the list to get promoted. It takes a lot more time and effort to score in the top 5 versus the top 25. Rollo recommends spending about an hour each day in test preparation while at the fire station. Kinders can then use his off-duty time for other activities: spending time with his family, taking a college class, running a part-time business, or working fire department overtime. According to Rollo, "Those activities provide a better return on your investment."

1. How much time do you really need to prepare for the exam?
2. If you pass the test, are you ready to be a supervising fire officer?
3. What is wrong with concentrating on being a good fire fighter for another 4 years?

Introduction

The purpose of this chapter is to provide a general description of the civil service promotional examination process that is used by most fire departments. Many variations of the testing procedures and promotional processes exist, so you need to understand the specific process that is used in your organization. Many organizations change elements or assessment tools from examination to examination.

Fire Officer I and II

The Origin of Promotional Examinations

Prior to the Civil War, most government jobs were awarded according to the patronage or spoils system—that is, those in power could appoint people to public office based on a personal relationship or political affiliation, rather than on merit. The best jobs went to political supporters and often required a payment to the individual who had the power to make appointments. Naturally, such a system was ripe for corruption.

Congress enacted the Pendleton Civil Service Reform Act in 1883 in response to the extensive corruption at Tammany Hall in New York City and the many other publicized abuses of the patronage system. The Pendleton Act established the civil service system within the federal government and provided a model for the civil service systems that were developed by many states and cities in subsequent years. The spoils system was gradually replaced by merit selection and promotion for most government employees. The process of developing promotional examinations for fire officer positions, based on testing for specific knowledge and skills, was derived directly from this act.

Fire Marks

Metropolitan Fire Department
When New York City created the paid Metropolitan Fire Department in 1865, Tammany Hall was the seat of political power. William March "Boss" Tweed was the chairman of the Democratic Party and the Grand Sachem (leader) of the Society of St. Tammany, which had been founded in 1789 as a club for patriotic and fraternal purposes. Between 1865 and 1871, Boss Tweed and Tammany Hall exploited the spoils system in New York City, swindling an estimated $75 million to $200 million from the city. In response to this widespread corruption, the public demanded political reform.

Getting It Done

Variations in Examinations
The fire officer promotional process remains an inexact science. Many variations of examination and testing procedures are in use, with differences in emphasis and on the weights assigned to particular dimensions (attributes or qualities that can be described and measured during a promotional examination). The rank (sergeant, lieutenant, or captain) and the classified job description determine the technical, theoretical, or behavioral emphasis of the examination.

Sizing up Promotion Opportunities

The 2008 recession resulted in many qualified fire officers not retiring and numerous agencies leaving vacant positions open. Promotional opportunities significantly slowed as departments decreased their size through attrition. A long recovery from the recession means that, as of this edition's publication date, public safety organizations have not returned to their 2006 staffing levels. This has reduced the number of promotional opportunities in most fire departments.

Firefighting is a lifestyle, like the military, medicine, and other public safety careers. The rotating shift work, the unique team-based workforce, and the hours of tedium mixed with moments of intense life-saving activity combine to create a special work environment. The rich tradition of service and sacrifice extends beyond the on-duty hours and becomes a major component of an individual's life. Some fire fighters represent the third or fourth generation of a firefighting family. Many fire fighters and officers enjoy their work so much that it is common to find them working a decade beyond their retirement eligibility dates or fighting to eliminate a mandatory retirement age FIGURE 2-1.

In most fire departments, completion of a promotional examination process creates an eligibility list that lasts from 2 to 6 years. Depending on local practice, the list may be either rank ordered or banded. On a rank-ordered list, the highest-scoring candidate is ranked number 1, the second-highest-scoring candidate is ranked number 2, and so forth. In most cases, individuals are promoted in the order in which they are placed on the eligibility list.

Other departments use a banded list, in which the candidates are placed into bands, or groups, of promotional candidates. The bands are usually identified as "Highly Qualified," "Qualified," and "Not Qualified." In this case, all of the candidates on the Highly Qualified list are considered equally qualified for promotional consideration, followed by all of the Qualified candidates.

■ Postexamination Promotional Considerations

Regardless of the eligibility-ranking scheme, the jurisdiction will make promotions to meet departmental and community needs. A basic requirement is that the candidate must be medically qualified to assume the new job. Most promotional job announcements require successful candidates to have an appropriate medical or physical performance rating. A candidate who is medically restricted or cannot complete the required physical ability assessment may not be considered for promotion until these issues are resolved FIGURE 2-2.

FIGURE 2-1 Many fire fighters and officers enjoy their work so much that it is common to find them working beyond their retirement eligibility.

FIGURE 2-2 Most promotional job announcements require that the candidate have an appropriate physical performance rating.

Assessment Center Tips

Walk the KSAs

Fire Fighter I training included an end-of-course written exam and demonstration of practical skills. All of the knowledge and psychomotor preparation was designed to enable the candidate to complete the certification exam successfully. The same approach should be taken when preparing for a promotional exam, because the candidate is demonstrating mastery of the knowledge, skills, and abilities identified for a managing or supervising fire officer.

In addition, the candidate must not be the subject of active formal discipline. Departments often identify a required time period that must pass between a formal disciplinary action and consideration for an officer promotion. This period, which varies from a month to a full year, will be described in the jurisdiction's personnel regulations, the department's administrative procedures, or the current labor contract. For example, a candidate who receives a 2-day suspension without pay for repeated tardiness in reporting for duty may not be considered for promotion for 365 days, even after ranking number 1 on a lieutenant eligibility list.

The increasing complexity of the fire department mission adds variables to the decision of who should be promoted. Some specialty supervisory positions require additional certifications or work experience. For example, to be promoted into the position of arson squad captain, the candidate also needs to have certification as an NFPA 1033 Fire Investigator and 2 years of experience as a fire investigator. Other assignments may include preferred qualifications. For example, the department may prefer promoting bilingual candidates in fire companies that serve diverse communities.

Each jurisdiction has a promotional process that evolved as a result of community needs, consent decrees, lawsuit settlements, arbitration decisions, grievance settlements, labor contracts, and memoranda of understanding. It is important to learn which rules, goals, or practices affect your department's promotional practice. Notably, results of the once-a-decade census document a community's diversity and may influence public safety promotion practices.

When Fire Officers Are Voted In

The American fire service started with neighbors helping neighbors using bucket brigades and long hooks. During that era, the community of fire fighters voted for their company foreman.

Today, some states do not require certification training for volunteer fire officers, instead depending on the authority having jurisdiction to establish experience and training requirements for fire company and command officers. Completing a fire officer training program will benefit the newly elected fire officer, providing information and knowledge of procedures that can assist the volunteer fire department in serving its community.

Preparing a Promotional Examination

The preparation of a promotional examination usually involves a combined effort between the fire department and the municipality's human resources section. Some jurisdictions may contract with a testing organization or a consultant to organize, develop, or deliver a promotional exam. Public safety promotional examinations are the most complex and detailed examinations conducted by a municipality.

The fire department usually establishes a test preparation committee consisting of three or more officers, chaired by a chief officer. These individuals are the subject-matter experts and are responsible for validating the technical content of the promotional examination. If the promotional examination is developed within the agency, the test committee members write the questions and establish its content **FIGURE 2-3**.

■ Charting the Required Knowledge, Skills, and Abilities

The municipality's personnel or human resources department uses two documents to define the <u>knowledge, skills, and abilities (KSAs)</u> that are required for every classified position within the municipality: a narrative job description and a technical class specification. Fire officers need to be familiar with both of these documents to prepare for a promotional examination.

A narrative <u>job description</u> summarizes the scope of the job and provides examples of the typical tasks a person holding that position would be expected to perform. The job description also lists the KSAs needed at the time of appointment and any special requirements that apply. Examples of special requirements could include possession of a valid driver's license and an Emergency Medical Technician certificate. Other requirements might include a Class "A" medical rating, passing a work performance test, or achievement of technical or professional certifications such as NFPA Fire Officer I. When the human resources department posts a promotional announcement, a full or partial narrative job description is usually included.

Human resources departments also prepare a technical <u>class specification</u> worksheet to quantify the KSA components of every classified municipal job **FIGURE 2-4**. The classification

FIGURE 2-3 If the examination is developed within the agency, test committee members establish its content.

NEW HANOVER COUNTY, NC
CLASS SPECIFICATION

CLASS TITLE: Fire Lieutenant

CLASS CODE:		
DEPARTMENT: Fire Services	ACCOUNTABLE TO: Fire Captain	FLSA STATUS: Non-exempt

CLASS SUMMARY:
Incumbents are responsible for shift supervision of County firefighters. Duties include: supervising and evaluating staff; overseeing responses to reports of fires; supervising rescue operations; preparing work schedules; writing reports of firefighting and rescue activities; training personnel; coordinating equipment and facility maintenance; presenting fire prevention programs; and, representing the department at special events.

DISTINGUISHING CHARACTERISTICS:
The Fire Lieutenant is the second level in a five level firefighter series. The Fire Lieutenant is distinguished from the Firefighter/Apparatus Operator in that it has shift supervisory responsibilities. The Fire Lieutenant is distinguished from the Fire Captain which has full supervisory authority.

DUTY NO.	TYPICAL CLASS ESSENTIAL DUTIES: (These duties are a representative sample; position assignments may vary.)	FRE-QUENCY
1.	Supervises two or more full-time staff to include: prioritizing and assigning work; conducting performance evaluations; ensuring staff are trained; and, making hiring, termination, and disciplinary recommendations.	Daily 25%
2.	Oversees response to reports of fires which includes: preparing pre-incident surveys; securing the scene; extinguishing the fire; salvaging structures and their contents; and providing emergency medical services to injured parties.	Daily 15%
3.	Supervises rescue operations by overseeing extrication activities and providing emergency medical services to injured parties.	Daily 15%
4.	Prepares daily and weekly work schedules for firefighters.	Daily 10%
5.	Prepares written reports of fire and rescue activities.	Daily 10%
6.	Provides training to shift personnel and students at the County fire academy on firefighting, rescue, and emergency medical topics.	Daily 10%
7.	Coordinates firefighter maintenance of vehicles, fire and rescue equipment, and facilities.	Daily 5%
8.	Presents fire prevention and education programs to businesses, schools, and community groups.	Weekly 5%
9.	Represents the department at special events including parades and open houses.	Occasionally 5%
10.	Performs other duties of a similar nature or level.	As Required
11.	Performs work during emergency/disaster situations.	As Required

FIGURE 2-4 Sample technical class specification worksheet. *(Continues)*

NEW HANOVER COUNTY, NC
CLASS SPECIFICATION

CLASS TITLE: Fire Lieutenant

POSITION SPECIFIC RESPONSIBILITIES MIGHT INCLUDE:
- Does not apply.

Knowledge (position requirements at entry):
Knowledge of:
- General principles of fire science;
- Emergency management techniques;
- Basic principles of rescue;
- Hazardous materials management techniques;
- Emergency medical practices;
- Departmental policies and practices;
- Local and state fire ordinances.

Skills (position requirements at entry):
Skill in:
- Performing fire suppression and rescue operations;
- Driving a vehicle;
- Preparing and making presentations;
- Preparing written incident reports;
- Using a computer and related software applications;
- Supervising and evaluating employees;
- Communication, interpersonal skills as applied to interaction with coworkers, supervisor, the general public, etc. sufficient to exchange or convey information and to receive work direction.

Training and Experience (positions in this class typically require):
High School Diploma or General Equivalency Diploma (GED) and five years of related firefighting experience, including two years of progressively responsible supervisory experience; or, an equivalent combination of education and experience sufficient to successfully perform the essential duties of the job such as those listed above; Associate's Degree in Fire Science preferred.

Licensing/Certification Requirements (positions in this class typically require):
- Class B Driver's License;
- Firefighter II Certification;
- Must be able to obtain EMT and Fire Instructor Level II certifications within one year and Level I Fire Inspector and ERT Certifications within two years.

Physical Requirements/Working Conditions:
Positions in this class typically require: climbing, balancing, stooping, kneeling, crouching, crawling, reaching, standing, walking, pushing, pulling, lifting, fingering, grasping, feeling, talking, hearing, seeing, and repetitive motions.

Heavy Work: Exerting up to 100 pounds of force occasionally, and/or up to 50 pounds of force frequently, and/or up to 20 pounds of forces constantly to move objects.

Incumbents may be subjected to moving mechanical parts, electrical currents, vibrations, fumes, odors, dusts, gases, poor ventilation, chemicals, oils, extreme temperatures, inadequate lighting, work space restrictions, intense noises, and travel.

NOTE:
The above job description is intended to represent only the key areas of responsibilities; specific position assignments will vary depending on the business needs of the department.

Classification History:
Draft prepared by Fox Lawson and Associates LLC (CC).
Date: 10/99

FIGURE 2-4 Continued.

system is a core component of the civil service system and is used to determine the compensation level for a position. For example, the classification details would show why the base salary for a battalion chief is set 35 percent higher than that for a lieutenant, based on the KSAs.

The worksheet also provides a map for the promotional examination to identify the factors that need to be evaluated. A promotional examination should focus on the unique, high-importance tasks that distinguish one position from another.

Periodically, a classification specialist from the human resources department surveys the individuals within a classification to rank the frequency and importance of a wide range of job tasks. This survey is performed to validate and update the job description and the KSA technical worksheet.

Promotional Examination Components

There is no perfect promotional exam. Through trial and error, research, grievances, and court decisions, the following components are frequently utilized components of a promotional exam.

The decision of which components to use is influenced by time requirements, expense, staff expertise, and past experience. These components can be locally developed or provided by a vendor, consultant, or assessment specialist.

■ Multiple-Choice Written Examination

Multiple-choice written examinations are widely used in the promotional process because they can be structured to focus on very specific subjects and factual information. There is no element of style or creativity in answering a multiple-choice question; the candidate simply has to select the appropriate answer. The scoring is equally straightforward because an answer is either right or wrong.

The multiple-choice written examination concentrates on facts that can be found within the materials on a reading list. The reading list generally includes textbooks, reference books, standard operating procedures manuals, department rules and regulations, and other locally developed reference materials.

Using the KSA technical worksheet, the test committee determines the number of questions to include from each knowledge area.

The supervising fire officer examination usually includes many technical questions covering engine, truck, and rescue company operations. These questions are directed toward a candidate's ability to demonstrate the basic knowledge that would be important for a lieutenant. In general, the first-level supervisory examination has the longest and most diverse reading list, and as much as 70 percent of this test focuses on the technical aspects of a supervising fire officer. Typical technical questions cover building construction, incident management, hydraulics, emergency medical care, and firefighting tactics. Many examinations also test the candidate's knowledge in specialized company tasks, such as technical/heavy rescue, truck company operations, and rapid intervention teams.

Supervisory questions tend to focus on the immediate requirements affecting fire company preparedness, such as a subordinate unfit for duty or injured on the job. For example, a supervising fire officer examination might include a question about how to handle a fire fighter reporting late for duty. The hierarchical nature of the fire department requires that the first-level supervisor identify the problem, stop the behavior, and report to or consult with a senior officer or chief before taking any significant supervisory action.

A managing fire officer examination usually includes fewer technical questions and more administration questions because the captain position involves more management responsibilities. The promotion process would assume that the candidate has already met the qualifications for the lower-level position. The second-level supervisor might also be expected to develop a fire fighter's work improvement plan, prepare a budget proposal, evaluate fire company performance, or develop a department-wide training plan. For example, a lieutenant candidate could be asked questions that relate to evaluating a fire fighter's reporting-time performance over a year. The supervisor should be able to identify patterns and determine the underlying causes for tardy reporting.

There are three options for constructing a multiple-choice written examination. Sometimes the local exam committee develops the test and the committee members write the questions. Alternatively, private companies may develop generic examinations that are used by many fire departments. The department has the third option of hiring a consultant to write a more specific examination that is directed toward local priorities.

Larger fire departments are more likely to develop an examination internally because they have the necessary resources. For example, the fire chief in a large city can probably assign a command officer and five to seven company officers to write the questions for a supervising fire officer examination.

The committee developing the examination first determines how many questions are needed to assess the candidate's knowledge within each particular area. For example, the committee might decide that nine building construction questions should be included in an examination. Two of the committee members are assigned to work together and develop 15 multiple-choice questions on building construction. The whole committee would then meet to select the nine best questions for the examination.

The selection process includes evaluating each question for job-content/criterion-referenced validity and reliability. Job-content/criterion-referenced validity means that the committee has certified that the knowledge being measured by the question is required on the job and referenced to known standards. The job requirement is identified through the KSA technical worksheet developed from the job description and class specification. The criterion comes from the appropriate NFPA standard and is linked to a published reference document. The best test questions selected by the committee are the ones with superior job-content/criterion-referenced validity.

Reliability is the characteristic where a test measures what it is intended to measure on a consistent basis. Test-and-item analysis can be performed to determine whether the more able candidates were less likely to select an incorrect answer. With a poor question, high scorers tend to answer the item incorrectly more frequently than low scorers.

A smaller department would be more likely to go to an external source for the examination questions, simply because of the time and effort that would be required to write and validate them internally. Writing good examination questions is difficult work. Each question has to be fully researched and carefully worded. A question can be challenged and invalidated if the wording is not clear, if there is no correct answer, or if more than one of the answers provided could be correct.

In many fire departments, multiple-choice written examinations are the only assessment tools used for promotional examinations. The simplicity of a multiple-choice examination can also be viewed as a weakness, however. A candidate who knows all of the facts can do very well on this type of examination, even if the individual is unable to apply the information in a real-life situation. The description "book smart, street dumb" has often been used in complaints about a process that relies solely on written multiple-choice questions to promote officers.

■ Assessment Centers

Some public safety agencies began to use assessment centers in promotional examinations in the late 1970s. An assessment center comprises a series of simulation exercises that are used to evaluate a candidate's competence in performing the actual tasks associated with a job. The assessment center process was developed in the officer corps of the German army in the 1920s. The goal of the program was to select future officers based on their predictive performance in a 2- to 3-day assessment procedure. The same type of process can be developed for fire officers.

In-Basket Exercises

The most common assessment center activity is the in-basket exercise **FIGURE 2-5**. It asks the candidate to deal with a stack of correspondence and related items that have accumulated in a fire officer's in-basket. This timed exercise measures the candidate's ability to organize, prioritize, delegate, and follow up on administrative tasks. Those individuals who do well on in-basket exercises tend to demonstrate successful managerial performance in real life.

A typical in-basket contains the following items:

- Instructions for the exercise
- A calendar
- An organizational chart or list of personnel
- Ten to 30 exercise items

An in-basket assessment generally begins by positioning the candidate as a newly promoted officer reporting to the fire station for the first time. The former officer suddenly retired or left the fire department, leaving a stack of issues that require attention in the in-basket. The candidate has a specified amount of time to go through the officer's in-basket and decide what to do with each item.

The situation is contrived so that the candidate is unable to call or contact anyone else directly for advice or assistance. The candidate might be provided with copies of the department's standard operating procedures and reference manuals. Hidden within the in-basket are surprises and time-critical issues, such as an important report that is overdue or an activity that has to be performed immediately. Frequently, the candidate encounters a scheduling conflict, such as an order from the operations chief to have the engine company report to a multiple-company drill at the same time the apparatus is scheduled for an annual pump test at the shop. Many in-baskets also include a writing exercise that could require the candidate to prepare a letter for the chief's signature or complete an incident report using correct formatting, spelling, and grammar.

From the test administrator's viewpoint, in-baskets are generally easy to organize and grade. This type of test can be difficult to develop for the lieutenant level, however, because many of the typical job tasks carried out by this fire officer in a fire station do not readily lend themselves to an in-basket assessment.

The following is a suggested method for handling promotional in-baskets:

- **Review.** Look at every item and determine which items are important (critical) and urgent (involving time constraints), which are important but not urgent, and which are unimportant.
- **Prioritize.** Handle the important and urgent items first, followed by the important and nonurgent items. When these are finished, you will have handled all items or will have only unimportant items left.
- **Identify resources/options/alternatives.** Determine who can handle the items, which options you may have in how they can be handled, and which alternative courses of action are available.
- **Follow up.** Provide some form of follow-up on all delegated items.
- **Make notifications.** Ensure that appropriate persons are notified of necessary and critical information.

Candidates can prepare for an in-basket exercise by practicing on sample exercises obtained from study materials publishers. Both lieutenant and captain in-basket exercises are

FIGURE 2-5 The in-basket exercise is a common assessment center event.

available; these include answers with explanations and descriptions of the dimensions assessed.

■ Emergency Incident Simulations

Emergency incident simulations are often included in a promotional examination to test the candidate's ability to perform in the role of officer at a fire or some other type of situation. Emergency incident management usually occupies a small amount of an officer's actual time, yet the required skill sets can be the most critical for evaluation in a promotional examination. An officer must be able to lead, supervise, and perform a set of essential skills at an emergency incident scene. Emergency incident simulations will take one of four formats.

In the first format, the candidate is provided with information concerning an emergency situation, usually in a written format that includes pictures or preincident plan information. The candidate has to explain which actions would be taken in the situation described and which factors would be considered in the making of those decisions. This kind of exercise is known as a "data dump" question because it provides an opportunity for the candidate to demonstrate a depth of knowledge about a particular subject. For example, the evaluators might find out how much expertise the candidate can demonstrate regarding basement fires in mercantile properties.

In the second format, the candidate is provided with a set of basic information concerning an emergency incident to begin the exercise. As the simulation progresses, the candidate is provided with additional updates of the situation, such as the following: "Immediately after you arrive on Engine 1, the store manager reports an acrid, unknown odor coming from a leaking package in the rear of the store. It has made a dozen people sick. Which actions will you take? What will you report to dispatch?" The candidate has to react to the unfolding situation, which can change based on the answers to the previous questions.

In the third format, the candidate participates in an interactive emergency scene simulation in a classroom or incident simulation trainer. The candidate is presented with a dynamic emergency incident through a multimedia format that typically includes full-color pictures depicting the incident. The graphics often include simulated smoke and flames superimposed on photos of an actual building in the community. Using a handheld radio, the candidate is expected to command the event using the local jurisdiction's standard operating procedures and incident command tools (accountability boards, command post clipboards, Incident Management System worksheets) that are appropriate to what a person at that rank would use on the street **FIGURE 2-6**. This type of exercise is complex and can be expensive because the test-site administrator might need to arrange for role players and technical support staff to run the simulation event for every candidate.

The fourth simulation format attempts to make the conditions as realistic as possible by actually having the candidate don protective clothing, climb into the officer's seat on an apparatus, and respond to a realistic scenario with a crew of fire fighters. This type of exercise is usually performed at a large urban or regional training academy, where there are buildings

FIGURE 2-6 The candidate is expected to command the simulated event.

and props that allow for a consistent simulation for all of the candidates. The real-life format allows a candidate to function fully in the role of an officer; however, it requires extensive preparations and a large supporting cast.

■ Interpersonal Interaction

An interpersonal interaction exercise is designed to test a candidate's ability to perform effectively as a supervisor. In a typical interpersonal interaction scenario, the candidate has to deal with a role player who has a problem, complaint, or dilemma. The candidate is usually provided with background information to prepare for the situation, followed by 10 to 20 minutes of face-to-face interaction with the role player **FIGURE 2-7**.

Many of these assessments require the candidate to deal with a poorly performing or troubled employee. The background information could include the employee's last evaluation

FIGURE 2-7 An interpersonal interaction exercise tests a candidate's ability to perform effectively as a supervisor.

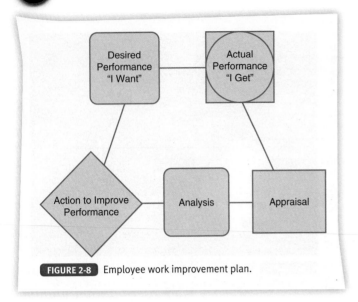

FIGURE 2-8 Employee work improvement plan.

report, a sick leave use printout, or recent disciplinary action. In many cases, the role player has important information related to the performance issue that is revealed only if the candidate asks the right questions during the interview phase. After the interview, the candidate may have to write up an employee improvement plan or a discipline letter FIGURE 2-8.

In an interpersonal interaction assessment, the key points are to meet the following criteria:

1. Maintain control of the interview.
2. Tell the employee the exact behavior you expect (e.g., arrive on time, wear your seat belt).
3. Give the employee a deadline to demonstrate a consistent behavioral change.
4. Specifically arrange for follow-up meetings. ("We will meet once a week to review your progress in meeting this goal.")
5. Attempt to get the employee to buy into or take personal responsibility for the improvement plan.
6. Be an empathetic listener, but remain focused on the reason for the interview.
7. Clearly explain the consequences if the employee's behavior does not change or improve.
8. Try to finish the session on a positive note ("You are a valuable member of this company and I am counting on you . . .").

A successful approach for an interpersonal interaction exercise is for the candidate to demonstrate the qualities of an "extreme supervisor." When this approach is used, any issue is handled in strict accordance with the rules and regulations of the organization.

An alternative scenario for this type of assessment is to have the role player confront the candidate as an angry or frustrated citizen. This exercise would assess how well the candidate can handle a customer service problem or a citizen complaint. This type of exercise is usually based on some situation that has actually occurred.

■ Writing or Speaking Exercise

A fire officer candidate should expect to deliver a short oral presentation or write a memo or report as part of the assessment center process. The oral presentation assesses oral communication, planning and organizing skills, and persuasiveness. Sometimes the candidate is required to write a report or a letter for a superior officer's signature. The written report assesses the candidate's written communication, problem analysis, and decision-making skills.

The oral presentation could be incorporated into one of the other exercises. For example, a candidate might be required to make a brief oral presentation to explain the size-up considerations of the emergency incident simulation.

Many assessment centers include an oral interview session in which the candidate is asked questions by an interview board FIGURE 2-9. One of the most popular questions at this type of interview is to ask the candidate why he or she should be selected for promotion. The response provides an opportunity for the candidate to demonstrate skills or knowledge in a particular area or even to correct an error made in a previous exercise. In some cases, a candidate's ability to recognize an error and correct a response could be viewed as equal to having the right answer to every question.

Videos are available from emergency service publishers and individual subject-matter experts to assist candidates in preparing to make a great presentation in an oral interview. The *Fire Officer Communications* chapter discusses fire officer communication skills in more detail.

■ Technical Skills Demonstration

Fire officers are expected to be working supervisors and to be skilled in both task- and tactical-level activities. For some supervising fire officer examinations, the candidates might also be required to demonstrate very specific technical skills. This type of exercise is often included in promotional tests for officers who would command highly specialized teams, such as urban search and rescue, hazardous materials, fire investigations, and paramedic teams.

FIGURE 2-9 Many assessment centers include an oral interview session.

Near-Miss REPORT

Report Number: 12-0000149

Event Description: During a rainstorm, heavy fire and smoke conditions were pushing from the second floor of a two-story commercial building. A nonoperational hydrant was located in front of building. After conferring with the store manager on the first floor, it was determined that there were no workers or occupants on the second floor. Heavy black smoke was banked down to the top of the second floor stair landing.

The first-due truck company had opened the scuttle over the stairs and a large hole over the main body of fire. This action did not alleviate conditions inside the building because the heavy rain prevented the smoke and heat from lifting. After seeing the color of the smoke change from dark gray to green and the smoke not lifting, I recommended to the incident commander (IC) that all fire fighters evacuate from the second floor in preparation for an exterior attack, and the IC concurred. A few minutes later, heavy fire was pushing out the front windows with stock to the ceiling. The life hazard was us (fire service). At street level we cooperated with a full roll call while aerial apparatus were simultaneously being supplied with large-diameter hose lines from nearby engines.

Lesson Learned: The safety chief (or designated officer) won't arrive for a while. It is incumbent that all supervisory personnel at the scene act as safety officers (don't wait for an invitation). Pick the brains of store managers or responsible people who know the building. If you have a bad hydrant, heavy rain, or other complicating factors, do not let the troops continue to occupy the building. If it doesn't seem right, it probably isn't. It is important to find out early if civilians are unaccounted for, as it's possible they got lost in the fire area.

Courtesy of the National Fire Fighter Near-Miss Reporting Systems (firefighternearmiss.com)

Preparing for a Promotional Examination

To prepare for an examination, the candidate must master both the content of the examination (technical and reference information) and the process (answering multiple-choice questions or role playing as an officer). Most of the content comes from the materials on the reference reading list; however, you should also be prepared for questions that relate to current issues and recent local history. For example, if your department recently had a problem with on-duty fire fighters taking emergency incident scene videos and posting them on a Web site, do not be surprised to find that issue somewhere in the examination. You should stay current by reading fire service publications and periodicals to keep up with trends and issues.

Mastering the content for an examination requires a personal study plan. Many tests require fire officer candidates to absorb a tremendous quantity of factual information, which can require a major investment of time and effort. It is not unusual for candidates to spend more than a year preparing for an examination, setting aside a period of time every day for reading and answering practice questions.

Fire fighters who participate in structured programs and study groups often perform better than fire fighters who study on their own **FIGURE 2-10**. Many fire departments have officially organized programs to help candidates prepare for examinations. In other cases, more informal study groups or preparatory programs are available for candidates who want to be well prepared. If your fire department does not have a promotional preparation program, you can set up your own study group. The group members support one another, share materials, and even develop their own multiple-choice questions.

FIGURE 2-10 Fire fighters who establish study groups often perform better than fire fighters who study on their own.

VOICES OF EXPERIENCE

Following a tough workday, and several months into my career as a promoted fire captain, the phone rang. The voice on the other end of the line was all too familiar as I heard the greeting, "Were you prepared for this?" I had heard this exact question asked in past firehouse stories and would continue to hear it as fire fighters prepare for promotion. Yes, I was prepared, and having learned how to handle situations to help others grow was extremely satisfying.

Preparing for promotion was one of the biggest investments in my career. After preparing for several promotional test and assessment interviews, and after watching others succeed, I have observed a great deal of what can successfully assist an individual in his or her own career enhancement. Proper preparation will take some monetary expense, but the quality time spent studying and gaining the necessary experience, though it can be overwhelming, will also award you with excellent leadership abilities. You will be prepared to handle the many situations that the dynamics of the fire service will place in front of you. You will be able to mitigate hazards, control company operations on scene, supervise and manage a firehouse responsibly, and understand the necessity of time and performance management. I have seen many prepared individuals walk out of the fire chief's office with a new bugle, yet fail to have that winning lottery ticket expression. Fast forward several months, however, and these individuals are steps beyond those who "got lucky." Sure, some can be successful at a test without putting forth the effort, but I have seen too many of these individuals sweat over the minor details that face an officer on a day-to-day basis.

Utilize your resources, practice officer skills, and invest in your future.

When your fire department provides the information for a future promotional process, you need to be "in the zone." Obtain personal copies of all of the promotional information and know the process. Fire fighters tend to listen to the rumor mill and end up with false information. Know what your department expects of you and all of its potential officers. Preparing yourself for the promotional process means obtaining all of the knowledge, skills, and abilities (KSAs) that you can, and utilizing as many resources as possible.

Take fire and educational courses outside of what your fire department has to offer. These courses will provide you with valuable information and skills that will remain with you throughout your career. Some of these courses may count toward a promotional process, but do not hang your hat on just those classes. Remember that this is an investment. Diversify your portfolio. A good company officer should provide you with the opportunity to "ride in the front seat" as the acting officer. Making decisions on incidents, and even throughout a station workday, will allow you to make decisions in a safe environment and with assistance from your officer.

Much of an officer's rank will lend itself to time management. Learning how to do the paperwork and company reporting will give you a chance to practice for a possible assessment center in-basket and also provide you with the knowledge needed for your written exam.

Obtaining textbooks and finding the time to study is what tends to be most overwhelming for fire fighters. Other than a few hundred dollars for textbooks and a few bucks for highlighters and note cards, it is the study time that will be your biggest investment. Explaining to family members and friends what you are trying to accomplish will assist you in your endeavors. Let them know what you want and what you are willing to sacrifice—for them and for your future.

Being able to answer "yes" to the question of whether you were prepared for promotion will end up being one of your most rewarding moments. This will be when you realize that all of your time, money, and sacrifice were worth it. Utilize your resources, practice officer skills, and invest in your future. The fire service is always looking for those who come prepared to lead their men and women.

Chris Riley
Fire Captain
Portsmouth Fire, Rescue, and Emergency Services
Portsmouth, Virginia

Building a Personal Study Journal

One valuable tool in studying for a promotional examination is a personal study journal. You can use this journal to set up your personal study schedule, keep track of your progress, and make notes about confusing, interesting, or important facts you discover. The first section could include all of the information about the promotional examination—a copy of the official announcement and copies of all of the documents the candidate was required to submit. Any test-related documents from the department can also be placed in this section.

The second section would contain a calendar, usually in the one-month-per-page format. The calendar covers the period from the start of the preparation process through the last promotional test activity. Important deadlines are highlighted, and the candidate can create a study plan, working backward from the examination date.

The third section would contain the written reference materials. Subdivided by reference, the parts of this section would cover the items used to master the written material. Some candidates outline each chapter; others make up practice tests. You should make a note of any questions you miss in practice tests to be sure that you go back and obtain the correct answers. This information can be extremely valuable during your final review before the examination date.

The fourth section would cover information about the announced or anticipated components of the promotional examination. Some departments provide information to the candidates about topics that may be emphasized or areas of the references that will be excluded. For example, the department may clarify that candidates are expected to conform to the mayday procedure that was in effect before July 1. In such a case, a new procedure may be in effect on the day of the promotional test, but the exam will test the candidate on the earlier procedure. The fourth section of your personal study journal is also where you would store information about the exam structure and process.

Preparing for Role Playing

The most effective candidates in role-playing exercises are the ones who act naturally during the examination, while making supervisory decisions that are strictly based on policies and

---

Getting It Done

Captain Jonah Smith Reflects on His First Officer

I had the good fortune to work for an officer who was one of the best in the department. He treated me just like a rookie should be treated: He gave me a hard time, pushed me to be better, mentored me, and most of all taught me about many parts of the service, especially tactically. I was assigned to him straight out of rookie school, and rode the back with two of the best fire fighters around. When I left that department for my current one, he was mad that I left but had done everything in his power to help me get the job with the new department. To me that showed me what kind of man he truly is.

When I was promoted to company officer a few years ago, I of course invited him because he had such a large influence on me over time. He had some issues and was unable to attend, but he did take the time to send me the e-mail shown here. He gave me permission to put this out there, but he will remain nameless.

> You are about to venture on the most challenging yet rewarding job you have faced in the fire service. There will be times when you will think, "Why didn't I stay a fire fighter?", but those days will soon disappear. I have this quote that I keep on my desk, and I look at it often: "A leader with great passion and few skills will always outperform a leader with great skills and no passion."
>
> You are lucky—you have the passion for the fire service and the knowledge, skills, and ability to be an outstanding captain. Not everyone has both. Keep that passion, work to develop those skills, and you will rise to the top. Remember this: If everyone below you is successful, then you are successful; and if they fail, it is because you failed them.
>
> Keep a humble attitude toward your men. Make sure they know that when the team is successful, they did it, and when the team fails, you did it. When they know you have their backs and will take hits for them, they are more likely to have yours.
>
> Always remember to hold the fire (where is it and where is it going—cut it off) until the cavalry arrives, and never forget the hook. You are truly missed. I hope you know that I think the world of you and desperately hate not being there.

Courtesy of Captain Jonah Smith

Getting It Done

Textbook Study Guides

A variety of publishers produce study guides with hundreds of multiple-choice questions taken from fire service textbooks and references. The answer to each question is keyed to a page number in the textbook or source material. Some book publishers release their own study guides, whereas others specialize in producing promotional study guides for specific examination levels.

Most publishers produce computer-based practice tests. In one version of computer-based assessment, a student can have the software assemble test questions based on the level of difficulty and subject area. Once the student has completed the test, the program generates the results, along with links back to textbook references.

regulations. This behavior is sometimes described as performing in the "extreme supervisor" role. Many successful candidates model their performance in the examination, as well as afterward, on the behavior of successful and respected officers. It is a good idea to pay attention to the supervisors and officers who are known for being efficient and effective.

One of the best ways to prepare for role playing is to experience working in one of the busier or larger fire stations within your department. More supervisory issues come up in a fire station with 14 fire fighters and 3 officers on each shift than in a station with 3 fire fighters and 1 officer. Similarly, a fire company that runs 15 calls per day encounters more problem-solving opportunities than a company that runs 3 calls per day. Working under one of the widely respected officers is also a good way to develop supervisory skills. Candidates who have worked for a supervisor who demonstrates excellent leadership and serves as a positive role model have a significant advantage over candidates who have never had that experience.

You Are the Fire Officer Conclusion

Kinders has been talking with fellow recruit school graduates about the upcoming exam. Some are creating a study group, including a Web site. A recruit school colleague says, "You can't be promoted if you don't take the exam." She suggests that you join the study group and give it your best shot.

Kinders asks Lieutenant Williams for study plan suggestions. Williams shows you the personal study journal he used for the last exam. Williams offers to spend some time every on-duty evening to assist the members who are studying for promotional exams.

Wrap-Up

Chief Concepts

- Promotional examinations were a product of the 1883 Pendleton Civil Service Reform Act.
- In most fire departments, completion of a promotional examination process creates an eligibility list that lasts from 2 to 6 years. Depending on local practice, this list may be either rank ordered or banded.
- Each jurisdiction has a promotional process that evolved as a result of community needs, consent decrees, lawsuit settlements, arbitration decisions, grievance settlements, labor contracts, and memoranda of understanding.
- The preparation of a promotional examination usually represents the combined effort of the fire department and the municipality's human resources section.
- The municipality's personnel or human resources department uses two documents to define the KSAs that are required for every classified position within the municipality: a narrative job description and a technical class specification.
- The decision of which components to use in a promotional examination is influenced by time requirements, expense, staff expertise, and past experience.
- Multiple-choice examinations concentrate on facts from the reading list.
- Assessment centers provide a variety of role-playing exercises.
- During an in-basket assessment, you should remember to review; prioritize; identify resources, options, and alternatives; follow up; and notify.
- Emergency incident simulations take one of four formats: the "data dump" question, a mock emergency incident, an interactive emergency scene simulation, or a full-scale incident simulation.
- During an interpersonal assessment, remember to remain in control, exactly state the desired behavior, give the employee a deadline, arrange follow-up meetings, get the employee to buy into or take personal responsibility for the improvement plan, be empathetic but focused, explain consequences, and finish the session on a positive note.
- Fire officer candidates may be required to deliver a short presentation or write a memo or report as part of the assessment center process.
- Technical skills may be evaluated during promotional tests for highly specialized positions.
- The candidate needs to develop a personal study plan to master the content for a promotional examination. Study techniques include keeping a study journal and participating in role-playing activities.

Hot Terms

<u>Assessment centers</u> A series of simulation exercises to identify a candidate's competency to perform the job that is offered in the promotional examination.

<u>Class specification</u> A technical worksheet that quantifies the knowledge, skills, and abilities (KSAs) by frequency and importance for every classified job within the local civil service agency.

<u>"Data dump" question</u> A promotional question that asks the candidate to write or describe all of the factors or issues covering a technical issue, such as suppression of a basement fire in a commercial property.

<u>Dimensions</u> Attributes or qualities that can be described and measured during a promotional examination. On average, 5 to 15 dimensions are measured on a promotional examination. The six most commonly addressed are oral communication, written communication, problem analysis, judgment, organizational sensitivity, and planning/organizing.

<u>In-basket exercise</u> A promotional examination component in which the candidate deals with correspondence and related items that have accumulated in a fire officer's in-basket.

<u>Job-content/criterion-referenced validity</u> A type of validity obtained through the use of a technical committee of job incumbents who certify that the knowledge being measured is required on the job and referenced to known standards.

<u>Job description</u> A narrative summary of the scope of a job. It provides examples of the typical tasks.

<u>Knowledge, skills, and abilities (KSAs)</u> The traits required for every classified position within the municipality. KSAs are defined by a narrative job description and a technical class specification.

<u>Personal study journal</u> A personal notebook to aid in scheduling and tracking a candidate's promotional preparation progress.

<u>Reliability</u> The characteristic where a test measures what it is intended to measure on a consistent basis.

<u>Spoils system</u> Also known as the patronage system; the practice of making appointments to public office based on a personal relationship or affiliation rather than because of merit. The spoils system scandals of the New York City "Tweed Ring" and the Tammany Hall political machine (1865–1871) resulted in Congress passing the Pendleton Civil Service Reform Act of 1883.

References and Additional Resources

NFPA reprinted material is not the complete and official position of the NFPA on the referenced subject, which is represented only by the standard in its entirety.

Blubaum, J. E. (2010). *Promotional Practices: Adaptive Change Issues Facing the Moscow Volunteer Fire Department*. Emmitsburg, MD: U.S. Fire Administration, Executive Fire Officer Program.

Compton, D., and G. Mack. (2004). *The Mental Aspects of Performance*. Stillwater, OK: IFSTA/FPP.

DeNavas-Walt, C., D. P. Bernadette, and J. C. Smith. (2012). *Income, Poverty, and Health Insurance Coverage in the United States: 2011*. U.S. Census Bureau, Current Population Reports, P60-243. Washington, DC: U.S. Government Printing Office.

Drucker, P. F. (1990). *From Mission to Performance: Managing the Nonprofit Organization*. New York, NY: HarperCollins, 53–106.

Filer, R. J., and R. R. Farr. (1982). *Analysis of Assessment Center Performances of Candidates for Captain and Assistant Chief, Montgomery County, MD*. Richmond, VA: Psychological Consultants, Inc.

Filer, R. J., and R. K. Filer. (1977). *Assessment Centers: Development and Use*. Richmond, VA: Psychological Consultants, Inc.

Fire and Emergency Service Image Task Force. (2013). *Taking Responsibility for a Positive Public Perception*. Fairfax, VA: International Association of Fire Chiefs.

Kastros, A. (2006). *Mastering the Fire Service Assessment Center*. Tulsa, OK: Pennwell/Fire Engineering.

Kiechel, W. K. III. (2012). "The Management Century." *Harvard Business Review* 90(11): 13.

Knowles, M. S., et al. (2005). *The Adult Learner: The Definitive Classic in Adult Education and Human Resource Development*. Burlington, MA: Elsevier, Butterworth, Heinemann.

Kreiger, A., and T. Masten. (2012). *Beyond the Consent Decree: Gender and Recruitment in the San Francisco Fire Department*. Emmitsburg, MD: U.S. Fire Administration, Executive Fire Officer Program.

LoSasso, C. (2012). *Mentoring Volunteer Officers Pre- and Post-promotion*. Emmitsburg, MD: U.S. Fire Administration, Executive Fire Officer Program.

Maher, P. T., and R. S. Michelson. (1992). *Preparing for Fire Service Assessment Centers*. Bellflower, CA: Fire Publications, Inc.

Mahoney, G. (2006). *Fire Department Interview Tactics*. Independence, KY: Delmar/Cengage.

McAniff, E. P., and J. J. Cunningham. (1974). *Leadership in the Fire Service*. Bayside, NY: McAniff Associates, Inc.

Metz, E. J. (2013). "New Venues for Discovering Fire and Emergency Services Literature." *Fire Technology* 49(2): 185–194.

Michelson, R. S., and P. T. Maher. (2009). *Assessment Centers for Public Safety*, 3rd ed. San Clemente, CA: LawTech.

Mittendorf, J. (2003). *Facing the Promotional Interview*. New York, NY: Fire Engineering.

Murtaugh, M. (1993). *Fire Department Promotional Tests. A New Direction: New Testing Components, New Testing Formats*. Pearl River, NY: Fire Tech Promotional Courses.

Myers, B. (2004). *Mentoring: Preparing Company Officers*. Emmitsburg, MD: U.S. Fire Administration Executive Fire Officer Program.

Reeder, F. F., and A. E. Joos. (2014). *Fire Service Instructor: Principles and Practice*, 2nd ed. Burlington, MA: Jones & Bartlett Learning.

Robinson, G. J. (2011). *Lasting Leadership: Preparing for the Transition to Company Officer*. Emmitsburg, MD: U.S. Fire Administration, Executive Fire Officer Program.

Roethlisberger, F. J., et al. (1939). *Management and the Worker: An Account of a Research Program Conducted by the Western Electric Company, Hawthorne Works, Chicago*. Boston, MA: Harvard University Press.

Smeby, L. C. Jr. (2014). *Fire and Emergency Services Administration: Management and Leadership Practices*, 2nd ed. Burlington, MA: Jones & Bartlett Learning.

Snook, J. W., et al. (2011). *A Leadership Guide for Volunteer Fire Departments*. Sudbury, MA: Jones & Bartlett Learning.

Sprouse, C. B. (2012). *When Good People Make Bad Decisions: Assessing Decision Fatigue in Las Vegas Fire and Rescue*. Emmitsburg, MD: U.S. Fire Administration, Executive Fire Officer Program.

Terpak, M. A. (2008). *Assessment Center Strategy and Tactics*. Tulsa, OK: Pennwell/Fire Engineering.

Thiel, A. K. (2012). Professional Development. In: *Managing Fire and Rescue Services*, A. K. Thiel and R. Jennings, eds. Washington, DC: International City/County Management Association, 247–272.

Tielsch, G. P., and P. M. Whisenand. (1978). *Fire Assessment Centers: The New Concept in Promotional Examinations*. Santa Cruz, CA: Davis Publishing.

Weinschenk, C., et al. (2008). "Analysis of Fireground Standard Operating Guidelines/Procedures Compliance for Austin Fire Department." *Fire Technology* 44(1): 39–64.

FIRE OFFICER in action

The Station 100 fire fighters have started a study group. Once every 2 weeks, the study group meets to review the material and take practice exams. Each member outlines a portion of a reference book and writes multiple-choice questions. A password-protected Web site includes a blog and an area to post the outlines, test questions, and other study material.

1. Which of the following actions is most valuable in the early phase of preparing for an announced promotional exam?
   A. Read more books on promotional exams.
   B. Ask individuals who scored high in the last promotional exam for preparation suggestions.
   C. Ask senior officers what they think will be on the promotional exam.
   D. Obtain old promotional exams from larger fire departments.

2. Which of the following is a key component in completing an in-basket exercise?
   A. Provide a "data-dump" response to each item.
   B. Demonstrate your mastery of a technical topic.
   C. Clearly show which items you consider "urgent."
   D. Properly document the follow-up activities.

3. A feature of an interpersonal interaction exercise is that the role player:
   A. has important information that is revealed only after detailed questioning by the candidate.
   B. reveals all of the essential information in the first 3 minutes of interaction.
   C. is expected to surprise or confuse the candidate through words or actions.
   D. does not know what the "correct" candidate response should be.

4. While preparing for the promotional exam, you learn about a firefighting technique that is superior to the method used by your department. You should:
   A. incorporate that technique into your response to the emergency management incident simulation.
   B. recommend to the test committee that they integrate this technique into the promotion process.
   C. identify locations within the promotional references that support the superior technique.
   D. make sure your exam responses reflect the current formalized practices of your department.

Fire Captain Activity

Lieutenant Williams meets with Captain Davis after the promotional process is completed: "I need your help to prepare for the next captain's exam."

NFPA 1021 Job Performance Requirement 5.2.3
Create a professional development plan for a member of the organization, given the requirements for promotion, so that the individual acquires the necessary knowledge, skills, and abilities to be eligible for the examination for the position.

Application of 5.2.3
1. Using a department you are familiar with, identify the education, certification, and experience requirements for a captain or managing fire officer.
2. Assume the candidate is a lieutenant with a high school degree and 30 hours of college credit. Describe a professional development plan so that the candidate will be well prepared before the next promotional process starts.

Fire Fighters and the Fire Officer

Fire Officer I

Knowledge Objectives

After studying this chapter, you will be able to:

- Describe the fire officer's basic tasks. (pp 42–44)
- Describe a typical fire station workday. (pp 44–45)
- Describe the transition from fire fighter to fire officer. (pp 45–46)
- Discuss the fire officer's role as a supervisor. (pp 46–47)
- Discuss the fire officer's role as a commander. (p 47)
- Discuss the fire officer's role as a trainer. (pp 47–49)
- Describe the activities a fire officer performs to maintain an effective working relationship with his or her supervisor (NFPA 4.1.1). (pp 49–50)
- Discuss the importance of integrity and ethical behavior (NFPA 4.1.1). (pp 50–51)
- Describe how to maintain workplace diversity (NFPA 4.1.1). (pp 51, 53–55)
- Describe the concept of the fire station as a business work location. (pp 55–56)

Skills Objectives

After studying this chapter, you will be able to:

- With a description of a fire station, work group, and schedule, prepare a beginning of shift report or activity plan (NFPA 4.2.2) (NFPA 4.4) (NFPA 4.4.2). (pp 42–43)
- Demonstrate the effective issuing of an unpopular order to a fire company. (pp 46–47)
- Demonstrate making a decision consistent with the department's core values, mission statement, and value statements given an ethical dilemma (NFPA 4.1.1). (pp 46–47, 51)
- Conduct an initial interview and notifications consistent with the department's policy, rules, and regulations given a harassment or hostile workplace complaint (NFPA 4.2.5). (pp 54–55)

Fire Officer II

Knowledge Objectives

There are no Fire Officer II knowledge objectives for this chapter.

Skills Objectives

There are no Fire Officer II skills objectives for this chapter.

CHAPTER 3

Principles of Fire and Emergency Service Administration (FESHE) Course Outcomes

1. Acknowledge career development opportunities and strategies for success. (pp 47–49)
6. Discuss the various levels of leadership, roles, and responsibilities within the organization. (pp 46–47, 49–50)
7. Describe the traits of effective versus ineffective management styles. (pp 49–50, 53–55)
8. Identify the importance of ethics as it relates to fire and emergency services. (pp 50–51)

You Are the Fire Officer

It is the first day for Lieutenant Taylor Williams on Quint 100. Arriving early, Captain Davis shows Williams around the shift officer work area and walks the lieutenant through the beginning of shift report on the department's intranet.

Davis explains that the quint officer is expected to submit the beginning of shift report by 7:15 A.M. and conduct the morning face-to-face line-up. This morning's line-up includes the following directive from the fire chief's office:

Effective immediately, the use of fire apparatus or ambulances to go to any establishment to purchase food or other station supplies is prohibited. A battalion chief is required to authorize any shopping trips.

This announcement makes for an uncomfortable morning meeting with the crew. It is clear that two of the fire fighters are very angry and start describing creative ways to combine a grocery store trip with an authorized activity. One plans to call the shop steward.

1. How do you respond to an order that you do not agree with?
2. How do you implement an unpopular directive?
3. What can you do to maintain the effectiveness of the work group?

Introduction

The fire officer is responsible for managing a work unit within the fire department. When we think about the duties of a fire officer, the most prominent vision is leading a team of fire fighters into a challenging emergency situation. In reality, much of what a fire officer does during a normal workday consists of routine administrative activities related to the work group, the assigned equipment, and the physical facility. The officer is expected to ensure that the work unit will be prepared to function effectively and efficiently when it is needed. That means managing personnel, resources, and programs. This chapter looks at a few overarching principles and examples of the day-to-day administration within a work group.

Supervising and managing fire officers usually report to higher-ranking chief officers. In a large fire department, the direct supervisor is usually an administrative fire officer, typically at the rank level of battalion or district chief. In a smaller organization, a fire officer might report directly to the fire chief or to a deputy or assistant chief. In most cases, this supervisor is an individual who has spent time coming up through the ranks, working as a supervising and managing fire officer.

Fire Officer I

The Fire Officer's Tasks

In a recent survey, chief officers working in cities and large counties responded to an e-mail request to "list the most important tasks you want a new fire officer to do well." This resulted in a list of four basic tasks that they consider vital: beginning of shift report, notifications, decision making, and problem solving. They believed that a new fire officer who meets these task expectations is on the right track.

■ The Beginning of Shift Report

Fire officers should provide a prompt and accurate report at the start of the workday **FIGURE 3-1**. This report is provided from each work location to the battalion or district chief within the first quarter hour of the reporting time. The format of the report may be electronic, paper, or verbal. Some departments use sophisticated online staffing systems, such as TeleStaff, that provide real-time scheduling that conforms to departmental operational requirements. The chiefs rely on this information to make staffing adjustments at the beginning of the shift. An accurate report is needed to ensure that adequate staffing and equipment are in place and ready for the balance of the shift **FIGURE 3-2**.

FIGURE 3-1 Providing the beginning of shift report is a vital company officer task.

Today's Date is: May 29

	Total	Paramedics	Fire Officer	EMS Officer	Prearranged Callback	Annual Leave	Vacancies	Detail Out-of-Operations	Injury Lv or Light Duty	LWOP	Fire OIC	EMS OIC
Minimums	10	3	2	1							Fire OIC	EMS OIC
Today's staffing	9	3	1	1	1	1	0	1	0	0	Smyth	Willow
Next day staffing	8	2	1	0							Smyth	????

Today's shortage	Why?	PM Surplus	Next Day's Shortage	Why?
Engine Officer	O/R	none	Engine Officer	Off Rep
			Medic Officer	Leave

Sick Leave	Detailed Out of Ops	Next Day APPROVED Leave
None	Capt. Johnson	FF Tolliver
		Lt. Willow

Injury Leave/Light Duty
None

		Vehicle Status	
		Engine 7746	Eng 46
Messages for the Chief:		Rescue 7099	Res 46
Vehicle 7234 overdue for preventative maintenance		Medic 6322	Med 46
Rescue 46 thermal imager broken		Reserve Engine 7234	Eng 35
Furnace malfunctioning		Reserve Medic 4276	Med 11
Fire Chief at 46 for dinner @ 1830		Battalion 9 5040	shop
		Reserve Suburban 5107	BC 09

FIGURE 3-2 Example of a municipal beginning of shift report.

Courtesy of Mike Ward.

The report provides the on-duty staffing information and sick leave list and identifies any positions that need to be filled for that shift. The positions are a priority because someone who worked the previous shift must remain on duty until a relief person shows up to fill any position that remains vacant. The battalion chief moves available staff to cover the vacancies noted on the beginning of shift reports, and any vacancies that remain have to be covered by fire fighters working overtime. It takes time to reassign on-duty personnel, call the overtime personnel, and get everyone to their work locations so the holdover personnel can be released. It is maddening when a new fire officer sheepishly calls the administrative fire officer 2 hours into a work shift to report that a fire fighter from the previous shift is still on involuntary holdover because someone failed to report the position vacancy on the beginning of shift report. By the time the position is covered, the fire fighter on involuntary holdover may have remained on duty for several more hours.

In addition, the beginning of shift report notes the location and condition of all of the apparatus or rolling stock, such as a reserve pumper that has been loaned to another station or an ambulance at the shop. Finally, the report provides the chief with any "must know" information that will require immediate attention.

Some reports include anticipated staffing for the next shift. This information allows the chief to make assignments in anticipation of the known vacancies in systems that lack an online staffing program.

■ Notifications

The second most important issue noted by the administrative fire officers in the previously mentioned survey was that the new supervising fire officer must make prompt notifications. Some types of information must be passed up the chain of command quickly. For example, all injury and infectious disease exposure reports must be processed without delay. If a fire fighter is exposed to a possible bloodborne pathogen early Saturday morning, the exposure report cannot sit on the fire officer's desk until Monday morning before it gets to the battalion chief. The designated infection control officer must be informed immediately so that he or she can get patient information while it can still be easily obtained. The same priority applies to any information the chief needs to know about when it is current, particularly before someone at a higher level calls to ask about the issue. Many chiefs call this the "no surprises" rule.

■ Decision Making and Problem Solving

The third and fourth issues cited by the administrative fire officers—decision making and problem solving—are of equal importance. Some chiefs complain that new supervising fire officers hesitate to make decisions. They seem to want the chief to make the hard decisions and to enforce unpopular rules. Chiefs typically want new officers to run their companies and make the decisions that are within their scope of responsibility. Chiefs are available for consultation, but they expect their officers to run the fire stations.

Fire officers should not complain about problems without proposing any solutions; they need to think through problems and contribute to solutions. The most valuable proposals from officers would consider the larger picture of how a possible solution would affect the rest of the department **FIGURE 3-3**.

■ Example of a Typical Fire Station Workday

A supervising fire officer is responsible for accomplishing the fire department mission through the efforts of the fire fighters under his or her command. At the company level, facilitating this outcome requires a balance of management and leadership skills. The officer must organize the work and provide the leadership to ensure that work gets done safely and effectively.

The fire department has an agency-wide mission that is translated into annual goals. These goals are used to develop annual, quarterly, and monthly objectives for each fire company. The administrative fire officer and the supervising or managing fire officer meet regularly to set these objectives and to review progress toward their achievement. The monthly goals show up as the planned activities on the fire company daily planner.

The following is an example of a 24-hour shift in a fire station. The starting point is a schedule that ensures all of the required tasks are completed in a logical order. The fire officer has to anticipate that emergency incidents will alter the workday and require adjustments in the schedule.

0700 Line-up and equipment check. Send beginning of shift report to chief. Clean quarters, empty trash, clean dishes.

0800 Dust and vacuum all carpeted areas. Sweep all tile floors.

0830 Physical training and outside skill drill. Go to store to pick up groceries.

1100 Heavy cleaning (while still in physical training clothes).
- Monday: Air out bunkroom and rotate mattresses; clean all windows.
- Tuesday: Clean utility rooms and shop area.
- Wednesday: Clean and inventory emergency medical service (EMS) equipment, self-contained breathing apparatus (SCBA), and decontamination areas.
- Thursday: Move recyclables outside for pickup, then clean weight room and lockers.
- Friday: Scrub kitchen and clean out refrigerators.

1130 Scrub bathrooms after fire fighters clean up from physical training.

Noon Lunch

1330 Scheduled productivity activity (e.g., fire safety inspections, school visits, inside or outside training).

1800 Dinner, followed by kitchen clean-up. Run dishwasher.

1930 Individual study time, occasional fire safety inspections (nightclubs) or drills.

2130 Remove all trash, tidy up day room, and make final pass through the kitchen.

Special station activity
- First and third Thursdays of the month: Scrub apparatus bay floor.
- Second and fourth Fridays of the month: Wax kitchen floor.
- February: Safety officer inspection of facility, apparatus, and personal protective equipment (PPE).
- March and September: Steam-clean carpeted areas.
- April: Fire chief's annual inspection of the fire station.
- May: EMS Week open house.
- October: Fire Prevention open house.

FIGURE 3-3 Discuss possible solutions to a problem with the chief.

Near-Miss REPORT

Report Number: 12-0000173

Event Description: An acting captain (a fire fighter usually assigned to an engine) was in charge of a tower ladder company that was dispatched to a house fire at a two-story Type V vacant house. While the company was en route to the scene, information was relayed that this was possibly an attic fire. Upon arrival, the company was given the assignment of securing utilities and asked to give a conditions report of the roof. Two fire fighters went to secure utilities while the driver/operator worked on setting up the aerial apparatus.

The acting captain was prompted to don his gear and make his way to the aerial apparatus. The fire fighters returned and gathered tools, and the two fire fighters and acting captain made their way to the bucket of the aerial apparatus. As the driver/operator maneuvered the aerial platform out of its cradle from the pedestal, the other crew members adjusted the tools and made a quick assessment of the scene. The driver/operator stopped maneuvering the aerial apparatus because he could no longer safely move them closer to the roof because of limited sight.

At this point, there was a delay in the decision of who would be taking over the final movements to get the aerial platform into place at the roof. Further, there was disagreement between the two fire fighters and the acting captain as to whether a parapet ladder should be used on the roof. Eventually, it was decided that they should use the parapet ladder; they proceeded to place this ladder in the bracket on the front of the platform bucket. The acting captain climbed down the parapet and broke the lightweight concrete tile with a trash hook, then sounded the roof decking. At this point, a roof report was relayed to command, and command asked the acting captain to watch the roof from the bucket.

This near miss occurred because the staffing allocation allowed a fire fighter who is not regularly assigned to the tower ladder to be the company officer. The fire fighter was not required to keep up his knowledge and not evaluated at any regular interval.

Lesson Learned: Personnel who act in positions of higher responsibility need to keep their knowledge base current. Personnel who are not assigned to specialty units (e.g., ladders, hazardous materials, technical rescue) should not be used to staff a position unless they are properly trained.

If a department has specialty apparatus and the expectation is that anyone can staff those apparatus, the department needs to provide regular training for those personnel who are not regularly assigned to those apparatus.

Courtesy of the National Fire Fighter Near-Miss Reporting Systems (firefighternearmiss.com)

■ Example of a Typical Volunteer Duty Night

The following example of a volunteer duty night is less detailed than the previous schedule, but incorporates the same essential tasks as a 24-hour shift in a municipal fire department, including equipment maintenance, training, and station maintenance. Many departments use a day book or monthly calendar to plan the duty crew activities.

- 1800 Evening duty crew starts. Equipment check.
- 1900 Dinner, followed by kitchen clean-up. Run dishwasher.
- 1930 Classroom session, skill drill, or community outreach activity.
- 2230 Remove all trash, tidy up, and make final pass through the kitchen. Take clean dishes and cups out of dishwasher and put away.

General station, company evolutions, and apparatus cleaning tasks are conducted on weekends. One section of the fire station gets a major cleaning two or three times each year (e.g., wax floors, scrub apparatus floors). Specialized heavy cleaning, such as steam-cleaning carpets and waxing apparatus, is scheduled throughout the year.

The Transition from Fire Fighter to Fire Officer

There are four times in a fire fighter's career when a major change occurs in how the individual relates to the formal fire department organization. The first change occurs when the fire fighter completes the probationary training period. Completing the initial training and probationary period is a major milestone—one that is marked in many departments by a change in helmet color or shield. The second change takes place when the fire fighter successfully completes a promotional process and starts working as a fire company commander. The third event is when the fire officer completes another level of training, advances through the promotional process, and starts working as a chief officer. The fourth event is when a fire fighter retires.

In all four situations, a significant change occurs in the individual's relationship to the organization and to the other members of the fire department. Part of this change is the individual's sphere of responsibility within the formal organization. As a full-fledged fire fighter, the individual shares the sense of mutual responsibility that prevails among the crew members on the rig. They work together to accomplish fire-ground tasks and look out for one another.

A promotion from fire fighter or driver/operator to company-level officer is a large step. The company-level officer is directly responsible for the supervision, performance, and safety of a crew of fire fighters. He or she has a sacred duty to ensure that all of the fire company team members remain safe when operating in a hostile or hazardous work environment. The company-level officer functions as a working supervisor, sharing the hazards and work conditions with the fire company team **FIGURE 3-4**.

This change in the sphere of responsibility often requires the new officer to change some on-duty behaviors or practices. The formal fire department organization considers a fire officer to be the fire chief's representative at the work location. This role creates an expectation that the fire officer will behave in a way that is appropriate for a first-line supervisor. Notably, behavior that was acceptable for a fire fighter may be unacceptable for a fire officer.

For example, consider a fire fighter who is known for developing elaborate practical jokes within the fire station environment. As a new fire officer, this individual would have to consider the impact that being seen as a practical joker would have on his or her ability to function as an effective supervising fire officer. Conversely, the new fire officer needs to consider how to respond to pranks and verbal jabs from fire fighters. What may have been an appropriate response from another fire fighter may no longer be the best response from an officer. Wearing the fire officer badge enhances the effect and consequences of any action or response.

Promotion to a chief officer rank changes the individual's relationship to the organization and the members to an even greater degree. A command officer has less of a hands-on role than an officer of a company. The command officer is typically directly responsible for several fire companies and must depend on company officers to provide direct supervision over crews performing fire-ground tasks, often in hazardous conditions. In many cases, the command officer works outside the hazardous area but is still responsible for whatever happens inside the site.

Fire Officer as Supervisor–Commander–Trainer

In his book *Effective Company Command*, James O. Page divided the company officer's duties into three distinct roles: supervisor, commander, and trainer.

■ Supervisor

In the supervisor role, the fire officer functions as the official representative of the fire chief. That means the fire chief expects that every fire officer will issue orders and directives and conduct business in a way that meets the chief's objectives. As part of that role, the fire officer is expected to supervise the fire company in a manner consistent with the rules and regulations of the fire department. For example, if all fire companies are expected to spend Tuesday afternoon conducting fire safety inspections in commercial properties, the fire officers are responsible for ensuring that their individual fire stations complete this task.

Unpopular Orders and Directives

On occasion, a fire officer may be required to issue and enforce unpopular orders **FIGURE 3-5**. Even if the officer disagrees with a particular directive, the formal organization requires and expects the officer to carry out that directive to the best of his or her ability. A fire officer can improve his or her effectiveness in handling an unpopular order by determining the story or history behind the order, which would enable the fire officer to put the directive in perspective.

When faced with an unpopular order, the fire officer should express any concerns and objections with his or her supervisor in private. This is the time for the fire officer to discuss suggestions for modifying or reversing the order. Occasionally, special circumstances may make the order difficult to implement, and the supervisor might be able to authorize adjustments. Once the meeting with the officer's supervisor is over, the formal organization expects the officer to enforce the order as issued or amended.

FIGURE 3-4 The company-level officer is responsible for the supervision, performance, and safety of a crew of fire fighters.

Courtesy of Captain David Jackson, Saginaw Township Fire Department

Fire Marks

James O. Page
In July 1971, James O. Page was promoted to battalion chief in the Los Angeles County Fire Department. In the same year he was admitted to the California Bar and finished the manuscript for his fire company supervision textbook, *Effective Company Command*. He became widely known as an author and publisher, the "father" of fire-based emergency medical services, and a practicing attorney.

FIGURE 3-5 An officer sometimes has to issue and enforce unpopular orders.

Telling the fire fighters that their officer does not agree with an order undermines the officer's authority and supervisory ability. The fire chief expects an officer to perform the required supervisory tasks, and the fire fighters must understand that the fire officer does not make all the rules or have a choice about which ones to enforce. Enforcing unpopular orders is part of the job.

■ Commander

When operating at the scene of an emergency incident, the fire officer is expected to function as a commander and to exercise strong direct supervision over the company members. In some cases, the fire officer could be responsible for directing the actions of additional resources, or he or she might function as the initial incident commander **FIGURE 3-6**.

Functioning as the initial incident commander on a major emergency is one of the higher-profile roles of a fire officer. The ability to bring order out of the chaos of an emergency incident is an art that requires a well-developed skill set. The fire officer needs to be clear, calm, and concise in the initial radio transmissions. The communication of incident size-up information must be consistent with the organization's requirements and Incident Management System (IMS).

FIGURE 3-6 At the scene of an emergency incident, the fire officer may function as the initial incident commander.

Fire Marks

Command Presence at Incidents

When retired Fire Department of New York (FDNY) Deputy Chief Vincent Dunn was a chief's aide in the 1960s, he drove for two different deputy chiefs on alternating days. Deputy chiefs were dispatched on multiple-alarm fires to function as the senior fire-ground commander. They directed the incident with the assistance of four to six battalion chiefs. Dunn, writing in *Command and Control of Fires and Emergencies*, compared the command presence of the two chiefs.

One of the deputy chiefs would arrive at the scene of a fire and not announce his arrival over the radio. He would don his helmet and fire coat, but stayed in his office shoes while he walked around the incident scene. He would stand next to the battalion chief who was running the incident and observe the activities. This deputy chief would have little impact on emergency operations.

The other deputy chief always maintained a high command presence when he responded to an incident. He would immediately announce his arrival on the fire ground over the radio. He would put on his fire boots, fire coat, and helmet and immediately report to the battalion chief who was running the incident. While the battalion chief provided a face-to-face status report to the deputy chief, Dunn would go to the rear of the fire building to observe conditions and report back to the deputy. Even if the battalion chief continued to function as the incident commander, everyone knew that the deputy chief was on the scene and providing direction. This deputy chief's command presence made him a much more effective leader of his team.

Developing a command presence is a key part of mastering the art of incident command. Command presence is the ability of an officer to project an image of being in control of the situation. To be a successful leader, the officer must convince others to follow by demonstrating the ability to take charge and make the right things happen. A fire officer who is going to establish command upon arriving at an emergency incident should have a detailed knowledge of the responding companies, a mastery of the local procedures, and the ability to issue clear direct orders. Fire fighters are aggressive, action-oriented people who also can be obsessive. It is important that a new fire officer develop a command presence to focus the efforts of this action-oriented team in a constructive direction **FIGURE 3-7**.

■ Trainer

The fire officer has the responsibility of making sure the fire fighters under his or her command are confident and competent in their skills. The company-level officer is responsible for the performance level of the fire company and must establish a set of expectations that the company will perform

FIGURE 3-7 It is important for the fire officer to develop a command presence.

at the highest level possible FIGURE 3-8. In a large department, one fire company may need to have a higher level of specialized skill in one area than in another. For example, a fire company may have a specialty assignment or a co-located specialized unit that the fire company staffs when it is requested. Technical rescue rigs, decontamination units, foam pumpers, mass-casualty units, and mobile command posts are examples of co-located specialized units.

In addition, the fire company's response district may require a higher level of fire fighter skill or knowledge. Consider hose handling skills. A fire company in a high-rise district would require different hose handling skills and competencies than those required of a company in a mountainous wildland interface area. Although both companies need to demonstrate familiarity with the basics of operating a hose line from a standpipe, the high-rise fire company should have a much higher level of expertise and competence in this skill. Conversely, the mountain fire company usually has a higher level of expertise and competence in rope rescue evolutions.

The company-level officer plays a key role in developing these competencies within the company. James O. Page makes three specific recommendations to assist fire officers in this task: develop a personal training library, know the neighborhood, and use problem-solving scenarios.

Developing a Personal Library

Page's personal library starts with a three-ring notebook with subject matter tabs. The subject tabs could match the topic headings in NFPA 1001, come from the recruit school curricula, or represent a personal list of important topics. Every time the fire officer attends a training event, the notes from that session are placed into the three-ring binder FIGURE 3-9. All related handouts, product information sheets, and other related items are also placed into the three-ring binder. When the officer is preparing to present a class that covers a specific topic, his or her personal training library is the first stop.

Leo D. Stapleton, a retired Boston fire commissioner writing in *Thirty Years on the Line,* advocates that officers maintain a personal journal where they record information about the incidents they run and the issues they handle. For working incidents, that would include what was encountered, what they learned, and how they would handle it the next time. Stapleton encourages officers to make these entries as soon as possible after the incident, while the information is vivid in the officer's memory.

Today, a fire officer can create an electronic version of Page's three-ring binder and Stapleton's journal with the use of a computer, scanner, and digital camera. As part of creating this personal electronic library, the fire officer can collect PowerPoint presentations and video clips, download files from the Internet, and convert various paper media into Adobe Portable Document Format (.pdf) files. A digital projector allows all these materials to be used in classroom sessions.

In addition to storing information in a three-ring notebook or on a laptop computer, Page and Stapleton recommend that the fire officer obtain personal copies of the textbooks and references used in fire fighter training and promotional examinations. You can highlight, tab, and write in this copy of the book until it becomes your personal reference. You can write notes in the margins and even note disagreements with the author's statement or point of view to personalize the learning experience. These kinds of annotations are especially valuable when you are reading a book in preparation for a promotional exam FIGURE 3-10.

FIGURE 3-8 The officer must ensure that fire fighters are confident and competent in their skills.

FIGURE 3-9 Notes from training events should be placed in a three-ring binder.

FIGURE 3-10 Highlight, tab, and write in your personal copies of textbooks used for training.

FIGURE 3-11 Area familiarization helps fire fighters get to know the properties in the neighborhood.

Know the Neighborhood

Fire fighters should have a detailed knowledge of the environment they protect FIGURE 3-11. That requires going out on inspections to walk through each nonresidential structure in the jurisdiction, from the roof through the subbasement; these walk-throughs reinforce the written preincident plan or diagrams. During the walk-through, the fire officer should take pictures to capture significant details. A detailed overhead view of many communities can be obtained from aerial photographs, digital maps, or Google Earth. This aerial view is especially valuable for apartment complexes, business parks, and retail areas. The maps and photos can be used to locate access routes, plan the placement of first-alarm apparatus, and identify exposure problems.

Knowing the neighborhood may include working with the building owners and occupants to practice incident action plans and procedures. Fire officers should strive to establish good working relationships with every building manager or emergency program manager in their response district.

Use Problem-Solving Scenarios

The fire officer can help the company members become more skilled and knowledgeable by providing opportunities to use their problem-solving skills. Instead of reading the code regulations to provide training for his company members, Page would present fact-based situations and require them to use the code to solve the problem. This technique forced the fire fighters to identify the occupancy use group, identify the issues, look up the applicable regulations, and make decisions. This is an excellent way for adults to learn concepts, regulations, and decision-making skills. The same technique can be used for fire-ground hydraulics, technical rescue scenarios, and incident management procedures.

In addition, this problem-solving approach can be used in reviews of preincident action plans. The fire officer can construct various emergency incident scenarios for training sessions, based on actual buildings and potential situations. Some computer-based fire incident simulators can accept digital pictures taken during walk-through visits and use them to create customized emergency incident scenarios.

The Fire Officer's Supervisor

Every fire officer has a supervisor. In municipal fire departments, a fire officer's supervisor is usually a command-level officer (a battalion chief, a district chief, or a battalion commander) who supervises numerous fire companies within a geographical area. The battalion chief's supervisor could be a deputy chief, who reports directly to the fire chief. In the same manner, the fire chief would report to the mayor, the city manager, the commissioner of public safety, or an individual at an equivalent level. Under the chain of command, the fire chief's orders and directives are passed down through the deputy chief to the battalion or district chief, who then ensures that the fire officers enact those orders or directives.

Regardless of the organization structure, every fire officer has an obligation to work effectively with a supervisor. Three activities are necessary to ensure a good working relationship:

- Keep your supervisor informed.
- Make appropriate decisions at your level of responsibility.
- Consult with your supervisor before making major disciplinary actions or policy changes.

FIGURE 3-12 Always keep your supervisor informed.

No supervisor likes surprises. Part of the supervisor–subordinate relationship is to make sure the supervisor is not surprised or blindsided FIGURE 3-12 . For example, if a fire company has a vehicle crash at 10:00 P.M., the chief needs to know about it as soon as possible. It would be poor form for the chief to find out about the accident from a story on the 11 o'clock news.

Supervising and managing fire officers should not hesitate to make decisions appropriate for their level of responsibility. This means that problems should be addressed and situations resolved where and when they occur. If a fire officer has the authority to solve a problem, he or she should not wait for the supervisor to arrive to solve it.

This level of authority usually covers fire station–level activities such as maintenance, training, and community outreach. At fire stations with rotating career work shifts, the expectation is that most issues between the shifts will be resolved at the supervising fire officer level. A fire officer who is working in a staff position should make the decisions appropriate for that position. Volunteer officers should make decisions appropriate for their level within the organization.

Some issues require the fire officer to consult with a supervisor before making a decision or taking action. Policy changes, for example, cannot be enacted within an administrative vacuum. If a decision will have an impact that goes beyond the fire officer's scope of authority, it is time to talk to the supervisor.

This policy also applies before major disciplinary actions are taken. Consultation with a supervisor is required by the municipal personnel regulations to ensure that all major discipline is delivered in a consistent and impartial manner. This is also a recommended practice in most volunteer fire departments.

Integrity and Ethical Behavior

The formal organization provides the new supervising fire officer with the symbols of power and authority that are associated with the badge, insignia, and distinctive markings on the officer's helmet. The individual, however, needs to provide the core values of integrity and ethical behavior that, combined with the formal symbols, create an effective fire officer. An unethical fire officer is ineffective and damages the department's reputation. If their behavior goes uncorrected, unethical and corrupt fire officers will corrode the department's ability to deliver services and maintain the public trust.

■ Integrity

Integrity refers to the complex system of inherent attributes that determine a person's moral and ethical actions and reactions, including the quality of being honest.

Getting It Done

One Captain's Expectations and Requirements to a New Fire Fighter

This document was provided to new fire fighters assigned to a senior captain's team:

> We expect you to be the best fire fighter on the best shift. We want other fire fighters at other stations to be envious of your position. Give 100 percent effort all of the time.
>
> Everyone will make mistakes. Own up to your blunder. Don't blame others. Always try to do the right thing. I will back you up if you can explain and justify why you did something wrong.
>
> When you go on a call, treat everyone like you would treat your Momma. I don't care if you have gotten up three times after midnight, even if it's a snivel call, treat the public with respect.
>
> This is a dangerous job. Use your common sense and experience to make decisions. If you don't know the answer, ask.
>
> We risk a lot to save a lot and risk little to save little. You will see many disturbing things on the street that will make your stomach turn. Learn how to deal with the stress of the job.
>
> Inside a burning building, if you think you are in trouble, call a mayday. Don't worry about peer pressure. If you think you are in trouble, you are. Get help; don't let your pride kill you.
>
> When given a difficult task, be able to grasp the detail of the order and put it into a safe practice to complete the task. This will apply under normal day-to-day routine situations and emergency conditions.
>
> Communicate properly and cordially, especially when dealing with the public. They are the ones who support us. Lead by example 24/7, not just at work.

FIGURE 3-13 The fire officer should demonstrate the behaviors that he or she says are important.

The fire officer should "walk the talk" and demonstrate the behaviors that he or she says are important **FIGURE 3-13**. If the company officer says that physical fitness is important, then the fire fighters should see their officer performing physical fitness training during the workday.

Integrity can be demonstrated by a steadfast adherence to a moral code. Such a code combines a fire officer's internal value system and the fire department's official organizational value system. Formal organizations publish their expectations as a code of ethics, a code of conduct, or a list of value statements.

■ Ethical Behavior

The fire officer position provides a wide range of opportunities to demonstrate ethical behavior. The fire officer demonstrating ethical behavior makes decisions and models behavior consistent with the department's core values, mission statement, and value statements.

Paragraph A.1.3 in the "Annex A: Explanatory Material" section of NFPA 1021, *Standard for Fire Officer Professional Qualifications*, makes the following statement:

> Fire officers are expected to be ethical in their conduct. Ethical conduct includes being honest, doing "what's right," and performing to the best of one's ability. For public safety personnel, ethical responsibility extends beyond one's individual performance. In serving the citizens, public safety personnel are charged with the responsibility of ensuring the provision of the best possible safety and service.
>
> Ethical conduct requires honesty on the part of all public safety personnel. Choices must be made on the basis of maximum benefit to the citizens and the community. The process of making these decisions must also be open to the public. The means of providing service, as well as the quality of the service provided, must be above question and must maximize the principles of fairness and equity as well as those of efficiency and effectiveness.

Reprinted with permission from NFPA 1021-2014, Fire Officer Professional Qualifications, Copyright © 2013, National Fire Protection Association, Quincy, MA.

Fire department activity tends to be high profile, regardless of the task. Even a trip to the grocery store to pick up dinner draws public attention. The fire officer should act as if someone is always documenting his or her actions when out of the fire station.

Workplace Diversity

The civil rights of Americans are established by federal laws, which are enforced in the workplace by the Equal Employment Opportunity Commission (EEOC). Title VII of the Civil Rights Act of 1964 covers state and local governments, schools, colleges, and unions. This law states that it is illegal for an employer to:

(1) fail or refuse to hire or discharge any individual, or otherwise discriminate against any individual with respect to compensation, terms, conditions, or privileges of employment because of such individual's race, color, religion, sex, or national origin, or (2) to limit, segregate, or classify employees or applicants for employment in any way that would deprive any individual of employment opportunities or otherwise adversely affect status as an employee because of such individual's race, color, religion, sex, or national origin.

The Equal Employment Opportunity Act of 1972 amended the Civil Rights Act of 1964 and expanded its coverage to include almost all public and private employers with 15 or more employees. In general, the 1972 act covers volunteer fire departments and other nonprofit emergency service organizations.

The Civil Rights Act of 1991 provides additional compensatory and punitive damages in cases of intentional discrimination under Title VII and the Americans with Disabilities Act of 1990. The changes introduced in the 1991 act increased the number of discrimination lawsuits filed against organizations.

Many fire departments have made changes to their recruitment, hiring, and promotion practices to comply with the various civil rights laws. Some departments took action to diversify their workforces without the prompting of the courts. Where this did not occur, the legal remedies have ranged from consent decrees, in which the fire department agrees to accomplish specific diversity goals within a specific time, to court orders outlining specific hiring practices.

Diversity, as applied to fire departments, means the workforce should reflect the community it serves. In all cases, the overarching goal is for the fire department to reflect the diversity of the community. Consider the example of a fire department that is 90 percent Caucasian in a community where the overall population is 40 percent African American, 20 percent Latino, and 20 percent Asian American. This fire department population is not reflecting the broader community.

If the department has agreed to an EEOC consent decree, it has promised the court that it will work to hire qualified individuals who reflect the community. A consent decree can require a variety of activities, including community outreach, job

VOICES OF EXPERIENCE

One of the most challenging times in one's career is promotion. I had worked hard to achieve my goal and then it was like the proverbial "dog chasing and catching a car." What was I going to do now?

There were mixed feelings now that I was a company officer: Would I be able to meet the expectations of my fire fighters? Of my fellow officers? Of my chief officers? Of myself?

Now that I was "the guy in charge" I had to come to grips with the fact the people were going to be coming to me for answers and those answers could make or break my reputation as a company officer and a leader. I was also going to have to give orders and I was expected to meld several individuals into a unit that could effectively manage details as minor as kitchen duties or as complex as multiple-alarm fires.

To work effectively with fire fighters, company officers should be consistent, fair, and resist micro-managing. If these goals are accomplished, management of the company should be easier.

> **Would I be able to meet the expectations of my fire fighters? Of my fellow officers? Of my chief officers? Of myself?**

One of the most important things a new officer can do is to be consistent. Do the same things the same way. Always strive to do the right thing—this will show your fire fighters that your intentions are to be a good company officer.

It is imperative to be fair. Resist the temptation to be everyone's buddy and also resist the temptation to be a task master. Remember, your fire fighters are looking to you to set the standards of the company.

It is also easy to micro-manage. Resist this at all costs. Let your fire fighters take on new responsibilities and help guide them. Your job as the company officer is to develop new company officers. To get the best out of your company you have to help the members of your company realize their strengths and help correct weaknesses. Never pass on an opportunity to teach and mentor. Don't forget to praise and don't forget to discipline. You will make mistakes, and how you handle those mistakes will go a long way in how you are perceived as a leader.

After a structure fire or any other incident, I would bring the company together and start by outlining the mistakes I felt I had made. This helped the other members open up, and productive discussions would follow. Show your fire fighters that you support and care for them. Take the time to get to know them. This helps when trying to motivate them.

Before my promotion I composed a list of "What I would do if I were an Officer". I made it a habit to refer to this list at least once a month to help keep myself in check. Here's my list:

- Stay real.
- Be fair.
- Do right.
- Listen.
- Don't ask anyone to do anything you wouldn't do.
- Be fast to praise; be slow to discipline.
- Teach.
- Never be satisfied.
- Lead by example.
- Get feedback.

My promotion (and subsequent promotions) were always challenging and rewarding personally and professionally but the biggest reward is having a former fire fighter tell you that it was a pleasure serving in your company.

Stephen McClure
Operations Chief (ret.), Charleston (WV) Fire Department
Director, Jackson County EMS
Ripley, West Virginia

fairs, pre-employment preparation, and peer group coaching. The court formally meets with the fire department representative periodically to see how well the department is progressing.

Some departments operate under a specific court-mandated hiring process. Using our earlier example, the court could require the department to hire two African Americans, one Latino, and one Asian American before it can hire one additional Caucasian. This requirement would remain in effect until the hiring diversity goal is accomplished.

Expiration of a court order does not relieve the fire department of its charge to maintain diversity. One large fire department had a qualified applicant pool that included 23 percent candidates representing a protected class when a consent decree was terminated in 1999. By 2008, that representation shrunk to 7 percent of the qualified applicant pool, resulting in another inquiry by the court.

The Fire Officer's Role in Workplace Diversity

Jack W. Gravely is a lawyer, subject-matter expert, and trainer on workplace diversity. He has been a frequent speaker at public safety agencies on issues of racial and cultural diversity.

In a presentation to fire department leaders, Gravely pointed out that today's fire officer has the benefit of four decades of equal employment opportunity (EEO)/affirmative action (AA) court decisions to guide decision making. When Gravely started teaching diversity training classes, the emphasis was on the language of the regulations and their potential impact. Today, the EEOC files approximately 400 lawsuits every year; hundreds of court decisions and a large body of case law present a clear and generally consistent policy on how a supervisor should behave in the workplace. Gravely recommends that a fire officer focus on actionable items and the definition of a hostile workplace.

Actionable Items

Actionable items are employee behaviors that require an immediate corrective action by the supervisor. Dozens of lawsuits have shown that failing to act when these situations occur is likely to create a liability and a loss for the department. The best example is the use of certain words in the workplace. An employee's use of derogatory or racist terms about people from other ethnicities, religions, or genders requires immediate corrective action by the supervisor. Regardless of the conditions, context, or situation, such words are inappropriate and represent a potential million-dollar liability to the organization.

The fire officer must act immediately in these situations. That means speaking with the offending fire fighter in private, and counseling the fire fighter that the use of such words is unacceptable in the fire station or in any situation where the individual is representing the department (either on duty or off duty but in uniform). The fire officer should provide the fire fighter with the fire department's or municipality's EEO/AA policy statement and, if applicable, the code of conduct. The fire officer should also maintain a formal or informal record of the counseling session **FIGURE 3-14**. If any doubt arises about whether the message was fully understood, the fire officer should ensure that a higher-level supervisor is informed of the action that has been taken.

The same policies should apply to fire fighters regularly assigned to the fire company, fire fighters detailed in or visiting

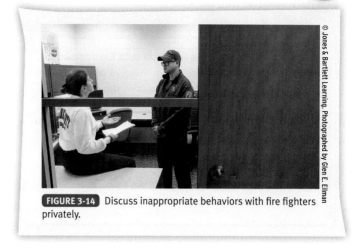

FIGURE 3-14 Discuss inappropriate behaviors with fire fighters privately.

the fire station, and other uniformed or civilian members of the fire department. Case law has shown that use of unacceptable language requires an immediate response. An officer's failure to act has been interpreted as official condoning or encouragement of such behavior.

Although enough case law is available to provide general guidelines and some specific examples, the subject of what constitutes harassment remains a dynamic aspect of the work environment. The fire officer needs to stay informed about the organization's EEO/AA and diversity policies. Most large organizations have a diversity or EEO/AA office that can provide up-to-date information and answer questions.

Hostile Workplace and Sexual Harassment

In 1999, the EEOC amended the guidelines on sexual harassment. The amended regulations broadened the types of harassment that are considered illegal and included a requirement that employers have a duty to maintain a harassment-free work environment. The standard for evaluating sexual harassment is what a "reasonable person" in the same or similar circumstances would find intimidating, hostile, or abusive. The 1999 guidelines clearly state that employers are liable for the acts of those who work for them if the organization knew or should have known about the conduct and took no immediate, appropriate, corrective action. That is why a fire officer must immediately respond to any utterances of offensive or derogatory language in the work environment.

Sexual harassment is unwanted, uninvited, and unwelcome attention and intimacy in a nonreciprocal relationship. The abuse of power is an essential component of sexual harassment. The EEOC guidelines state that verbal and physical conduct of a sexual nature is harassment when the following conditions are present:

- The employee is made to feel that he or she has to endure such treatment to remain employed.
- Whether the employee submits to or rejects such treatment is used when making employment decisions.
- The employee's work performance is affected.
- An intimidating, hostile, or offensive work environment is present.

The term "hostile work environment" can be used to describe a broad range of situations in which an employee is subject to discrimination in the workplace. The generalized hostile

> **Getting It Done**
>
> **The Fire Company Defines Its Nonhostile Environment**
> A fire officer who unilaterally imposes new and severe requirements that affect the language fire fighters use in the workplace would encounter much resistance. One method of fostering the concept of on-duty speech or appropriate workplace behavior is to have the fire fighters develop their own set of rules or behaviors. Provide the fire fighters with the source documents (federal and municipal codes and regulations), and ask them to consider which words or behaviors are inappropriate while at work.

workplace definition can apply to a variety of circumstances that do not necessarily include a specific abuse of supervisory power.

Gravely believes that hostile workplace complaints will shape workplace diversity. There are few quid pro quo sexual harassment cases—cases in which a supervisor specifically promises a work-related benefit in return for a sexual favor. Instead, the trend is toward more complaints about a hostile workplace, which can involve sexual issues.

Many examples can be cited of hostile workplace issues that resulted in large court-directed settlements for the complainant. In one municipality, a formal investigation was initiated by the EEOC after the fire department settled multiple cases that cost the city more than $10 million. The investigation took more than a year, and its outcome included a federal requirement to provide antiharassment training throughout the department.

The Tipping Point of Harassment and Hostile Workplace Issues

Complaints settled by a court decision represent less than 3 percent of all of the harassment, hostile workplace, and discrimination complaints filed against fire departments. The company officer plays a vital role in establishing a work environment that is free from conditions that might lead to such a complaint.

Social media and a 24-hour news cycle can transform a minor firehouse incident into an international news item in hours. This process irrevocably changes the public's perception of the department and may lead to costly and lengthy investigation and legal action. Failing to respond to actionable items or ignoring elements leading to a hostile environment creates organizational liability and places the fire officer in professional peril.

Handling a Harassment or Hostile Workplace Complaint

Fire fighters who want to initiate a harassment complaint have a choice of three methods for doing so. They can start with the federal government, they can start with the local government, or they can start within the fire department—it is their choice. If the process starts within the fire department, a fire officer may be the first formal point of contact. In this case, the fire officer would function as the first step of a multistep procedure.

The fire officer should know the department's procedure for handling a harassment or hostile workplace complaint. The fire officer's designated role in conducting an investigation of an EEO complaint depends on the procedures adopted by the jurisdiction or the fire department. In some cases, an officer's role is limited to starting the process by making appropriate notifications. Many local governments have specially trained EEO staff who conduct an investigation. In other cases, the fire officer might be required to perform the initial investigation and submit a report. Following are general guidelines:

- **Keep an open mind.** Many fire officers have a difficult time believing that discrimination or harassment could be happening right under their noses. Failure to investigate a complaint, however, is the most common reason a local government is found liable. Every complaint must be investigated and documented. Do not come to any conclusions until your investigation is complete. Follow your organization's procedures and inform your supervisor.
- **Treat the person who files the complaint with respect and compassion.** Employees often find it extremely difficult to complain about discrimination or harassment. When an employee comes to you with concerns about discrimination or harassment, be professional, but also be understanding.
- **Do not blame the person filing the complaint.** Case law and current practice dictate that it is the complainant who determines whether the situation is hostile or is harassment. Blaming the person bringing the complaint to you is the second most common way that local governments lose harassment or hostile workplace complaints.
- **Do not retaliate against the person filing the complaint.** It is against the law to punish someone for complaining about discrimination or harassment. The most obvious forms of retaliation are termination, discipline, demotion, or threats to do any of these things. More subtle forms of retaliation could include changing the shift hours or work location of the accuser, even if the intent is to remove the alleged victim from the problem.
- **Follow established procedures.** Local government personnel regulations generally provide a detailed procedure on how to handle harassment or hostile workplace complaints. Follow the procedure used in your jurisdiction.
- **Interview the people involved.** An initial investigation usually involves conducting interviews with the people who are involved in the situation, typically starting with the person who made the complaint. The interviewer needs to find out exactly what the employee is concerned about. Get details: what was said or done, when and where, and who was present. Then talk to any employees who are being accused of discrimination or harassment. Get details from them as well. Be sure to interview any witnesses who may have seen or heard any problematic conduct. Take notes about your interviews and gather any relevant documents.

- **Look for corroboration or contradiction.** Discrimination and harassment complaints often involve "he said/she said" situations. The accuser and the accused may offer different versions of an incident, leaving you with no way of knowing who is telling the truth. In such a case, you may have to turn to other sources for clues. Witnesses may have seen part of an incident. In some cases, documents, such as e-mails and posted notes, may prove one side to be right.
- **Keep it confidential.** A discrimination complaint can polarize a workplace. The unique team nature of 24-hour fire service work creates a close environment where it is difficult to maintain confidentiality. The fire officer must insist on and enforce confidentiality during the investigation.
- **Write it all down.** Take notes during all interviews. Before the interview is over, go back through your notes with the interviewee to ensure accuracy. Keep a journal of the investigation. Write down the steps you have taken to get at the truth, including dates and places of interviews. Keep a list of all documents that are reviewed. Document any action taken against the accused or the reasons for deciding not to take action. Anticipate that all of this written record will be used in any subsequent civil service or court actions.
- **Cooperate with government agencies.** If the fire fighter files a complaint with another government agency (either the federal EEOC or an equivalent state agency), that agency may investigate. Notify your supervisor as soon as you receive a call or visit from an EEOC investigator. You will probably be asked to provide certain documents, to give your side of the story, and to explain any efforts you made to deal with the complaint yourself. Be cautious, but cooperative.

Regardless of where the complaint is filed, the fire chief is required to take corrective action if the investigation confirms that the complaint has merit. If the department concludes that some form of discrimination or harassment occurred, formal corrective action could include mandatory training, work location transfer, or demotion. Termination may be proposed for more egregious kinds of discrimination and harassment, such as threats, stalking, or repeated and unwanted physical contact.

■ The Fire Station as a Business Work Location

The fire officer needs to consider the fire station or other fire department facility as a business work location. This perspective represents a drastic change from the concept of a fire station as a home away from home for a group of fire fighters, but it is necessary to ensure that the fire station maintains a professional work environment. In the eyes of the law and in the opinion of administrators and elected officials, the same rules of behavior apply to an office in city hall at 2 P.M. and a fire station at 2 A.M.

The fire officer can help maintain an appropriate work environment by encouraging and enforcing acceptable behavior whenever fire fighters are on duty. The fire officer accomplishes this by taking the following steps:

- Educate the employees on the workplace rules and regulations that define expected behavior. Start with the local government's "code of conduct" or other documents that outline the chief administrative officer's expectations for all municipal employees. This information can usually be found in the municipality's mission statement, core values, or personnel regulations.
- Promote the use of "on-duty speech." The goal is not to change the thoughts or feelings of individual fire fighters, but rather to establish a workplace environment where certain behaviors and words are not used. Fire fighters can think what they want. However, while they are on duty, in the fire station, or in uniform, they cannot use certain words or phrases or act out certain behaviors.
- Be the designated adult. This action requires the fire officer to model appropriate behavior as well as encourage and enforce the same behavior by the fire fighters. The fire officer must identify and correct unacceptable workplace behavior whenever it is observed. Ignoring a problem is, in reality, permitting it to continue. The fire officer who is a candidate for promotion is expected to identify, explain, and enforce the limits of unprofessional behavior **FIGURE 3-15**.

A company-level officer should make it a practice to walk around the fire station at various times during the workday to observe what is going on. This walk-around is more important

FIGURE 3-15 The company officer must identify and correct unacceptable behavior.

Assessment Center Tips

Know Your Organization's Procedures

Even if the normal practice for your department is to have a specialized or designated person conduct an investigation of a harassment or hostile workplace complaint, the promotional candidate must know both the local and federal procedures that must be followed when an employee files a complaint. Harassment and hostile workplace complaints require a specific and detailed response by the first-line supervisor.

An assessment center scenario may involve a fire fighter coming to the candidate for advice. During the interview process, the fire fighter reveals that he or she may be a victim of harassment or a hostile workplace. The candidate would be expected to explain the options available to the fire fighter in filing either a federal or a local complaint.

when the officer is in a big station with multiple companies and in combination career–volunteer departments, where there is a constant flow of people coming into and out of the fire station. This practice is not designed to catch someone doing something wrong, but rather is intended to make sure that everything is functioning properly. The officer should routinely determine what the crew members are doing and check on the safety of the facility and equipment. At the same time, the officer should look out for unexpected situations and surprises. Having the reputation of knowing what is going on in the station and reacting to inappropriate situations goes a long way toward encouraging appropriate workplace behaviors **FIGURE 3-16**.

FIGURE 3-16 The officer should routinely check in with crew members.

You Are the Fire Officer Conclusion

Lieutenant Williams has three choices: enforce the chief's directive completely, modify compliance after consulting with the captain, or circumvent the directive and help the fire fighters create shopping work-arounds. Circumventing the ban may make Williams popular with the fire fighters, but it could have significant career consequences. Ignoring a directive from the fire chief is not a career-enhancing activity.

The most appropriate response is to enforce the shopping ban as outlined by the fire chief. Sudden directives emerge from a situation several levels above the fire officer. The formal organization expects that the fire fighters will immediately comply.

Williams and Davis meet with Chief Johnson later in the day. The chief says that this directive represents a response to an issue that is about to hit the media. No fire department apparatus can be parked near grocery stores or shopping centers unless they are handling an incident. You cannot go into a store once you clear an incident next to the store. There is no room for adjustment at this time.

After the chief leaves, Williams has a follow-up meeting with the crew, explaining the details of the directive, stating that Station 100 will comply, and asking the members how to change the daily routine to make this work.

Wrap-Up

Chief Concepts

- A fire officer is responsible for accounting for the people and resources at a fire station and work location. This role may require development of a report at the beginning of duty, outlining staffing and equipment status.
- The beginning of shift report is provided from each work location to the battalion or district chief within the first quarter hour of the reporting time.
- The new supervising fire officer must make prompt notifications, as some types of information must be passed up the chain of command quickly.
- Chiefs typically want new officers to run their companies and make the decisions that are within their scope of responsibility. Chiefs are available for consultation, but they expect their officers to run the fire stations.
- Each fire department should have an agency-wide mission that is translated into annual goals. These goals are used to develop annual, quarterly, and monthly objectives for each fire company.
- Transitioning from fire fighter to fire officer changes how the individual relates to the formal fire department organization and which role the fire officer plays with fellow fire fighters.
- A fire officer has a larger sphere of responsibility when supervising a work group than he or she had as a fire fighter.
- A company officer's duties can be divided into three distinct roles: supervisor, commander, and trainer.
  - As a supervisor, the fire officer functions as the official representative of the fire chief.
  - As a commander, the fire officer is expected to exercise strong direct supervision over the company members when operating at the scene of an emergency incident.
  - As a trainer, the company-level officer is responsible for the performance level of the fire company and must establish a set of expectations that the company will perform at the highest level possible.
- Every fire officer has a supervisor. Keep your supervisor informed, make appropriate company-level decisions, and consult with your supervisor before making major disciplinary or policy changes.
- A fire officer needs to model the core values of integrity and ethical behavior that, combined with the formal symbols, create an effective fire officer. An unethical fire officer is ineffective and damages the department's reputation.
- A fire officer should "walk the talk" and demonstrate integrity by behaving ethically.
- The Equal Employment Opportunity Commission is the federal agency empowered to enforce compliance with the Civil Rights Act of 1964, the Equal Employment Opportunity Act of 1972, and the Civil Rights Act of 1991. Current fire department recruitment, hiring, and promotion practices are guided by the EEOC to achieve workplace diversity.
- The EEOC files approximately 400 lawsuits every year; hundreds of court decisions and a large body of case law present a clear and generally consistent policy on how a supervisor should behave in the workplace.
- The fire officer should follow the local procedures when he or she encounters a harassment or hostile workplace complaint. Take notes, make appropriate notifications, and do not take sides.
- The fire officer needs to consider the fire station or other fire department facility as a business work location. This perspective represents a drastic change from the concept of a fire station as a home away from home for a group of fire fighters, but it is necessary to ensure that the fire station maintains a professional work environment.

Hot Terms

<u>Actionable items</u> Employee behavior that requires an immediate corrective action by the supervisor; dozens of lawsuits have shown that failing to act in the face of such behavior will create a liability and a loss for the department.

<u>Administrative fire officer</u> IAFC description of a person who has worked as a managing fire officer for 3 to 5 years, is certified at the NFPA Fire Officer III level, and has accomplished formal education equivalent to a bachelor's degree.

<u>Diversity</u> A characteristic of a fire workforce that reflects differences in terms of age, cultural background, race, religion, sex, and sexual orientation.

<u>Ethical behavior</u> Decisions and behavior demonstrated by a fire officer that are consistent with the department's core values, mission statement, and value statements.

References and Additional Resources

NFPA reprinted material is not the complete and official position of the NFPA on the referenced subject, which is represented only by the standard in its entirety.

Blubaum, J. E. (2010). *Promotional Practices: Adaptive Change Issues Facing the Moscow Volunteer Fire Department*. Emmitsburg, MD: U.S. Fire Administration, Executive Fire Officer Program.

Carter, H. R., and E. Rausch. (2008). *Management in the Fire Service*, 4th ed. Sudbury, MA: Jones and Bartlett.

Davis, L., and D. Colletti. (2002). *The Rural Firefighting Handbook*. Royersford, PA: Lyon's Publishing.

Drucker, P. F. (1990). *Managing the Nonprofit Organization*. New York, NY: HarperCollins.

Dunn, V. (1999). *Command and Control of Fires and Emergencies*. Tulsa., OK: Pennwell/Fire Engineering.

Floryan, J. P. (2009). "Post-Pierce Program: Using IDR to Improve the Los Angeles Fire Department's Current Complaint and Disciplinary Procedure." *Pepperdine Dispute Resolution Law Journal* 9(1): 147–166.

Kreiger, A., and T. Masten. (2012). *Beyond the Consent Decree: Gender and Recruitment in the San Francisco Fire Department*. Emmitsburg, MD: U.S. Fire Administration, Executive Fire Officer Program.

LoSasso, C. (2012). *Mentoring Volunteer Officers Pre- and Post-promotion*. Emmitsburg, MD: U.S. Fire Administration, Executive Fire Officer Program.

Myers, B. (2004). *Mentoring: Preparing Company Officers*. Emmitsburg, MD: U.S. Fire Administration, Executive Fire Officer Program.

Page, J. O. (1973). *Effective Company Command*. Alhambra, CA: Borden Publishing.

Robinson, G. J. (2011). *Lasting Leadership: Preparing for the Transition to Company Officer*. Emmitsburg, MD: U.S. Fire Administration, Executive Fire Officer Program.

Smeby, L. C. Jr. (2014). *Fire and Emergency Services Administration: Management and Leadership Practices*, 2nd ed. Burlington, MA: Jones & Bartlett Learning.

Snook, J. W., et al. (2011). *A Leadership Guide for Volunteer Fire Departments*. Sudbury, MA: Jones & Bartlett Learning.

Stapleton, L. D. (1983). *Thirty Years on the Line*. Dover, NH: DMC Associates.

Title VII of the Civil Rights Act of 1964. (Pub. L. 88-352) *U.S. Code*, Volume 42, section 2000e. Available at http://www.eeoc.gov/laws/statutes/titlevii.cfm. Accessed May 23, 2013.

FIRE OFFICER
in action

The first day for Lieutenant Williams seems filled with administrative and supervisory tasks that are more complex and time consuming than anticipated. Some tasks require immediate individual action, others need coordination, and a few make the fire officer feel like a schoolteacher in the middle of a playground spat. Here are three events in Williams's first tour as the Quint 100 supervisor.

1. Quint 100 is handling an "assist citizen" incident, a situation that none of the crew has experienced. The crew devises a solution that falls into the gray area of authorized fire department activities. What should Lieutenant Williams do?
    A. Do nothing that is out of the clear scope of fire department–approved activities.
    B. Follow through with the solution even though it is contrary to his personal ethics.
    C. Follow through with the solution if it is consistent with the department's and the officer's own ethics.
    D. Review the proposed solution with the assistant chief of operations and request approval.

2. Engine and Medic 100 are on an incident. A contractor working in the Fire Station 100 kitchen is using derogatory and racist terms. What should Lieutenant Williams do?
    A. Ignore him and move the crew out of the kitchen.
    B. Wait for Captain Davis to return and report the issue.
    C. Notify Battalion Chief Johnson and the fire department EEO/AA representative.
    D. Immediately meet with this person in private and state that the use of such words is unacceptable at this work location.

3. While operating at a 2 A.M. traffic accident, a Quint 100 fire fighter comes into contact with blood and body fluids from a patient. What should Lieutenant Williams do?
    A. Write a memo about the exposure and send it to Chief Johnson.
    B. Notify the appropriate supervisor as soon as possible.
    C. Make a notation in the beginning of shift report on the next scheduled workday.
    D. Finish clearing the incident, and then have Quint 100 proceed to the hospital to obtain treatment.

Fire Officer Communications

Fire Officer I

Knowledge Objectives

After studying this chapter, you will be able to:

- Describe the steps in the communication cycle (NFPA 4.2.2). (pp 62–63)
- List the basic skills for effective communication (NFPA 4.2.2). (pp 63, 65–66)
- Identify ways to improve listening skills. (pp 63, 65)
- Describe the ways to counteract environmental noise (NFPA 4.2.2). (pp 65–66)
- Identify the key points for emergency communications (NFPA 4.2.1) (NFPA 4.4.5). (pp 66–68)
- Identify types of reports and discuss their use (NFPA 4.1.2) (NFPA 4.4.5). (pp 68–72)

Skills Objectives

After studying this chapter, you will be able to:

- Perform emergency communications with fire officers, fire fighters, and dispatch. (pp 66–68)
- Give a verbal report. (p 68)
- Write reports based on report purpose and requirements. (pp 68–71)

Fire Officer II

Knowledge Objectives

After studying this chapter, you will be able to:

- Explain the difference between formal and informal communications. (pp 72–73, 75)
- Identify different types of formal communications and discuss their use (NFPA 5.6.3). (pp 72–73, 75)
- Describe the considerations and requirements to keep in mind when writing a report (NFPA 5.4.5) (NFPA 5.6.3). (p 75)
- Describe how to give an oral presentation of a written report (NFPA 5.3). (pp 75–76)
- Identify the elements to include in a news release and the best format in which to put these elements (NFPA 5.3) (NFPA 5.4.4). (p 76)
- Discuss the effects of social media on fire department communications (NFPA 5.3). (pp 76–79)

Skills Objectives

After studying this chapter, you will be able to:

- Write a formal report (NFPA 5.4.5) (NFPA 5.6.3). (p 75)
- Give an oral presentation of a written report (NFPA 5.3). (pp 75–76)
- Write a news release (NFPA 5.3) (NFPA 5.4.4). (p 76)
- Develop a social media policy for your department. (pp 76–79)

CHAPTER 4

Principles of Fire and Emergency Service Administration (FESHE) Course Outcomes

2. Recognize the need for effective communication skills, both written and verbal. (pp 62–63, 65–66, 68–73, 75)
6. Discuss the various levels of leadership, roles and responsibilities within the organization. (pp 68–71, 73, 75)
9. Identify the roles of the National Incident Management System (NIMS) and Incident Management System (ICS). (pp 66–68)

You Are the Fire Officer

Quint 100 is returning from an early morning water-flow alarm at a shopping mall. Lieutenant Williams is reflecting on the after-dinner conversation with Captain Davis about the lack of standard operating procedures (SOPs) for quint companies. The result of that conversation is a new assignment to develop procedures for Quint 100 that could be presented to the battalion chief. Williams's reflection is interrupted, however, when the quint crew encounters a two-story mid-row townhouse with flames coming out of a basement window.

Williams sees an adult leaning out of a smoky second-floor window and picks up the microphone to give dispatch a size-up and incident report. The lieutenant tells the crew to grab the man at the second-floor window. Returning from a 360-degree size-up, Williams sees the man still at the window and the quint crew deploying an attack line through the front door. They enter the townhouse and immediately fall through the floor. As Williams runs up to the front door, dispatch asks, "Is the fire out or do you need additional assistance?"

1. What should you do now?
2. Why is it important to complete the entire communications process when you are in an emergency situation?
3. What can you do in the future to improve information exchange?

Introduction

Many fire officers wear a rank insignia that features bugles, representing the fire officer's speaking trumpet. At the turn of the 20th century, such a trumpet was used to shout orders on the fire ground. Today, this symbol emphasizes the fire officer's requirement to communicate. Although the technology has certainly advanced, communication skills remain critically important in the fire service. An officer must be able to communicate effectively in many different situations and contexts.

A fire officer must be able to process several types of information to supervise and support the fire company members effectively. An almost overwhelming volume of information is available today, particularly regarding newly identified hazards. A fire officer should regularly access publications and social media to stay aware of trends and issues within public safety. In addition, the officer should share significant information with the company members.

Clearly, effective communication skills are essential for a fire officer. These skills are required to provide direction to the crew members, review new policies and procedures, and simply exchange information in a wide range of situations. Effectively transmitting radio reports requires a unique skill set. Communication skills are equally important when working with citizens, conducting tours, releasing public information, and preparing reports.

Fire Officer I

The Communication Process

Communication is a repetitive circular process. Successful communication occurs whenever two people can exchange information and develop mutual understanding. When information flows from one person to another, the process is truly effective only when the person receiving the information is able to understand what the other person intended to transmit.

Effective communication does not occur unless the intended message has been received and understood. This message must make sense in the recipient's own terms, and it must convey the thought that the sender intended to communicate. The sender cannot be sure that this outcome has occurred unless the recipient sends some confirmation that the message arrived and was correctly interpreted.

■ The Communication Cycle

The communication cycle consists of five parts:
1. Message
2. Sender
3. Medium (with noise)
4. Receiver
5. Feedback

The Message

The message represents the text of the communication. In its purest form, the message contains only the information to be conveyed. Messages do not have to be in the form of written or spoken words, however. For example, a stern facial expression with purposeful eye contact can convey a very clear message of disapproval, whereas a smile can convey approval. These messages are clear, yet no words accompany them.

The Sender

The sender is the person or entity who is sending the message. We think of the sender as a person, but it could also be a sign, a sound, or an image. The message needs to be properly targeted to the right person and formulated so that the receiver will understand the meaning. The sender is responsible for the receiver properly understanding the message.

A fire officer must be skilled if he or she is to transmit information and instructions to subordinates and co-workers effectively. The tone of voice or the look that accompanies a spoken message can profoundly influence the receiver's interpretation. Body language, mannerisms, and other nonverbal cues may all affect the interpretation.

Senders may sometimes convey messages that are not intended, not directed to anyone in particular, or not even meant to be messages. For example, a person can outwardly express personal disappointment; however, others may interpret this look as a message of disapproval aimed at them.

The Medium

The medium refers to the method used to convey the information from the sender to the receiver. The medium can consist of words that are spoken by the sender and heard directly by the receiver. Spoken words or sounds can also be transmitted as electromagnetic waves through a radio system. Written words, pictures, symbols, and gestures are all examples of messages transmitted through a visual medium. The sender should consider the circumstances, the nature of the message, and the available methods before choosing a particular medium to send a message.

When information has to be transmitted to subordinates, a fire officer can post a notice on a bulletin board, announce it at a formal line-up, or mention it during a firehouse meal. The medium that is chosen influences the importance that is attached to the message. If the information is really important, the fire officer might announce the key points at line-up, and then direct all members to read a written document and sign a sheet to acknowledge that they have read and understood it. When choosing the circumstances for transmitting a personal message, remember this guideline: Praise in public; counsel, coach, or discipline in private.

The Receiver

The receiver is the person who receives and interprets the message. Unfortunately, there are many opportunities for error in the reception of a message. Although it is up to the sender to formulate and transmit the message in a form that should be clearly understandable to the receiver, it is the receiver's responsibility to capture and interpret the information. The same words can convey different meanings to different individuals, so the receiver might not automatically interpret a message as it was intended. In the fire service, the accuracy of the information that is received can be vital, so both the sender and the receiver have responsibilities to ensure that messages are properly expressed and interpreted. Many messages are directed to more than one person, so there can be multiple interpretations of the same message.

Feedback

The sender should never assume that information has been successfully transferred unless some confirmation is provided that the message was received and understood. Feedback completes the communication cycle by confirming receipt and verifying the receiver's interpretation of the message. The importance that is attached to feedback depends on the nature of the message.

When relaying critical information during a stressful event, the sender should have the receiver repeat back the key points of the message in his or her own words. Without feedback, the sender cannot be confident that the message reached the receiver and was properly interpreted.

Effective communication should contain all five components of the communications process; if one or more is missing, communication does not occur.

Effective Communication Skill Basics

Almost every task tackled by the fire officer depends on the ability to communicate effectively. An officer must be effective as both a sender and a receiver of information and, in many cases, as a processor of information that has to flow within the organization.

■ Active Listening

Your success as a supervisor depends on how freely your subordinates talk to you, keep you informed, and tell you what is bothering them. That success also depends on your effectiveness in communicating with your superiors and keeping them informed.

The ability to listen becomes increasingly more important as one advances through the organization. A fire fighter must listen effectively to information that is coming from a higher level. A fire officer must also be able to listen to company members or other subordinates and accurately interpret their comments, concerns, and questions.

Listening is a skill that must be continually practiced to maintain proficiency. A typical listening situation for a fire officer could be a meeting with a fire fighter who is expressing a problem or a concern. Listening, in a face-to-face situation, is an active process that requires good eye contact, alert body posture, and frequent use of verbal engagement **FIGURE 4-1**.

FIGURE 4-1 Listening is an active process that requires good eye contact, alert body posture, and frequent use of verbal engagement.

Near-Miss REPORT

Report Number: 12-0000194

Event Description: The involved crew arrived on Engine 1, which was the second-arriving fire apparatus behind Engine 2, at the scene of a fire involving a two-and-a-half-story single-family dwelling. The crew of Engine 1 laid a large-diameter hose (LDH) supply line from a hydrant a half-block away from the scene and found a moderate to heavy amount of dark gray smoke issuing from Side A of the first floor as well as Side D, with reports of possible persons trapped inside.

The driver/operator and one fire fighter established a water supply to Engines 1 and 2 while the remaining crew of three (the company officer and two fire fighters) prepared to enter. It was noted that no attack line was deployed into the house, and our crew could not determine the location of the Engine 2 crew. It was assumed that Engine 2 entered to conduct a primary search. (Note: No one from the first-arriving company had formally established command at this point, nor were we able to establish radio contact with them prior to our arrival.)

Engine 1 then deployed a 1¾-inch preconnected attack line from our apparatus and entered the first floor for fire attack. Shortly after this company's entry, Engine 2 crew members entered with another 1¾-inch attack from their apparatus and started attacking the fire in the C/D quadrant. Engine 1 personnel then left the attack line and began a primary search. After quickly searching the first floor, the crew went to the second floor and found heavy fire blowing into the hallway from a room in the A/D quadrant.

Engine 1 retrieved its attack line from the first floor and began to knock down the fire in that room so the search could be completed without being cut off by the fire. The crew was able to knock down the fire in the room when water pressure was lost to the attack line. The officer of Engine 1 radioed the apparatus operator about the pressure loss. Within seconds, the fire intensified and began blowing into the hallway, nearly engulfing the crew of Engine 1 in the hallway. With no water, we quickly self-rescued down the stairs and out of the house.

The fire fighter at the nozzle suffered first-degree burns to the left shoulder, upper left arm, and left ear and was treated on scene by EMS. The second fire fighter (backup man) was uninjured; the company officer was at the bottom of the stairs awaiting control of the fire before ascending. All personnel were properly and fully equipped with SCBA and structural firefighting turnout gear.

It was noted later that Engine 2's crew had switched attack lines while our crew was searching. Furthermore, their initial attack line had approximately 12 kinks in it, as it appeared to have been laid on the ground and then charged without properly flaking (stretching) it out. We believe this action caused the loss of water pressure in the hose line, as the operators at both engines maintained proper discharge pressures (according to the pump panel gauges). Other contributing factors were the failure of Engine 2 to establish incident command, the failure to announce the strategic mode to be used, and the failure to advise the other companies which tasks/actions they were performing.

Lesson Learned: Several lessons were learned and/or reinforced. First, incident command must be established by the first-arriving officer or company. Failure to do so jeopardizes the safety of fire fighters and risks the overall success of the operation for fire fighters, owing to a loss of control of the scene and failure to establish the strategic goals and action plan.

Second, and probably the most important factor in this incident, training needs to be reinforced and practiced routinely so that fire fighters are proficient in the proper deployment of attack lines to prevent kinks and to allow for the proper pressure and flow to be realized from the attack line.

In this case, the involved crew's training and experience permitted them to realize the need to retrieve the attack line to knock down the fire so the search could be completed without being cut off by the fire and to self-rescue before they sustained further injury.

An important factor that minimized the injuries sustained in this incident was that all members of the involved crew were properly wearing all components of their structural firefighting turnout gear and SCBA. Failure to have done so would have resulted in severe burns or possibly death.

Courtesy of the National Fire Fighter Near-Miss Reporting Systems (firefighternearmiss.com)

The purpose of active listening is to help the fire officer understand the fire fighter's viewpoint to solve an issue or a problem.

The following techniques may help improve your listening skills:

- Do not assume anything. Do not anticipate what someone will say.
- Do not interrupt. Let the individual who is trying to express a point or position have a full say.
- Try to understand the need. Often, the initial complaint or problem is a symptom of the real underlying issue.
- Look for the real reason the person wants your attention.
- Do not react too quickly. Try not to jump to conclusions. Avoid becoming upset if the situation is poorly explained or if an inappropriate word is used. The goal is to understand the other person's viewpoint.

■ Stay Focused

It is easy to get sidetracked and bring unrelated issues into a conversation. Directed questioning is a good method to keep a conversation on topic. If the speaker starts to ramble, ask a specific question that moves the conversation back to the appropriate subject. For example, if you are attempting to find out why a fire fighter failed to wear a uniform shirt to roll call and he starts talking about how he does not like the fire department patch, you could ask, "How does the patch affect whether you wear your uniform shirt to roll call?"

■ Ensure Accuracy

A fire officer needs to have up-to-date information on the fire department's standard operating guides, policies, and practices. An officer should also be familiar with the personnel regulations, the approved fire department budget, and, if applicable, the current union contract.

If a fire fighter is misinterpreting a factual point or a departmental policy, the fire officer is obligated to clarify or correct the information. Ignoring an inaccurate statement may simply foster erroneous information that is transmitted along the fire service "grapevine" (discussed later in this section). If necessary, obtain the accurate information from a chief officer or headquarters staff and correct the misunderstanding.

A fire officer sometimes has to exercise control over what is discussed in the work environment. Fire station discussions can easily encroach on subjects that are intensely personal, such as politics, religion, or social values, and can quickly escalate into confrontations. To address potential problems proactively, the fire officer needs to establish some ground rules about the topics and level of intensity when discussing issues. Rumor control is useful in de-escalating the spread of inaccurate information that can harm an individual, the department, or the fire service.

■ Keep Your Supervisor Informed

The administrative fire officer depends on you to share information about what is happening at the fire station or in your work environment. In particular, a fire officer needs to keep the chief officer informed about three areas:

- **Progress toward performance goals and project objectives.** A fire officer needs to keep his or her chief apprised of work performance progress, such as training, inspections, smoke detector surveys, and target hazard documentation. It is especially important to let the chief know about anticipated problems early enough to get help and to keep the projects running on time.
- **Matters that may cause controversy.** The chief should be informed about conflicts with other fire officers or between shifts, or conflict that extends outside the organization. The chief also needs to know about any disciplinary issues or a controversial application of a departmental policy. When conveying these kinds of messages, sooner is better than later. Contact your supervisor to discuss a potential disciplinary issue and the best approach to enforcement and resolution.
- **Attitudes and morale.** A fire officer spends the workday with a group of fire fighters at a single fire station, whereas the command officer spends much of the workday in meetings or on the road. The fire officer should communicate regularly with the chief about the general level of morale and fire fighter response to specific issues.

■ The Grapevine

Every organization has an informal communication system, often known as the "grapevine." The flow of informal and unofficial communications is inevitable in any organization that involves people. The grapevine flourishes in the vacuum created when the official organization does not provide the workforce with timely and accurate information about work-related issues. Much of the grapevine information is based on incomplete data, partial truths, and sometimes outright lies.

A fire officer can often get clues about what is going on but should never assume that grapevine information is accurate and should never use the grapevine to leak information or stir controversy. In addition, a fire officer needs to deal with grapevine rumors that are creating stress among the fire fighters by identifying the accurate information and sharing it with subordinates.

■ Overcoming Environmental Noise

<u>Environmental noise</u> is a physical or sociological condition that interferes with the message. Within this definition, "noise" includes anything that can clog or interfere with the medium that is delivering the message.

Physical noise includes background conversations, outside noises, or distracting sounds that make it difficult to hear. For example, siren noise makes it difficult to communicate over the radio. The squeal of portable radio feedback competes with the clarity of information exchange. Digital radios and cell phones can suffer from poor reception or static. A rapid flow of incoming messages can overload the receiver's ability to deal with the information, even if the words come through clearly **FIGURE 4-2**. Darkness or bright flashing lights make it difficult to see clearly and interpret a visual message.

FIGURE 4-2 Trying to talk on a cellular phone with poor reception is one example of physical noise interfering with communication.

Another form of noise interference occurs when the receiver is distracted or thinking about something else and blocks out most or all of an incoming message. The human brain has difficulty processing more than one input source at a time. For example, you could be having a conversation on the telephone with one individual when someone else calls your name. In such a case, your attention is diverted while you try to determine who called you and why. During that moment, you are likely to miss part of the telephone conversation. A similar situation occurs if the sender or receiver is unable to concentrate on or respond to the message due to fatigue, boredom, or fear.

Sociological environmental noise is a more subtle and difficult problem. Prejudice and bias are examples of sociological environmental noise. If the receiver does not believe that the sender is credible, the message will be ignored or result in an inadequate response. The following list provides suggestions to improve communication by minimizing this type of environmental noise:

- **Do not struggle for power.** Focus attention on the message, not on whether the sender has the authority to deliver it to the receiver. The situation—not the people involved—should drive the communication and the desired action.
- **Avoid an offhand manner.** If you want your information to be taken seriously, you must deliver it that way. Be clear and firm about matters that are important.
- **Keep emotions in check.** Emergency operations often deal with events that invoke intense feelings that can interfere with focus and attention to the facts of the situation at hand.
- **Remember that words have meaning.** Select words that clearly convey your thoughts, and be mindful of the impact of the tone of voice.
- **Do not assume that the receiver understands the message.** Encourage the receiver to ask questions and seek clarification if the intent of the message is unclear. A good technique of confirming understanding is for the receiver to repeat the key points of the message back to the sender.
- **Immediately seek feedback.** If the receiver identifies an error or has a concern about the message, encourage that individual to make the statement sooner rather than later. It is always better to solve a problem or identify resistance while there is still time to make a change.
- **Provide an appropriate level of detail.** Think about the person who is receiving the message and how much information that individual needs. Consider a fire officer telling a fire fighter to set up the annual hose pressure test. The information needs are different for a fire fighter who has 22 years of experience as a pump operator versus a rookie who has 22 days on the job.
- **Watch out for conflicting orders.** Make sure that your message is consistent with information coming from other sources. A fire officer should develop a network of peers to consult when dealing with unfamiliar or confusing situations. A subordinate should also be expected to inform the fire officer of a conflicting order so that you can consider the difference and then give the direction that is most appropriate at that moment.

These eight suggestions are intended to improve communications dealing with administration and supervisory activities. The fire ground or emergency scene requires different communications practices.

Emergency Communications

A fire officer needs to communicate effectively during emergency incidents. There is no time to be elegant when communicating within the emergency environment. The direct approach requires asking precise questions, providing timely and accurate information, and giving clear and specific orders.

Under the time pressure of an emergency incident, management of the communications process can be as important as communicating effectively. The incident commander has to establish and maintain a command presence to manage the exchange of information so that the most important messages go through and lower-priority or unnecessary communications do not get in the way. The incident commander should be the gatekeeper for information exchange via radio communications. Additional control can be added by stating, "Unit(s) stand by." In severe situations, requesting emergency radio traffic only will reduce communications to immediate necessary communications. This is critical in the event of a mayday situation or substantial change in fire conditions that must be communicated without delay or interference.

Key points for emergency communications are as follows:

- Be direct **FIGURE 4-3**.
- Speak clearly.
- Use a normal tone of voice.
- If you are using a radio, hold the microphone about 2 inches (5 cm) from your mouth.
- If you are using a repeater system, allow for a time delay after keying the microphone before speaking.

FIGURE 4-3 Emergency communications require a direct approach.

- Use plain English rather than "10 codes."
- Use common terminology that is recognized by the National Incident Management System (NIMS), especially when interacting with multiple agencies and disciplines.
- Try to avoid being in the proximity of other noise sources, such as running engines.

■ "Unit Calling, Repeat..."

Radio messages must be accurate, brief, and clear. An officer should be as consistent as possible when sending verbal messages over the radio. The performance goal should be to sound the same and communicate just as effectively when reporting a minor incident as when communicating under intense stress. An overly excited commander is difficult to understand and gives the listener(s) the impression that the incident is out of control. In contrast, calm and confident orders from the person in charge set the tone for more efficient and effective incident management.

Recordings of radio messages transmitted during emergency incidents are an effective training tool. Listening to others allows a fire officer to identify and emulate techniques that are clear and precise, leaving no doubt about the intent of the message. Listening to recordings of your own radio communications also provides valuable feedback, allowing you to compare the actual communications output with your thought process at that moment.

■ Initial On-Scene Radio Report

Size-up is the mental process of gathering and considering all of the pertinent details of a given incident. The initial radio report follows the size-up. During this report, the fire officer puts the key information into words clearly and concisely and shares it with all of the personnel responding to the incident. The initial report should describe what you have, state what you are doing, and provide direction for other units that will be arriving. Routinely using the same terminology and format helps ensure that nothing is missed. An example might sound like this:

Engine 1: Dispatch, from Engine 1.

Dispatch: Engine 1, this is Dispatch.

Engine 1: Engine 1 at 2345 Central Avenue. We have a two-story, wood-frame, single-family dwelling with heavy smoke showing on side Charlie and fire showing from the first floor on side Delta. Engine 1 has a supply line and will be making an offensive attack through side alpha. Engine 1 is establishing Central Avenue Command on side Alpha.

Dispatch: Engine 1 at 2345 Central Avenue with a two-story, wood-frame, single-family dwelling, heavy smoke showing on side Charlie and fire showing on side Delta first floor. Engine 1 is making an offensive attack through side Alpha and establishing Central Avenue Command on side Alpha.

Through this simple exchange of information, everyone who needs to know is informed of the situation and the action being taken by the first-arriving unit, and the stage is set for all other units to take action based on SOPs. The dispatcher's repeating of the key information provides solid confirmation that it was received and understood.

■ Using the Communications Order Model

The communications order model is a standard method of transmitting an order to a unit or company at the incident scene. It is designed to ensure that the message is clearly stated, heard by the proper receiver, and properly understood. It also confirms that the receiver is complying with the instruction.

Command: Ladder 2, from Command.

Ladder 2: Ladder 2, go ahead Command.

Command: Ladder 2, come in on side Charlie and conduct a primary search on the second floor. Also advise if there is any fire extension to that level.

Ladder 2: Ladder 2 going to side Charlie to do a primary search on the second floor and check for fire extension.

Command: Ladder 2, that is correct.

■ Radio Reports

Company fire officers frequently provide reports over the radio. Radio communications are essential for emergency operations because they provide an instantaneous connection and can link all of the individuals involved in the incident to share important information. During an emergency incident, both the sender and the receiver should strive to make their radio messages accurate, clear, and as brief as possible. Conditions are often stressful, with several parties competing for air time and the receiver's attention. There may be only a limited time to transmit an important message and ensure that it is received and understood.

The individual who is transmitting a radio report often feels an intense pressure to get the message out as quickly as possible. The realization that a large audience could be listening often adds to the sender's anxiety level.

When given the option, many fire officers prefer to use a telephone or face-to-face communications instead of a radio to transmit complicated or sensitive information. These options allow a more private exchange of information between two individuals, including the ability to discuss and clarify the information. A good practice is to think first; position the microphone; depress the key; take a breath; and then send a concise, specific message in a clear tone.

Reporting

Some reports are prepared on a regular schedule, such as daily, weekly, monthly, or yearly. Other types of reports are prepared only in response to specific occurrences or when requested. To create a useful report, the fire officer must understand the specific information that is needed and provide it in a manner that is easily interpreted.

Types of Reports

Some reports are presented orally, whereas others are prepared electronically and entered into computer systems. Some reports are formal, whereas others are informal. This chapter covers some of the widely used elements of reporting and provides examples of the most common types of reports that a fire officer might have to prepare. More experienced fire officers can also provide assistance in mastering the reporting requirements of a new position.

■ Verbal Reports

The most common form of reporting is verbal communication from one individual to another, either face-to-face or via a telephone or radio. To be effective, the transfer of information must be clear and concise, using terminology that is appropriate for the receiver.

Face-to-face conversation is the most effective means of conveying many types of information. The sender and the receiver can engage in a two-way exchange of information that incorporates body language, facial expressions, tone of voice, and inflection. When verbal communication must be conducted over radio or telephone, however, these supplementary expressions are sacrificed. During emergency operations, key information from a face-to-face encounter may need to be transmitted over the radio so everyone has access to it.

A CAN report is an effective method of exchanging information when the magnitude or speed of the event requires maximum efficiency. The acronym CAN stands for "conditions–actions–needs." Suppose incident command contacts the Ventilation Group and requests a CAN report. The officer assigned to the Roof Group would report back to command the conditions of the roof and ventilation process, actions taken, and any additional needs that must be met to complete the task. Radio reports during an emergency incident should be directed back to the incident commander once the task is completed, when a progress update is necessary, or when additional resources are required.

■ Written Reports

Routine Reports

Routine reports provide information that is related to fire department personnel, programs, equipment, and facilities. Most fire departments also require company officers to maintain a company journal or log book **FIGURE 4-4**. A fixture in fire stations since the 19th century, the company journal provides an extemporaneous record of all emergency, routine, and special activities that occur at the fire station. Some fire departments have advanced from keeping the journal in a handwritten, hard-covered book to maintaining it as a database on the station computer.

The company journal serves as a permanent reference that can be consulted to determine what happened at that fire station at any particular time and date, as well as who was involved. The general orders usually list all of the different types of information that must be entered into the company journal. The company journal is also the place to enter a record of any fire fighter injury, liability-creating event, and special visitors to the station.

Morning Report to the Administrative Fire Officer

Most career fire officers are required to provide some type of morning report to their superior officer. This report is made by telephone or on a simple form that is transmitted by fax or e-mail.

One purpose of the morning report is to identify any personnel or resource shortage as soon as possible. For example, if the driver/operator calls in sick, the driver/operator from the off-going shift might have to remain at work to keep the engine company in service until a replacement arrives. The administrative fire officer needs to know that the company is short a driver/operator and has a person working involuntary overtime or holdover, or that a replacement needs to be found.

Monthly Activity and Training Report

The monthly activity and training report documents the company's activity during the preceding month. Such a report typically includes the number of emergency responses, training activities, inspections, public education events, and station visits that were conducted during the previous month. Some monthly reports include details such as the number of feet of hose used and the number of ladders deployed during the month.

In many cases, the officer delegates the preparation of routine reports that do not involve personnel actions or supervisory responsibilities to subordinates. In such a case, the officer is always responsible for checking and signing the report before it is submitted.

Two other routine reports are the annual fire fighter performance appraisals and the fire safety inspection. These reports are covered in detail in the *Evaluation and Discipline* and *Preincident Planning and Code Enforcement* chapters.

Some fire companies or municipal agencies post a version of a monthly report on their public Web site. The reports often include digital pictures or other information of the incidents and the personnel involved.

Special events or unusual situations that occurred during the month might also be included in the monthly report. For example, this report might note that the company provided standby coverage for a presidential visit or participated in a local parade.

Incident Reports

An incident report is required for every emergency response. The nature and complexity of the report depend on the situation:

Fire Station 100 Company Journal

out	in	inc#	address / comments			sign
	0700		Thursday, February 12 - A shift on duty. Captain Davis OIC			
			Engine	Quint	Medic	
			Cpt. Davis	Lt. Williams	Lt/PM Turner	
			AO Anders	AO Rollo	FF/PM Olliges	
			FF/PM Thompson	FF Grynski		
			FF Sorce (hold)	FF/HM Schultz		
			Status:			
			FF Kinders on Sick Leave			
			FF Source on involuntary hold over			
			FF Wirth on Exchange of Shift until 1900 with FF/HM Schultz			
			FF/PM O'Brien on Annual Leave			
			Vehicle #130654 at Maintenance			
			Vehicle #030112 running as Engine 100			
			Vehicle #110065 running as Medic 44			
			Portable Radio / Keys / Controlled drugs	Anders - 4 / Davis / Thompson	Rollo - 4 / Williams / xxxxxx	Olliges - 2 / Turner / Turner
	0730		Paramedic Intern Bartlow (Community College) on Medic 100			Turner
0813	1011	0345	7111 Hogarth St. - Injury		Medic	Olliges
	0820		FF Cegar detailed in from Fire Station 47			Ceg
	0830		FF Sorce off duty			Sorce
0855	0914	0456	9103 Cross Chase Rd - Trouble Breathing		Engine	Anders
1122	1255	0499	4110 Green Springs Dr. #101 - sick		Medic	Olliges
1141	1217	0512	8902 Harrivan Ln - Kitchen fire		Quint	Rollo
1230	1645		Quint to Academy			Williams
	1300		12-lead monitor #1314 damaged on incident #0499			Turner
1314	1655		Facilites working on furnace			Davis
	1345		EMS3 delivers loaner 12-lead #1133 and picks up #1314			Turner
1355	1506	0701	10614 Hampton Rd - STEMI		Medic	Olliges
1355	1422	0710	10614 Hampton Rd -STEMI		Engine	Davis
1430	1450	0715	I-95 North at Franconia exit - MVC		Engine	Davis
1455	1511	0733	6600 Springfield Mall - Injury		Engine	Davis
	1815		Assistant Chief Arrow and City Manager in quarters			
1824	1930	1135	10400 Richmond Hwy - ALS		Medic	Olliges
	1830		FF Wirth on duty			Wirth
1844	0020	1148	7401 Eastmoreland Ave - 2nd Alarm		Quint	Williams
1909	2253	1148	7401 Eastmoreland Ave - 3rd Alarm		Engine	Davis
	1922		Engine 44 filling Fire Station 100			Jones
			Friday, February 13			
0112	0223	0078	9788 Whispering Meadow Ln - stroke		Medic	Olliges
0249	0311	0101	6600 Springfield Mall - Alarm Bells		Quint	Williams

FIGURE 4-4 Example of a company journal.

Minor incidents generally require simple reports, whereas major incidents require extensive reports.

The National Fire Incident Reporting System (NFIRS) is a nationwide database managed by the U.S. Fire Administration that collects data related to fires and other incidents. Generally, the first-arriving officer (or the incident commander) completes an NFIRS report for each response, including a narrative description of the situation and the action that was taken. The full incident report includes a supplementary report from the officer in charge of each additional unit that responded.

Many departments use a narrative format when reporting on incidents FIGURE 4-5. Some incidents require an expanded incident report narrative, in which all company members submit a narrative description of their observations and activities during an incident. The fire officer should anticipate the need to provide expanded incident reports for the following types of incidents:

- The fire company was one of the first-arriving units at a fire with a civilian fatality or injury.
- The fire company was one of the first-arriving units at an incident that has become a crime scene or an arson investigation.
- The fire company participated in an occupant rescue or other emergency scene activity that would qualify for official recognition, an award, or a bravery citation.
- The fire company was involved in an unusual, difficult, or high-profile activity that requires review by the fire chief or designated authority.
- A fire company activity occurred that may have contributed to a death or serious injury.

> **Civilians Rescued From Burning Home**
>
> Twenty-eight fire fighters responded to an early morning house fire in the Grandview district. The first 9-1-1 call was at 10:47 pm on Monday December 10, 2014, reporting smoke coming from a three-story townhome.
>
> Metro County fire fighters arrived at 11:07 pm, encountered smoke and flames coming from the first floor windows at 1928 Braniff Boulevard. Two elderly females were found on the third floor. They were taken out of the building through a window via an aerial tower, treated by paramedic/fire fighters, and transported to University Hospital.
>
> It took 17 minutes of aggressive fire suppression before Battalion Chief Devon Jones declared the fire "under control." The first floor of the townhouse was extensively damaged, with heat and smoke damage to the adjacent townhouses.
>
> The cause of the fire remains under investigation. Preliminary results show that the fire started in the kitchen. Smoke detectors were in the home, but batteries were removed.
>
> The monetary loss has not been calculated.
>
> Submitted by T. L. Gaines, Metro County Fire Department
>
> MCFDPIO@metrocounty.gov (111) 555-3473

FIGURE 4-5 Sample incident report.

- A fire company activity occurred that may have created a liability.
- A fire company activity occurred that has initiated an internal investigation.

The author of a narrative should describe any observations that were made en route to the incident or at the scene, and should fully document his or her actions related to the incident. The narrative should provide a clear mental image of the situation and the actions that were taken, and may include pertinent negatives—elements noted not to be present or actions not taken. For example, if no gas-powered tools were used on the fire floor, this fact may influence the effectiveness of an arson investigation and be cited in future court proceedings.

Infrequent Reports

Infrequent reports usually require a fire officer's personal attention to ensure that the report's information is complete and concise. These special reports include the following documents:

- Fire fighter injury report
- Citizen complaint
- Property damage or liability-event report
- Vehicle accident report
- New equipment or procedure evaluation
- Suggestions to improve fire department operation
- Response to a grievance or complaint
- Fire fighter work improvement plan
- Request for other agency services

Some situations require two or more different report forms to cover the same event. For example, if a fire engine collides with a private vehicle at an intersection while responding to an emergency call, the fire officer will spend significant time completing a half-dozen forms, even if no injuries occur. A supervisor's report must be filed, and the apparatus driver must complete an accident report. Another form identifies the damage to the fire apparatus and any repairs that are needed. The local jurisdiction's risk management division may also require a report. The insurance company that covers the fire apparatus usually requires the completion of another form.

A supervisor's report is required by state worker's compensation agencies whenever an employee is injured. This report serves as the control document that starts the state file relating to an injury or a disability claim **FIGURE 4-6**.

The supervisor's first report of an injury must be submitted within 24 to 72 hours of the incident or in line with company policy, after conducting an investigation into the facts.

Safety Zone

Knowing Reporting Procedures

A fire officer is responsible for the timely and accurate completion of property damage and injury reports. That does not mean that the fire officer must personally fill out every form. All fire fighters should know how to fill out the required administrative paperwork for a minor injury, bloodborne pathogen exposure, or damage to fire department property. Departments using electronic patient care reports must make note of the presence of known environmental or health hazards, as this information may affect future worker's compensation coverage or long-term healthcare issues for an individual or an entire crew.

The late notification of the duty officer regarding infectious disease encounters and the tardy or incomplete documentation of injuries or toxic exposures are continuing challenges to fire fighter health and safety. All fire fighters should know the reporting procedures to be followed to protect them throughout their careers.

SUPERVISOR'S ACCIDENT REPORT
WORKERS' COMPENSATION CLAIMS

Claims Management Inc.
PO Box 342
Sacramento, CA 95812-3042
(916) 631-1250
FAX (916) 635-6288

DATE & TIME REPORTED:

OSHA CASE NO:

COMPANY	LOCATION	LOCATION CODE NO:
Scotts Valley Fire Protection District	*7 Erba Lane, Scotts Valley, CA, 95066*	*1100*

A. EMPLOYEE

NAME | JOB TITLE

DEPARTMENT *Scotts Valley Fire Protection District 1100* | ☐☐ LOST TIME ☐☐ NO LOST TIME | ☐☐ FIRST AID

B. TIME AND PLACE OF ACCIDENT

DATE	HOUR	DEPARTMENT	IMMEDIATE SUPERVISOR
		1100	

IDENTIFY EXACT LOCATION WHERE ACCIDENT OCCURRED *(Be specific)*

JOB OR ACTIVITY AT TIME OF ACCIDENT *(Be specific)*

C. WITNESSES - *List of Names and Addresses*

Name	Address

D. DESCRIBE THE ACCIDENT/ACCIDENT CAUSE - *Please be specific*

E. UNSAFE ACT/CORRECTIVE ACTION TAKEN - *Include both employee and supervisor corrective actions to prevent future occurrences.*

EMERGENCY - WENT TO THE DOCTOR
☐☐ YES
☐☐ NO

If yes, please fill out the following information:
Name of Doctor: ____________________
Address of Doctor: ____________________

☐☐ NON-EMERGENCY, BUT PLAN ON SEEING A PHYSICIAN
Physician's Name: ____________________

Courtesy of Claims Management, Inc.

FIGURE 4-6 Supervisor's accident report for worker's compensation.

This report often identifies the injured person, the nature of the injury, the means by which the incident occurred, practices or conditions that contributed to the incident, any potential loss, the typical frequency of occurrence, and actions that will be taken to prevent the same incident from occurring again. A chronological statement of events is a detailed account of activities, such as a narrative report of the actions taken at an incident or accident, and should be included in the supervisor's report. Emotional statements, opinions, and hearsay have no place in a formal report that will remain on permanent record and potentially be subject to subpoena as a public record.

Using Information Technology

Most reports are completed using a computer and software. The role of the fire officer will range from selecting preformatted information in an interactive online form, as with an NFIRS report, to composing a narrative and following a guide sheet to make a suggestion to improve fire department operations. The physical resources available within a fire station for reporting purposes typically include a computer, printer, and network connection.

The computer includes a keyboard, display screen, mouse or touchpad, central processing unit, memory, hard drive, and a disk drive.

Two types of software are found on the computer: the operating system and applications. The operating system handles the input, manages files, and manages all of the activity. It is like the drivetrain of a quint. Commonly used operating systems include Microsoft Windows, Apple's OS, and Linux. Application programs are designed to provide a specific task, much like a pump panel controls the delivery of water. Fire officers commonly use word processing, spreadsheet, and presentation applications.

Word processors are used to produce memos, reports, and letters. Many programs provide templates that serve as premade layouts for routine reports. Most provide spell-checking and grammar guides. Popular word processing programs include Microsoft Word, OpenOffice Writer, GoogleDocs, WordPerfect, and Page.

Spreadsheets are used to tabulate numbers and information in columns and rows, often as a means to analyze data and develop charts. The display resembles a ledger sheet, with columns represented by letters and rows identified by numbers. The intersection of a column and row, called a cell, can contain numbers, letters, or a formula. Formulas are used to make calculations. For example, in cell A20 you might see "=A19*D03," which means the software multiplies the number in cell A19 with the number in cell D03 and places the result in cell A20. Spreadsheets have immense calculation capabilities. Widely used spreadsheet applications include Microsoft Excel, Open Office Calc, Google Spreadsheet, Apple Numbers, and Libre Office Calc.

Presentation software is used to display information in a format that allows the audience to understand visually the message presented. The images can be displayed on a screen or Web site, and images, videos, and sounds can be integrated as desired. Commonly used presentation software includes Microsoft PowerPoint, OpenOffice Impress, and Apple Keynote.

Fire Officer II

Written Communications

Written communication is needed to document both routine and extraordinary fire department activities. The resulting documents establish institutional history and serve as the foundation of any activity that the fire department wishes to accomplish. As a senior chief explained, "If it is not written down, it did not happen."

Informal Communications

Informal communications include internal memos, e-mails, instant messages, and messages transmitted via mobile data terminals. This type of communication is used primarily to record or transmit information that may not be needed for reference in the future. Even so, all communications from a company officer are retrievable within an information management system. Consequently, each message and report needs to be appropriate, accurate, and ethical.

Some informal reports are retained for a period of time and may become a part of a formal report or investigation. For example, a written memo could document an informal conversation between a fire officer and a fire fighter regarding a performance issue. The memo might note that the employee was warned that corrective action is required by a particular date. If the performance is corrected, the warning memo may be discarded after 12 months. If the performance remains uncorrected, the memo becomes part of the formal documentation of progressive discipline.

Formal Communications

A formal communication is an official fire department document printed on business stationery with the fire department letterhead. If it is a letter or report intended for someone outside the fire department, it is usually signed by the fire chief or a designated staff officer to establish that the document is an official communication. Subordinates often prepare these documents and submit them for an administrative review to check the grammar and clarity. A fire chief or the staff officer who is designated to sign the document performs a final review before it is transmitted. When sending electronic messages with attachments, it is a good practice to double-check exactly what you are sending by opening the attachment(s) to make sure it is what you intended.

The fire department maintains a permanent copy of all formal reports and official correspondence from the fire chief and senior staff officers. Formal reports are usually archived.

Standard Operating Procedures

Standard operating procedures (SOPs) are written organizational directives that establish or prescribe specific operational or administrative methods to be followed routinely for the performance of designated operations or actions. They are

Assessment Center Tips

Write a Letter for the Fire Chief
A popular assessment center exercise requires the candidate to prepare a letter for the fire chief's signature. This represents a common assignment for a supervising fire officer in many fire departments. In the scenario, the candidate is provided with all of the necessary information and must demonstrate the ability to prepare a properly formatted and worded letter. The letter is typically a response to a complaint or inquiry by a citizen or community interest group.

This exercise has two primary goals:
1. Assess the candidate's writing skills.
2. Assess the candidate's decision-making skills.

Preparing for this part of the assessment center requires the candidate to learn the proper format for an official letter or formal report. Generally, the assessment center exercise requires a business letter format. One method of preparing for this exercise is to contact the fire chief's administrative assistant and ask how this type of letter is typically prepared for the fire chief.

---

intended to provide a standard and consistent response to emergency incidents as well as personnel supervisory actions and administrative tasks. SOPs are a prime reference source for promotional exams and departmental training.

SOPs are formal, permanent documents that are published in a standard format, signed by the fire chief, and widely distributed. They remain in effect permanently or until they are rescinded or amended. Many fire departments conduct a periodic review of all SOPs so they can revise, update, or eliminate any that are outdated or no longer applicable. Any changes in the SOPs must also be approved by the fire chief (See the *Introduction to the Fire Officer* chapter for more information on SOPs.)

Standard Operating Guidelines

Standard operating guidelines (SOGs) are written organizational directives that identify a desired goal and describe the general path to accomplish the goal, including critical tasks or cautions. Like SOPs, SOGs are formal, permanent documents that are published in a standard format and remain in effect until they are rescinded or amended. SOGs are often used in multijurisdictional activities, such as an automatic aid response plan, that provide flexibility for each responding agency to achieve a goal through various methods and resources.

The U.S. Fire Administration's *Developing Effective Standard Operating Procedures for Fire and EMS Departments* looked at the difference between SOPs and SOGs. Its review of legal proceedings indicates that terminology is less important than content and implementation of SOPs/SOGs. The review revealed that the courts tend to assess liability based on factors such as the following:

- Systems in place to develop and maintain SOPs/SOGs
- Compatibility with regulatory requirements and national standards
- Consideration of unique departmental needs
- Adequacy of training and demonstration of competence
- Procedures used to monitor performance and ensure compliance

General Orders

General orders are formal documents that address a specific subject, policy, condition, or situation. They are usually signed by the fire chief and can remain in effect for various periods of time, ranging from a few days to permanently. These documents may also be called *executive orders*, *departmental directives*, *master memos*, or other terms. Many departments use general orders to announce promotions and personnel transfers. Copies of the general orders should be available for reference at all fire stations.

Announcements

The formal organization may use announcements, information bulletins, newsletters, Web sites, or other methods to share additional information with fire department members. The identification of these communications is different for each department. These types of announcements are used to distribute short-term and nonessential information that is of interest.

Legal Correspondence

Fire departments are often required to produce copies of documents or reports for legal purposes. Sometimes the fire department is directly involved in the legal action, whereas in other cases the action involves other parties but relates to a situation in which the fire department responded or had some involvement. It is not unusual for a fire department to receive a legal order demanding all of the documentation that is on file pertaining to a particular incident. Collecting and transmitting this information is a task that can require extensive time and effort. In any situation in which reports or documentation is requested, the fire department should consult with legal counsel.

The fire officer who prepared a report may be called on—sometimes years later—to sign an affidavit, appear at a deposition, or appear in court to testify that the information provided in the report is complete and accurate. The fire officer might also have to respond to an interrogatory, which is a series of written questions asked by someone from an opposing legal party. The fire department must provide written answers to the interrogatories, under oath, and produce any associated documentation. If the initial incident report or other documents provide an accurate, factual, objective, complete, and clear presentation of the facts, the response to the interrogatory is often a simple affirmation of the written records. The task can become much more complex and embarrassing if the original documents are vague, incomplete, false, or missing.

Recommendation Report

A recommendation report is a document that suggests a particular action or decision. Some reports incorporate both a chronological section and a recommendation section. For

VOICES OF EXPERIENCE

Effective communication is crucial not only for our success on the fire ground, but also for the safety of our fire fighters. When reviewing line-of-duty deaths or near-miss reports, it seems a breakdown in communication is a contributing factor in the majority of them. Think about all the times within your own organization that miscommunication has occurred on your fire ground. Now ask yourself, why does this occur?

> **The incident commander attempted to call the ladder crew on the radio, but no response came back.**

Several years ago, I served in the Training Division for a neighboring department. We were in the middle of conducting annual live fire scenarios. In one particular scenario, crews arrived at a three-story burn building and reported smoke evident from the C and D sides on the second and third floors. The incident commander (IC) assigned his engine crew to fire attack (Attack Group); the rescue crew was assigned to search and rescue (Search Group). During his walk-around, the captain also noticed a victim (mannequin) positioned at a window on the C/D corner of the third floor. The arriving ladder crew was assigned by Command (via radio) to throw a 35-foot ladder to the third floor and attempt to rescue the victim. The ladder crew positioned the ladder at the window; while they were doing so, the tip of the ladder bumped the victim, knocking it down away from view. After the ladder was in position, the ladder crew returned to their rig to retrieve tools and stage for further orders.

During this time, the IC was moving around the exterior of the building, continuing to assess the situation and give assignments to the approaching medic unit and engine. When he came back around to the C side, he noticed the ladder and also noticed the victim was no longer visible. He radioed to the Search Group inside that they should meet up with the ladder crew who had the victim. The Search Group advised they had made access to the third floor via the stairwell and had not met the ladder crew. The incident commander attempted to call the ladder crew on the radio, but no response came back. After a second attempt to reach the ladder crew via radio with no response, the IC ordered the evacuation signal to sound and transmitted the evacuation message over the radio to all units. As units evacuated, it was quickly discovered that the ladder crew was standing on the other side of their rig outside of the view from Command. The senior fire fighter on the ladder never heard the order for them to make entry, only to throw the ladder. When the assignment was complete, he did not inform Command because he assumed the IC would see the ladder in position, and he returned to the rig to gather more tools and await further orders.

The captain in charge quickly realized something was not right and made the right call in evacuating all crews to determine the issue. Had it not been for the quick discovery of the problem, the IC was already prepared to issue a mayday call on the ladder crew's behalf.

Feedback in this situation would have prevented this incident from occurring at all. It was a great eye opener for the crews on scene as to how quickly miscommunication could happen. I have used this scenario as an example in many classes to demonstrate the importance of effective communication and how it impacts everything we do.

Billy Floyd, Jr.
Division Chief of Fire Training
North Myrtle Beach Department of Public Safety
North Myrtle Beach, South Carolina

example, the fire fighter line-of-duty death investigations produced by the National Institute for Occupational Safety and Health (NIOSH) include a chronological report of the event, followed by a series of recommendations aimed at preventing the occurrence of a similar situation.

The goal of a decision document is to provide enough information and make an effective persuasive argument so that the intended individual or body accepts your recommendation. This type of report includes recommendations for employee recognition or formal discipline, for a new or improved procedure, or for adoption of a new device. A decision document usually includes the following elements:

- **Statement of problem or issue:** One or two sentences.
- **Background:** Brief description of how this became a problem or an issue.
- **Restrictions:** Outline of the restrictions affecting the decision. Factors such as a federal law, state regulations, local ordinances, budget or staff restrictions, and union contracts are all restrictions that could affect a decision.
- **Options:** Where appropriate, provide more than one option and the rationale behind each option. In most cases, one option is to do nothing.
- **Recommendation:** An explanation of why the recommended option is the best decision. The recommendation should be based on considerations that would make sense to the decision maker. If the recommendation is going to a political body and involves a budget decision, it should be expressed in terms of lower cost, higher level of service, or reduced liability. The impact of the decision should be quantified as accurately as possible, with an explanation of how much money it will save, which new levels of service will be provided, or how much the liability will be reduced.
- **Next action:** A clear statement of the action that should be taken to implement the recommendation. Some recommendations may require changes in departmental policy or budget. Others could involve an application for grant funds or a request for a change in state or federal legislation.

Writing a Report

Reports should be accurate and present the necessary information in an understandable format. The report could be intended to brief the reader, or it might provide a systematic analysis of an issue. If the fire chief wants only to be briefed on a new piece of equipment and the fire officer delivers a fully researched presentation, both parties would be frustrated. An executive summary would help in this case.

The fire officer must consider the intended audience. Consider a report that proposes replacing four engine and three ladder companies with three quints. If the report is internal, the technical information could include fire department jargon. In contrast, if the report will be submitted to the city council, with copies going to the news media, many of the terms and concepts would have to be explained in detail.

When preparing the report, the appropriate format must be selected. Options range from an internal memo format with one or two pages of text to a lengthy formal report that includes photographs, charts, diagrams, and other supporting information.

Some reports require a recommendation. In such a case, the fire officer may need to analyze and interpret data that are maintained in a database or a records management system. The fire department's records management system typically contains NFIRS run reports, training records, occupancy inspection records, preincident plans, hydrant testing records, staffing data, and company activity records.

A computer-aided dispatch system uses a combination of databases to verify addresses and determine the units that should be dispatched; it also maintains information that can be retrieved relating to individual addresses. All of the transactions performed by a computer-aided dispatch system are recorded in another database. This information can be retrieved to examine the history of a particular incident or to identify all of the incidents that have occurred at a particular location.

When developing some reports, the fire officer might be looking at variances, such as differences between the projected budget and actual expenses in different accounts in a given year. When a variance is encountered, the analysis should consider both the cause and the effect. If the budget included an allocation of $10,000 for training and the expenditure records indicate that only $500 was spent, there would be a variance of $9500 that could be reallocated for some other purpose. This variance could also indicate that very little training was actually conducted.

Presenting a Report

A fire officer should be prepared to make an oral presentation of a decision document or to speak to a community group on a fire department issue. Using the written report as a guide, a verbal presentation would consist of four parts:

1. *Getting their attention.* Have an opening that entices the audience to pay attention to your message.

Assessment Center Tips

Verbal Presentations

A promotional candidate should be prepared for two different types of verbal presentations. The first type is a meeting with a citizen or community group to explain a fire department practice or issue. This is equivalent to a verbal presentation of a decision document.

The second type of presentation could follow an emergency incident scenario. After the incident management portion of the exercise is completed, the candidate could be required to make a presentation of the incident as if he or she is speaking to the press or reporting to a senior command officer.

2. *Interest statement.* Immediately and briefly explain why listeners should be interested in this topic.
3. *Details.* Organize the facts in a logical and systematic way that informs the listeners and supports the recommended decision.
4. *Action.* At the close of your presentation, ask the audience to take some specific action. The most effective closing statements are actions that relate directly to the interest statement at the beginning of the presentation.

Preparing a News Release

Fire organizations need to be able to communicate effectively through the local mass media. A news release allows a fire department to reach a large audience at virtually no cost. Such a release could be structured to promote a public education message, draw attention to the community's risk problems, announce a new program or the opening of a new facility, or simply develop good public relations. A well-prepared news release encourages the media to cover your story.

The first step in preparing a news release is to formulate a plan. Consider what the release is intended to accomplish. Is it an urgent fire safety message or an invitation to the fire department picnic? Who is the target audience? Most messages are directed toward some particular group or audience. A safety message about a smoke alarm program, for example, is directed toward those residents who do not have smoke alarms. Every contact with the community should be viewed as an opportunity to teach or reinforce some safety message. You already have community members' attention due to the event, so here is the opportunity to teach them something that can prevent an injury or death in the future.

During this first step, you should also identify what makes the story interesting and worthy of the time and energy the media source must expend to present it. The media may have hundreds of releases that they must prioritize to decide which ones should get attention. If the news release is about smoke alarms, the release should emphasize why this subject is important.

The second step is to develop the concept and write the release. The format should be clear, concise, and well organized. The first paragraph or two should cover the "who, what, when, where, and why" of the story. Successive paragraphs should provide further information, progressing from the most important information to the least important. Do not be wordy in a press release, and stick to the facts. The idea is to give the media enough information to convince them that the story warrants coverage. Each media outlet will consider the release in the context of whether the information would be of interest to its audience. If the release is on target, a reporter will follow up. It is very important that sensitive information, such as names or identifiers, is not released to protect the individuals as well as the department from Health Insurance Portability and Accountability Act (HIPAA) violations.

The use of the department's letterhead will help draw the attention of the news agency. Formatting can also help bring attention to the release. Use at least 1-inch (2.5-cm) margins, double-space the text, and make the news release fit on one page. Keep the document neat and clearly organized, and make sure that it is free from typographical or grammar errors.

At the top of the page, print "NEWS RELEASE" in all capital letters. The time and date should appear on the following line. Separate this heading from the text by a series of five to seven pound signs (#######). Pound signs should also be used to separate the text from the ending, which should include a contact name and number for further information.

The last step is to get the news release out to the media. It is important to distribute the release to all media outlets in an equitable manner.

Social Media

The term social media describes a continually changing utilization of digital communications in which users create online communities to share information, ideas, personal messages, videos, pictures, and other content. Social media differ from traditional or corporate media in the areas of quality, reach, frequency, usability, immediacy, accuracy, reliability, and permanence.

The proliferation of social media has drastically changed the way news is reported. Several years ago, much of the reporting fell to news reporters who collected information from a variety of sources before presenting what they had learned to the public. The impact of social media on news reporting is that anyone with a social media account can report what they have seen and heard, and news spreads much faster. This news, however, is not always accurate or well researched.

■ Establishing a Fire Department Social Media Policy

Fire departments need to establish and regularly review their social media policies. Technology, court decisions, and high-profile incidents continue to shape the development of best practices.

The Cumberland Valley Volunteer Firemen's Association's *Fire Service Reputation Management White Paper* includes this observation on departmental and personal information technology:

> Fire departments should also be mindful of department-related material/content that members may post to their individual or other websites or blogs, including "unofficial" departmental websites. Acting within the bounds of the law and, in particular, respecting individuals' freedom of speech and expression, departments should have policies to ensure that members do not post to their own or any other website inappropriate or offensive department-related material. Such policies should, among other things, explicitly aim to prevent

any disclosures that might reveal sensitive Homeland Security–related information or that might constitute a potential HIPAA or other privacy-related breach or violation.

Courtesy of Cumberland Valley Volunteer Firemen's Association.

The fire chief identifies who will develop a social media policy, including the process of submitting, reviewing, and approving content. Such a policy should cover use of personal social media sites as related to the member's fire department duties as well as official fire department social media sites.

The fire department has an obligation to actively protect itself from creation and sharing of inappropriate images. Using a pumper as a background for a sexy photo session is an example of an inappropriate image. Posting racist and X-rated messages about the department or the job are actions that have terminated the careers of both volunteer and career fire fighters. Posting images of dead, intoxicated, or seriously injured victims can destroy decades of trust built up between the department and the citizens they serve. In fact, Connecticut and New Jersey have made it a criminal offense for an emergency responder to take a photo of a victim or patient; posting that image online generates a second criminal offense.

Using images or discussing incidents or issues identified as "fire department related" on personal social sites can also be controlled by the department, even when the information is posted by off-duty personnel. Developing and enforcing the department's social media policy will require coordination among the authority having jurisdiction, human resources, public information agencies, and law enforcement **FIGURE 4-7**.

■ Helmet Cams and On-the-Job Digital Images

Digital technology provides an opportunity to provide a dramatic first-person perspective while a fire fighter is operating at a structure fire, vehicle crash extrication, or life-threatening medical emergency. High-definition picture and sound can create a compelling recording.

Recordings also create significant risk to the department. Posting such videos on social media could interfere with an arson investigation, illegally identify a medical patient, or expose a crime victim. Any of these outcomes may subject the department to legal sanctions and penalties. Deleting such recordings could be considered destruction of evidence.

No incident is handled perfectly. A real-time digital documentation of a fire-ground misadventure could be used to point out reckless behavior, poor procedures, or a practiced pattern of incompetence. Some departments now prohibit the use of any video recording device when responding to or operating at an incident.

Issues related to fire service imagery extend beyond on-duty members responding to incidents. Utilizing images or logos that identify the department is prohibited when members post private pictures or videos on social media. For some agencies, this is a digital extension of the rules against wearing elements of the fire fighter's uniform or personal protective clothing when off-duty.

> **Fire Marks**
>
> **LAFD Is Early Social Media Adopter**
> The Los Angeles Fire Department uses social media to assist during emergencies as well as to engage the community in emergency preparedness. As reported in a 2007 article in *Computerworld,* Public Information Officer Brian Humphrey was able to obtain first-person accounts of fire development at a major emergency brush fire in Griffith Park. Humphrey provided that information to the incident commander as well as public information and instructions through Twitter. LAFD utilizes approximately 80 different social media applications or services.

■ Engaging the Community Through Social Media

In *The Path Forward*, the American Red Cross presented research at the 2010 Emergency Social Data Summit showing that nearly half of respondents affected by a disaster asked for help on social media and 75 percent expected help to arrive within the hour. FEMA Administrator W. Craig Fugate, the keynote speaker at the summit, points out, "As social media becomes more a part of our daily lives, people are turning to it during emergencies as well. We need to utilize these tools, to the best of our abilities, to engage and inform the public, because no matter how much federal, state, and local officials do, we will only be successful if the public is brought in as part of the team."

Social media adds immediacy that empowers citizens to share information with government, first responders, media, and one another, thereby complementing messages transmitted by radio, broadcast or cable TV, printed notices, and ham radio networks. A key to successful fire department use of social media is establishing the public's trust.

Outward social media efforts provide the public with timely and appropriate awareness of active emergency incidents that may affect traffic or their neighborhood, alerts regarding impending natural or human-made significant events, emergency directions during a major weather event, and emergency communications during a disaster. In addition, outward social media can supplement traditional fire prevention and injury prevention campaigns, increase awareness of fire department community activities, and provide a constantly updated digital presence.

A federal homeland security group recruited first responders to develop best practice recommendations in engaging communities. The resulting 2012 Department of Homeland Security report, *First Responder Communities of Practice Virtual Social Media Working Group Social Media Strategy*, made the following recommendations:

> Establishing and maintaining brand standardization, including social media profiles, logos, messaging styles,

SAN JUAN COUNTY FIRE DEPARTMENT

SOCIAL MEDIA POLICY

1. **PURPOSE:** The purpose of this document is to define and regulate the use of social media by San Juan County Fire Department employees and volunteers.

2. **DEFINITIONS**

    a. Social media: forms of electronic communication through which users create online communities to share information, ideas, personal messages, and other content. The term social media includes, but is not limited to, social networking sites such as Facebook, Myspace, LinkedIn, Twitter, and YouTube.

    b. SJCFD social media site: a social media site created, maintained and controlled by SJCFD.

    c. Personal social media: social media content maintained and controlled by an individual employee or volunteer member of SJCFD.

3. **SCOPE** This policy applies to the use of personal social media relating to an employee's or volunteer's duties, and to social media on SJCFD social media sites.

4. **SJCFD SOCIAL MEDIA SITES:**

    a. SJCFD social media sites shall not be created without the approval of the SJCFD Fire Chief or the SJCFD Fire Chief's designee.

    b. All content posted on SJCFD social media sites shall be approved by the Fire Chief or the Fire Chief's designee.

    c. Social media content on SJCFD social media sites shall adhere to all applicable laws, regulations and policies including the records management and retention requirements set by law and regulation.

5. **PERSONAL SOCIAL MEDIA**

    a. No information, videos or pictures gathered while on SJCFD business (including emergency calls, meetings, drills, details, trainings or anything obtained on organization property or at organization functions) may be shared or posted in any format without the approval and written consent of the Fire Chief or the Fire Chief's designee.

    b. Speech that impairs the performance of SJCFD, undermines discipline and harmony among co-workers, or negatively affects the public perception of SJCFD or San Juan County is prohibited and may be sanctioned.

    c. Social media content shall adhere to all applicable laws, regulations and SJCFD policies.

6. **GUIDELINES FOR USE OF PERSONAL SOCIAL MEDIA**

    a. Do not share confidential or proprietary information of SJCFD or San Juan County.

    b. Do not violate SJCFD or San Juan County policies and procedures.

    c. Do not display SJCFD or San Juan County logos, uniforms or similar identifying items without prior written permission.

    d. Do not publish any materials that could reasonably be considered to represent the views or positions of SJCFD or San Juan County without authorization.

7. **OWNERSHIP OF DATA AND MONITORING**

    a. San Juan County owns the right to all data files in any San Juan County owned computer, network, cell phone or other information system.

    b. San Juan County also reserves the right to monitor electronic mail messages (including text and instant messaging systems) and their content created, viewed or accessed on San Juan County computers, networks and cell phones.

8. **NONCOMPLIANCE**

    a. Inappropriate use of social media may result in disciplinary actions, up to and including termination as an employee or volunteer member of SJCFD.

    b. SJCFD employees and volunteers must comply with San Juan County Computer Use Policy in addition to the San Juan County Fire Department Social Media Policy.

FIGURE 4-7 Sample social media policy.

contact information, etc. across all agency-related social media profiles is essential to awareness, recognition, and familiarity. Additionally, developing a social media presence before the onset of a crisis or emergency event will establish a response agency as a credible and authoritative source amidst a sea of voices. It is also essential that agencies maintain both on and offline communications, and ensure that all communications are consistent across all information channels.

The report provided best practice examples of the following social media techniques. Details of these practices are covered in the *Working in the Community* chapter:

- Crowdsourcing for creative problem solving
- Online collaboration and multimedia information sharing
- Developing creative and engaging content
- Relationship building and community partnerships
- Volunteer networks
- Text campaigns
- Incentification (causing someone to be excited about what you are offering)
- Gamification (use of game-oriented thinking and game mechanics to engage users and solve problems)

The frequency of social media outward messages should change depending on the current conditions. The public may not wish to receive multiple messages on a daily basis. During an emergency, however, the public may wish to receive updates as often as possible. Asking the community for their preferences for messaging frequency, and adjusting message frequency based on the feedback and target demographics, is a trust-building activity.

Social media provide fire departments with a creative means to engage the community through a variety of channels and activities. Social media tools also provide the community with a voice and a means to participate in their own preparedness.

Getting It Done

Getting in Front of a Controversial Issue
Dave Statter, a retired TV reporter who runs STATter911 Communications, has warned about trends in news stories that end up becoming issues for fire departments across the United States. Most of these stories have dealt with budget concerns such as claims of excessive overtime, sick leave, and shift swapping. Clear patterns are evident in how those stories evolved in multiple jurisdictions across the country. Often these reports have hurt the image of fire fighters even when there has been no wrongdoing. In the presentation "Effective Communications in a Digital Age," Statter provides this advice about facing down controversial issues:

- Get your house in order now—before a political leader makes it an issue or a reporter starts asking questions.
- Take corrective action on any abuses you uncover.
- If you believe any problems you discovered are likely to become news, consider breaking the news yourself.
- Be able to defend your policies publicly.
- Change the policy if you are unable to defend it publicly.

You Are the Fire Officer Conclusion

Lieutenant Williams realizes that dispatch did not understand the first radio message. With a slightly quavering voice, Williams reports that Quint 100 has a working basement fire in a townhouse with an occupant showing at a second-floor window. Fire fighters have fallen into the basement. Using clear and concise words, Williams calls for a second, or greater, alarm and an EMS mass-casualty response. The lieutenant confirms that the dispatcher understands and is sending help.

Williams then calls the fire attack team. One of the members responds and provides a location, unit, name, assignment, and resource (LUNAR) report. The team is in the basement and appears to be in a laundry room. Most of the fire in the basement has been knocked down. One fire fighter appears to have a broken leg. The 360-degree size-up noted that the structure was two stories in the front and three stories in the rear. The rear includes a sliding glass door. Williams and the apparatus operator proceed to the rear, force open the sliding door, and are removing the injured fire fighter when a rising cacophony of sirens, air horns, and Jake brakes announces the impending arrival of help.

Wrap-Up

Chief Concepts

- A fire officer must be able to process several types of information to supervise and support the fire company members effectively.
- Successful communication occurs when two people can exchange information and develop mutual understanding.
- The communication cycle includes five components: message, sender, medium, receiver, and feedback.
- An officer must be effective as both a sender and a receiver of information and, in many cases, as a processor of information that has to flow within the organization.
- To improve your listening skill, do not assume, do not interrupt, try to understand the need, and do not react too quickly.
- Directed questioning is a good method to keep a conversation on topic.
- A fire officer needs to have up-to-date information on the fire department's standard operating guides, policies, and practices.
- Fire officers should keep their superior officers informed about progress toward goals and projects, potential controversial issues, fire fighter attitude, and morale.
- A flow of informal and unofficial communications is inevitable in any organization that involves people. The grapevine flourishes in the vacuum created when the official organization does not provide the workforce with timely and accurate information about work-related issues.
- Environmental noise can be counteracted by avoiding a power struggle, communicating clearly and firmly, carefully choosing words, confirming that the receiver understands the message, and providing consistent and appropriately detailed messages.
- The direct approach to emergency communications entails asking precise questions, providing timely and accurate information, and giving clear and specific orders.
- An officer should be as consistent as possible when sending verbal messages over the radio. The performance goal should be to sound the same and communicate just as effectively when reporting a minor incident as when communicating under intense stress.
- Following the on-scene size-up, you as the fire officer should provide an initial radio report that describes what you have, states what you are doing, and provides direction for other units that will be arriving.
- Radio communications are essential for emergency operations because they provide an instantaneous connection and can link all of the individuals involved in the incident to share important information.
- To create a useful report, the fire officer must understand the specific information that is needed and provide it in a factual manner that is easily interpreted.
- The most common form of reporting is verbal communication from one individual to another, either face-to-face or via a telephone or radio. To be effective, the transfer of information must be clear and concise, using terminology that is appropriate for the receiver.
- Written reports vary in purpose, formality, and frequency. Examples include company journals, morning reports to the administrative fire officer, monthly activity and training reports, and incident reports.
- Informal reports are not considered official fire department permanent records.
- Formal reports are official fire department documents and are usually archived.
- When writing a report, the appropriate format should be used. It can range from an internal memo format with one or two pages of text to a lengthy formal report that includes photographs, charts, diagrams, and other supporting information.
- A verbal presentation of a written report should consist of four parts: an introduction that gets the audience's attention, an interest statement, report details, and the action the audience should take.
- A news release allows a fire department to reach a large audience at virtually no cost. The release should always be structured to effectively promote a public education message, draw attention to the community's risk problems, announce a new program or the opening of a new facility, or simply develop good public relations.
- Social media differ from traditional or corporate media in the areas of quality, reach, frequency, usability, immediacy, accuracy, reliability, and permanence.
- Each fire department should establish and regularly review its social media policy. Technology, court decisions, and high-profile incidents continue to shape the development of best practices.
- Video recordings create significant risk to the department. Posting videos on social media could interfere with an arson investigation, illegally identify a medical patient, or expose a crime victim.
- The effective use of social media introduces an element of immediacy that empowers citizens to share information with government, first responders, media, and one another, thereby complementing messages transmitted by radio, broadcast or cable TV, printed notices, and ham radio networks.

Hot Terms

<u>Chronological statement of events</u> A detailed account of the fire company activities as related to an incident or accident.

<u>Company journal</u> A log book at the fire station that creates an extemporaneous record of the emergency, routine activities, and special activities that occurred at the fire station. The company journal also records any fire

Wrap-Up, continued

fighter injuries, liability-creating events, and special visitors to the fire station.

<u>Environmental noise</u> A physical or sociological condition that interferes with the message in the communication process.

<u>Expanded incident report narrative</u> A report in which all company members submit a narrative on what they observed and which activities they performed during an incident.

<u>Formal communication</u> An official fire department communication. Such a letter or report is presented on stationery with the fire department letterhead and generally is signed by a chief officer or headquarters staff member.

<u>General orders</u> Short-term directions, procedures, or orders signed by the fire chief and lasting for a period of days to 1 year or more.

<u>Health Insurance Portability and Accountability Act (HIPAA)</u> Enacted in 1996, federal legislation that provides for criminal sanctions and civil penalties for releasing a patient's protected health information in a way not authorized by the patient.

<u>Informal communications</u> Internal memos, e-mails, instant messages, and computer-aided dispatch/mobile data terminal messages. Informal reports have a short life and may not be archived as permanent records.

<u>Interrogatory</u> A series of formal written questions sent to the opposing side of a legal argument. The opposition must provide written answers under oath.

<u>National Fire Incident Reporting System (NFIRS)</u> A nationwide database at the National Fire Data Center under the U.S. Fire Administration that collects fire-related data in an effort to provide information on the national fire problem.

<u>Recommendation report</u> A decision document prepared by a fire officer for the senior staff. Its goal is to support a decision or an action.

<u>Social media</u> Digital communications through which users create online communities to share information, ideas, personal messages, videos, pictures, and other content.

<u>Standard operating guidelines (SOGs)</u> Written organizational directives that identify a desired goal and describe the general path to accomplish the goal, including critical tasks or cautions.

<u>Standard operating procedures (SOPs)</u> Written organizational directives that establish or prescribe specific operational or administrative methods to be followed routinely for the performance of designated operations or actions.

<u>Supervisor's report</u> A form that is required by most state worker's compensation agencies and that is completed by the immediate supervisor after an injury or property damage accident.

References and Additional Resources

NFPA reprinted material is not the complete and official position of the NFPA on the referenced subject, which is represented only by the standard in its entirety.

Baron, G. (2006). *Now Is Too Late 2: Survival in an Era of Instant News*. Bellingham, WA: Edens Veil Media.

Benoit, J., and K. B. Perkins. (2001). *Leading Career and Volunteer Firefighters: Searching for Buried Treasure*. Halifax, NS: Henson College, Dalhousie University.

Boyd, B. (2012). *Social Media in Emergency Management*. Bellingham, WA: Agincourt Strategies.

Bramble, D. (2011). *Facebook and the Fire Department: Who Is Using It and How?* Emmitsburg, MD: U.S. Fire Administration, Executive Fire Officer Program.

Brunacini, A. V. (1996). *Essentials of Fire Department Customer Service*. Stillwater, OK: Fire Protection Publications.

Burke, M., et al. (2012, October 29). "Don't Worry: Indian River Inlet Bridge Is There. Fake Photo Causes Flurry of Excitement." *DelawareOnline*. http://www.delawareonline.com

Carpenter, Q. (2002, December 26). "Kid Drownings: After Decades of Failure, the Well-Intentioned Still Don't Get It." *Phoenix (Arizona) New Times*.

Caulfield, H. J., and D. Benzaia. (1985). *Winning the Fire Service Leadership Game*. New York, NY: Fire Engineering.

Center for Homeland Defense and Security. (2009). *Ogma Workshop: Exploring the Policy and Strategy Implications of Web 2.0 on the Practice of Homeland Security*. Monterey, CA: Naval Postgraduate School.

Coombs, W. T. (2012). *Ongoing Crisis Communication: Planning, Managing, and Responding*, 3rd ed. Thousand Oaks, CA: Sage Publications.

Fire and Emergency Service Image Task Force. (2013). *Taking Responsibility for a Positive Public Perception*. Fairfax, VA: International Association of Fire Chiefs.

Fire Chief's Association. (2012, August 15). *San Juan County Fire Department: Social Media Policy*. Aztec, MN: San Juan County Fire Department.

First Responder Communities of Practice Program and Virtual Social Media Working Group. (2012). *First Responder Communities of Practice Virtual Social Media Working Group Social Media Strategy [Final]*. Washington, DC: Department of Homeland Security, Science and Technology Directorate.

Gormley, W. T., and S. J. Balla. (2012). *Bureaucracy and Democracy: Accountability and Performance*, 3rd ed. Washington, DC: CQ Press.

Graham, D. H. (1999). *The Missing Protocol: A Legally Defensible Report*. Ashton, MD: Clemens Publishing.

Harman, W., and G. Huang. (2011). *The Path Forward: A Follow-up to The Case for Integrating Crisis Response with Social Media and a Call to Action for the Disaster Response Community*. Washington, DC: American Red Cross.

Havenstein, H. (2007, August 3). "LA Fire Department All 'aTwitter' over Web 2.0." *Computerworld*.

Hawkins, D. (2007, May). *Communications and the Incident Management System. Issue Brief*. Washington, DC: U.S. Department of Justice.

Herrman, J. (2012, October 30). "Twitter Is a Truth Machine. During Sandy, the Internet Spread—Then Crushed—Rumors at Breakneck Speed." *BuzzFeed FWD*.

Wrap-Up, continued

Howard, A. (2011, March 7). *Social Media in a Time of Need: How the Red Cross and the Los Angeles Fire Department Integrate Social Tools into Crisis Response.* Sebastopol, CA: O'Reilly Radar.

Howitt, A. M. and H. B. Leonard, Eds. (2009). *Managing Crises: Response to Large-Scale Emergencies.* Washington, DC: CQ Press.

International Association of Fire Chiefs. (2011). *Fire and EMS Department Social Media Policy.* Fairfax, VA: International Association of Fire Chiefs.

Keith, B. D. (2012). *Improving Communication with Non-English Speaking Populations in the City of Dalton, Georgia.* Emmitsburg, MD: U.S. Fire Administration, Executive Fire Officer Program.

Lesperance, A., et al. (2010). *Social Networking for Emergency Management and Public Safety.* Oak Ridge, TN: Pacific Northwest National Laboratory, 95.

Levy, J. M. (1998). *Take Command of Your Writing: A Comprehensive Guide to More Effective Writing.* Campbell, CA: Firebelle Productions.

Mills, S. E. (2012). *Effective Emergency Operations Preparedness for Recurring Planned Public Events.* Emmitsburg, MD: U.S. Fire Administration, Executive Fire Officer Program.

Murphy, M. (2013). "Social Media and the Fire Service." *Fire Technology* 49(1): 175–183.

Rakestraw, M. R. (1992). *Management Communications Program.* Emmitsburg, MD: Executive Fire Officer Program.

Reid, D. B. (2012). *Developing an Internal Communications Plan for Strathcona County Emergency Services.* Emmitsburg, MD: U.S. Fire Administration, Executive Fire Officer Program.

Sonderman, J. (2012). *Hurricane Sandy Tests Twitter's Information Immune System.* Poynter.

Sprouse, C. B. (2012). *When Good People Make Bad Decisions: Assessing Decision Fatigue in Las Vegas Fire & Rescue.* Emmitsburg, MD: U.S. Fire Administration, Executive Fire Officer Program.

Tobias, E. (2011). "Using Twitter and Other Social Media Platforms to Provide Situational Awareness During an Incident." *Journal of Business Continuity & Emergency Planning* 5(3): 208.

U.S. Fire Administration. (1999). *Developing Effective Standard Operating Procedures for Fire and EMS Departments.* Washington, DC: Federal Emergency Management Agency.

U.S. Fire Administration. (2008). *Voice Radio Communications Guide for the Fire Service.* Washington, DC: Federal Emergency Management Agency.

Walker, D. (2011). *Mass Notification and Crisis Communications: Planning, Preparedness, and Systems.* Boca Raton, FL: CRC Press.

Watson, C. (2012). *Using Social Media to Communicate with the Occupants of Large Residential Buildings During Fire Emergencies.* Emmitsburg, MD: U.S. Fire Administration, Executive Fire Officer Program.

Weider, M., Ed. (2010). *Fire Service Reputation Management White Paper.* Berkeley Springs, WV: Cumberland Valley Volunteer Firemen's Association.

Weinschenk, C., et al. (2008). "Analysis of Fireground Standard Operating Guidelines/Procedures Compliance for Austin Fire Department." *Fire Technology* 44(1): 39–64.

Wenger, D. H., and D. Potter. (2011). *Advancing the Story: Broadcast Journalism in a Multimedia World,* 2nd ed. Washington, DC: CQ Press.

FIRE OFFICER in action

The dispatch center provides Battalion Chief Johnson with a copy of the radio transmissions related to the townhouse fire. The initial on-scene radio report by Quint 100 was garbled. The dispatcher asked for a repeat of the message, but did not receive a reply. The next transmission came in response to the dispatcher's question about whether the fire was out. Effective communication requires a coordination of verbal and written procedures, consideration of environmental factors, and awareness of information technology.

1. _____ is (are) a series of formal written questions sent to the opposing side of a legal argument. The opposition must provide written answers under oath.
   - **A.** General orders
   - **B.** Arbitration
   - **C.** Interrogatory
   - **D.** Chronological statement of events

2. The townhouse fire resulted in a fire fighter mayday and a serious civilian injury. In addition to the NFIRS report, which documentation will the supervising or managing fire officer complete?
   - **A.** An expanded incident report narrative
   - **B.** A PowerPoint presentation
   - **C.** An informal report to the administrative fire officer
   - **D.** A summary from a counseling session

FIRE OFFICER in action (continued)

3. Lieutenant Williams is tasked with providing an after-action presentation to the command staff. What is the first step in presenting a verbal presentation?
   A. Provide the presenter's professional background.
   B. Paint a picture of what happens if nothing is done.
   C. Introduce your proposed action.
   D. Get the audience's attention.

4. _____ is not part of the communication cycle.
   A. The medium
   B. Translation
   C. Feedback
   D. The receiver

Fire Captain Activity

Battalion Chief Johnson meets with Captain Davis to share the results from a recent senior command staff meeting. "We need to maintain our ability to handle life-threatening calls, serious transportation crashes, and structural fires with a smaller workforce. To make smart risk management decisions, we need to get a better understanding of which types of events Station 100 handles."

As the captain, Davis is the formal fire department representative to the communities served by Station 100.

NFPA Fire Officer II Job Performance Requirement 5.4.4
Prepare a news release, given an event or topic, so that the information is accurate and formatted correctly.

Application of 5.4.4
1. Prepare a news release for the community newspaper explaining how Station 100 will maintain its ability to serve the neighborhood after nearby Fire Station 44 (engine and paramedic ambulance) is closed and Truck 112 is moved 15 miles farther away from the community at Fire Station 47.

NFPA Fire Officer II Job Performance Requirement 5.4.5
Prepare a concise report for transmittal to a supervisor, given fire department record(s) and a specific request for details such as trends, variances, or other related topics.

Application of 5.4.5
1. Using information from a fire department you are familiar with or have data access for, provide a five-year analysis of one of the following department responses:
   - Emergency medical first responder calls
   - Activated fire alarms without smoke or fire
   - Cooking fires
   - Water-flow alarms without sprinkler activation.
   - "Fuel in the creek" or other petroleum spills
   - Lift assist or similar calls where the fire department is requested to move a patient after EMS is on the scene
   - Suspicious packages or "white powder" calls
   - Dumpster fires

NFPA Fire Officer II Job Performance Requirement 5.6.3
Prepare a written report, given incident reporting data from the jurisdiction, so that the major causes for service demands are identified for various planning areas within the service area of the organization.

Application of 5.6.3
1. Using the annual statistical report of a fire department you are familiar with, evaluate one of the five most common types of incidents that the department responds to, as divided by geographic planning area, such as the central business district. Which factors contribute to this workload? What can the fire department do to reduce the cost of handling those incidents by 25 percent?

Safety and Risk Management

Fire Officer I

Knowledge Objectives

After studying this chapter, you will be able to:

- Discuss how to develop an incident action plan (NFPA 4.2.1). (pp 86, 89)
- Describe the initiatives that have been implemented to reduce fire fighter injuries and deaths. (pp 86–92)
- List the most common causes of personal injury and deaths to fire fighters (NFPA 4.7.1) (NFPA 4.7.3). (pp 87–89)
- Describe methods for reducing the risk of personal injury and death to fire fighters (NFPA 4.2.1) (NFPA 4.7) (NFPA 4.7.3). (pp 86–99)
- Discuss the role and requirements of an incident safety officer. (pp 92–94)
- Describe safety policies and procedures and basic workplace safety (NFPA 4.2.3) (NFPA 4.7) (NFPA 4.7.1). (p 95)
- Describe principles to prevent emergency incident injuries (NFPA 4.2.1) (NFPA 4.7) (NFPA 4.7.1). (pp 95–96)
- Describe safety considerations for the fire station (NFPA 4.7) (NFPA 4.7.1). (pp 96–97)
- Describe the components of an infectious disease control program (NFPA 4.7) (NFPA 4.7.1). (pp 97–99)
- Describe procedures for conducting and documenting an accident investigation (NFPA 4.7.2). (p 99)
- List the elements of a postincident analysis (NFPA 4.6.3). (p 99)

Skills Objectives

After studying this chapter, you will be able to:

- Develop an incident action plan (NFPA 4.2.1). (pp 86, 89)
- Implement safety policies and procedures for basic workplace safety (NFPA 4.2.3) (NFPA 4.7) (NFPA 4.7.1). (p 95)
- Conduct and document an accident investigation (NFPA 4.7.2). (p 99)
- Develop a postincident analysis (NFPA 4.6.3). (p 99)

Fire Officer II

Knowledge Objectives

After studying this chapter, you will be able to:

- Discuss fire fighter death and injury data (NFPA 5.7). (pp 101–102)
- Discuss the impact of sudden cardiac arrest on the fire service (NFPA 5.7). (pp 101–102)
- Discuss the role that traumatic injuries play in fire fighter injuries and fatalities (NFPA 5.7). (p 102)
- Discuss the risk of asphyxiation and burns to fire fighters (NFPA 5.7). (p 102)
- Discuss near-miss report analysis using the Human Factors Analysis and Classification System (HFACS) tool (NFPA 5.7). (pp 103–104)
- Discuss the role of data analysis and risk management in fire departments (NFPA 5.7) (NFPA 5.7.1). (pp 104–105)
- Describe how to use postincident analysis to mitigate hazards (NFPA 5.7.1). (p 105)

Skills Objectives

After studying this chapter, you will be able to:

- Use the Human Factors Analysis and Classification System (HFACS) tool to analyze near misses (NFPA 5.7). (pp 103–104)
- Use postincident analysis to mitigate hazards (NFPA 5.7.1). (p 105)

CHAPTER 5

Additional NFPA Standards

- NFPA 472: *Standard for Competence of Responders to Hazardous Materials/Weapons of Mass Destruction Incidents*
- NFPA 1002: *Standard for Fire Apparatus Driver/Operator Professional Qualification*
- NFPA 1407: *Standard for Training Fire Service Rapid Intervention Crews*
- NFPA 1451: *Standard for a Fire and Emergency Services Vehicle Operations Training Program*
- NFPA 1500: *Standard on Fire Department Occupational Safety and Health Program*
- NFPA 1521: *Standard for Fire Department Safety Officer*
- NFPA 1581: *Standard on Fire Department Infection Control Program*
- NFPA 1583: *Standard on Health-Related Fitness Programs for Fire Department Members*
- NFPA 1600: *Standard on Disaster/Emergency Management and Business Continuity Programs*
- NFPA 1710: *Standard for the Organization and Deployment of Fire Suppression Operations, Emergency Medical Operations, and Special Operations to the Public by Career Fire Departments*
- NFPA 1720: *Standard for the Organization and Deployment of Fire Suppression Operations, Emergency Medical Operations, and Special Operations to the Public by Volunteer Fire Departments*
- NFPA 1975: *Standard on Station/Work Uniforms for Emergency Services*

Principles of Fire and Emergency Service Administration (FESHE) Course Outcomes

4. Select and implement the appropriate disciplinary action based on the employee's conduct. (pp 88, 95, 99, 103–105)
6. Discuss the various levels of leadership, roles, and responsibilities within the organization. (pp 89–92, 94, 105)
8. Identify the importance of ethics as it relates to fire and emergency services. (pp 99, 103–105)
9. Identify the roles of the National Incident Management System (NIMS) and Incident Management System (ICS). (pp 89–90, 92–94)

You Are the Fire Officer

Engine and Quint 100 are part of a greater alarm assignment to a restaurant where there has been an explosion and partial collapse. The incident started as an investigation of an odor. As the fire crew searched for the source, an explosion collapsed the front of the restaurant and partly buried the first-arriving pumper. Captain Davis is appointed to run the Rescue Group in Division A, which is assigned to search the front of the restaurant. Davis has Engine 100, Quint 100, and two additional fire suppression companies.

1. What are the common causes of fire fighter deaths and injuries on the fire ground?
2. Whose responsibility is it to conduct a risk–benefit analysis, and what are the key elements of that process?
3. What are the primary responsibilities of a rapid intervention crew?
4. What are the primary responsibilities of an incident safety officer?
5. What are the typical tasks expected of an incident safety officer?

Introduction

Fire department operations often include high-risk situations that can occur under any weather conditions at any time of the day or night. The fire officer is responsible for ensuring that every fire fighter completes every incident without injury, disability, or death. This idea is expressed as a fire officer's special obligation to ensure that "everyone goes home" at the end of a workday.

Fire officers have a personal and professional obligation to prevent deaths and injuries to the fire fighters who are working under their supervision. The responsibilities of a fire officer include identifying hazards and mitigating dangerous conditions to provide a safe work environment for fire fighters. The fire officer must also identify and correct behaviors that could lead to a fire fighter's injury or death. The officer should set a good example because the crew members will follow their officer's lead. Safe practices must be the only acceptable behavior, and good safety habits should be incorporated into all activities.

Fire Officer I

Fire Fighter Death and Injury Trends

As part of his or her duties, the fire officer develops an incident action plan (IAP) that addresses and minimizes the chances of harm by identifying and controlling the factors that might lead to fire fighter injury or death. Fire fighters must be fully prepared to work safely in high-risk situations.

Understanding the causes of fire fighter deaths and injuries is the first step in developing an incident action plan. Prevention depends on the ability to halt the cascade of events that leads to a serious injury or death. The National Fire Protection Association (NFPA), the U.S. Fire Administration (USFA), the National Institute for Occupational Safety and Health (NIOSH), and the International Association of Fire Fighters (IAFF) all publish reports and statistical analyses that provide information about the causes and circumstances of these events.

■ Everyone Goes Home®

Everyone Goes Home is a program developed by the National Fallen Firefighters Foundation (NFFF) to prevent line-of-duty death and injuries. The NFFF held a Firefighter Life Safety Summit in 2004 that resulted in 16 initiatives:

1. Define and advocate the need for a cultural change within the fire service relating to safety, incorporating leadership, management, supervision, accountability, and personal responsibility.
2. Enhance the personal and organizational accountability for health and safety throughout the fire service.
3. Focus greater attention on the integration of risk management with incident management at all levels, including strategic, tactical, and planning responsibilities.
4. Empower all fire fighters to stop unsafe practices.
5. Develop and implement national standards for training, qualifications, and certification (including regular recertification) that are equally applicable to all fire fighters based on the duties they are expected to perform.
6. Develop and implement national medical and physical fitness standards that are equally applicable to all fire fighters based on the duties they are expected to perform.

7. Create a national research agenda and data collection system that relates to the initiatives.
8. Utilize available technology wherever it can produce higher levels of health and safety.
9. Thoroughly investigate all fire fighter fatalities, injuries, and near misses.
10. Create grant programs that support the implementation of safe practices and/or mandate safe practices as an eligibility requirement.
11. Develop and champion national standards for emergency response policies and procedures.
12. Develop and champion national protocols for response to violent incidents.
13. Ensure that fire fighters and their families have access to counseling and psychological support.
14. Ensure that public education receives more resources and is championed as a critical fire and life safety program.
15. Strengthen advocacy for the enforcement of codes and the installation of home fire sprinklers.
16. Ensure that safety is a primary consideration in the design of apparatus and equipment.

A follow-up summit was held in 2007 to develop key recommendations for each of these initiatives. A standard approach to safety incorporates best practices that should be part of every operational situation.

Fire fighters must work in teams at emergency incidents, and fire officers must maintain accountability at all times for the location and function of all members working under their supervision. Every company must operate within the parameters of an incident action plan, under the direction of the incident commander.

Reliable two-way communications must be maintained through the chain of command. Also, adequate backup lines must be in place to ensure that a safe exit path is maintained for crews working inside a fire building and that any sudden flare-ups can be controlled. Rapid intervention crews must be established to provide immediate assistance if any fire fighter is in danger. In addition, air supplies must be monitored to ensure that fire fighters leave the hazardous area before their low-air alarm activates and they run out of air. All fire fighters must watch for, recognize, and communicate any indications of impending building collapse.

National Fire Fighter Near-Miss Reporting System

In August 2005, the International Association of Fire Chiefs (IAFC) launched a Web-based system to report near misses. The goal of www.firefighternearmiss.com is to track incidents that avoided serious injury or death, to identify trends, and to share the information with other fire fighters in a confidential and nonpunitive way. This program is based on the Aviation Safety Reporting System, which has been gathering reports of close calls from pilots, air traffic controllers, and flight attendants since 1976. Examples of near-miss reports are provided throughout this book.

Getting It Done

Habits to Improve Safety
A fire officer can develop four simple habits to improve the safety of his or her crew:
- Be physically fit.
- Wear seat belts.
- Practice safety through training and personal example.
- Maintain fire company integrity at emergency incidents.

Reducing Deaths from Sudden Cardiac Arrest

The NFFF Life Safety Summit noted that a disproportionate number of fire fighters older than the age of 49 die of cardiac arrest while on duty. NFPA's *Firefighter Fatalities in the United States—2011* noted that heart attacks are the leading cause of death for fire fighters, accounting for 41 percent of all line-of-duty deaths from 2007 to 2011. The NFPA report further noted that postmortem medical documentation was available for 22 of the 31 cardiac arrest deaths in 2011 and showed that 13 of the stricken fire fighters were hypertensive, six had coronary artery disease, five had diabetes, and four had a history of cardiac problems, such as prior heart attacks, bypass surgery, or angioplasty/stent placement. Some of the victims had more than one condition. Other risk factors noted were obesity, high cholesterol levels, smoking, and family history.

Every fire fighter candidate should undergo a medical examination before he or she is allowed to respond to incidents. Regular physical examinations should be scheduled for as long as the fire fighter is engaged in performing emergency duties, with an emphasis on identifying risk factors that could lead to a heart attack under stressful conditions. Fire officers should look for indications that a member is in poor health or is unfit for duty and, if necessary, arrange for a special evaluation by a fire department–approved physician.

Physical fitness activities should be considered an essential component of every fire fighter's training regimen. Changes in lifestyle can often reduce the risk of a fatal heart attack; such changes include stopping smoking, lowering high blood pressure, reducing high blood cholesterol level, maintaining a healthy weight, and managing diabetes. Fire officers must understand that these changes are as important to the body as wearing full protective equipment.

The NFPA, IAFC, NFFF, National Volunteer Fire Council (NVFC), and IAFF have developed resources to help the fire officer encourage healthy living. Fire officers should advocate methods of making positive lifestyle changes to increase fire fighter safety. NFPA 1583, *Standard on Health-Related Fitness Programs for Fire Department Members*, provides a structure and resources to help the fire officer develop a health-related fitness program. The IAFC partnered with the IAFF to develop the Fire Service Joint Labor Management Wellness Fitness Initiative (WFI), which in turn produced the Candidate Physical Aptitude Test as well as a peer fitness training certificate program with the American Council on Exercise.

Regardless of any fire department mandates that are handed down, every fire officer should strive to be physically

FIGURE 5-1 The fire officer should strive to be physically fit.

fit **FIGURE 5-1**. Fitness should be a personal priority and an expression of leadership.

Safety Zone

Physical Fitness Tips

The NFFF summarizes the importance of physical fitness with the following recommendations that support fire fighter maintenance:

- Regular medical check-ups: Yes, they can be a pain, but if you do not do it for you, do it for those who need you.
- Regular exercise: Even walking makes a *big* difference! Walk a mile a day and watch the changes.
- Eat healthy: Think about what you are eating, and then picture operating interior at a working fire 30 minutes later. Now, what do you want to eat?

■ Reducing Deaths from Motor Vehicle Collisions

Collisions account for the largest percentage of traumatic fire fighter deaths. NFPA's *Firefighter Fatalities in the United States—2011* notes that the number of fire fighter deaths resulting from motor vehicle collisions averaged 15 per year over the past decade (range: 9 to 25). Volunteer fire fighters have an increased risk from motor vehicle collisions. In *What's Changed Over the Past 30 Years?*, NFPA's review of data from 1977 to 2007 show that three-fourths of the 406 fatalities stemming from such collisions were volunteer fire fighters. Almost 40 percent of the fire fighters died in their personal vehicles. More than three-fourths of the fire fighters who died in these incidents were not wearing a seat belt. Excessive speed given the road conditions and operator error are frequently cited causes of these fatal collisions. Obeying traffic laws, using seat belts, driving sober, and controlling speed would prevent most of the fire fighter fatalities in road collisions.

Section 6.2.1 of NFPA 1500, *Standard on Fire Department Occupational Safety and Health Program,* states that "Fire apparatus shall be operated only by members who have successfully completed an approved driver training program commensurate with the type of apparatus the member will operate or by trainee drivers who are under the supervision of a qualified driver." Putting an untrained driver behind the wheel of an emergency vehicle places both the occupants of that vehicle and the general public in immediate danger. The driver of an emergency vehicle is legally authorized to ignore certain restrictions that apply to other vehicles, but only when operating the emergency vehicle in a manner that provides for the safety of everyone using the roadways. Specific procedures for safe emergency response must be practiced. The fire officer is responsible for ensuring that the driver consistently follows the rules of the road for emergency response.

Driver minimum qualifications are established in NFPA 1002, *Standard for Fire Apparatus Driver/Operator Professional Qualifications*; NFPA 1451, *Standard for a Fire and Emergency Services Vehicle Operations Training Program*; and NFPA 1500, *Standard on Fire Department Occupational Safety and Health Program*. The fire officer should set high expectations for driver training and performance. In particular, the apparatus operator should be required to have nonemergency driving experience with a specific piece of apparatus before being assigned to emergency response driving duties. There are tremendous differences between the handling characteristics of fire apparatus versus automobiles or light trucks. Stopping distances are directly related to the weight of the vehicle, so fire apparatus take a relatively longer distance (and time) to come to a complete stop. In addition, heavy fire apparatus can turn over easily in a collision. Driving a ladder truck is vastly different from driving an engine, and a tender has a different set of handling characteristics from either of these apparatus. As this brief summary of handling characteristics suggests, driving fire apparatus can be quite challenging, and all of the necessary skills must be learned and practiced under controlled conditions before an operator is qualified to drive under emergency response conditions.

Requiring fire fighters to wear seat belts is a simple requirement that could prevent 8 to 12 fatalities every year. Many of the fire fighters killed in motor vehicle crashes are thrown from the vehicle. A fire officer who allows company members to respond while unrestrained in the fire apparatus has no excuse if a fatal accident results. The mandatory use of seat belts by fire fighters, like many other safety procedures, may require a change in the culture. A fire officer must be prepared to accept the responsibility and provide the leadership to bring about positive changes.

Getting It Done

Developing an Incident Action Plan

An IAP provides a concise, coherent means of capturing and communicating the overall incident priorities, objectives, and strategies in the contexts of both operational and support activities. Every incident must have an action plan.

Most initial response operations are not captured with a written IAP. However, if an incident involves hazardous materials or is likely to extend beyond one **operational period**, become more complex, or involve multiple jurisdictions and/or agencies, a written IAP will be required to maintain effective, efficient, and safe operations.

A single-unit IAP includes the following basic elements:

- Specifies the incident objectives
- States the activities to be completed
- Covers a specified time frame, known as an operational period

Consider a single fire company response to an activated automatic fire alarm. The company officer verbalizes the incident action plan. The incident objectives are first to investigate the source of the alarm and then to restore the alarm system. The corresponding activities are for the crew to proceed to the alarm panel to identify the location of the alarm. With one radio-equipped member remaining at the alarm panel, a crew of two, wearing full personal protective equipment (PPE) and carrying a radio and hand tools, will proceed to the location of the activated alarm. The crew will report back their findings when they reach the activated device. If the alarm is not due to an incipient fire, the member at the alarm panel will attempt to reset the alarm system.

■ Reducing Deaths from Fire Suppression Operations

Asphyxia and burns are prime factors that directly cause death while operating in burning buildings, followed by trauma. Even though they may wear protective clothing and self-contained breathing apparatus (SCBA), fire fighters routinely operate in situations where a problem, a procedural error, or an equipment failure could result in a fatality.

Fire officers must fully understand all local policies and procedures that should guide their actions. The established standard operating procedures (SOPs) and safety practices should always be followed in situations that meet the criteria for their use. If a situation does not fall under an established SOP, the fire officer must determine an appropriate course of action and give specific directives to subordinates to indicate clearly how the situation is to be handled. In these situations, the fire officer must provide an even greater level of supervision to ensure that the crew members understand and follow the plan. While doing so, the fire officer must always be prepared for changing conditions and unanticipated hazards.

Maintaining Crew Integrity

A consistent challenge is to maintain the integrity of the fire company while operating at an emergency incident. The officer must know the location and function of every crew member at all times. The U.S. Fire Administration's *Firefighter Fatality Retrospective Study* indicated that 82 percent of the fatal fire suppression incidents tallied in the report involved the death of a single fire fighter. Many of these fire fighters became lost or disoriented and died before the fire officer or incident commander was aware that a fire fighter needed help.

NIOSH investigates fire fighter line-of-duty deaths as part of its Fatality Assessment and Control Evaluation program. The lack of an effective Incident Management System that includes a fire fighter accountability component is a common scenario in many of the single fire fighter deaths investigated by NIOSH.

Operating in an IDLH Environment

The Occupational Safety and Health Administration (OSHA) establishes federal workplace safety regulations in the United States, including 29 CFR 1910.134 (Respiratory Protection), which applies to the use of SCBA by the fire service. This rule applies when members are operating in an environment that is immediately dangerous to life and health (IDLH). NFPA 1670, *Standard on Operations and Training for Technical Search and Rescue Incidents*, defines IDLH as "Any condition that would pose an immediate or delayed threat to life, cause irreversible adverse health effects, or interfere with an individual's ability to escape unaided from a hazardous environment." The interior of a fire building, where fire fighters are using SCBA, is considered to be an IDLH atmosphere.

The OSHA regulation and NFPA 1500 establish specific requirements for fire fighters operating in an IDLH environment. This "two-in, two-out rule" includes the following elements:

- A designated officer-in-charge
- Two fire fighters who enter the IDLH area together and remain in visual or voice contact with each other at all times, while wearing SCBA **FIGURE 5-2**

FIGURE 5-2 Two fire fighters should enter an area together and remain in visual or voice contact with each other at all times, while wearing SCBA.

Getting It Done

Vehicle Safety Tips

The NFFF summarizes the issue of vehicle safety by pointing out the following items that support the driver response plan:

- It's not a race.
- Safe is more important than fast.
- Stop at red lights and stop signs! *No excuses*!
- If they do not get out of your way, do not run them over! *Think* and *react carefully*!

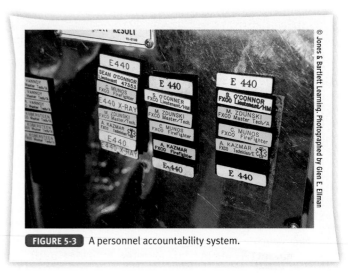

FIGURE 5-3 A personnel accountability system.

- Two properly equipped and trained fire fighters who must:
  - Be positioned outside the IDLH environment
  - Account for the interior team
  - Remain capable of rescue of the interior team

The two-in, two-out rule leads to the requirement for a <u>rapid intervention crew (RIC)</u>. NFPA 1710, *Standard for the Organization and Deployment of Fire Suppression Operations, Emergency Medical Operations, and Special Operations to the Public by Career Fire Departments*, Section 3.3.44 and NFPA 1720, *Standard for the Organization and Deployment of Fire Suppression Operations, Emergency Medical Operations, and Special Operations to the Public by Volunteer Fire Departments*, Section 3.3.31 define RIC as "A dedicated crew of firefighters who are assigned for rapid deployment to rescue lost or trapped members." Most fire departments add another fire company to the first alarm assignment for a structure fire, either as part of the initial dispatch or as soon as a working fire is reported, to function as the four-person RIC.

To comply with OSHA regulations, an initial rapid intervention crew (IRIC) is required to be assembled prior to operations within IDLH environments. Two fire fighters from the initial attack crew will be assigned as the IRIC until the RIC company is deployed at the incident. Both the IRIC and RIC members should meet the requirements of NFPA 1407, *Standard for Training Fire Service Rapid Intervention Crews*.

To ensure that the RIC can recognize in a timely manner that a fire fighter is missing, a <u>personnel accountability system</u> is needed to track the identity, assignment, and location of all fire fighters operating at the incident scene **FIGURE 5-3**.

An accountability system typically provides the following components:

- A method to identify all personnel who are on the scene
- A method to identify personnel who are in the hazard area
- A process to account for personnel quickly when an evacuation order or unusual event occurs (e.g., a flashover, backdraft, or collapse)
- A process to ensure that personnel do not go unaccounted for an extended period of time

An accountability system must function at multiple levels at an incident scene. Although every fire officer must be accountable for all assigned company members, command officers are accountable for the location and function of full companies. An officer who is assigned to manage a sector for a division at an incident must know where every assigned company is operating and what they are doing at all times. The incident commander must know which companies are assigned to each sector or division.

To ensure an effective accountability system, crews must routinely receive training on the system and follow the procedures. During an emergency situation, fire fighters will revert to habit. The proper use of an accountability system must be fully integrated as one of those habits.

A personnel accountability system is required by NFPA 1500, *Standard on Fire Department Occupational Safety and Health Program*. Some departments require a personnel accountability report (PAR) roll call every 20 minutes during fire attack or other high-hazard operations. Each sector chief or company officer must account for the personnel assigned to that sector. A PAR is also undertaken after a sudden change in operating conditions, after a flashover or collapse, or when moving from offensive to defensive fire attack.

Air Management

Self-contained breathing apparatus provides a safe and reliable air supply that allows a fire fighter to operate in an IDLH atmosphere for a finite length of time. The length of time that a SCBA air tank will last depends on the rate at which the individual consumes the available air supply, which varies

Fire Marks

Are You a RAT?

Departments use different titles or terms to identify the resources that will be committed to meet the two-in, two-out requirement. The terms FAST truck (fire fighter assist search team), RIT (rapid intervention team), RICO (rapid intervention company operations), and RAT (rapid assist team), among others, are used to identify resources that cover the same assignment.

Company Tips

Overriding the Two-In, Two-Out Rule
The OSHA rule includes an exception that would allow a first-arriving fire company to override the two-in, two-out rule and initiate operations in an IDLH environment if there is a chance that a victim will be rescued alive. The company officer must communicate this action to the other companies that are responding to the incident. The mere fact that someone might be in danger is not sufficient justification to ignore the two-in, two-out rule; this exception exists only to avoid the situation where a procedural rule could stand in the way of saving a life.

FIGURE 5-4 Thermal imaging devices are a critical component of a scene safety program.

considerably depending on the user and the nature of the work being performed. The rated time is based on a standard consumption rate for an individual at rest and does not reflect typical experience in firefighting operations. Running out of air in an IDLH atmosphere can result in an unconscious fire fighter who will die of asphyxiation.

Some departments have replaced their 30-minute rated SCBA air tanks with 45- or 60-minute air tanks to increase the amount of time a fire fighter can operate in IDLH atmospheres. Other departments have established or strengthened incident management procedures to monitor individual fire fighter air use more closely.

Low-pressure warning devices, which provide an indication when the remaining air supply reaches a set point, are effective only if the fire fighter heeds the warning and is able to exit to a safe atmosphere in the time that is provided. The low-pressure alarm might not provide sufficient exit time for a fire fighter who is deep inside a burning building or unable to exit immediately. Many departments monitor the air levels of fire fighters on entry and exit. If a fire fighter fails to exit within the expected time period, an accountability roll call is taken. Fire fighters should know the amount of air in their cylinders at all times and develop an exit strategy before the low-air alarm activates.

Teams and Tools
The minimum size of an interior work team is two fire fighters. Every team member must have full personal protective equipment, including SCBA and a personal alert safety system (PASS) device unit. At least one member (if not every member) of every team should have a radio.

If such equipment is available, every interior operating team should also have a thermal imaging device, which is a critical component of a comprehensive safety program. The thermal imaging device allows crews to navigate, locate victims, evaluate fire conditions, locate hazards, and find escape routes in smoke-filled buildings or total darkness **FIGURE 5-4**.

NIOSH's investigation of line-of-duty deaths revealed that some departments with thermal imaging devices did not use them. Either the fire fighters did not know how to use the imager, the device was inoperable, or the device was not used.

Thirty years ago, the same observation could have been made regarding breathing apparatus, as SCBA units were often left sitting in the apparatus compartments while fire fighters made interior attacks. The fire officer must provide frequent training to ensure that the company members are fully prepared to integrate the thermal imager into interior firefighting operations.

Situational Awareness
Every fire officer is expected to maintain a continual connection between the functions being performed by the company and the overall situation. This is one of the most important reasons for establishing and maintaining an effective Incident Command Structure (ICS) at every incident. The fire officer maintains situational awareness by staying oriented, making observations, providing and receiving regular updates within the ICS, listening to fire-ground radio communications, and continually performing risk–benefit analysis.

It is all too easy to become distracted or preoccupied with a particular task or function and, in doing so, lose track of the larger situation that is occurring and changing simultaneously. This can be a problem at a dynamic emergency incident, particularly for an officer who is supervising a crew that is performing a complex task deep inside a smoke-filled building. Conditions can change quickly, and crews might have no way of observing what is going on around them.

Risk–Benefit Analysis
A risk–benefit analysis is based on a hazard and situation assessment that weighs the risks involved in a particular course of action against the benefits to be gained from taking those risks. Fire fighters sometimes use the axiom, "Risk a little to save a little; risk a lot to save a lot"—a simplified description of the risk–benefit process. Risk–benefit analysis must always be approached in a structured and measured manner.

Life safety, including the lives of fire fighters, is a paramount goal. If the situation requires placing fire fighters in extreme danger with little chance of success, the operation should not be undertaken. There is no justification for risking the lives of fire fighters to save property that is already lost

Fire Marks

Rapid Intervention Is Not Rapid

In 2001, two fire fighters became disoriented when the Phoenix Fire Department moved from offensive to defensive operations at a supermarket fire. The incident commander sent two companies into the building as the RIC. Despite heroic efforts by the on-scene crews, one of the disoriented fire fighters died in the building. Twelve fire fighters involved in the rescue effort ran into problems and declared their own maydays during the 53 minutes it took to remove Firefighter/Paramedic Bret Tarver from the building.

The Phoenix Fire Department embarked on a comprehensive recovery effort to learn what must be done to avoid another fire fighter death in a commercial structure fire. Assistant Chief Steve Kreis coordinated the evaluation of the rapid intervention effort. To validate the proposed changes in fire fighter rapid intervention and rescue procedure, Phoenix conducted more than 200 multiple-company drills in three large single-story commercial buildings. Arizona State University was a partner in this venture, with Dr. Ron Perry making sure that the drills were valid for academic research purposes.

The training drill scenario was similar to the supermarket fire and designed to reinforce RIC procedures. Two fire fighters were located within 100 feet (30.5 meters) of a building exit and not entrapped in any way. The fire attack hose line was not obstructed by debris or convoluted in any way. There was no smoke, heat, or fire in the drill building. One fire fighter was alert, separated from his partner, and running out of air. The other fire fighter was unconscious, about 40 feet (12 meters) away from the first fire fighter, with his or her PASS alarm activated.

During the drills, the vision through participants' SCBA face pieces was obscured by a laminated piece of window tint fitted within the mask. Depending on ambient conditions and hand lights, the participants could see 5 to 20 feet (1.5 to 6 meters) ahead of them. A recording with fire-ground sounds was played during the rescue effort. At that time few thermal imagers were available in the Phoenix fire service, and they were not used in these drills.

The drill objectives were twofold: (1) find the alert fire fighter, assess his or her level of air, and transfill the fire fighter's SCBA air (buddy breathing) while assisting the fire fighter out of the building; and (2) find, transfill SCBA air for, and remove the unconscious fire fighter. The results from the 2002 drills showed that it takes 21 minutes to rescue a downed fire fighter using a team of 12 fire fighters.

During the rescue effort drills, 20 percent of the RIC team got into some trouble themselves, usually running out of air. A 3000-psi SCBA bottle in the Phoenix drill provided 18.7 minutes of air, plus or minus 30 percent. In this straightforward and noncomplex scenario, the RIC members were running out of air before completing the removal of a downed fire fighter. Add stairs, heat, adrenaline, or entrapment to this scenario, and it becomes apparent that most of the RIC team will have profound difficulties in rescuing and removing a downed fire fighter.

---

or has no real value. For example, conducting an interior attack on a fire in an unoccupied abandoned building needlessly places fire fighters at risk. Similarly, entering a burning building to search for missing occupants who could not possibly be alive cannot be justified.

Every emergency incident exposes fire fighters to a set of unavoidable inherent risks. These risks are managed within a measured and controlled system that is based on training, coordination, and the use of protective clothing and equipment. The nature of the mission requires fire fighters to be able to work safely in situations that are inherently dangerous. Fire officers keep this system in balance.

The only situation that truly justifies exposing fire fighters to a high level of risk is one where there is a realistic chance that a life can be saved. Even in this circumstance, fire fighters must use all of the resources at their disposal to limit the risks. All of the fire department's training, equipment, and systems are designed to enable fire fighters to be effective in the face of such challenging and dangerous situations.

Operating within the National Incident Management System (NIMS) provides the framework for commanding high-risk tasks under a dynamically changing environment. NIMS incorporates a process to move from strategy to tactics to task in an organized fashion.

The fire officer starts the risk–benefit analysis by preparing a preincident plan, which is a written document that provides information that can be used by responding personnel to determine the appropriate actions in the event of an emergency at a specific facility. Building construction, occupancy, use, contents, and condition are all factors that should go into the development of the risk–benefit analysis.

When operating at an emergency incident, the fire officer reviews the preincident plan and makes observations about current conditions. These two inputs are combined to produce an incident action plan, which is developed by the incident commander and incorporates the overall incident strategy, tactics, risk management evaluation, and organization structure for that particular situation. Incident action plans are updated throughout the incident, based on progress updates from operating crews and observations by the incident management team. More information on this process can be found in the chapter *Rules of Engagement*.

Incident Safety Officer

An incident safety officer is a designated individual at the emergency scene who performs a set of duties and responsibilities that are specified in NFPA 1521, *Standard for Fire Department Safety Officer*. The incident safety officer functions as a member of the incident command staff, reporting directly to the incident commander. The incident commander is personally responsible for performing the functions of the incident safety officer if this assignment has not been assigned or delegated to another individual.

Many fire departments assign a designated officer to respond to emergency scenes to fill this position. In the absence of a designated safety officer, this duty may be

> ### Company Tips
>
> **Situational Awareness**
>
> In *Understanding and Implementing the 16 Firefighter Life Safety Initiatives*, the NFFF summarizes situational awareness by specifying the following items that support the interior firefighting plan:
> - Work as a team.
> - Stay together.
> - Stay oriented.
> - Manage your air supply.
> - Get off the apparatus with tools and a thermal imager for *every* interior operating team.
> - Provide a radio for *every* member.
> - Provide regular updates.
> - Constantly assess the risk–benefit model.

assigned to a qualified individual by the incident commander. NFPA 1500 and NFPA 1521 also specify that the fire department must have a standard operating procedure to define the criteria for the response or appointment of an incident safety officer. The same principles should be applied to any situation, whether a designated safety officer has been dispatched or an officer has been assigned to fulfill this function by the incident commander.

The fact that a safety officer has been assigned does not relieve any officer or fire fighter of the responsibility to operate safely and responsibly. The incident safety officer is simply an additional resource to ensure that the safety priorities of the situation are being addressed. Every fire officer shares the responsibility to act as a safety officer within his or her scope of operations.

■ Incident Safety Officer and Incident Management

The incident safety officer is a key component of the Incident Command System. The Incident Command System (ICS) is the standard organizational structure that is used to manage assigned resources so as to accomplish the stated objectives for an incident. Every incident requires someone, known as the incident commander, to be in charge at all times to coordinate resources, strategies, and tactics. This requirement comes into force with the initial arriving officer, who functions as the incident commander until he or she is relieved by a higher-ranking officer. The incident management structure can become larger and more complex, depending on the nature and the magnitude of the incident.

The incident safety officer reports to the incident commander and is required to monitor the scene, to identify and report hazards (the potential for harm to people, property, or the environment) to the incident commander, and, if necessary, to take immediate steps to stop unsafe actions and ensure that the department's safety policies are followed. In most situations, the exchange of information between the incident commander and the incident safety officer is conducted verbally and quickly at the command post.

Safety officers have specific authority and a special set of responsibilities under the ICS. One of the primary responsibilities of an incident safety officer is to identify hazardous situations and dangerous conditions at an emergency incident and recommend appropriate safety measures. In most cases, the safety officer acts as an observer, monitoring conditions and actions and evaluating specific situations. When an unsafe condition is observed that does not present an imminent danger, the safety officer consults with the incident commander and with other officers to determine a safe course of action.

If a situation creates an imminent hazard to personnel, the incident safety officer has the authority to immediately suspend or alter activities. When this special authority is exercised, the incident safety officer must immediately inform the incident commander of the hazardous situation and his or her actions. It is ultimately up to the incident commander to either approve or alter the action taken by the safety officer.

■ Qualifications to Operate as an Incident Safety Officer

NFPA 1521 outlines the criteria for an incident safety officer. Every fire officer should be trained to perform the basic duties of an incident safety officer and be prepared to act temporarily in this capacity if he or she is assigned to this position by the incident commander.

According to NFPA 1521, the incident safety officer must be a fire department officer and at a minimum must meet the requirements for Fire Officer I as specified in NFPA 1021, *Standard for Fire Officer Professional Qualifications*. The incident safety officer also must be qualified to function in a sector officer position under the local Incident Management System.

The general knowledge requirements for an effective incident safety officer are as follows:

- Safety and health hazards involved in emergency operations
- Building construction
- Local fire department personnel accountability system
- Incident scene rehabilitation

Incident safety officers at a special operations incident require additional specialized knowledge and experience. Special operations are emergency incidents to which the fire department responds that require specific and advanced training and specialized tools and equipment, such as water rescue, extrication, confined-space entry, hazardous materials situations, high-angle rescue, and aircraft rescue and firefighting. For example, the incident safety officer at a hazardous materials incident needs to have an advanced understanding of this situation, perhaps by training to the Hazardous Materials Technician level of NFPA 472, *Standard for Competence of Responders to Hazardous Materials/Weapons of Mass Destruction Incidents*.

In many fire departments, specialized teams have their own designated safety specialists who are trained to work directly with the incident management team.

Typical Incident Safety Officer Tasks

The specific duties that an incident safety officer must perform at an incident depend on the nature of the situation. The following is a partial listing of functions that may need to be addressed at incidents:

- Ensure that safety zones, collapse zones, and other designated hazard areas are established, identified, and communicated to all members present on scene.
- Ensure that hot, warm, decontamination, and other zone designations are clearly marked and communicated to all members.
- Ensure that a rapid intervention crew is available and ready for deployment.
- Ensure that the personnel accountability system is being used.
- Evaluate traffic hazards and apparatus placement at roadway incidents.
- Monitor radio transmissions and stay alert to situations that could result in missed, unclear, or incomplete communication.
- Communicate to the incident commander the need to appoint assistant incident safety officers because of the need, size, complexity, or duration of the incident.
- Immediately communicate any injury, illness, or exposure of personnel to the incident commander and ensure that emergency medical care is provided.
- Initiate accident investigation procedures and request assistance from the health and safety officer in the event of a serious injury, fatality, or other potentially harmful occurrence.
- Survey and evaluate the hazards associated with the designation of a landing zone and interact with helicopters.
- Ensure compliance with the department's infection control plan.
- Ensure that incident scene rehabilitation and critical incident stress management are provided as needed.
- Ensure that food, hygiene facilities, and any other special needs are provided for members at long-term operations.
- Attend strategic and tactical planning sessions and provide input on risk assessment and member safety.
- Ensure that a safety briefing, including an incident action plan and an incident safety plan, is developed and made available to all members on the scene.

Additional duties that the incident safety officer must perform when fire has involved a building or buildings are as follows:

- Advise the incident commander of hazards, collapse potential, and any fire extension in such buildings.
- Evaluate visible smoke and fire conditions to advise the incident commander, tactical-level management unit officers, and other officers on the potential for flashover, backdraft, or any other fire event that could pose a threat to operating teams.
- Monitor the accessibility of entry and egress of structures and the effect it has on the safety of members conducting interior operations.

Getting It Done

Incident Safety Plan

For the incident safety officer to perform these duties and responsibilities adequately at a large or complex event, the incident commander must define an incident action plan. The incident safety officer can then provide the incident commander with a risk assessment of incident scene operations. The incident safety officer can also develop an **incident safety plan**, which outlines the actions that are required to provide for safety at the scene, based on the incident commander's incident action plan and the type of incident encountered. At a large or extended-duration operations event, the incident action plan and incident safety plan are often compiled as written documents and updated at the beginning of every work period.

Assistant Incident Safety Officers at Large or Complex Incidents

Some incidents, based on their size, complexity, or duration, require more than one safety officer. Assistant incident safety officers can be assigned to subdivide responsibilities for different areas and functions at events such as high-rise fires, hazardous materials incidents, and special rescue operations. In these cases, the incident safety officer should inform the incident commander of the need to establish a safety unit as a component of the incident management organization. Under the overall direction of the incident safety officer, assistant incident safety officers can be assigned to various functions, such as scene monitoring, action planning and risk management, interior operations, or special operations teams. During extended-duration incident operations, a relief rotation can be established to ensure that safety supervision is maintained at all times.

Incident Scene Rehabilitation

Rehabilitation is the process of providing rest, rehydration, nourishment, and medical evaluation to members who are involved in strenuous or extended-duration incident scene operations **FIGURE 5-5**. Part of the incident safety officer's role is to ensure that an appropriate rehabilitation process is established. Incident scene rehabilitation is the tactical-level management unit that provides for medical evaluation, treatment, monitoring, fluid and food replenishment, mental rest, and relief from climatic conditions of the incident.

Fire fighters are aggressive by nature. They want to do a good job and want to be where the action is occurring. This action orientation makes them susceptible to exceeding the physical limitations of their own bodies. When they overdo it, dehydration may contribute to sudden cardiac arrest on the fire ground. As a fire officer, you must constantly monitor the health and welfare of your crew. The incident safety officer is a third line of defense, after the individual and the fire officer, for ensuring that fire fighters obtain appropriate rehabilitation.

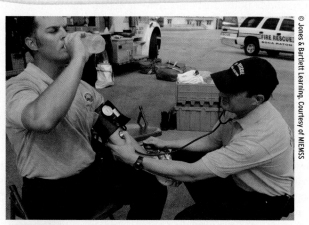

FIGURE 5-5 Rehabilitation is the process of providing rest, rehydration, nourishment, and medical evaluation to members who are involved in extended-duration or extreme incident scene operations.

Creating and Maintaining a Safe Work Environment

Although the highest priority of a fire officer is to ensure that every member goes home alive from a call, the reality is that for every fire fighter death in the line of duty, nearly 1000 fire fighter injuries occur. A safety program that is based on preventing fatalities is far from adequate. Every injury, or near miss, should be viewed as a potentially fatal or disabling situation, and injury prevention should be an equally important concern of the fire officer. The fire officer must model good behavior to help develop the subordinates' attitudes about injury prevention.

■ Safety Policies and Procedures

Most fire departments have policies that regulate safety practices at the company level. These policies are designed to address routine circumstances, and many have been developed in reaction to a previous accident or injury. The fire officer is on the front line in ensuring compliance with all safety policies. The fire officer needs to fully understand each policy, follow all safety policies and procedures, and ensure that all subordinates fully understand and follow them.

A number of methods may be used to ensure that fire fighters understand safety policies and procedures. Many departments require all members to sign a document acknowledging that they have read and understand each policy and any new or amended policy. Some departments leave it to the officer to read and explain each policy to the crew members. This can be done at the morning briefing; however, there may be little retention of information at this time. A more effective method would be to have the members individually read the policy and then lead a group discussion to ensure that it is understood.

One method to reinforce safety policies is to watch videos of incidents and critique them based solely on safety policies. This exercise often reveals significant differences between the way things are supposed to be done and what actually occurs at an emergency incident. It can be a significant learning experience for everyone involved.

Five other sources that may be accessed to review safety policies are the "Report of the Week" from the National Fire Fighter Near-Miss Program, "The Secret List" sent by the Firefighters Close Calls Web site, information posted by the Emergency Responder Safety Institute (ERSI), incident videos collected by STATter911.com, and NIOSH case studies. In addition to reporting on incidents and trends, the NFFF Near-Miss Program, Firefighters Close Calls, and ERSI provide training programs and teaching resources on their Web sites.

The fire officer must conscientiously ensure that all safety policies are followed in training activities. Training provides the luxury of time to learn and correct errors. On emergency scenes, when there is only one opportunity to do things properly, fire fighters will perform the way they have been trained.

It is impossible to write a policy to cover every conceivable hazard. For this reason, the fire officer should use good judgment to identify hazardous situations and implement mitigating measures. These efforts might include requiring fire fighters to use safety glasses while mowing the lawn or ensuring that food preparation surfaces are properly cleaned to prevent the spread of germs. The fire officer should make sure that company members understand the need for safe work practices and develop an attitude that internalizes safety, rather than relying on the fire officer to be a "safety police officer."

■ Emergency Incident Injury Prevention

Many of the same techniques that are used to prevent fire fighter deaths also help prevent fire fighter injuries; however, they are not necessarily the same. The following principles should be implemented to prevent injuries as well as deaths.

Physical Fitness

Fire fighters who are in good physical condition are less prone to injury as well as at reduced risk for heart attack. Through strength and flexibility training, the body becomes more resistant to sprains, strains, and other injuries and is more capable of responding to critical situations. Sprains and strains are the leading type of fire fighter injury. A physical fitness program that includes cardiovascular, strength, and flexibility training is needed to reduce fire fighter deaths and injuries.

Getting It Done

Fire Fighter Rehabilitation

The NFFF summarizes incident scene rehabilitation with the following items that support the fire fighter rehabilitation guidelines:

- Stop before you drop.
  - Cool down when hot.
  - Warm up when cold.
  - Dry off when wet.
- Stay hydrated.
  - With noncaffeinated drinks.
- Monitor vital signs.

Personal Protective Equipment

Personal protective equipment (PPE) performance has improved tremendously over the past 40 years. The protective ensemble for structural firefighting includes a turnout coat and pants, boots, gloves, hood, helmet, SCBA, and a PASS. NFPA standards define the minimum performance standards for each item.

Of course, the best protective equipment is of no use if it is not worn. The fire officer should monitor the proper use of protective clothing and take immediate action to correct any deficiencies. In some departments, fire fighters routinely enter buildings to investigate fire alarm activations and other seemingly minor situations without fully donning their PPE and SCBA, and sometimes without bringing tools, hose packs, or extinguishers. This practice has proved deadly when the situation turned out to be more serious than anticipated. The officer must set a good example and require every crew member to treat every call—no matter how seemingly benign—like an actual fire.

The need for protective clothing and equipment does not end when the fire is extinguished. SCBA must be used until carbon monoxide and particulate matter levels have been reduced. Coats, pants, gloves, goggles, and helmets are needed during overhaul to prevent wounds. The incident commander and the safety officer are responsible for determining whether it is safe to reduce the level of PPE that will be used at any time.

In addition to the protective clothing and equipment, fire fighters operating in hazardous areas should carry several safety-related items. A radio is needed to maintain contact, to receive instructions, or to request assistance in an emergency; at least one member of each team should have a radio, and preferably every fire fighter should carry one. A flashlight is needed to help prevent fire fighters from becoming lost or disoriented and as a signal for help. A forcible entry tool should be carried to provide a method of escape in an emergency. Personal wire cutters can also prove valuable if a fire fighter becomes entangled in wires. A personal escape rope should be carried by every fire fighter as well. Each item provides an additional measure of safety, and the fire officer should always set the example by making sure he or she carries this equipment.

Many nonfire incidents also require the use of appropriate protective equipment. The *2008 Near Miss Annual Report*, for example, reviewed 590 submitted reports dealing with the operation of hydraulic rescue tools and power saws. Fire fighters must pay attention to the task at hand and use proper eye protection during forcible entry or vehicle extrication.

Although station uniforms are not included in most definitions of personal protective equipment, the clothing that is worn under turnout gear can provide an additional level of protection. NFPA 1975, *Standard on Station/Work Uniforms for Emergency Services*, provides a set of performance requirements for station clothing. The use of clothing that meets this standard under turnout gear will further reduce the risk of burn injuries in a severe situation.

With emergency medical services (EMS) responses accounting for more than 70 percent of responses in many fire departments, the fire officer must also understand the requirements for medical PPE. In many cases, this requirement is as simple as wearing appropriate gloves—healthcare providers wear gloves for all patient contact situations. When blood or other body fluids are present, eye and face protection and full-body protection may also be needed. An effective education program is essential for fire fighters to understand the importance of wearing the appropriate protection for every situation.

■ Fire Station Safety

In addition to emergency situations, safety considerations apply to the fire station environment. The fire station and other fire department facilities are a workplace, and the fire department is fully responsible for maintaining a safe work environment. Every fire department should have a comprehensive workplace safety program that applies to all of its facilities. The station setting allows the fire officer to be even more proactive in enforcing safety policies than is possible at an incident scene.

Safety hazards at the fire station can have the same consequences as hazards encountered at emergency incidents. The most important difference is that the fire department has control over the fire station environment at all times and has the ability, as well as the responsibility, to identify and correct any safety problems that are present. The fire department does not have the same type of control over the conditions that are typically encountered at the scene of an emergency incident.

Clothing

Protective clothing can become contaminated at incident scenes and should never be worn in the living quarters of the fire station. Turnout gear that is known to be contaminated with fire residue, chemicals, or body fluids should be cleaned as soon as possible. When washing turnout gear, be sure to follow the manufacturer's recommendations. Such cleaning requires the use of a commercial extractor-type washer rather than a residential washer. Many fire departments provide special washing machines for protective clothing at the fire station or send gear out for professional cleaning and decontamination. Duty uniforms should also be washed at the fire station or sent out to a commercial laundry to prevent cross-contamination of the family laundry.

Protective clothing should be inspected regularly by the fire officer, and any items that are worn or damaged should be repaired or replaced. Properly fitting gear is also important, both for comfort and for protection. Gear inspection should include an exterior evaluation while the clothing is being worn and a close examination of the condition of all interior layers.

Housekeeping

General housekeeping around the fire station is important in injury and accident prevention. Apparatus bay floors that become slippery when wet have caused many fire fighters to slip and fall. A squeegee should be used to remove standing water, and "wet floor" warnings should be used. The same is true for interior station floors that require mopping.

Leaving equipment lying around can present a tripping hazard to fire fighters when they are in a hurry to respond to an alarm. The walking traffic flow areas should always be kept clear.

The same fire hazards that exist in other types of occupancies can be found in many fire stations. Leaving food unattended on the stove and improperly storing flammable liquids can lead to a publicly embarrassing situation if they lead to a fire at the facility. Every fire station should have the appropriate number and types of fire extinguishers, which must be properly maintained. The fire officer should ensure that working smoke alarms are present and checked regularly. Automatic fire suppression sprinklers are highly recommended.

Food preparation activities are responsible for many relatively minor injuries to fire fighters. Inappropriate use of kitchen knives has caused injuries resulting in lost time and worker's compensation claims. Proper decontamination of all food preparation surfaces is often overlooked. Regular hand washing while cooking prevents the transmission of illness.

Diesel exhaust systems installed in apparatus bays reduce the exposure of fire fighters to hazardous gases. If the fire station does not have an exhaust system, the fire officer should ensure that vehicles are not left running inside the building.

Lifting Techniques

Back injuries can occur while lifting and moving patients, dragging fire hoses, setting ladders, or working at the station. No matter what the cause, back injuries are serious and potentially career ending.

Proper lifting techniques should be used to reduce the risk of back injuries. In particular, a fire fighter should never bend at the waist to pick up items or victims. Instead, the fire fighter should bend at the knees and lift by standing straight up. The back should be kept in a natural position rather than locked in a hyperextended manner. When an object is too bulky, in an awkward position, or too heavy for one fire fighter, additional help should be sought to lift it. Fire officers should always reinforce the use of proper techniques and procedures to avoid injuries.

Some injuries have resulted from horseplay around the station. This is an area where the fire officer must exercise supervisory control to reduce the risk of unjustifiable injuries.

Infection Control

Every fire department should have an infection control program that meets NFPA 1581, *Standard on Fire Department Infection Control Program*. This standard identifies six components of a program:

1. A written policy with the goal of identifying and limiting exposures
2. A written risk management plan to identify risks and control measures
3. Annual training and education in infection control
4. A designated infection control officer
5. Access to appropriate immunizations for employees
6. Instructions for handling exposure incidents

Proper decontamination procedures are followed after emergency medical incidents or any situation where equipment could have become contaminated. Disposable items, gloves, and contaminated expendable equipment should be disposed of in a specially marked bag that is designed for that use.

Patient compartments in ambulances should be properly decontaminated after every transport. Likewise, patient cots should be decontaminated after every call. Equipment that is designed to be decontaminated and reused should be cleaned only in an approved decontamination sink, typically in a designated area of the fire station. Contaminated equipment should *never* be taken into the living area of the station or cleaned in a sink where food is prepared.

Always follow written procedures and the manufacturer's guidelines for decontamination of medical equipment. A 1 percent bleach and water solution is typically used for this purpose; however, this solution should *not* be used on turnout gear. Many of the materials commonly used in protective clothing can be seriously damaged or destroyed by this type of cleaning agent.

NFPA 1581 provides specific information on establishing an infection control program, including guidelines on equipment cleaning and storage, facility requirements, and methods of protection. The fire officer should also consult departmental guides and policies for further details.

■ Infectious Disease Exposure

NFPA 1581 also provides a model program for situations where a fire fighter has been exposed to an infectious or contagious disease. Most fire departments have established policies that govern postexposure procedures. The fire officer must be prepared to fulfill the duties of the initial supervisor when a member has experienced an infectious disease exposure.

The most important first step with any exposure is to wash the exposed area immediately and thoroughly with soap and running water. If soap and running water are not available, waterless soap, antiseptic wipes, alcohol, or other skin cleaning agents can be used until soap and running water are obtained.

The fire department infection control officer should be immediately notified after an exposure incident has occurred. The designated infection control officer should arrange for the member who has experienced an exposure to receive medical guidance and treatment as soon as is practical. An exposure can be a stressful event for a fire fighter, so it is important to provide confidential postexposure counseling and testing. Because the fire fighter may not seek out this information on his or her own, the department should proactively inform the individual of this service.

Document all exposures as soon as possible using a standardized reporting form **FIGURE 5-6**. At a minimum, the record should include the following information:

- Description of how the exposure occurred
- Mode of transmission
- Entry point
- Use of personal protective equipment
- Medical follow-up and treatment

The record of an exposure incident will become part of the member's confidential health database. A complete record of the member's exposure incidents must be available to the member on request. Data on exposure incidents should also be maintained by the department but without personal identifiers to allow for analysis of exposure trends and to develop strategies to prevent such incidents.

Exposure Event Number_____

Sample Blood and Body Fluid Exposure Report Form

Facility name: _____

Name of exposed worker: Last _____ First: _____ ID #: _____

Date of exposure: _____/_____/_____   Time of exposure: _____:_____ AM   PM (Circle)

Job title/occupation: _____ Department/work unit: _____

Location where exposure occurred: _____

Name of person completing form: _____

Section I. Type of Exposure *(Check all that apply.)*

☐ **Percutaneous** (Needle or sharp object that was in contact with blood or body fluids)
  (Complete Sections II, III, IV, and V.)

☐ **Mucocutaneous** *(Check below and complete Sections III, IV, and VI.)*
  ___ Mucous Membrane   ___ Skin

☐ **Bite** *(Complete Sections III, IV, and VI.)*

Section II. Needle/Sharp Device Information
(If exposure was **percutaneous**, *provide the following information about the device involved.)*

Name of device: _____   ☐ Unknown/Unable to determine

Brand/manufacturer: _____   ☐ Unknown/Unable to determine

Did the device have a sharps injury prevention feature, i.e., a "safety device"?

☐ Yes   ☐ No   ☐ Unknown/Unable to determine

If yes, when did the injury occur?

☐ Before activation of safety feature was appropriate   ☐ Safety feature failed after activation

☐ During activation of the safety feature   ☐ Safety feature not activated

☐ Safety feature improperly activated   ☐ Other: _____

Describe what happened with the safety feature, e.g., why it failed or why it was not activated: _____
_____

Section III. Employee Narrative *(Optional)*

Describe how the exposure occurred and how it might have been prevented:
_____
_____
_____

NOTE: This is not a CDC or OSHA form. This form was developed by CDC to help healthcare facilities collect detailed exposure information that is specifically useful for the facilities' prevention planning. Information on this page (#1) may meet OSHA sharps injury documentation requirements and can be copied and filed for purposes of maintaining a separate sharps injury log. Procedures for maintaining employee confidentiality must be followed.

FIGURE 5-6 Exposure report form.

Courtesy of CDC

Due to the hazardous nature of some communicable diseases, individuals who have been exposed to these risks are often required to report to the infection control officer. This information must be maintained with the strictest of confidence. It is up to the fire department physician to determine the member's fitness-for-duty status after reviewing documentation of his or her exposure.

Accident Investigation

The fire department health and safety officer is charged with ensuring that all injuries, illnesses, exposures, fatalities, or other potentially hazardous conditions and all accidents involving fire department vehicles, fire apparatus, equipment, or fire department facilities are thoroughly investigated. The initial investigation of many situations is often delegated to a local fire officer.

An accident is any unexpected event that interrupts or interferes with the orderly progress of fire department operations. This definition includes personal injuries as well as property damage. An accident investigation should determine the cause and circumstances of the event and identify any corrective actions that are needed to prevent another injury, accident, or incident from occurring.

The fire officer must ensure that all required federal, state, and local documentation is complete and accurate. The result of an accident investigation should always include recommended corrective actions that are presented to the fire chief or the chief's designated representative.

■ Accident Investigation and Documentation

Most fire departments have established procedures for investigating accidents and injuries. The fire officer is usually responsible for conducting an initial investigation and for fully investigating many minor accidents. Accidents that involve serious injuries, fatalities, or major property damage are usually investigated by the health and safety officer or by other qualified individuals.

An officer could be expected to conduct the full investigation for a simple ankle sprain, a broken pike pole, or a dented rear step on the apparatus. In each case, it is the fire officer's responsibility to protect the physical and human resources of the department. To accomplish this goal, the officer must understand the basic principles of investigation.

An investigation normally consists of three phases:

1. Identification and collection of physical evidence
2. Interviews with witnesses
3. Written documentation (at the end of an investigation)

Each of these steps is essential for the development of a comprehensive report that can be used to initiate safer work practices and conditions. The investigation report can also provide evidence that could be used to support or refute future claims. If a department has a standardized procedure for investigating accidents, the fire officer should always use the recognized procedure.

Examination of the physical evidence includes documenting the time, day, date, and conditions that existed at the time of the incident. Such factual information should be verifiable and should not include any interpretation. The weather conditions should be noted. The scene of the accident should be fully documented, with drawings noting the locations of relevant artifacts that could provide information about how or why the accident occurred. These details are needed to help determine prevention methods.

Witnesses to an accident should also be interviewed, including all of the individuals who were directly or indirectly involved, as well as anyone who simply observed what happened. When interviewing a witness, explain the rationale for the interview and note that this is a fact-finding process. Advise the person that it is okay not to know the answer to every question. Begin by asking the individual to give a verbal chronology of whatever occurred, being sure not to interrupt. Watch for nonverbal communication by the witness. Once the opening report has been given, ask questions to clarify the details, but do not ask leading or misleading questions. If the level of PPE being worn at the time of the incident has not been mentioned, determine what was being worn, how it was being worn, and whether it performed as designed. Last, restate previous questions in a different order and from a different perspective to verify previous answers.

The last step in an accident investigation is the documentation of the findings and the conclusions and recommendations. The facts should be presented in a logical sequence with the results of the interview or interviews attached. A determination as to the most likely cause or causes of the accident should be included, along with a recommendation on how this type of accident could be prevented. Some departments require a determination about the likely frequency as well as the likely severity of this type of accident—an assessment intended to optimize prevention efforts.

As a fire officer, you have a duty to perform an initial accident investigation that is fair and unbiased. Both the department and the employee have a vested interest in the final report. An employee who applies for a disability pension based on an accident that occurred several years earlier could be denied the pension if adequate documentation is not maintained. For this reason, most departments require a report to be filed whenever there is a potential exposure to an infectious disease, when injury has occurred to an employee or a citizen, and when property damage has occurred to either private property or departmental property.

Postincident Analysis

The incident safety officer provides a written report for the department that includes pertinent information relating to safety and health issues involved with the incident. This would include information about the use of protective clothing and equipment, personnel accountability system, rapid intervention crews, rehabilitation operations, and other issues that directly affect the safety and welfare of members at the incident scene.

VOICES OF EXPERIENCE

A fire officer needs to understand that the crew's health and safety should be viewed as just as important, if not more important, than any other assignment the officer will be given. Without a healthy crew ready to perform when the tone goes off for an emergency, the crew will not respond as an effective unit, possibly leading to injury or death. Most fire officers think of only the physical needs of their crew when thinking about health and safety, not the emotional wellbeing. The emotional wellbeing of the fire fighter can have an impact on the fire fighter, the crew, the fire fighter's family, and the community.

Throughout my career as a fire fighter, I have responded to numerous incidents involving loss of life, but one incident had a profound impact on how I viewed the need for ensuring the emotional needs of the fire fighters was taken into consideration during and after the call. We received a call early in the morning for a structure fire with possible civilians inside the home. When the first units arrived on scene, the home was fully involved and conditions unsafe to make entry to search for possible civilian victims. The fire fighters had responded to numerous past calls at the address and recognized the family's vehicles in the yard, so we knew there would be a chance of finding victims once the fire was knocked down and entry was safe.

> **In the days after the incident, one fire fighter quit the fire department and other fire fighters started showing signs of emotional trauma.**

When crews entered the remains of the structure, the four victims were discovered and it was quickly determined by crews that this was a crime scene due to the trauma the victims had received that was not caused by the fire. Fire fighters were asked by the county coroner and state medical examiner to assist in digging through the debris by hand to remove the victims. It took hours to ensure nothing was missed that could help solve the crime. Once the last victim was removed, crews returned to the station to clean equipment to return to service. The officers quickly asked the fire fighters if they were okay, and then returned to deal with the press because the story had grabbed the attention of the national news agencies. There were no SOGs to deal with critical incident stress, and no one was trained in debriefing or how to get help for the fire fighters to prevent long-term emotional trauma from dealing with the victims at the scene. In the days after the incident, one fire fighter quit the fire department and other fire fighters started showing signs of emotional trauma, stating they were on the verge of leaving the fire service due to what they experienced on scene.

The fire officers decided there was a need for a critical incident stress management plan to ensure the emotional health of the fire fighters. SOGs were developed to debrief fire fighters after scenes where there is potential for emotional trauma and to activate the state crisis response team, who are first responders trained to assist with Critical Incident Stress Debriefing (CISD) and refer fire fighters to mental health professionals when they need more emotional care beyond the debriefing.

Since this incident I have been involved in several debriefings and activations of the state crisis response team after critical incidents. The fire fighters I have spoken with have stated the CISD was a positive experience and they were glad it was offered because they wouldn't have asked for it prior to the SOGs being put into place due to the fear of looking weak in the eyes of the other fire fighters. A well-executed CISD can have enormous potential to alleviate the emotional feelings and potentially dangerous physical symptoms, thus extending the careers of personnel and saving great outlays of resources to replace fire fighters who have seen too much human misery.

Jennifer Henson
Director, Fire Protection Technology and Firefighter Certification
Catawba Valley Community College
Hickory, North Carolina

Fire Officer II

Analyzing Death and Injury Data

In preparing the report *What's Changed Over the Past 30 Years?*, NFPA's Fire Analysis and Research Division conducted a 30-year retrospective analysis of fire fighter fatalities. According to this research, the number of fire fighters dying on duty declined 30 percent over a 30-year period, from an average of 151 deaths annually in the 1970s to an average of 99 deaths per year in the 2000s. The two primary drivers of this decline are a reduction in the number of fire fighters dying of sudden cardiac death and the decreasing number of structure fires. However, the rate of fire fighters dying while operating within structure fires is higher than it was in the 1970s. Given the significant improvements in protective clothing, breathing apparatus, training programs, and incident management, fire fighter safety within burning structures is an area that needs focused attention from the fire officer.

The U.S. Fire Administration, through the National Fallen Firefighters Foundation (NFFF), has adopted a goal of 50 percent reduction in fatalities within a decade. The U.S. Fire Administration's *Firefighter Fatalities in the United States in 2011* notes that the "2011 total of 64 firefighter fatalities was, for the second year in a row, the lowest number of firefighter losses on record" **FIGURE 5-7**. There were 108 fire fighter fatalities in 2004, when the "Everybody Goes Home" program was started.

The data indicate that fewer fire fighters are dying on the fire ground, both in actual numbers and in relative terms; however, more are being killed while responding to emergency incidents or performing duties other than fighting fires **FIGURE 5-8**. Each time a company rolls out of the station responding to a call, the fire fighters are putting their lives at risk.

NFPA reports that 70,090 fire fighters were injured in the line of duty in 2011. Approximately 43.5 percent of these injuries occurred during fire-ground operations, with the balance split between nonfire emergency incidents and other on-duty activities, such as training, physical fitness, cleaning, and maintenance. The most frequently reported type of injury was strain, sprain, or muscular pain (50.7 percent), followed by wound, cut, bleeding, or bruise (14.5 percent); thermal stress (6.9 percent); and burns (6.2 percent).

■ Sudden Cardiac Arrest

As an occupational group, fire fighters are more likely to die of a heart attack while on duty than other U.S. workers. Although the term "occupation" is used to refer to fire fighters, this category includes a full range of affiliations, including career, volunteer, contract, and wildland fire fighters.

The relatively high heart attack rate for fire fighters is related to the nature of the work, which can suddenly change from low activity to an episode of high stress and intense exertion. This type of sudden stress can trigger a heart attack, particularly in an individual who is predisposed to cardiovascular disease. Security guards also have a high rate of heart attacks—a cause that accounts for 25 percent of their on-duty deaths. Construction workers, who routinely perform strenuous physical labor, and police officers, who experience similar

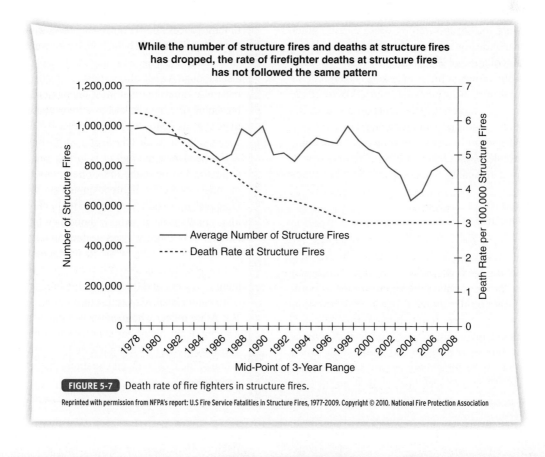

FIGURE 5-7 Death rate of fire fighters in structure fires.

Reprinted with permission from NFPA's report: U.S Fire Service Fatalities in Structure Fires, 1977-2009. Copyright © 2010. National Fire Protection Association

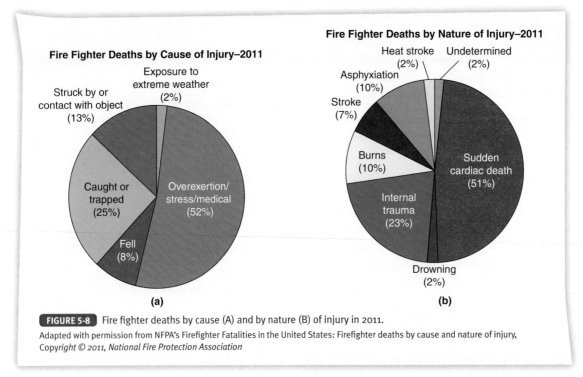

FIGURE 5-8 Fire fighter deaths by cause (A) and by nature (B) of injury in 2011.
Adapted with permission from NFPA's Firefighter Fatalities in the United States: Firefighter deaths by cause and nature of injury, Copyright © 2011, National Fire Protection Association

variations between low-intensity and high-stress periods, both have lower heart attack rates than fire fighters. Among workers in all occupational categories, only 15 percent of on-duty deaths are caused by heart attacks.

Regular medical examinations and physical fitness programs are the most significant factors in preventing heart attacks. Other risk factors were represented among the victims of sudden cardiac death, including obesity, high cholesterol, smoking, and family history.

■ Struck by or Contact with an Object

Traumatic injuries are the second leading cause of fire fighter fatalities. Most of these deaths are caused by vehicle accidents, falls, or structural collapse of a burning building. Trauma deaths resulting from motor vehicle collisions accounted for 19 percent of annual line-of-duty deaths from 2008 to 2012. The most common fatal motor vehicle collision scenario involves a fire fighter responding to an emergency incident in a personal vehicle. The second most common type of fatal collision is a tender (mobile water tanker) rollover. Tenders have a higher, disproportionate fatality rate compared to other types of fire apparatus.

■ Caught or Trapped

The third most frequent category of fire fighter fatalities includes asphyxiation and burns. In these situations, the fire itself is the direct cause of the fire fighter's death. The fatal injury can involve either burns or asphyxiation, or a combination of both.

The risks of respiratory injuries and burns are inherent in a fire fighter's work environment. Firefighting is routinely performed in situations that involve high temperatures and contaminated atmospheres. Protective clothing and SCBA are designed to reduce the risk of injury from these causes, but they have limited effectiveness. When the capabilities of the protective ensemble are exceeded, a serious or fatal injury can occur quickly.

Fire Marks

National Fallen Firefighters Memorial Weekend

The National Fire Academy in Emmitsburg, Maryland, has hosted the National Fallen Firefighters Memorial Weekend every year since 1981. This national event is held to honor and remember the fire fighters who died in the line of duty during the previous year. The memorial lists the names of all of the fire fighters who have died in the line of duty each year.

The memorial service, like the image of a fire fighter's funeral procession, is not easily forgotten. A fallen fire fighter is honored as a hero, buried with great pomp and circumstance, and long remembered on plaques and memorials. Although we will always honor our fallen comrades, we must also remember that most fire fighter fatalities can be prevented. The fire service has a tremendous responsibility to its own members, to their families and friends, and to our communities to protect the lives of all fire fighters.

Whenever a fire fighter is lost, a painful and lasting impact is felt by many individuals, beginning with the members of the fire fighter's immediate family. A spouse must pick up a shattered life and try to make sense of the loss. Children will grow up without their father or mother. Parents have to bury a son or daughter. Personal friends and co-workers of the fallen fire fighter feel a similar type of loss. The loss is also felt throughout the fire department or agency and throughout the fire service whenever a fire fighter is killed or seriously injured.

A fire officer who experiences the loss of a member working under his or her supervision often feels a sense of responsibility for having allowed the loss to occur. This is one of the most difficult situations that a fire officer can experience.

Near-Miss REPORT

Report Number: 12-0000233

Event Description: This incident involved an investigation of an odor of natural gas in a commercial strip mall at approximately 1900 hours. I responded on the engine with a crew of three. Upon arrival, we discovered the caller was in a busy pizza parlor restaurant. We entered the business and investigated the kitchen. We checked the ovens for leaks and found no odors or obvious leaks.

We were leaving and took off our SCBA and were putting them away when the captain noticed one of the blacked-out windows next to the pizza parlor was hot. We called for the manager to unlock the business and after 5 minutes the front door was opened. There was no fire or smoke, but upon entry we realized this was a remodel job and the entire interior had just been sheetrocked, mudded, and taped. Two electric, industrial propane heaters were running full speed, each hooked up to a large, 20-gallon propane tank; one had an orange hot element and the other had malfunctioned and was freely releasing propane into the room. The whole room was completely charged with propane.

One of the crew reached up to where the extension cord went into a temporary electrical junction box up to the ceiling 10 feet up. He said "Here, I will unplug it." Before any of us could say or do anything, he pulled the plug. Because there were no lights on and the windows were taped over, I clearly saw the electrical arc when he pulled that cord. That room could have had just enough oxygen with the front door wide open to have combusted. The explosion would have been deadly for the fire crews and the civilians next door as well as the civilian manager in the doorway.

Lesson Learned: First, in regard to scene safety, when we went to investigate the space with the hot windows, all safety equipment (SCBA) should have been re-donned. The door should have been opened by fire fighters as we kept the civilian at a safe distance. Upon discovery of this potentially catastrophic condition, the pizza parlor should have been evacuated and response for a full commercial structure fire should have been requested, as well as assistance from PD code three. Once the area was evacuated, crews should have pulled attack lines and had them charged until the rest of the alarm responders showed up. Then the back door should have been opened and ventilation procedures started. I cannot comment on the crew member who pulled the extension cord; that is a training issue.

Courtesy of the National Fire Fighter Near-Miss Reporting Systems (firefighternearmiss.com)

Analyzing Near-Miss Reports

Every August, a working group of fire fighters and officers assemble to analyze near-miss reports using a tool modified from the U.S. Navy's Human Factors Analysis and Classification System (HFACS). The results of the analysis are provided in an annual report.

■ HFACS Level 1: Unsafe Acts

Level 1 in the HFACS includes two categories: errors and violations. Errors are considered unintentional and can be based on decisions, skills, or perceptions. Decision-based errors include flaws in communicating information to decision makers. Skill-based errors include attention failure (lack of situational awareness), memory failure (forgotten or missed step in a procedure), or technique failure (lack of training). Perception-based errors may include visual illusions.

Violations are considered intentional and are classified as either routine or exceptional. Routine violations include failure to use safety equipment, failure to follow a recommended tactile best practice (sounding the floor before entering), or failure to follow recommended cerebral best practices (conducting a risk–benefit analysis). Exceptional violations include not being qualified to perform an action.

■ HFACS Level 2: Preconditions to Unsafe Acts

Level 2 analyzes substandard conditions and practices of the individuals involved in the incident. Substandard conditions include factors contributing to adverse mental states, psychological states, and physical limitations. Substandard practices include failure to use elements of crew resource management and personal readiness.

■ HFACS Level 3: Unsafe Supervision

Unsafe supervision is broken down into four categories: inadequate supervision, allowing inappropriate operations, failure to correct known problems, and supervisory violations. Level 3 is intended to examine the role (if any) of supervision in a near-miss event TABLE 5-1.

■ HFACS Level 4: Organizational Influences

The most difficult HFACS level to analyze in a near-miss report is Level 4. The operating culture of the fire department is often as significant to the near miss as the individual's action. This

Fire Marks

FDNY Deaths in Structure Fires: 1994 to 2008

While commanding Division 3 in mid-town Manhattan, Deputy Chief Vincent Dunn wrote his first book, *Collapse of Burning Buildings*. The book was dedicated to "the forty-six FDNY chiefs, company officers and fire fighters who have been killed by burning buildings which collapsed 1956–1986." Dunn provides an analysis on the 28 Fire Department of New York (FDNY) members who died in burning buildings from 1994 to 2008:

- Seventy-five percent of them were members of ladder companies. Three were from rescue companies and four were from engine companies.
- The fatalities included 17 fire fighters, 7 lieutenants, and 4 captains. Eighty-six percent of the deaths occurred while searching the structure. Four fire fighters were killed while operating a hose line.
- Multiple-family dwellings accounted for 15 of the fatalities, followed by eight commercial structures, three vacant buildings, and two single-family homes. Thirteen FDNY members were killed operating above the fire floor, 10 on the fire floor, and three in the basement.
- There were more deaths during the fire growth stage (64 percent) than when the fire was fully developed.

Company Tips

Enhancing Fire Fighter Safety

Gordon Graham is a risk management expert. He has shared his success as a supervisor in a large police agency, where his systematic approach to risk management reduced organizational liability and injuries.

Graham advocates that emergency service organizations concentrate on high-risk/low-frequency events, when the fire officer or fire fighter does not have any discretionary time to evaluate the situation or consult with others. Clear policies and frequent training for these incidents are essential, so that the response is instinctive when a high-risk/low-frequency event is encountered with no time to think about the course of action.

Graham applied a systems approach in considering safety in public safety agencies, known as *Graham's Rules for Enhancing Firefighter Safety* (GREFS). Here is a condensed version of the rules:

1. People are the most critical element of the safety process. Select the best. Screen out those with a history of serious medical conditions and inappropriate vehicle operations.
2. Management has leadership responsibility for preventing injuries. Identify those high-risk/low-frequency events that have a probability of causing the greatest problems. Have a workable fire fighter accountability system in place.
3. Employees must be continuously trained to work safely.
4. Safety should be a condition of employment. Every fire fighter must recognize that he or she has an ongoing obligation to work safely. Supervisors and managers must both monitor safety performance in the workplace and follow the same rules themselves.
5. When safety rules are not followed, address the issue with prompt, fair, and impartial discipline.
6. Answer this question: "Does the benefit of fighting this fire justify the risks involved in fighting this fire?"
7. Women and men who work safely are more likely to be productive.
8. Safety extends beyond the job to be part of every person's life.
9. Safety is a business responsibility. Ethically, all management teams have an affirmative obligation to make sure that fire fighter safety is being taken seriously.
10. Most things that go wrong are highly predictable. Predictable is preventable.

level examines both resource management (staffing, training, budget resources, and equipment/facility resources) and organizational climate (chain of command, delegation of authority, risk management programs, and safety programs).

Data Analysis

The fire department is required to maintain records of all accidents, occupational deaths, injuries, illnesses, and exposures in accordance with federal, state, and local regulations. In addition to meeting regulations, these reports can assist the fire officer in identifying trends and safety-related issues.

Risk management is the identification and analysis of exposure to hazards, selection of appropriate risk management techniques to handle exposures, implementation of chosen techniques, and monitoring of results, with respect to the health and safety of members. Accident and injury reduction should be a major concern of every fire officer. The fire officer should review all injury, accident, and health exposure reports to identify unsafe acts and work conditions.

Table 5-1 HFACS Level 3: Unsafe Supervision Factors

Inadequate Supervision	Allowing Inappropriate Operations	Failure to Correct Known Problems	Supervisory Violations
• Guidance not provided • Training not provided • Qualifications not tracked	• Personnel not adequately briefed • Understaffed • Unnecessary hazards permitted	• Unidentified and unqualified • Failure to provide training to unqualified personnel • Failure to correct inappropriate or unsafe behaviors	• Authorization of unnecessary hazards • Failure to enforce department rules • Authorization of unqualified personnel to perform tasks

Reprinted with permission of Elsevier Public Safety, copyright 2008.

When reviewing an accident report, it is critical to determine the root cause of the accident. Were unsafe acts being committed? If so, what must be done to change the behavior or attitude? If the cause was an unsafe condition, how can this condition be corrected? An evaluation of all unsafe acts and conditions and identification of the means to address each is required to prevent future occurrences. Some corrective actions can be made at the company level, whereas others require action at a higher level.

Once the cause has been identified, a report should be filed with your supervisor that outlines the problem, actions taken to correct the problem, and any additional actions that should be implemented. Some departments have these reports sent directly to the department's health and safety officer.

The fire officer may be required to complete a longitudinal study of accidents, injuries, and exposures for the company. The officer must review the data to determine which areas are causing the greatest number of incidents as well as which problems have the greatest potential for loss. This review should look for trends, such as an increase in the number of slips, trips, and falls on the apparatus bay floor. This pattern might point to the condition of the floor surface being a contributing factor, which could be corrected. Similarly, a series of exposures to blood by a few individuals might indicate a need for additional training for the entire company.

Fire companies should train on every aspect of the job—but the reality is that there is not enough time to do it all. Risk management principles suggest that fire officers should focus attention on the high-risk activities that are performed infrequently. An example would be training on locating and removing lost or trapped fire fighters.

Mitigating Hazards

The real value in a postincident analysis is the learning process that results from the information obtained during the process. Here are a few tips to help a fire officer to maximize the value of a postincident analysis:

1. Determine what the preferred and safer response or activity should be.
2. Develop a procedure, practice, or equipment list needed to deliver the preferred and safer response or activity.
3. Consult with supervisors for approval and formal adoption of the revised response or activity.
4. Train your peers and subordinates in the preferred and safer response or activity. Training should address recommendations arising from the investigation of accidents, injuries, occupational deaths, illnesses, and exposures, as well as the observation of incident scene activities.
5. Make the revision part of an updated standard operating procedure or directive.
6. Integrate the new response or activity into the existing continuing education and basic program.

The fire officer must fulfill four roles to provide a safe environment for the fire company and ensure that everyone goes home:

1. Identify unsafe and hazardous conditions.
2. Mitigate or reduce as many problems as possible.
3. Train and prepare for the remaining hazards.
4. Model safe behavior.

You Are the Fire Officer Conclusion

Division A is covering the front of the restaurant and the street. The explosion broke the columns supporting the front of the restaurant and brought the roof down. A lean-to collapse area includes the officer's side of the first-arriving engine. Captain Davis subdivides the Rescue Group into three teams, each with a lieutenant or captain, and divides the search area into sectors. Part of the Rescue Group incident action plan includes securing adequate lighting and ventilation in the search area and setting up medical care and rehabilitation areas in the cold zone. Captain Davis works with the incident safety officer to identify and isolate an area where the roof remains unstable. Two fire fighters and three civilians are found in the debris and moved into medical treatment. In addition, the Rescue Group locates two civilian fatalities. They are left in place until the investigation is completed and the unsafe areas are secured.

The primary function of a rapid intervention crew is to rescue any fire fighter who is lost, disoriented, trapped, or injured. The RIC should be strategically positioned, standing by ready for immediate action and monitoring fire-ground radio traffic. The incident or sector safety officer is responsible for verifying that operations are conducted safely. This includes identifying imminent and potential hazards, providing a risk assessment to the incident commander, and ensuring that all safety policies and procedures are followed.

Wrap-Up

Chief Concepts

- The fire officer is responsible for ensuring that every fire fighter completes every incident without serious injury, disability, or death. This concept is expressed as a fire officer's special obligation to ensure that "everyone goes home" at the end of a workday.
- The fire officer must develop an incident action plan (IAP) that addresses and minimizes the chances of harm by identifying and controlling the factors that lead to fire fighter injury or death.
- Everyone Goes Home is a program developed by the National Fallen Firefighters Foundation to prevent line-of-duty death and injuries; it includes the 16 Firefighter Life Safety Initiatives.
- The goal of www.firefighternearmiss.com, launched by the International Association of Fire Chiefs in 2005, is to track incidents that avoided serious injury or death, to identify trends, and to share the information with other fire fighters in a confidential and nonpunitive way.
- Heart attacks are the leading cause of death for fire fighters, accounting for 41 percent of all line-of-duty deaths from 2007 to 2011. A disproportionate number of fire fighters older than the age of 49 die of cardiac arrest while on duty.
- Motor vehicle collisions account for the largest percentage of traumatic fire fighter deaths. More than three-fourths of the fatalities from this cause were found to not be wearing a seat belt.
- Asphyxia and burns are prime factors as a direct cause of death while fire fighters are operating in burning buildings.
- An incident safety officer is a member of the incident command staff who monitors the scene, identifies and reports hazards to the incident commander, and takes immediate steps to stop unsafe actions and ensure that the department's safety policies are followed.
- Every fire officer should be trained to perform the basic duties of an incident safety officer and prepared to act temporarily in this capacity if he or she is assigned to fill this position by the incident commander.
- Some incidents, based on their size, complexity, or duration, require more than one safety officer. Assistant incident safety officers can be assigned to subdivide responsibilities for different areas and functions at incidents such as high-rise fires, hazardous materials incidents, and special rescue operations.
- Part of the incident safety officer's role is to ensure that an appropriate rehabilitation process is established. Incident scene rehabilitation provides for medical evaluation, treatment, monitoring, fluid and food replenishment, mental rest, and relief from climatic conditions of the incident.
- A safety program that is based on preventing fatalities is far from adequate. Every injury, or near miss, should be viewed as a potentially fatal or disabling situation, and injury prevention should be an equally important concern of the fire officer.
- The fire officer needs to fully understand each policy, follow all safety policies and procedures, and ensure that all subordinates fully understand and follow them.
- Principles that should be implemented to prevent injuries and deaths include physical fitness and consistent use of personal protective equipment.
- Safety hazards at the fire station can have the same consequences as hazards encountered at emergency incidents. The fire department, however, has control over the fire station environment at all times and has the ability, as well as the responsibility, to identify and correct any safety problems.
- Every fire department should establish an infection control program that meets NFPA 1581, *Standard on Fire Department Infection Control Program*.
- The most important first step with any infectious disease exposure is to wash the exposed area immediately and thoroughly with soap and running water. If soap and running water are not available, waterless soap, antiseptic wipes, alcohol, or other skin cleaning agents can be used until soap and running water are obtained.
- The fire department health and safety officer ensures that all injuries, illnesses, exposures, fatalities, or other potentially hazardous conditions and all accidents involving fire department vehicles, fire apparatus, equipment, or fire department facilities are thoroughly investigated.
- The fire officer is usually responsible for conducting an initial investigation and for fully investigating many minor accidents.
- The incident safety officer provides a postincident analysis for the department in the form of a written report that includes pertinent information relating to safety and health issues involved with the incident.
- The number of fire fighters dying on duty in the United States has declined 30 percent over the last 30 years, from an average of 151 deaths annually in the 1970s to an average of 99 deaths per year in the 2000s.
- The relatively high heart attack rate among fire fighters reflects the nature of the work, which can suddenly change from low activity to an episode of high stress and intense exertion.
- Traumatic injuries are the second leading cause of fire fighter fatalities. Most of these deaths are caused by vehicle accidents, falls, or structural collapse of a burning building.
- The third most frequent category of fire fighter fatalities includes deaths from asphyxiation and burns. In these situations, the fire itself is the direct cause of the fire fighter's death.
- Near-miss reports are analyzed annually using a tool modified from the U.S. Navy's Human Factors Analysis

and Classification System (HFACS) that identifies four categories of risks:
- HFACS level 1: unsafe acts
- HFACS level 2: preconditions to unsafe acts
- HFACS level 3: unsafe supervision
- HFACS level 4: organizational influences

■ The fire department is required to maintain records of all accidents, occupational deaths, injuries, illnesses, and exposures in accordance with federal, state, and local regulations.

■ The real value in a postincident analysis is the learning process that results from the information obtained during the evaluation.

■ The fire officer has four roles in providing a safe environment for the fire company and ensuring that everyone goes home:
- Identify unsafe and hazardous conditions.
- Mitigate or reduce as many problems as possible.
- Train and prepare for the remaining hazards.
- Model safe behavior.

Hot Terms

Accident An unplanned event that interrupts an activity and sometimes causes injury or damage; a chance occurrence arising from unknown causes; an unexpected happening due to carelessness, ignorance, and the like.

Hazard Any arrangement of materials and heat sources that presents the potential for harm, such as personal injury or ignition of combustibles.

Health and safety officer The member of the fire department assigned and authorized by the fire chief as the manager of the safety and health program.

Immediately dangerous to life and health (IDLH) Any condition that would do one or more of the following: (1) pose an immediate or delayed threat to life, (2) cause irreversible adverse health effects, or (3) interfere with an individual's ability to escape unaided from a hazardous environment.

Incident action plan (IAP) The objectives reflecting the overall incident strategy, tactics, risk management, and member safety that are developed by the incident commander. Incident action plans are updated throughout the incident.

Incident Command System (ICS) A system that defines the roles and responsibilities to be assumed by personnel and the operating procedures to be used in the management and direction of emergency operations; also referred to as an Incident Management System (IMS).

Incident safety officer An individual appointed to respond to or assigned at an incident scene by the incident commander to perform the duties and responsibilities specified in NFPA 1521, *Standard for Fire Department Safety Officer*.

Incident safety plan The strategies and tactics developed by the incident safety officer based on the incident commander's incident action plan and the type of incident encountered.

Incident scene rehabilitation The tactical-level management unit that provides for medical evaluation, treatment, monitoring, fluid and food replenishment, mental rest, and relief from climatic conditions of an incident.

Operational period A term used in a written incident action plan identifying a period of time during a long-term incident that a specific incident action plan covers. For federally funded incidents, the operational period is 12 hours; local incidents may use operational periods of 8 hours.

Personnel accountability system A method of tracking the identity, assignment, and location of fire fighters operating at an incident scene.

Preincident plan A written document resulting from the gathering of general and detailed data to be used by responding personnel for determining the resources and actions necessary to mitigate anticipated emergencies at a specific facility.

Rapid intervention crew (RIC) A dedicated crew of four fire fighters who are assigned for rapid deployment to rescue lost or trapped members.

Rehabilitation The process of providing rest, rehydration, nourishment, and medical evaluation to members who are involved in extended or extreme incident scene operations.

Risk–benefit analysis A decision made by a responder based on a hazard and situation assessment that weighs the risks likely to be taken against the benefits to be gained for taking those risks.

Risk management Identification and analysis of exposure to hazards, selection of appropriate risk management techniques to handle exposures, implementation of chosen techniques, and monitoring of results, with respect to the health and safety of members.

Safety unit A member or members assigned to assist the incident safety officer; the tactical-level management unit that can be composed of the incident safety officer alone or with additional assistant safety officers assigned to assist in providing the level of safety supervision appropriate for the magnitude of the incident and the associated hazards.

References and Additional Resources

NFPA reprinted material is not the complete and official position of the NFPA on the referenced subject, which is represented only by the standard in its entirety.

Beaty, D. (1995). *The Naked Pilot: The Human Factor in Aircraft Accidents.* Shrewsbury, UK: Airlife Publishing.

Compton, D., and G. Mack. (2004). *The Mental Aspects of Performance.* Stillwater, OK: IFSTA/FPP.

Ezell, G. S. (2010). *Firefighter Safety in Abandoned and Vacant Structures Within the City of Joplin.* Emmitsburg, MD: U.S. Fire Administration, Executive Fire Officer Program.

Fahy, R. F. (2010). *U.S. Firefighter Fatalities in Structure Fires, 1997–2009.* Quincy, MA: National Fire Protection Association.

Fahy, R. F., P. R. LeBlanc, and J. L. Molis. (2007). *What's Changed Over the Past 30 Years?* Quincy, MA: National Fire Protection Association.

Fahy, R. F., P. R. LeBlanc, and J. L. Molis. (2012). *Firefighter Fatalities in the United States—2011.* Quincy, MA: National Fire Protection Association.

Gasaway, R. B. (2013). *Situational Awareness for Emergency Response.* Tulsa, OK: Pennwell/Fire Engineering.

International Association of Fire Chiefs and National Fire Protection Association. (2012). *Live Fire Training: Principles and Practice.* Burlington, MA: Jones & Bartlett Learning.

Karter, M. J. Jr. (2012). *Patterns of Firefighter Fireground Injuries.* Quincy, MA: National Fire Protection Association.

Karter, M. J. Jr., and J. L. Molis. (2012). *U.S. Firefighter Injuries—2011.* Quincy, MA: National Fire Protection Association.

Karter, M. J. Jr., and G. P. Stein. (2012). *U.S. Fire Department Profile Through 2011.* Quincy, MA: National Fire Protection Association.

Kolomay, R., and R. Hoff. (2003). *Firefighter Rescue and Survival.* Tulsa, OK: PennWell.

Marsar, S. (2009). *Can They Be Saved? Utilizing Civilian Survivability Profiling to Enhance Size-up and Reduce Firefighter Fatalities in the Fire Department, City of New York.* Emmitsburg, MD: U.S. Fire Administration, Executive Fire Officer Program.

National Fallen Firefighters Foundation. (2011). *Understanding and Implementing the 16 Firefighter Life Safety Initiatives.* Stillwater, OK: Fire Protection Publications.

National Fire Data Center and National Fallen Firefighters Foundation. (2012). *Firefighter Fatalities in the United States in 2011.* Emmitsburg, MD: U.S. Fire Administration.

National Fire Protection Association. Reprinted with permission from NFPA 1500-2013, Fire Department Occupational Safety and Health Program, Copyright © 2012.

National Fire Protection Association. Reprinted with permission from NFPA 1670-2009, Standard on Operations and Training for Technical Search and Rescue Incidents. Copyright © 2008.

National Fire Protection Association. Reprinted with permission NFPA 1710, Standard for the Organization and Deployment of Fire Suppression Operations, Emergency Medical Operations, and Special Operations to the Public by Career Fire Departments. Copyright © 2009.

National Volunteer Fire Council and National Fire Protection Association. (2012). *Understanding and Implementing Standards NFPA 1500, 1720, and 1851.* Greenbelt, MD: National Volunteer Fire Council.

Phoenix Fire Department. (2002). *Final Report: Southwest Supermarket Fire, 35th Avenue and McDowell Road.* Phoenix, AZ: Phoenix Fire Department.

Salmon, M. A. (2012). *Body Mass Index in Relation to Job Performance.* Emmitsburg, MD: U.S. Fire Administration, Executive Fire Officer Program.

Smeby, L. C. Jr. (2014). *Fire and Emergency Services Administration: Management and Leadership Practices,* 2nd ed. Burlington, MA: Jones & Bartlett Learning.

Tri Data Corporation. (2002). *Firefighter Fatality Retrospective Study.* Washington, DC: U.S. Fire Administration.

Zotti, T. J. (2011). *Can I Take This Off Yet: A Breathing Apparatus Policy for the Wolfeboro Fire-Rescue Department.* Emmitsburg, MD: U.S. Fire Administration, Executive Fire Officer Program

FIRE OFFICER in action

After the injured fire fighters and civilians are removed, command has Division A evacuated because the roof continues to shift and crumble. The Rescue Group is assigned to assist in overhauling the garden shop that has been damaged by the collapse. The incident safety officer requires that the group wear medical exam gloves under their fire gloves and remain on SCBA air while working in the shop. Upon completion of this task, the Rescue Group goes through technical decontamination and proceeds to rehabilitation.

The fire officer has three roles when it comes to safety and risk management. The first role is to model appropriate behavior and work habits. The second role is to ensure that the members under the officer's supervision are not subjected to avoidable risks and unsafe work conditions. The third role is to participate fully in the local, regional, and national effort to reduce fire fighter deaths, disabilities, and injury.

1. What is the most common characteristic of fire fighters who died from a sudden heart attack while at work?
   A. Preexisting cardiovascular disease or cardiac surgery history
   B. Cigarette smoker for more than 5 years
   C. Body fat representing more than 20 percent of total body weight
   D. Family history of high cholesterol and diabetes

2. "Failure to correct known problems" describes which level of HFACS classification?
   A. Level 1: Unsafe Acts
   B. Level 2: Preconditions to Unsafe Acts
   C. Level 3: Unsafe Supervision
   D. Level 4: Organizational Influences

3. Which of the following is part of the general knowledge requirements to operate as an incident safety officer?
   A. History of the incident management process
   B. Safety and health hazards involved in emergency operations
   C. Certification as a hazardous materials response team technician
   D. Human anatomy and physiology

4. Who is authorized to submit a report to the National Fire Fighter Near-Miss Reporting System?
   A. Chief of department or union president
   B. Incident commander
   C. Supervising fire officers or higher
   D. Anyone who experienced a near miss

Fire Captain Activity

Injured fire fighters and civilians have been removed from the restaurant. The building official and technical rescue chief agree that a heavy crane will be needed to remove the collapsed roof elements before a secondary search can be conducted.

There may be additional viable victims trapped deeper within the rubble. The restaurant includes a basement that has been inaccessible.

NFPA Fire Officer II Job Performance Requirement 5.7.1
Analyze a member's accident, injury, or health exposure history, given a case study, so that a report including action taken and recommendations made is prepared for a supervisor.

Application of 5.7.1
Pick one of these situations and prepare a recommendation for your chief. Should the member continue to work as a fire fighter?
1. A member under your command suffers a heart attack and receives a stent.
2. After receiving a head injury during physical training, a member now suffers from seizures when staring at the front of the pumper when the emergency lights are on.
3. A member who is a recovering alcoholic shows up for work intoxicated.

Understanding People: Management Concepts

Fire Officer I

Knowledge Objectives

After studying this chapter, you will be able to:

- Discuss the principles of supervision and basic human resources management (NFPA 4.1.1) (NFPA 4.2). (pp 112–113)
- Explain the history and principles of scientific management (NFPA 4.2.6). (pp 113–114)
- Explain the history and principles of humanistic management (NFPA 4.2.6). (pp 114–118)
- Discuss the function of human resources management (NFPA 4.2) (NFPA 4.2.5). (pp 118, 120–121)
- Coordinate the completion of assigned tasks and projects (NFPA 4.2.6) (NFPA 4.2). (pp 120–121)

Skills Objectives

There are no Fire Officer I skills objectives for this chapter.

Fire Officer II

Knowledge Objectives

After studying this chapter, you will be able to:

- Discuss the importance of daily training and performance management (NFPA 5.2) (NFPA 5.2.1). (pp 121–122)
- Discuss the human resources function of compensation and benefits (NFPA 5.2). (pp 121–122)

Skills Objectives

There are no Fire Officer II skills objectives for this chapter.

CHAPTER 6

Principles of Fire and Emergency Service Administration (FESHE) Course Outcomes

3. Identify and explain the concepts of span and control, effective delegation, and division of labor. (pp 112–113, 117–118, 120–122)
4. Select and implement the appropriate disciplinary action based on the employee's conduct. (pp 114–115, 117–118, 122)
5. Explain the history of management and supervision methods and procedures. (pp 113–120)
6. Discuss the various levels of leadership, roles, and responsibilities within the organization. (pp 114–118, 120–121)
7. Describe the traits of effective versus ineffective management styles. (pp 113–115, 117–118)
8. Identify the importance of ethics as it relates to fire and emergency services. (pp 114, 122)

You Are the Fire Officer

Battalion Chief Frank Johnson is conducting a six-month probationary assessment of both Captain Jean Davis and Lieutenant Taylor Williams at Fire Station 100. The assessment includes a review of fire company performance as well as in-station activities. The chief notes very few target hazard walk-throughs in the district and reduced company performance on standardized evolutions at a recent evaluation at the fire academy.

The evaluation revealed that fire fighters are taking too much time and missing critical elements of fire hose deployment. Lieutenant Williams explains that the crew has been focusing on advanced forcible entry and ladder drills in preparation for the Fire Fighter Olympics. According to Captain Davis, this effort is part of a "Circle of Consensus" process that leads to a holistic development of internal satisfaction. Station 100 fire fighters want to win the Olympic truck company challenge. Johnson is irritated by this response, and he suggests that Williams and Davis focus on the external satisfaction of their battalion chief by complying with departmental goals: perform a standardized skill drill every day and a target hazard walk-through every week.

1. Why are there so many management concepts that provide contradictory information?
2. How do these theories assist you in executing supervisory tasks?
3. Is there a way for you to evaluate a new management concept or procedure before using it at the fire station?

Introduction

The start of the Management Century was marked by a May 1886 address by Yale Lock Manufacturing founder Henry R. Towne to the American Society of Mechanical Engineers: "the *management of works* has become a matter of such great and far-reaching importance as perhaps to justify its classification also as one of the modern arts." Management science is the systematic pursuit of practical results, using available human and knowledge resources in a concerted and reinforcing way.

A fire officer is a manager who has been given the responsibility of directing and supervising a group of fire fighters, as well as apparatus, equipment, facilities, and other resources, to achieve certain outcomes. In the fire service, the desirable outcomes begin with protecting people and property from a variety of hazardous situations. Additional desirable outcomes include ensuring that the work is performed safely, efficiently, promptly, and in accordance with a long list of rules, regulations, procedures, and additional concerns.

Fire Officer I

Managing People

Most fire officers will find that their greatest challenge relates to managing people. It is the workers who get the job done. The managing or supervising fire officer is responsible for performing a set of functions that direct and coordinate these workers' efforts, provide them with the necessary tools and resources, and ensure that the outcome meets the standards. Fire department operations are labor-intensive ventures, with tasks being completed by skilled workers **FIGURE 6-1**. To be effective as a manager, a fire officer must develop skills that are directly related to managing human resources.

The concept of management emerged from the Industrial Revolution. The introduction of steam power in the late 1700s led to the creation of large factories in Europe that were staffed with agricultural and small-town residents who needed direction and supervision.

Human resources management is built upon the concepts introduced by two generalized schools of management thought: scientific management and humanistic management. Each school developed a set of theories on how supervisors can manage people to accomplish tasks in the work environment. You can think of these theories and the related practices as

FIGURE 6-1 Managing people can be the fire officer's greatest challenge.

tools in the fire officer's management toolbox. The fire officer will not necessarily use all of the tools or the same tools for every situation. Instead, selection of the appropriate management tools must consider the situation, the task, and the individuals or team that will be involved in producing a desired outcome. Some of these theories are directly applicable for use by a fire officer in assigning tasks or responsibilities to crew members.

Scientific Management

Adam Smith noted that it took several hundred years for a region to develop a tradition of labor and the expertise in manual and managerial skills needed to support the Industrial Revolution. That was far too long to meet the needs of rapidly growing textile factories, however. Enter the engineering approach to management, education, and training. When New York City established the Metropolitan Fire Department in 1865, the Industrial Revolution was in full swing. Engineers were the movers, shakers, and creators of the product-based business world, and their influence spilled over into the fire service.

■ Frederick Winslow Taylor

In the 19th century, Frederick Winslow Taylor decided to forsake his educational pursuits at Harvard in favor of a career in industry. In 1874, the skills he needed were not taught in universities; they were learned on the shop floor. Taylor was hired as a laborer in the machine shop of Midvale Steel.

At Midvale, Taylor developed and put into place the basic elements of what later came to be known as scientific management. Scientific management is based on the breaking down of work tasks into their constituent elements; the timing of each element based on repeated stopwatch studies; the fixing of piece-rate compensation based on those studies; the standardization of work tasks on detailed instruction cards; and generally, the systematic consolidation of the shop floor's brain work in a "planning department."

Taylor published *The Principles of Scientific Management* in 1911, describing how the application of the scientific method to the management of workers could greatly improve productivity. Application of scientific management called for optimizing the ways tasks were performed and simplifying the jobs so workers could be trained to perform a specialized sequence of motions in the one "best" way.

Prior to scientific management, work was performed by skilled craftsmen who served lengthy apprenticeships to learn their trades. Individual workers made their own decisions about how their job was to be performed. Scientific management took away much of this autonomy and converted skilled crafts into a series of simplified jobs that could be performed by unskilled workers. To determine the optimal way to perform a job, Taylor performed experiments that he called time studies (also known as time and motion studies). These studies featured use of a stopwatch to measure a worker's sequence of motions, with the goal of determining the most efficient way to perform each task. That worker and others could then be trained to perform those same specific tasks in the same efficient manner.

Taylor's Four Principles of Scientific Management

After years of various experiments to determine optimal work methods, Taylor proposed four principles of scientific management:

1. Replace "rule-of-thumb" work methods with methods based on a scientific study of the tasks.
2. Scientifically select, train, and develop each worker, rather than passively leaving workers to train themselves.
3. Cooperate with the workers to ensure that the scientifically developed methods are being followed.
4. Divide work nearly equally between managers and workers so that the managers apply scientific management principles to planning the work and the workers actually perform the tasks.

Factories that implemented these principles, including Henry Ford's Model T automobile factories, achieved impressive productivity. Versions of scientific management can still be found in the 21st-century workplace.

■ Multiphase Fire Fighter Safety and Deployment Study

A research partnership of the Commission on Fire Accreditation International (CFAI), International Association of Fire Chiefs (IAFC), International Association of Firefighters (IAFF), National Institute of Standards and Technology (NIST), and Worcester Polytechnic Institute (WPI) was formed to conduct a multiphase study of the deployment of resources as it affects fire fighter and occupant safety. This study investigated the effects of varying crew size, first apparatus arrival time, and

---

Fire Marks

A Paramilitary Legacy

Organizational structure and development both influence management concepts, as can be seen in the history of the U.S. fire service. The typical municipal fire department structure was created at the end of the Civil War, when city leaders asked West Point–educated generals to organize the individual volunteer fire companies into a municipal fire department. The first formal fire company leaders were the individual fire company foremen. When the company foreman was off, an assistant or acting foreman was in charge of the fire company.

Individual fire companies were organized into battalions consisting of four to eight fire stations. Battalions were organized into divisions, reflecting the area's geographic boundaries. The city created a department and appointed a chief officer who oversaw all of the fire companies in the municipality.

The city fire department supplemented, replaced, or evicted the volunteer fire companies that were established before 1860. Unlike other city employees, career fire fighters were on continuous duty, with a couple of hours off for meals and one or two days off each month. They lived in the fire station.

> **Fire Marks**
>
> **Fire Fighter Combat Challenge**
> When the first Combat Challenge was held at the Maryland Fire and Rescue Institute in May 1991, the team from Prince William County, Virginia, won with a time of 10:08 minutes. Within a decade, the winning time to complete the same course had dropped to less than 2 minutes. By closely analyzing every movement, competitors learned how to perfect their techniques and shave seconds from every step.

response time on fire fighter safety, overall task completion, and interior residential tenability using realistic residential fires. More than 60 laboratory and residential fire-ground experiments were designed to quantify the effects of various fire department deployment configurations on the most common type of fire—a low-hazard residential structure fire. Experiments measured 22 fire-ground tasks.

The four-person crews operating on a low-hazard structure fire completed all the tasks on the fire ground an average of 7 minutes faster—nearly 30 percent faster—than the two-person crews. The four-person crews completed the same number of fire-ground tasks an average of 5.1 minutes faster—nearly 25 percent faster—than the three-person crews.

Humanistic Management

One problem with the scientific management method was that people were treated like carbon-based cogs in a scientific management production line. Taylor considered workers as cheap, stupid, and interchangeable. Each worker was trained to perform just one task, a small portion of a production process, and to repeat that task at a mind-numbing rate. The humanistic management school shifted the focus to pay more attention to the workers and to the working conditions that would make those workers more productive.

The humanistic management school started with Professor George Elton Mayo, a Harvard University industrial psychology professor. From 1924 until 1933, Western Electric's Hawthorne Works plant was the site of a series of experiments in which changes in the working conditions were evaluated against productivity in an electronic relay assembly line. These experiments yielded the term *Hawthorne effect*—a phenomenon in which people improve their performance or behavior not because of any specific condition being tested, but simply because of the extra attention they receive as part of the study.

Humanistic management research continued through the work of Douglas McGregor and Abraham H. Maslow.

■ McGregor: Theory X and Theory Y

Douglas McGregor, a social psychologist and professor of management at the Massachusetts Institute of Technology (MIT), studied the work environment from the mid-1930s through the mid-1950s. McGregor's study period covered the Depression, World War II, and the booming prosperity of the 1950s that created the U.S. middle class. McGregor's 1960 publication, *The Human Side of Enterprise*, summarized the results of his research and added the Theory X and Theory Y concepts to the manager's motivational toolbox.

Whereas immigrants to the United States in the early 20th century worked to survive, most workers in the mid-20th

Near-Miss REPORT

Report Number: 10-0000880

Event Description: The morning of my shift, the temperature was 101 degrees; heat index 110; humidity more than 80 percent; clear, sunny, and dry; and no rain in the past 12 to 24 hours. My company was dispatched for a working structure fire in a subdivision. Upon arrival at the scene, we found a 2000-square-foot home with 25 percent involvement, with exposures on the two sides. I hand-jacked 400 feet of 5-inch large-diameter hose (LDH) to the hydrant and then proceeded to charge two 1.75-inch cross-lays. I assisted with pulling a 2.5-inch preconnected portable master stream to the rear of the building where the fire was located. I performed interior operations for approximately 12 minutes and went to rehab with the rest of my crew. I was in rehab for about 10 minutes when I began to feel chest palpitations, weakness, and dizziness. According to medical crews on the scene, I was also lethargic, disoriented, and hypotensive. I was moved to the medic unit, and an IV was established with normal saline. The cardiac monitor showed sinus tachycardia. All my gear was removed, and active cooling took place. I was in the hospital for several hours and was administered three 1-liter bags of saline.

Lesson Learned: The entire shift (two on the engine and two on the ambulance) was hydrating prior to and during the shift to prevent this very problem. Proper staffing would have meant that some of the tasks I did would be performed by someone else.

century enjoyed much better retirement and financial resources. McGregor observed that in spite of these improvements, many mid-20th-century workers were dissatisfied with their jobs. He concluded that worker motivation is directly related to autonomy and responsibility; workers with greater autonomy are more likely to be motivated in their jobs. McGregor also felt that modern employment often stifled human creativity and impaired motivation. McGregor developed the Theory X and Theory Y concepts to define this problem in terms of the manager's or the organization's view of the workers.

A Theory X manager believes that people do not like to work, so consequently they need to be closely watched and controlled. Theory X makes sense when you look at the Industrial Age working conditions of the late 1800s. The assembly-line factories were horrible, characterized by scant light, poor ventilation, and unsafe and unsanitary working conditions. Some workers had to relieve themselves at their workstations. Many of the factory workers were recently arrived immigrants, with no industrial work skills and limited ability to read or write English. Many were children. No wonder they had to be coerced and threatened to keep up their productivity!

A Theory Y manager has an entirely different view of employee creativity and motivation. A Theory Y manager believes that people do like to work and that they need to be encouraged, rather than controlled.

These two distinctly different philosophical models of worker motivation and supervisory strategies can be observed in many organizations today. The working world has changed significantly since 1960, especially with the increased participation of female managers, and Theory Y is the prevailing trend. Nevertheless, academic and business writers point out that job satisfaction and employee loyalty are far lower now than they were in 1960. Theory X remains a commonly encountered managerial style.

Theory X Versus Theory Y for the Fire Officer

McGregor's Theory X and Theory Y models can be valuable tools for a fire officer. Individuals seek to become fire fighters because they are attracted to the work. Indeed, few people are forced by economic conditions to become fire fighters. The fire service requires a strong personal commitment, beginning with the initial investment in physical and technical training. Many fire fighter activities are difficult, physically demanding, and unpleasant, yet most fire fighters love their work. They tend to operate from a different worldview than most workers in other fields, such as those in the information technology or service industries.

The fire officer is supervising a self-selected workforce whose members are dedicated and enthusiastic about their responsibilities as fire fighters. The officer's challenge is often to steer these members' efforts in the right direction and to create an effective team. The Theory Y concepts can be used effectively in many situations to encourage fire fighter creativity.

Theory Y does not work for every situation, however. In particular, there are three situations in which the fire officer must temporarily behave as a Theory X manager. The first situation is when operating at a fire or other high-risk activity—the fire officer must provide close and autocratic supervision. The second situation is when the fire officer must take control

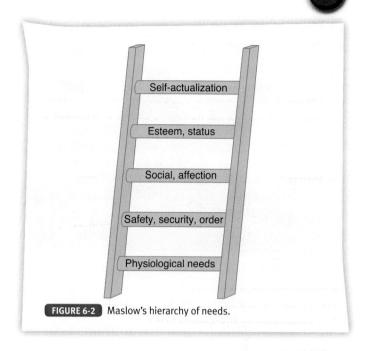

FIGURE 6-2 Maslow's hierarchy of needs.

of a workplace conflict and issue specific directions to defuse the situation. The third situation is when a fire officer is near the end of a series of negative disciplinary measures; for example, the Theory X concept and techniques are applicable in an involuntary work performance meeting with a subordinate.

Maslow: Hierarchy of Need

Abraham H. Maslow was a psychologist who researched mental health and human potential. He is most often associated with the concepts of a hierarchy of needs, self-actualization, and peak experiences. Maslow's concept of the hierarchy of needs, for example, depicted human needs as being arranged like a ladder or a pyramid **FIGURE 6-2**.

Level 1: Physiological Needs

The most basic human needs, at the bottom of the ladder, are physical—air, water, food, and shelter. It is difficult to be creative or productive when you are exhausted and hungry. For the fire officer, this point emphasizes the importance of staying alert for conditions where the fire fighters are hungry, exhausted, dehydrated, too hot, or too cold. Sometimes fire fighters have to work at the extremes of discomfort and personal inconvenience due to the nature of emergency activities, but these situations should be the exceptions and not a normal way of getting the job done. The practice of making a fire company work without a rest and rehydration period will fail to meet their most basic needs **FIGURE 6-3**. Meals may be delayed, but food is needed to keep fire fighters working, just as fuel is needed to keep apparatus operating.

Level 2: Safety, Security, and Order

keeping FF's safe

Safety, security, and order needs are next in Maslow's hierarchy of needs. Safety is obviously a primary concern in the fire service, although a fire fighter's perceptions of safety are probably quite different from those of a typical office or factory worker. If the fire fighters feel that their fire officer is leading them in an unsafe manner or a particular policy or practice in the fire department is exposing them to an avoidable risk, safety is likely

FIGURE 6-3 Fire fighters need periods of rest and rehydration to meet their most basic physiological needs.

FIGURE 6-4 Fire fighters often engage in social and recreational activities with their co-workers.

to become a very significant issue. Years of deferred fire station maintenance or postponed equipment replacement may result in unsafe work conditions, for example.

Security is more closely associated with maintaining employment or status within the organization. It could take a form as subtle as a fear that a personality clash with a particular company officer will destroy or impede a fire fighter's career. An individual who feels threatened may do what is necessary to survive, but will not feel highly motivated to advance the fire officer's or fire department's objectives.

Imagine the impact on your fire company if a notice from fire administration suddenly announced that all incumbent fire fighters would be required to undergo paramedic training, and that any fire fighter who does not pass the National Registry certification exam within 1 year will be terminated. Although this would be a great example of Theory X coercion, such a policy would also create tremendous fire fighter stress.

The need for security and order could be a factor when a fire department is undergoing a significant reorganization, such as the appointment of a new fire chief or the combination of departments into a regional authority. Any change in the established order is likely to arouse insecurities. The anticipation of significant changes in the organization or budget-related cutbacks can easily impede fire fighter creativity and motivation.

Level 3: Social Needs and Affection

At the third level of Maslow's hierarchy of needs are psychological or social needs related to belonging to a group and feeling acceptance by the group. Group acceptance and belonging are particularly strong forces within the fire service. The majority of fire fighters have some type of identification on their personal vehicles that visibly declares their local fire department affiliation and membership in the fire service at large. Fire fighters wear their company patches proudly and often engage in social and recreational activities with their co-workers **FIGURE 6-4**. Transferring a fire fighter away from a "good" assignment can be a strong demotivator, just as suspending a volunteer from riding can have a significant impact at the social and affection levels.

Level 4: Esteem and Status

Promotions, gold badges, special awards, and take-home fire department vehicles are all symbols that apply to the esteem and status level **FIGURE 6-5**. Membership in an elite

Getting It Done

Consistency Counts

Given a choice, fire fighters prefer a company officer who is consistent. The brilliant officer who is inconsistent creates concern and doubt in the fire fighter.

Consider the dilemma that arises when the first-arriving engine company lays a supply line if the department lacks a standard operating procedure. Instead of focusing on the upcoming fire problem, the fire fighter will be looking for clues that the rig is slowing down to deploy a supply line. A consistent response plan reduces confusion and allows supervisors and fire fighters to concentrate on the emergency scene issues.

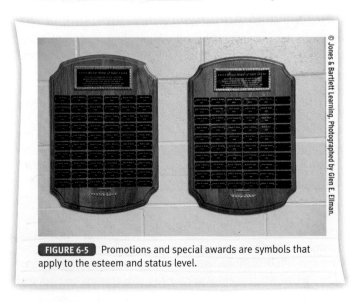

FIGURE 6-5 Promotions and special awards are symbols that apply to the esteem and status level.

Company Tips

Building Trust in the New Fire Officer

It is vital for a new supervisor to make an effort to build trust with the workers in his or her company. The failure to establish trust from the beginning of the relationship can quickly undermine a supervisor's career—especially when that manager is a new fire officer. Fire fighters make or break a new officer, and they will make every effort to protect themselves from an officer who appears to be unsafe, unstable, or unprepared for the job.

How can the new fire officer build trust? Here are five suggestions:

1. Know the fire officer job, both administrative and tactical.
2. Be consistent. Strive to provide a measured response to any problem, emergency, or challenge.
3. Walk your talk. Actions speak much louder than words.
4. Support your fire fighters. Make sure you help meet their physiological, safety, security, and order needs.
5. Make fire fighters feel strong. Help them become competent and confident in their emergency service skills. Show how they can control their destiny.

---

fire department unit is often an indicator of special qualifications or achievements. Many fire fighters are willing to invest both money and off-duty time to compete for a high-esteem or high-status position. For example, fire fighters may spend months preparing for a fire officer promotional exam or taking courses to qualify for a rescue company. Most elite fire units have more applicants than available positions.

Level 5: Self-Actualization

Peak experiences are profound moments of love, understanding, happiness, or rapture, when a person feels more whole, alive, self-sufficient, and yet a part of the world—that is, more aware of truth, justice, harmony, and goodness. Self-actualizing people have many such peak experiences. These individuals tend to focus on problems outside of themselves and have a clear sense of what is true.

Maslow felt that unfulfilled needs at the lower levels of the "needs" ladder would inhibit a person from climbing to the next step. As he pointed out, a person who is dying of thirst quickly forgets that thirst when deprived of oxygen.

Maslow's thinking was original. Most psychology research before him, including Maslow's first textbook, had been concerned with the abnormal and the ill. Maslow, however, wanted to know what constituted positive mental health. As he observed, "Human nature is not nearly as bad as it has been thought to be."

■ Blake and Mouton's Managerial Grid

In the early 1960s, the Exxon Corporation hired behavioral scientists Robert Blake and Jane Mouton to perform a series of experiments designed to increase leadership effectiveness.

In the 1990s, the grid theory that these researchers developed was applied to crew management of aircraft and aerospace teams—a concept later adopted by the fire service as "crew resource management" to improve incident safety.

The grid theory assumes that every decision made and every action taken in the workplace is driven by people's values, attitudes, and beliefs. At the individual level, these values are based on two fundamental concerns that influence behavior—a concern for people and a concern for results. Blake and Mouton developed a survey document with 35 questions that measures a person's level of concern in each area. The results are plotted on an X–Y chart. Blake and Mouton described five behavioral models based on a person's position on this grid.

Indifferent: Evade and Elude

The indifferent style represents the lowest level of concern for both results and people. The key word for this style is "neutral." An individual who demonstrates this style is the least visible person in a team; he or she is a follower who maintains distance from active involvement whenever possible. An indifferent person carefully goes through the motions of work, doing enough to get by, but rarely making a deliberate effort to do more.

The stereotypical image of this personality is the minion in a bureaucratic government agency where everyone is treated like a number. This sort of workplace allows the person to blend in without attracting attention. In fact, he or she often seeks work that can be done in isolation so as to carry on without being disturbed or noticed.

The indifferent manager relies heavily on instructions and process, depending on others to outline what needs to be done. Reliance on instructions avoids the need to take personal responsibility for results. If problems arise, the indifferent person is often content to ignore or overlook them, unless the instructions specify how to react to that particular problem. The indifferent person might point out the problem to someone else, but would be unlikely to offer a solution. With no instructions, he or she simply carries on with the attitude that "This is not my problem."

Controlling: Direct and Dominate

A controlling person demonstrates a high concern for results, but a low concern for others. The high concern for results brings determination, focus, and drive for success. Consequently, such a person is usually highly trained, organized, experienced, and qualified to lead a team to success. Unfortunately, the low concern for others prevents the controlling person from being aware of others involved in an activity, beyond what is expected of them in relation to results. The controlling person expects everyone else to "keep up" with his or her efforts, and so moves ahead, intensely focused on results, often leaving others lost in the wake of his or her forceful initiative.

Accommodating: Yield and Comply

The accommodating person demonstrates a low concern for results and a high concern for other people. Such an individual maintains a heightened awareness of the personal feelings, goals, and ambitions of others, and always considers how proposed actions will affect them. He or she is approachable, fun, friendly, and always ready to listen with sympathy and encouragement.

Some cornerstone phrases of the accommodating attitude are "Let's talk about it," "What can I do to help?", and "Let me know what you think." The main weakness in this behavior lies in the focus of the ensuing discussions. Discussions with an accommodating person tend to include an overwhelming emphasis on personal feelings and preferences, while avoiding concrete issues. The discussion itself becomes the goal, so conversations can meander in any direction instead of concentrating on solving the problem.

The controlling and accommodating styles are diametrically opposed in their perspectives. Each of these orientations leads in a narrow and singularly focused manner by ignoring the other primary concerns in the workplace.

Status Quo: Balance and Compromise

The status quo person believes there is an inherent contradiction between the concerns for results and for people, but does not value one concern over the other. Instead, the status quo person sees a high level of concern for either people or results as too extreme and tries to moderate both in the workplace.

The objective of the status quo person is to play it safe and work toward acceptable solutions that follow proven methods. Such a politically motivated approach seeks to avoid risk by maintaining the tried-and-true course, following popular opinion and norms without pushing too hard in any direction.

Another key aspect of the status quo approach is the emphasis on maintaining popular status within the team and organization. This type of manager must be intelligent and informed enough to persuade people and companies to settle for a compromise—often less than they want and less than they could achieve. This requires being well liked, keeping well informed, and effectively convincing people that the consequences are not worth the risk. On the surface, this might make the status quo appear unbiased and impartial, but the status quo approach actually represents a narrow view that underestimates people, results, and the power of change.

Sound: Contribute and Commit

The sound person sees no contradiction in demonstrating a high concern for both people and results at the same time. He or she feels no need to restrain, control, or diminish the concerns for both people and results in a relationship. The consequence is a freedom to test the limits of success with enthusiasm and confidence. The sound attitude leads to more effective work relationships based on "what's right" rather than "who's right."

The sound behavioral model is preferred for a candidate who seeks to become a successful fire officer. The full integration of concerns for both people and results stands in contrast to the outcomes associated with the other behavioral styles. Each of the other models represents a weakness in one or both critical areas. The controlling person feels that a high concern for results is more important than a high concern for people. The accommodating person feels the reverse—namely, that a high concern for people is more important than results. The person employing the status quo approach feels that a high concern for either people or results is too risky, and prefers to maintain the current situation, holding the safe middle ground. The indifferent person sees any major concern for people or results as unrealistic and too demanding.

Human Resources Management

Human resources management focuses on the task of managing people using physical, financial, and time assets. Although some human resources functions affect career departments more than volunteer departments, all functions are required. Typical human resources management functions include the following:

- Human resources planning
- Employee (labor) relations
- Staffing
- Human resources development
- Performance management
- Compensation and benefits
- Employee health, safety, and security

Human resources planning is the process of having the right number of people in the right place at the right time who can accomplish a task efficiently and effectively. Most often, this includes forecasting future staffing needs and determining how those needs can be met. When a fire department projects the number of retirements for the next year and determines that a new recruit class should begin prior to those vacancies, the department is fulfilling the human resources planning function. Typically, this is not an activity that is conducted at the company level, other than determining needs while developing an incident action plan.

Employee relations include all activities designed to maintain a rapport with the membership. Most often, these actions are associated with working with labor organizations. Fire officers must know and understand all agreements between labor and the fire department. The fire officer should also be aware of the vast number of laws that regulate the relationship between employees and their employer. Ignorance of the law is no excuse, and fire officers who violate the law may quickly find themselves and their department faced with a federal complaint. (The *Organized Labor and the Fire Officer* chapter focuses on this humans resource function.)

Staffing is the process of attracting, selecting, and maintaining an adequate supply of labor. The fire service traditionally has been fortunate in its ability to attract applicants. However, with the increase in the services provided and the greater educational requirements, recruitment is becoming more difficult. Although some fire departments have

---

Getting It Done

Assigning a Mentor for Each Rookie

One effective method of orienting a rookie is to select a fire fighter who exemplifies good performance to act as a mentor to the new member. The fire officer asks that experienced fire fighter to become the "big brother" or "big sister" to the recruit. If the fire fighter accepts the responsibility, the recruit is then brought into the discussion so he or she knows that the mentor fire fighter is there to help him or her with any questions he or she might have. This gives the recruit an experienced fire fighter to use as a resource for questions.

VOICES OF EXPERIENCE

I have been blessed in being exposed to a variety of officers in my career, some of whom displayed positive traits and others not so good. I will always remember a particular lieutenant—not for what he did for me, but for what I saw him do for others.

In many fire departments, there are those who never seem to fit into the organization. In some cases, those who cannot seem to get along or make grade are shuffled around like checkers. We were no different: We had two fire fighters who were always being shifted around. It was as though no one would address concerns about performance or team interaction. This did finally change due to some fine work of a lieutenant who made sure that every opportunity was provided for those working under his leadership to be successful.

I remember observing this lieutenant at a fire while he was off duty. During the fire, he observed that one of the fire fighters who never seemed to fit in did not perform well. Based on his observation at the incident, the lieutenant could have had some ammunition to discredit the fire fighter when he later was assigned to his shift, but he did not. He had the integrity to know that he had to provide the fire fighter a clean slate when he was assigned under his supervision. His ability and insight to manage a difficult situation resulted in tremendous value to the fire fighter, the shift, and the department. A sign of a great leader is the ability to understand how to manage others despite their past or differences.

> *If you do not share what you and the department expect for performance, how can you ever expect a fire fighter to make a valuable contribution?*

What the lieutenant did to make those under his leadership successful was not magic; it basically came down to communications. The lieutenant did not have a secret—he did what every leader should do. It was a simple act of management when a new member was assigned to his company; he would sit down at the start of the first shift and share his expectations of that individual. An important element in his message was the emphasis to the new shift member on his or her personal value to the shift and department. The bottom line was that his expectations were always met and the entire shift benefited through his clear parameters. If the expectations were not met, the new shift member and department were not successful. Fortunately, these individuals were able to meet the expectations and to make contributions to the entire fire department.

The lesson from this lieutenant is an essential nugget for those managing fire companies: If you do not share what you and the department expect for performance, how can you ever expect a fire fighter to make a valuable contribution?

Bill Guindon
Director
Maine Fire Service Institute
Brunswick, Maine

developed programs to actively seek out qualified women and minorities, most fire departments have not.

Staffing also includes labor force reductions. There are many methods of shrinking the fire department's size; most fire departments have started with furloughs and attrition rather than terminations and focused on increasing work hours without increasing pay. The staffing function is typically accomplished at the organizational level.

Human resources development includes all activities to train and educate the employees. This function is heavily dependent on the fire officer at the company level. The development of employees begins when they first arrive at the department. The fire department orients the recruits to the fire department's methods of operation through a recruit class. The process continues once they arrive at their fire station assignment. The fire company officer will usually sit down with each recruit and orient him or her to the job and the ways things are done at the fire station.

Performance management; compensation and benefits; and employee health, safety, and security are covered in the Fire Officer II section of this chapter.

Utilizing Human Resources

■ Mission Statement

One basic principle is for the fire officer to know and understand the fire department's mission. Frequently, the fire department's mission is expressed through a written statement. The mission statement is a formal document that outlines the basic reason for the organization's existence and states how it sees itself. It is designed to guide the actions of all employees.

■ Getting Assignments Completed

One of the greatest demands on the fire officer is to make effective use of time. Fire officers will find that a great number of demands on the company's time are made. These demands include conducting public education, doing inspections, and undertaking other fire prevention efforts. They also include training and education of the crew members, and routine duties such as cleaning the station, doing paperwork, and maintaining the apparatus. And, of course, the company must respond to calls.

Some of these activities are known months in advance; others may require the immediate response of the crew. The daily schedule of some crews is in a constant state of flux due to the call volume. Other crews may be interrupted only occasionally. No matter what the situation, the fire officer can ensure maximum efficiency by applying good time management skills.

The first duty is to determine which activities are to be completed, when they must be completed, and how long it will take to complete them. The fire officer can then identify what needs to be done during the shift, the week, the month, and the year. Items that must be completed during the shift are a higher priority than those that must be completed next week. For example, completing the daily log is more important than completing a weekly inspection.

Occasionally, there is not sufficient time to complete all the required tasks during the shift. When this occurs, the fire officer must determine the fire department's priorities. This step allows the fire officer to determine which activities must be completed and which will have to wait. For example, meeting the deadline for the employee's timesheets may be a higher priority than getting the fire engine to the service center for an oil change.

Because many activities, such as an emergency call, are not known until they occur, the fire officer must plan ahead and build in the expected interruptions. The fire officer must not wait until the last minute to have inspections completed because he or she may have calls that prohibit this from happening. The sooner the scheduled activity is completed, the more flexibility the fire officer has.

Once the activities have been prioritized and it is determined when they must be accomplished, the fire officer must develop a plan that lays out how the activities will be accomplished. Some activities might require the entire company to

> **Assessment Center Tips**
>
> **Practice Techniques**
> An effective way to ensure that all tasks are properly handled in an assessment center is to practice the techniques while working as a fire fighter or supervising fire officer. A common error is not documenting a follow-up date to check on progress or use as a milestone target. Follow-up is documented in the response to an in-basket exercise by establishing a due date for the item as well as on the calendar.

> **Getting It Done**
>
> **Company Officer Delegation**
> There are seven steps in effective delegation:
> 1. Define your desired results.
> 2. Select the appropriate fire fighter.
> 3. Determine the level of delegation.
> 4. Clarify expectations and set parameters.
> 5. Give authority to match the level of responsibility.
> 6. Provide background information.
> 7. Arrange feedback during the process.
>
> Determining the level of delegation in Step 3 relates to the amount of decision-making authority provided to the fire fighter. The company officer has five options at his or her disposal:
> 1. Take action independently—no need to report back to the officer.
> 2. Take action and report back to the officer when done (as with a task assignment within the Incident Command System).
> 3. Recommend action that the company officer must approve.
> 4. Provide two or more recommended actions from which the company officer will choose.
> 5. Provide information about the pros and cons of different recommendations.

be involved in the activity, whereas others might require only a part of the crew to complete them. An example of splitting up the crew to accomplish multiple goals simultaneously would be sending one crew member to the store while a second member cleans the fire station kitchen and a third member completes the weekly inventory of supplies.

Having many tasks to coordinate can easily become overwhelming. One method to assist in making sure that activities are accomplished is to place all scheduled events on a monthly calendar. Inspections, public education, and special training should be noted, along with employee leave time. The calendar provides a visual method of tracking upcoming events.

Another method is to create a "daily" file. Within this file, a page describes each activity and specifies when it is to be completed. The file is organized from the earliest-due activity to complete to the latest. This method allows the fire officer to determine what needs to be done quickly without having to look through a pile of papers.

One of the best tools to improve time efficiency is delegation. As discussed previously, delegation allows subordinates to complete tasks they are capable of performing. These duties should be ones that allow the subordinate to grow. An example is the fire officer delegating the responsibility for ordering supplies to a fire fighter. Delegation allows the fire officer to focus on duties that cannot be delegated, such as performance appraisals.

Once a task is assigned, the fire officer must provide the fire fighter with regular follow-up. This can take many forms. At the station level, a verbal progress report is most often given. On more formal projects, the progress report may be in writing to provide long-term documentation.

Fire Officer II

Human Resources for Managers

The Fire Officer I section of this chapter listed seven typical functions of human resources. Three of the seven are covered in this section of the chapter:

- Performance management
- Compensation and benefits
- Employee health, safety, and security

The fire service is unlike most other private and public organizations in its ability to spend vast amounts of time solely devoted to developing its employees while at work. Ensuring daily training is one of the most basic responsibilities of the fire company officer.

<u>Performance management</u> is the process of setting performance standards and evaluating performance against those standards. Generally, the standards are set at the fire organizational or fire departmental level; however, the evaluation of the fire fighter's performance is a primary function of the fire company officer. This issue is covered in greater detail in the *Evaluation and Discipline* chapter.

Human resources management also includes the setting of <u>compensation and benefits</u>. Although the managing fire officer may not determine the compensation policy, the officer must understand how the system is designed and to which benefits the employees are entitled. Most fire departments operate on a step-and-grade pay system. In such an approach, the position of fire fighter is established on a particular pay grade level that comprises a number of steps. If the fire fighter demonstrates satisfactory performance, he or she will progress from one pay step up to the next until he or she eventually reaches the top step of the pay grade. The fire fighter will remain at this pay step and grade until he or she is promoted. He or she will then move to a new pay grade and progress up through the steps until he or she again reaches the top step.

The other most commonly used compensation systems include merit-based pay and skill-based pay. In merit-based pay systems, the fire fighter is typically paid a base amount and then receives additional compensation for good performance. For example, a fire fighter who receives an outstanding evaluation might receive a bonus of 5 percent of his or her annual pay. In contrast, skill-based pay systems typically pay a base amount and then give additional compensation for any skills the fire fighter can demonstrate. For example, a fire fighter who is also a hazardous materials technician would be paid an additional amount because of those special skills.

The 2009 recession drew attention to and placed pressure on public safety benefits programs because many fire departments provide more generous packages than private-sector employers. Many fire fighter benefits programs include a defined benefit retirement where the employee contributes a set percentage of his or her pay and receives a guaranteed benefit amount. Some include defined contribution retirement plans where the employee pays in a percentage of his or her pay and receives a benefit amount that is dependent on the performance of the account. Typically, these accounts take the form of a 457 deferred compensation plan. Other benefits include paid holidays; vacation leave; sick leave; and health, dental, vision, and life insurance.

<u>Health, safety, and security</u> include all activities to provide and promote a safe environment for fire fighters. The fire officer is responsible for many of these activities, such as ensuring that seat belts are utilized, proper procedures are followed, and personal protective equipment is used correctly. The fire department or organization may provide other health activities such as an employee assistance program, which is covered in detail in the *Evaluation and Discipline* chapter. The fire department may also offer tobacco cessation classes and incentives or diet and exercise information.

The fire officer's ability to utilize the human resources that are assigned is an essential function of the position. The fire officer must be able to accomplish the department's work with and through other people. In doing so, he or she must ensure that duties are carried out in accordance with departmental policies. Normally, this is done through direct supervision.

<u>Direct supervision</u> requires that the fire officer directly observe the actions of the crew. If the crew makes an interior attack, the fire officer is present. When a hose is loaded, the fire officer oversees the reloading. Direct supervision allows the fire company officer to ensure that departmental safety policies are used. The more dangerous the activity, the more direct supervision is required. For nonemergency activities, less direct supervision is needed.

The more direct supervision that is needed, the less efficient the crew. The goal of the fire officer is to reduce the need for direct supervision and increase the utilization of the fire company. Frequently, this goal is accomplished through a series of policies and procedures. Human resources policies and procedures guide the fire officer's decisions as they relate to personnel issues.

The Four Borders of Managing Human Resources

The fire officer must be familiar with the locations and topical areas that are covered in the fire department's human resources policies and procedures. Specifically, the organization may be subject to federal laws, a union contract, city regulations, and departmental policies that must all be followed. These four borders define the fire officer's human resources arena.

When presented with an issue, the fire officer must first determine which laws apply. The laws most commonly involved are the Fair Labor Standards Act (FLSA), the Civil Rights Act, the Age Discrimination in Employment Act (ADEA), the Americans with Disabilities Act (ADA), the Family and Medical Leave Act (FMLA), the Uniformed Services Employment and Reemployment Rights Act (USERRA), the Freedom of Information Act (FOIA), and the Health Insurance Portability and Accountability Act (HIPAA). The U.S. Department of Labor provides a great deal of information about the various laws that affect fire fighters. The fire officer may also want to discuss issues that are affected by these laws with the human resources department and legal counsel.

Once it is determined which laws affect a decision, the fire officer should review the fire department's policies to ensure compliance. If a labor contract exists, the fire officer must determine whether the activity is addressed in the contract. If the fire organization has a human resources department, it may prove a valuable asset in determining the proper action by the fire officer.

The fire officer must apply his or her actions with fairness and equity to the parties involved. For example, if an employee desires to schedule leave time in short increments, the fire officer must determine whether any laws, organizational policies, or departmental policies address the issue. If not, the fire officer must consider the impact of the request on the achievement of the department's mission. Lastly, the fire officer must consider whether the action is fair to all employees.

You Are the Fire Officer Conclusion

It has been 2 weeks since Chief Johnson directed Captain Davis to focus on meeting the chief's external satisfaction by complying with departmental goals. Station 100 has completed three target hazard walk-throughs, and the captain has paid closer attention to the daily skill drills.

The chief meets privately with Davis: "A challenge when considering new management concepts is making sure you are still meeting the departmental needs. The Circle of Consensus may be a useful tool as long as it results in meeting departmental goals. I was disappointed that your delegation of the in-station training did not appear to include monitoring the activity. Lieutenant Williams had the authority, but you retained the responsibility. The lieutenant is sharp but inexperienced. I need you to structure a professional development plan for Williams."

The chief has Lieutenant Williams join the meeting with the captain: "I appreciate the efforts to meet the departmental goals. Our challenge is to get the best from our people as effectively as possible. You know how you feel when you first arrive at a working structure fire, with smoke showing from many windows and a report of people trapped? You want to do everything at once and use every tool carried on the rig. There are not enough people or time to use all of the tools to complete critical fire-ground tasks. The smart company officer sizes up the situation and selects the best tool to accomplish the task."

Chief Johnson continues: "Management concepts are like the tools on the quint. You have collected a large set of management ideas, concepts, and practices in your management toolbox. You need to pick the best idea, concept, or practice that accomplishes the objective without pummeling fire fighters or eroding your authority."

You will see new management tools during your tenure as a company officer. Like new firefighting tools, they will have high promise. Until you evaluate how each new management tool works by understanding its history and development and trying it out in a training session, it remains an unknown resource.

Wrap-Up

Chief Concepts

- Management science is the systematic pursuit of practical results, using available human and knowledge resources in a concerted and reinforcing way.
- Scientific management breaks down work tasks into constituent elements. The timing of each element is based on repeated stopwatch studies that seek to standardize work tasks into simple, repeatable tasks.
- Frederick Winslow Taylor developed four principles of scientific management:
  - Replace rule-of-thumb work methods with scientific study.
  - Scientifically select, train, and develop each worker.
  - Cooperate with workers to ensure methods are being followed.
  - Enforce division of work: managers think, workers work.
- Humanistic management shifted the focus from scientific methods in favor of paying more attention to the workers, and to the working conditions that would make them more productive.
- A McGregor Theory X manager believes that people do not like to work, so they need to be closely watched and controlled.
- A McGregor Theory Y manager believes that people do like to work and that they need to be encouraged, not controlled.
- Maslow's hierarchy of needs is a ladder comprising five need levels. The supervisor's role is to meet the employee's needs so that the employee can proceed to the next level.
  - Level 1: Physiological
  - Level 2: Safety, security, and order
  - Level 3: Social needs and affection
  - Level 4: Esteem and status
  - Level 5: Self-actualization
- The grid theory of management assumes that every decision made and every action taken in the workplace is driven by people's values, attitudes, and beliefs. Blake and Mouton described five behavioral models based on a person's position on this grid:
  - Indifferent
  - Controlling
  - Accommodating
  - Status quo
  - Sound
- Managing fire fighters requires physical, financial, human, and time resources.
- Human resources planning is the process of having the right number of people in the right place at the right time.
- One of the greatest demands on the fire officer is to ensure the effective use of time.
- Performance management is the process of setting performance standards and evaluating performance against those standards. Although the standards are set at the fire organizational or fire departmental level, the evaluation of the fire fighter's performance is a primary function of the fire company officer.
- The fire officer's ability to utilize the human resources that are assigned is essential.
- Direct supervision requires the fire officer to directly observe the actions of the crew.

Hot Terms

Compensation and benefits Human resources system to identify and determine the pay, leave, and fringe benefits for each position in the organization.

Direct supervision A type of supervision in which the fire officer is required to observe the actions of a work crew directly; it is commonly employed during high-hazard activities.

Health, safety, and security Human resources activities intended to provide and promote a safe work environment.

Hierarchy of needs Maslow's description of human needs as a pyramid or ladder that starts with physiological needs and ends with self-actualization.

Human resources development All activities to train and educate employees.

Human resources planning The process of having the right number of people in the right place at the right time who can accomplish a task efficiently and effectively.

Humanistic management A management strategy that emphasizes human need and attitude; motivation comes from within the employee and not from authoritarian control. It leads to Maslow's hierarchy of needs.

Performance management The process of setting performance standards and evaluating performance against those standards.

Scientific management The breakdown of work tasks into constituent elements. The timing of each element is based on repeated stopwatch studies; the fixing of piece-rate compensation based on those studies; standardization of work tasks on detailed instruction cards; and generally, the systematic consolidation of the shop floor's brain work.

Staffing The process of attracting, selecting, and maintaining an adequate supply of labor, as well as reducing the size of the labor force when required.

Theory X McGregor's description of the management assumption that people do not like to work and must be closely watched and controlled.

Theory Y McGregor's description of the management assumption that people like to work and need to be encouraged, not controlled.

Wrap-Up, continued

References and Additional Resources

NFPA reprinted material is not the complete and official position of the NFPA on the referenced subject, which is represented only by the standard in its entirety.

Carter, H. R., and E. Rausch. (2008). *Management in the Fire Service,* 4th ed. Quincy, MA: National Fire Protection Association.

Drucker, P. F. (1995). "Really Reinventing Government." *Atlantic Monthly.* 50–52.

Edwards, S. T. (2009). *Fire Service Personnel Management,* 3rd ed. Upper Saddle River, NJ: Brady/Prentice Hall Health.

England, R. E., et al. (2012). *Managing Urban America,* 7th ed. Washington, DC: CQ Press.

Gentry, J. (2012). *Understanding Leadership's Role with Employee Engagement.* Applied Research Project. Emmitsburg, MD: U.S. Fire Administration, Executive Fire Officer Program.

Gilbreth, F. (1912). *Primer of Scientific Management.* New York, NY: Van Nostrand.

Greenberg, A. S. (1998). *Cause for Alarm: The Volunteer Fire Department in the Nineteenth-Century City.* Princeton, NJ: Princeton University Press.

Haboush, D. G. (2011). *Adopting Organization Values in the Carmel Fire Department.* Applied Research Project. Emmitsburg, MD: U.S. Fire Administration, Executive Fire Officer Program.

Heil, G., et al. (2000). *Douglas McGregor, Revisited: Managing the Human Side of the Enterprise.* New York, NY: John Wiley & Sons.

Kanigel, R. (1997). *The One Best Way: Frederick Winslow Taylor and the Enigma of Efficiency.* New York, NY: Viking.

Kiechel, W. K. III. (2012). "The Management Century." *Harvard Business Review.* 90(11): 13.

Maslow, A. H. (1954). *Motivation and Personality.* New York, NY: Harpers.

Maslow, A. H. (1998). *Maslow on Management.* New York, NY: John Wiley & Sons.

McGregor, D. (1960). *The Human Side of Enterprise.* New York: McGraw-Hill.

Mintzberg, H. (2011). *Managing.* San Francisco, CA: Berrett-Koehler.

Nelson, D. (1980). *Frederick W. Taylor and the Rise of Scientific Management.* Madison, WI: University of Wisconsin Press.

Osborne, D., and T. Gaebler. (1992). *Reinventing Government: How the Entrepreneurial Spirit Is Transforming the Public Sector.* Cambridge, MA: Perseus.

Peters, T., and R. Waterman. (1982). *In Search of Excellence: Lessons from America's Best-Run Companies.* New York, NY: HarperCollins.

Roethlisberger, F. J., et al. (1939). *Management and the Worker: An Account of a Research Program Conducted by the Western Electric Company, Hawthorne Works, Chicago.* Boston, MA: Harvard University Press.

Scott, G. (2006). *Human Resources Management: Managing Volunteer and Combination Emergency Service Organizations.* York, PA: W. F. Jenaway. VFIS: 95–113.

Smith, K. B., and A. Greenblatt. (2013). *Governing States and Localities,* 4th ed. Washington, DC: CQ Press.

Snook, J. W., et al. (2011). *A Leadership Guide for Volunteer Fire Departments,* 4th ed. Sudbury, MA: Jones & Bartlett Learning.

Taylor, F. W. (1911). *The Principles of Scientific Management.* New York, NY: Harper.

Wilson, W. (1887). "The Study of Administration." *Political Science Quarterly* (2): 197–222.

FIRE OFFICER in action

Promotions and transfers have placed a new driver and rookie fire fighter on Quint 100. The driver is an experienced pump operator but never worked on a ladder company or quint. He is inconsistent and not confident when operating the aerial. The rookie is enthusiastic and prone to overexcitement. Lieutenant Williams is working with both of them to improve performance.

1. Lieutenant Williams directs the driver to have the rookie practice donning fire gear and SCBA, grab the preassigned tools, and exit the quint while in the station. This is an example of which management principle?
- **A.** Scientific management
- **B.** Theory X management
- **C.** Theory Y management
- **D.** Hierarchy of needs

2. Quint 100 is operating at a commercial structure fire and is assigned to open the roof. What is the best management principle for Lieutenant Williams to use during this task?
- **A.** Scientific management
- **B.** Theory X management
- **C.** Theory Y management
- **D.** Hierarchy of needs

FIRE OFFICER *in action* (continued)

3. Quint 100 has been performing difficult overhaul operations, with each member going through two SCBA bottles during this task. Lieutenant Williams tells the group officer that the crew needs relief. This is an example of which management principle?
   A. Theory Y management
   B. Maslow's safety, security, and order level
   C. Maslow's physiological needs level
   D. Blake and Mouton's balance and compromise

4. Station 100 winning the Fire Fighter Olympics competition is an example of which management principle?
   A. Blake and Mouton's direct and dominate
   B. Blake and Mouton's contribute and commit
   C. Maslow's esteem and status level
   D. Maslow's self-actualization level

Fire Captain *Activity*

Captain Davis delegated the in-station training program to Lieutenant Williams. A 1-hour skill drill is scheduled for every workday, focused on performing 30 standardized fire-ground suppression activities. At the 6-month assessment by the battalion chief, Station 100 fire fighters are delivering below-standard performance on fire company evolutions involving hose deployment. The crews take too much time and miss critical elements. Williams has been spending the daily skill drill time to prepare for the Fire Fighter Olympics advanced forcible entry and ladder contests.

NFPA Fire Officer II Job Performance Requirement 5.2.1
Initiate actions to maximize member performance and/or to correct unacceptable performance, given human resource policies and procedures, so that member and/or unit performance improves or the issue is referred to the next level of supervision.

Application of 5.2.1
1. Using human resources policies and procedures from a fire department familiar to you, develop an action plan for Lieutenant Williams that will improve Station 100 fire fighter performance in 10 hose deployment and water supply standardized evolutions.
2. The battalion chief has scheduled a formal reevaluation of Station 100 fire fighters in 7 weeks. Lieutenant Williams has 16 workdays to prepare the crew. Prepare a memorandum to the battalion chief describing the action plan.

NFPA Fire Officer II Job Performance Requirement 5.2.2
Evaluate the job performance of assigned members, given personnel records and evaluation forms, so each member's performance is evaluated accurately and reported according to human resource policies and procedures.

Application of 5.2.2
1. Using human resources policies and procedures from a fire department familiar to you, appropriately document the 6-month performance of Lieutenant Williams in running the in-station training program.

NFPA Fire Officer II Job Performance Requirement 5.2.3
Create a professional development plan for a member of the organization, given the requirements for promotion, so that the individual acquires the necessary knowledge, skills, and abilities to be eligible for the examination for the position.

Application of 5.2.3
1. Using human resources policies and procedures from a fire department familiar to you, provide a professional development plan to Lieutenant Williams to improve fire fighter training outcomes and prepare for a captain promotional process. Prepare a memorandum for Lieutenant Williams describing the plan and available resources.

Leading the Fire Company

Fire Officer I

Knowledge Objectives

After studying this chapter, you will be able to:

- Describe the role of the fire officer as both a leader and a follower. (pp 128–129)
- Identify leadership styles used in the fire service (NFPA 4.2.1) (NFPA 4.2.2). (p 129)
- Identify types of power used in leadership. (pp 129–130)
- Describe leadership in routine situations (NFPA 4.2.2). (p 130)
- Describe emergency scene leadership (NFPA 4.2.1). (pp 130, 132–134)
- Describe the leadership challenges related to the fire station work environment. (pp 134–135)
- Describe the leadership challenges related to the volunteer fire service. (p 135)

Skills Objectives

After studying this chapter, you will be able to:

- Demonstrate planning and setting priorities (NFPA 4.2.6). (pp 129–130)
- Demonstrate clear and concise communication (NFPA 4.4.4). (pp 130, 132–134)
- Demonstrate allocation of resources (NFPA 4.6.1). (p 134)
- Supervise and account for assigned personnel under emergency conditions (NFPA 4.6.1) (NFPA 4.2.2). (pp 130, 132–133)

Fire Officer II

Knowledge Objectives

After studying this chapter, you will be able to:

- Identify the principles of motivation (NFPA 5.2.1). (pp 135–136)
- Describe the reinforcement theory of motivation. (pp 135–136)
- Describe the motivation–hygiene theory of motivation. (p 136)
- Describe the goal-setting theory of motivation. (p 136)
- Describe the equity theory of motivation. (p 136)
- Describe the expectancy theory of motivation. (p 136)

Skills Objectives

After studying this chapter, you will be able to:

- Demonstrate an ability to increase teamwork (NFPA 5.2.1). (pp 135–136)

CHAPTER 7

Principles of Fire and Emergency Service Administration (FESHE) Course Objectives

2. Recognize the need for effective communication skills both written and verbal. (pp 132–134)
3. Identify and explain the concepts of span and control, effective delegation, and division of labor. (pp 130–134)
5. Explain the history of management and supervision methods and procedures. (pp 135–136)
6. Discuss the various levels of leadership, roles, and responsibilities within the organization. (pp 128–130, 132–134)
7. Describe the traits of effective versus ineffective management styles. (pp 128–130, 132–133, 135–136)
8. Identify the importance of ethics as it relates to fire and emergency services. (pp 134–135)

You Are the Fire Officer

An early-morning response to an activated smoke detector in an industrial park became a greater alarm structure fire. Quint 100's crew was rocked by a fire roll-over when members opened an interior door in the building. As the crews clean up from the fire, Battalion Chief Johnson meets privately with Captain Davis and Lieutenant Williams.

The chief is angry that the Engine 100 crew entered the building with no tools or self-contained breathing apparatus (SCBA) and did not follow the standard operating procedure (SOP) on commercial fire responses. The chief notes that after the quint discovered the working fire, there was significant delay in getting water to the fire. Captain Davis is directed to improve engine company response to commercial fires, and Lieutenant Davis is ordered to improve the quint's capabilities on attack line deployment.

1. How can you improve company performance?
2. How can you improve fire fighter response to a changing incident?

Introduction

Leadership has been described by Gary Yukl as "the process by which a person influences others to understand and agree about what needs to be done and how to do it, and the process of facilitating individual and collective efforts to accomplish shared objectives" (Yukl 2013). A person carries out this process by applying his or her leadership attributes, such as beliefs, values, ethics, character, knowledge, and skills. Although your position as a fire officer gives you the authority to accomplish certain tasks and objectives in the organization, this power does not make you a leader—it simply makes you the supervisor. Becoming a leader starts from within and builds from your first childhood experience when learning to share with others. Years of observation, effort, and experience shape leadership skills and abilities. Bosses tell people to accomplish a task or objective, whereas leaders make them want to achieve high goals and objectives. Your goal should be to lead, not just to be a boss.

Fire Officer I

The Fire Officer as a Follower

Leaders can be effective only to the extent that others are willing to accept their leadership. This characteristic is sometimes described as followership. It is self-evident that a leader cannot lead if others will not follow. An effective leader uses persuasiveness and motivation to overcome resistance. In some cases, for a variety of reasons, others simply refuse to follow.

The fire officer has to be both a leader and a follower. On the one hand, the officer leads the fire company to achieve the goals and objectives that have been established by the department or the jurisdiction. On the other hand, the officer has to follow leadership that comes from a higher level, even if it is not always pointing in a direction where the officer would prefer to go. In many cases, the officer is the messenger who has to deliver unpopular news to the company members to ensure compliance with instructions that came from a higher level. The *Fire Fighters and the Fire Officer* chapter discusses how the fire officer handles an unpopular order.

Followership is particularly important for a fire officer because subordinates are always aware of what the fire officer does. If the fire officer demonstrates selective following of orders from the fire chief, deciding which ones to follow and

> ### Fire Marks
>
> **The Role of the Officer Is to Keep Calm**
> David Halberstam wrote *Firehouse*, a book about his Upper West Side neighborhood fire station, Engine 40 and Ladder 35. Only one of the 13 fire fighters from this station returned from the World Trade Center collapse on September 11, 2001.
>
> During a June 3, 2002, online interview conducted by *The Washington Post*, Halberstam discussed the role of the fire officer. It is an observation not found in the book (The Washington Post 2002):
>
> > One of the key things in a firehouse is the role of the officer and the need on their part in whatever danger, in the worst kind of crisis, to stay calm. Because if they're not, and if they show fear, it will ripple right through the men. And so the men will say of an officer, "He's a very good officer. He always stayed calm." . . . the captain is the first in, by tradition, and the last out.

Getting It Done

Making a Decision

Leaders make decisions. Here is a five-step approach:
1. **Clear your mind.** Take a deep breath or pause to concentrate on the issue.
2. **List options.** Brainstorm to develop a long list.
3. **Weigh the outcomes.** Is the best possible outcome worth the worst possible risk? What is the likelihood of the outcome? Which physical, legal, organizational, or moral barriers exist with the choice?
4. **Make a choice.** Accept responsibility for the decision and have a "Plan B" identified.
5. **Evaluate decision.** Did the decision have the desired outcome?

FIGURE 7-1 An autocratic approach to leadership is required in a high-risk, emergency scene activity.

which ones to ignore, he or she sends a clear message to the company members: It is acceptable for the fire fighters to be selective in following orders from their fire officer. Followership is an important character trait that will serve the officer well in the future when dealing with others who have not followed the rules because they do not seem fair.

Leadership Styles

There are many ways to look at leadership, including how to describe a leader, what leaders do, and how leaders act. One approach to better understanding a leader is to look at the possible styles of leadership. Most sources agree that an effective leader changes his or her style of leadership based on the specific situation. The three major leadership styles are traditionally identified as autocratic, democratic, and laissez-faire.

■ Autocratic

An iron-hand approach is used when the fire officer needs to maintain high personal control of the group. In such a case, the fire officer is telling subordinates what to do and is expecting immediate and complete adherence to the issued instructions.

The autocratic style of leadership is required in two situations. The first situation occurs when the fire company is involved in a high-risk, emergency scene activity, such as conducting a primary search during a structure fire **FIGURE 7-1**. There is no time for discussion, and this is not the situation to experiment with alternative approaches.

The second situation occurs when the fire officer needs to take immediate corrective supervisory activity, such as during a "control, neutralize, command" response to a confrontation. In this scenario, the officer must be firmly in control of the situation.

■ Democratic

A consultative approach takes advantage of all of the ingenuity and resourcefulness of the group in determining how to meet an objective or complete a task **FIGURE 7-2**. The officer should use the democratic style of leadership when planning a project or developing the daily work plan of the company. This approach can also be used in some low-risk emergency scene operations.

Specialized and highly technical fire companies often use the democratic approach when faced with a complex or unusual emergency situation. They depend on the skills and experience of the individual team members to analyze the situation, consider the alternatives, and develop the incident action plan. Execution of the plan often involves an autocratic command style, however.

■ Laissez-Faire

A free-rein style of leadership moves the decision making from the fire officer to the individual fire fighters. The fire officer depends on the fire fighters' good judgment and sense of responsibility to get things done within basic guidelines. The laissez-faire approach is an effective leadership style when working with experienced fire fighters and when handling routine duties that pose little personal hazard.

Power

Power is the capacity of one party to influence another party. One of the first works on the management of power was *The Prince*, Niccolo Machiavelli's masterpiece posthumously published in 1532.

FIGURE 7-2 The democratic approach uses the resourcefulness of the group in determining how to meet an objective.

Company Tips

Delegating Routine Activities
A goal of an effective fire officer is to push decision making to the lowest possible level. Because a fire officer gains experience as a leader and builds confidence and trust with a team, many routine activities can be delegated. For example, one fire fighter could be assigned to manage in-station training activities, another could oversee routine station maintenance, and a third could keep preincident plan files updated. The fire officer continues with the morning line-up and evening check-up, but the individual fire fighters perform their assignments autonomously. This sort of delegation allows the officer to focus on activities that require his or her personal attention.

French and Raven describe types of power in their 1959 *Studies of Social Power* as the result of the "target person's" response from the "agent" making a request.

- **Legitimate power.** The target person believes that the agent has the right to make the request and the target person has the obligation to comply. For example, under the Incident Management System, the incident commander has the legitimate power to reassign the ventilation sector.
- **Reward power.** The target person complies to obtain rewards believed to be controlled by the agent.
- **Expert power.** The target person complies due to a belief that the agent has special knowledge.
- **Referent power.** The target complies due to admiration of or identification with the agent and seeks approval.
- **Coercive power.** The target person complies to avoid punishment believed to be controlled by the agent.

Professor Gary A. Yukl at the University of Albany researches leadership, power, and influence. He updated the French and Raven taxonomy to define two types of power: personal and positional. Personal power, which includes expert and referent power, reflects the effectiveness of the individual. Positional power, in contrast, is defined by the role an individual has within the organization. Legitimate, reward, and coercive power are the three examples of positional power. Yukl provides two additional position-based power descriptions:

Getting It Done

Transitioning from Democratic to Autocratic Leadership
The increased deployment of specialized and highly technical emergency services creates a unique challenge for the fire officer. The best way to size up a complicated situation and develop an incident action plan is to use the democratic style of leadership, using the knowledge and experience of all responders to develop the best approach. Once the operation begins, however, the fire officer needs to assume the autocratic role to ensure the safe execution of the plan.

- **Information power:** Control over information. Unlike expert power, information power is based on the target person's assessment of the agent's ability to discover or obtain relevant information rapidly and efficiently, usually through a cultivated network of sources.
- **Ecological power:** Control over the physical environment, technology, or organization of work. The target person's behavior is based on perceptions of opportunities and constraints.

Leadership in Routine Situations

Most fire officer leadership activity is directed toward accomplishing routine organizational goals and objectives in nonemergency conditions. This includes being well prepared to perform in emergency situations. The fire officer accomplishes most of these goals and objectives through the efforts of fire fighters. The role of the officer involves influencing, operating, and improving to ensure that those efforts achieve the desired results.

Fifty years ago, fire officers used an autocratic style of leadership in both emergency and nonemergency duties. The officer decided what was to be done and who was to do it, and the fire fighters responded without question. Today, the workforce is very different. Employees demand to be included in the decision-making process. Consequently, effective fire officers provide a more participative form of leadership in routine activities.

Although the fire officer is commonly given specific assignments that the company must complete, much discretion may apply regarding when, how, and by whom activities are carried out. Some officers sit down with their crews at the start of the shift and cover the assignments, if any, that the administrative fire officer has indicated need to be completed for the day. The officer may also outline the status of other duties that will need to be completed over the long term but do not need to be undertaken immediately. An open question is then addressed to the crew as to what they think should be done, including any areas that they believe should be addressed. Depending on the outcome of the discussion, the officer decides or has the group decide on the plan for the day to accomplish the activities.

Some tasks must be completed every day. In most departments, station cleaning is performed daily, making a trip to the grocery store is required, and equipment must be checked. The fire officer may decide that these duties are easiest to accomplish if they are assigned in advance. Typically, in conjunction with the crew, a determination is made about which duties will be completed every day based on the work position of the individual. The driver is usually assigned to check all equipment and clean the apparatus bay. The rookie fire fighter may be assigned to the bathroom and kitchen detail, whereas the senior fire fighter is assigned to sweeping and vacuuming floors.

Emergency Scene Leadership

A core responsibility of a fire officer is to handle emergencies effectively. The chapter *Managing Incidents* covers structural components of incident management in more detail. This chapter examines some of the leadership issues that apply specifically to a fire officer.

VOICES OF EXPERIENCE

Leadership, including how it relates to "leading the company," is as mysterious and confrontational a topic as the firehouse conversation of smooth-bore nozzle versus fog nozzle. Leadership is getting the troops to do what we need to attain the goals of the organization or department. How do we get them to do the assigned tasks—and, better yet, how do we get them to *enjoy* those tasks? We as leaders are tasked with getting the job done, which is more easily accomplished on the fire ground than it is around the firehouse. In our paramilitary structure, the fire ground is the battlefield. Order, unity, honor, and a strong command presence are a must. We lead by example; we lead from the front. We listen to what the troops tell us. As command officers, we must trust our troops. As company officers, we must trust our troops. Where does this trust come from?

> **Our challenge as leaders is to find that desirable trait in everyone we work with.**

Many more decisions are made *away* from the fire ground than *on* the fire ground. In training, it is important to allow the troops to make a mistake (as long as it does not result in injury). If you, as their leader, make a mistake, do not be afraid to say, "I blew that." In making a mistake, you learn a very important lesson, and can then relay that lesson to the troops. You might be surprised of the extent to which your creditability as an officer will flourish.

We ask fire fighters every day to place their lives in harm's way to potentially save someone they do not know. When was the last time we told our troops "thank you"? I firmly believe that, both on and off the fire ground, the best leaders focus on what their people are doing right and reward them for it. A simple "thank you" can have an immense impact. Treating the troops with respect and honor will pay dividends in the long run.

Good leaders know their people—both their strengths and their weaknesses. A good leader can take the most inexperienced employee and determine the positive elements that person can bring to the table, expand on the positive traits, and motivate that employee to become a critical and vital member of the team. Our challenge as leaders is to find that desirable trait in everyone we work with. The true leader can develop this skill and apply it to all ranks and members of their organization.

Stephen K. Lovette
Captain
High Point Fire Department
High Point, North Carolina

A fire officer always has direct leadership responsibility for the company that he or she is commanding. The first-arriving fire officer has additional important responsibilities to establish command of the incident and to provide direction to the rest of the responding units. The first-arriving fire officer also has leadership responsibilities that relate to the communications center.

■ Methods of Assigning Tasks

The fire officer's primary responsibility is to the team of fire fighters under his or her direct supervision. The officer is responsible for their safety, their actions, and their performance at the incident scene. The fire officer should develop a consistent approach to emergency activities, in which all of the department's operating procedures are followed.

Most departments have some form of standard operating procedures (SOPs) that specify what is expected of each company and what to do in a given situation. Some SOPs are very specific and detailed, whereas others are more general. For departments with very detailed procedures, the task that each fire fighter is to perform may already be determined. The SOPs may indicate which tools the fire fighter is to carry based on whether the company includes three or four members. In these cases, the fire officer is responsible for ensuring that the fire fighters know and follow the policy.

For departments that have broad SOPs or none at all, the fire officer has two choices in assigning tasks: preassigning them or assigning them as needed on the scene. The advantage of preassignment is that it relieves the officer of the burden of having to make some decisions during the emergency. The advantage of assigning tasks on the scene is that it reduces unnecessary effort.

When giving assignments at the scene, the fire officer must be clear and concise and ensure that all fire fighters understand their assigned tasks. In this situation, the fire officer uses an autocratic style of leadership because the emergency scene does not allow for participative decision making. For this method of operation to be effective, the fire fighters must clearly understand the reason for the use of different leadership styles before the incident. One method that may help a fire officer make assignments under pressure is to develop a standard method of handling situations. Making standard decisions in a consistent manner will assist in the decision-making process. If you cannot use a written checklist, then developing a mental checklist that is rapidly completed at every incident will create a predictable outcome. Because so many factors come into play when determining tactics, it is crucial that officers be trained in the critical decision-making process. For example, if you need to assign someone to rescue responsibilities, always consider first whether you have a rescue company available; if not, then a truck company and lastly an engine company can be used to handle this task. This understanding allows a sequential thought process to develop, making assignments easier.

One method of assigning tasks that allows for more participation in the process yet reduces the number of decisions the officer must make at the scene is to rely on the broad SOPs, such that the fire officer discusses with the crew the various "routine" emergencies that they respond to. The officer and crew can discuss the needs of the incident and the steps that must be taken to mitigate it. Finally, the officer and crew can decide who will be responsible for doing what.

For example, the crew may decide that at any life-threatening medical incident, the fire fighter is responsible for checking the victim's respirations, securing the airway, and then providing ventilation. The driver is responsible for checking for a pulse and uncontrolled bleeding and then providing compressions. The officer is responsible for setting up and using the automated external defibrillator. This plan ensures that the crew covers all the basics of care without fumbling and waiting to be directed.

■ Critical Situations

Dangerous situations may develop suddenly during incident operations. A fire officer must use the autocratic leadership style when immediate action is required. Orders to evacuate a building and a fire fighter mayday are two critical events that require immediate autocratic action.

When command issues an evacuation order, all members of the fire company should immediately leave the fire building.

Safety Zone

Practice Like You Play

A key to consistent performance at emergency incidents is to "practice like you play." Every training activity and every response should be approached with the same degree of professionalism and strict adherence to SOPs. This approach leads to the company performing as expected when the situation demands maximum effort.

Most structure fire responses address relatively minor situations, such as activated fire alarms and fires that involve less than a room and contents. To prepare the fire company for larger fires and critical situations, each response should require the same consistent approach. Following are some possible examples of response actions:

- All crew members don their protective clothing, sit down, and fasten their seat belts before the apparatus moves, and they remain seated and belted as long as the apparatus is moving.
- The first-arriving engine company always establishes a water supply.
- The first-arriving ladder company positions and prepares to deploy the aerial apparatus.
- Crews entering the building wear full protective gear, including self-contained breathing apparatus (SCBA); carry their tools; and perform their assignments for a working fire.

If there is any doubt about whether a fire is present in the structure, all operations should be performed with the assumption that a fire truly exists. Without creating a customer service problem, the fire company should search the entire building, including attics and basements. This approach allows each fire fighter to develop awareness of the built environment. More importantly, when the company encounters a working fire, the standard fire-ground tasks have been practiced and are familiar.

Near-Miss REPORT

Report Number: 11-0000414

Event Description: Early this morning, fire fighters responded to assist a neighboring department at a fire involving a multiple-occupancy dwelling. Visibility was poor. An officer led a crew of three to the fourth floor, where they located a victim who had become lost and disoriented in the thick smoke. The officer and fire fighter [A] led the victim to the stairwell, while fire fighter [B] continued to search for another victim, who was said to be in one of the apartments on that floor. The victim who was being led out fell unconscious and pinned the officer in the stairwell. Fire fighter [A] worked to free the officer (no mayday was called and command was not notified of a found victim), and they were able to drag the victim down to an exit. Having located a second victim, fire fighter [B] notified command and requested help to remove the second victim to safety.

Hearing the call for help, the officer and fire fighter [A] returned to the fourth floor, along with another officer. The second officer had been part of the initial attack crew and was out of air. Together, the four fire fighters carried the unconscious victim from the fourth floor to awaiting medics at the rear entrance of the building. Three out of the four fire fighters involved in the rescue were hospitalized for smoke inhalation and heat exhaustion. No accountability system was in place, and command was not aware of the crew splitting while conducting a search.

Brackets [ ] denote reviewer de-identification.

Lessons Learned: Strong incident command, freelancing, and the lack of accountability played a big part in the second officer entering an immediately dangerous to life and health (IDLH) area without an ample air supply to do the work that had to be done. Lack of experience and lack of situational awareness played a role in the first officer splitting the crew and leaving one fire fighter alone to conduct a search. Neglecting to call a mayday when becoming trapped by a victim or notifying command of the situation was also an issue.

Become aware of a mayday and when to call a mayday. A strong accountability system in place can make a difference. Review the department SOPs for radio communication. Discipline among company officers and attitude reflect leadership.

Evacuation may mean that the company leaves a fire hose in the building and the fire fighters quickly exit the building. The fire officer is responsible for accounting for the fire fighters under his or her supervision. The fire officer provides a head count to the group command officer or the incident commander.

A mayday requires a complex response from the company operating within a hot zone or burning structure. The first obligation is to maintain radio discipline so that command can determine the mayday location and situation. The second obligation is to maintain company or group integrity. Incident control evaporates if every fire fighter rushes to assist the fire fighter in trouble. The fire officer needs to maintain company integrity to facilitate the mayday rescue and continue to fight the fire or control the hazard. Changes in assignment come from the group incident commander.

The fire officer may directly encounter a dangerous situation, such as a collapsing building, a person with a gun, or a careening vehicle. In such a case, the officer must make an immediate, clear, and autocratic command to protect the fire fighters under his or her supervision by removing them from the danger area. Then, the fire officer needs to inform command.

Assistant Chief Joseph Pfeifer, when reviewing a rescue of a driver trapped in a sanitation truck that crashed through a fourth-floor exterior wall in New York City, made this observation: "As a novel incident increases in scale and complexity, it is important for incident commanders to separate management and leadership functions. Leadership is about stepping back and detaching from the management of the incident to analyze what is taking place and project future actions; then the leader reconnects to guide management in adapting to novelty" (Pfeifer 2012).

After every incident, the fire officer should briefly review the event while still at the scene or as soon as possible after returning to quarters. This is an opportunity to clarify issues and answer questions. The fire officer can reinforce good practices and immediately identify any unacceptable performance.

■ The Dispatch Center

The first-arriving company at an incident needs to describe conditions to the dispatch center concisely. The content of the initial radio report needs to meet the departmental operating procedures. In addition, this verbal picture establishes the tone of the incident. A fire officer who provides a calm and complete description of the situation on arrival demonstrates leadership and ability to control the action that will occur. An officer who gives the impression of confusion, indecision, or uncertainty fails the first test of leadership.

If a specific requirement for the initial communications has been established, a radio report should meet those criteria. Otherwise, the report should include the following information:

- Identification of the company arriving at the scene.
- A brief description of the incident situation. This may include providing information about the building size, occupancy, hazardous chemical release, or multiple-vehicle accident.
- Obvious conditions, such as a working fire, multiple victims, a hazardous materials spill, or a dangerous situation.
- Brief description of action to be taken: "Engine 2 is advancing an attack line into the first floor."
- Declaration of strategy to be used: offensive or defensive.
- Any obvious safety concerns.
- Assumption, identification, and location of command.
- Request or release resources as required.

A calm, concise, and complete radio report helps the dispatch center, the chief officer, and other responding companies to understand the situation at hand. It is important to use radio terminology that meets the department's operating procedures and is clear to everyone who is listening. Close compliance with the communications SOPs is an important component of effective fire-ground command.

A visit to the dispatch center is a valuable experience for every fire officer. It is important to understand the dispatch center's operating conditions and to appreciate how dispatchers depend on fire officers at the incident scene to provide good information.

■ Other Responding Units

The first-arriving fire officer provides leadership and direction to responding units by implementing the incident management plan. The fire officer must demonstrate the ability to take control of the situation and provide specific direction to all of the units that are operating and arriving. This requires mastery of the autocratic style of leadership.

FIGURE 7-3 Wearing a uniform alerts the fire fighters that a formal supervisory task is at hand.

Getting It Done

Looking Official When Performing Formal Fire Officer Duties

Many departments provide two different uniforms that can be worn by the fire officer. The station work uniform could be a golf shirt, or "job" shirt. The more formal "Class A" uniform generally includes a long-sleeve collared shirt with insignia, badge, and tie.

One method to help fire fighters understand when the officer is performing an official supervisory task is to wear the uniform shirt. If the officer always puts on a dress shirt and tie to conduct employee evaluations or disciplinary actions, the serious and official nature of these activities is clearly visible. This increases the fire officer's effectiveness **FIGURE 7-3**.

Leadership Challenges

The fire officer works in a dynamic environment with changing conditions and evolving organizational needs. The basic leadership concepts prevail when considering two unique challenges: the fire station as a work location and leading a volunteer fire company.

■ Fire Station as Municipal Work Location Versus Fire Fighter Home

Fire fighters work rotating shifts together and prepare and share meals. There are days of dull routine, punctuated by episodes of intense excitement. These factors create a powerful and special community among fire fighters that is difficult to compare with any other workplace environment. The fire station becomes a "home away from home," and fire fighters become part of an extended family.

Although this workplace environment produces a special type of bonding among fire fighters, it can easily lead to a variety of productivity problems, as well as behavioral and personality traits that are often associated with a dysfunctional family. The resulting situations can be quite different from the problems that occur in a "normal" work environment, and the behavioral issues are often more severe. Case law and administrative actions reinforce the notion that the fire station is a local government facility, subject to the same rules and expectations as any other workplace. The formal organization expects fire fighters who are in the station at 3 A.M. to behave the same way as the administrative services staff behaves at 3 P.M.

A fire officer must balance the expectations of the employer with the realities of a fire station work environment and the desire to create an effective team. A certain amount of spirited behavior can be healthy, as long as basic rules are observed. The fire officer has to maintain order and ensure that whatever occurs can be explained and would be acceptable to a rational observer.

Two general rules for nonemergency activities are as follows:

1. Do not compromise the ability of the fire company to respond to emergencies in its district.
2. Do not jeopardize the public's trust in the fire department.

Responding to emergencies and preserving the public trust are two values that every fire fighter needs to demonstrate. Within those values is wide latitude for team-building extracurricular activities.

Leadership in the Volunteer Fire Service

Fire officers leading volunteer fire companies have to rely on their leadership skills even more than their municipal counterparts do. Pride, group identity, and personal commitment are key factors that keep volunteers active and loyal to the organization—there is no paycheck that compels a volunteer to endure an unpleasant situation. Consequently, the volunteer fire officer must pay attention to the satisfaction level of every member and be alert for issues that create conflict or frustration.

The relationship between the volunteer fire officer and fire fighters is more like that of a mom-and-pop family business than that of a municipal agency. Effective leadership is often the strongest force that influences members' performance and commitment to the organization. If the negative aspects outweigh the positive aspects, an individual can step aside or drop out.

Among the unique issues that a volunteer fire officer must consider are the following:

- Changes in employment, family situations, or child/elder care profoundly affect the time a volunteer is able to devote to the fire department.
- Extensive training requirements can have an impact on volunteer availability. It is easy to assign a municipal employee to attend a 40-hour training class that occurs during the regular workweek, but it can prove difficult to accomplish the same training when the opportunities to do so are restricted to evenings and weekends.
- Interpersonal conflicts can develop between members. Conflicts can drive away members and erode fire company preparedness. The volunteer officer must act quickly when such problems are identified. The rights and reasonable expectations of individual members must be considered, without compromising the mission or the good of the organization.

A fire commissioner serving in a large county observed four phases of volunteer participation:

1. **Large loss of applicants during initial fire fighter training.** The candidates could not make the time commitment, were physically unprepared, or could not pass the cognitive certification exam.
2. **Small loss during the probationary period.** During the first year after joining the department, members are typically consistent participants in emergency and administrative activities.
3. **Moderate to high loss of fire fighters between the third and sixth years of membership.** The twenty-somethings are finishing college, getting married, having children, and building their careers. The ones who do not leave cannot devote as much time to the volunteer service as they once did.
4. **Recommitment between the 15th and 18th years of membership.** Those volunteer fire fighters who never left significantly increase their time with the department, and many who left start returning. The children are grown; the career is established. The commissioner noted that this group forms the core of the volunteer fire department.

According to this fire commissioner, his county reported working with a total of 1000 volunteers annually. It processed 250 applications per year, and lost 250 members per year. Six hundred members had certification to operate as fire fighters, apparatus operators, company commanders, and chief officers. That group averaged 36 hours of service per month. Within that group were 125 individuals who were the pillars of the volunteer organization, providing 80 to 200 hours of service per month.

A special concern in volunteer organizations is the political balance that results from electing officers. A conscientious volunteer officer has to use strong leadership and the courage of conviction to implement an unpopular policy.

Fire Officer II

Motivation

One of the key components of leading is the ability to motivate. The chapter *Understanding People: Management Concepts* introduces some of the concepts related to motivation. The fire officer must be able to apply methods to inspire subordinates to achieve their maximum potential. Although each method has a slightly different focus, five overriding principles apply:

- Recognize individual differences.
- Use goals.
- Ensure goals are perceived as attainable.
- Individualize rewards.
- Check for system equality.

Reinforcement Theory

Probably the best-known motivational theory is the reinforcement theory. This theory suggests that behavior is a function of its consequences. To motivate employees to perform, the officer must provide reinforcers to encourage the employee to act in the desired manner. Reinforcement must immediately follow an action to increase the probability that the desired behavior will recur. Four types of reinforcers exist:

- **Positive reinforcement:** Giving a reward for good behavior
- **Negative reinforcement:** Removing an undesirable consequence of good behavior
- **Extinction:** Ignoring bad behavior
- **Punishment:** Punishing bad behavior

Whereas positive and negative reinforcement increase the likelihood of good behavior, extinction and punishment decrease the likelihood of bad behavior. Using punishment and extinction does not guarantee good behavior, but rather simply reduces the specific bad behavior. It may be replaced with a different undesirable behavior.

Most fire officers quickly learn to use positive reinforcement. It is easy to pat someone on the back for a good job. Positive reinforcement must be sincere and deserved.

Most people are also familiar with and may use punishment to correct an employee who does not perform as desired. Most parents can identify with the use of extinction, as this strategy is frequently used when a child is misbehaving. The parent simply ignores the behavior to decrease the chance of the child doing it again.

Negative reinforcement is often underutilized as management strategy. It involves removing the negative consequences of good performance. For example, if a company completes its yearly inspection list early, many departments may increase the number of inspections the company is required to complete the following year. In effect, the crew is punished for good behavior. Negative reinforcement would remove the undesirable consequence.

Motivation–Hygiene Theory

One method of motivation was described by Frederick Herzberg, a psychologist who developed the concepts of job enrichment and the motivation–hygiene theory. His 1968 *Harvard Business Review* article "One More Time, How Do You Motivate Employees?" sold millions of copies and remains the magazine's most requested article.

Motivation–hygiene theory breaks the motivational process into two parts: hygiene factors and motivation factors. Hygiene factors are conditions that are external to the individual, such as pay and work conditions. Motivation factors are the individual's internally determined motivators, such as the desire for recognition, achievement, responsibility, and advancement.

Hygiene factors do not motivate individuals, but if the person is not satisfied with any of the external conditions, he or she will not be motivated. To achieve maximal performance, the officer must ensure that the fire fighters are satisfied with their work environment, pay, supervision, and company policy, and then motivate them by giving recognition for performance, providing promotional opportunities, and allowing for added responsibilities.

Probably the most significant lesson to be learned from this theory is that employees who are dissatisfied with the external conditions will not be motivated to achieve maximal performance for the company. Alternatively, once fire fighters are satisfied with the external conditions, improving these factors will not increase performance. No two fire fighters will consider the same conditions to be at the same satisfaction level. The officer must consider the individual's perception. The fire officer must have open and honest communication with the individual fire fighters.

Goal-Setting Theory

Another method of motivating fire fighters is by goal setting, which relies on the natural competitiveness of people. According to this theory, the key to motivation is for the officer to set specific goals that will increase performance. The goals must be difficult, but attainable. If they are easy to achieve, there is little motivation to work hard. Conversely, if the goals are perceived to be impossible to achieve, the fire fighter will not try hard because failure is inevitable. The fire fighter must honestly believe that performing well can result in success.

Clear, specific, and measurable goals are essential for motivation. Vague goals, such as "get to the apparatus as quickly as you can," do not give the fire fighter clear enough expectations to determine success. Instead, a goal might be to "reduce your turnout time to less than 80 seconds." The fire fighter in the latter example would be much more motivated than the fire fighter in the former example.

The most significant lesson in goal-setting theory that a fire officer can use to motivate fire fighters is to consider carefully which actions are needed to improve the organization and then to develop clear, specific, and challenging but attainable goals for the individual fire fighter.

Equity Theory

The motivational process known as equity theory suggests that employees evaluate the outcomes they receive for their inputs and compare them with the outcomes others receive for their inputs. Outcomes range from pay and benefits to recognition, achievement, and promotion. Inputs include educational level, performance level, risks taken, and special skills.

Equity theory explains why the fire chief is paid more than a fire fighter and why a fire fighter is paid more than a restaurant dishwasher. When this theory is applied, however, most fire fighters would not understand why they ranked among the lowest one third in pay in a survey of salaries across comparable cities, whereas the fire chief's pay ranked in the top one third in the same survey.

If a fire officer wants to motivate the fire fighter, he or she must determine where the fire fighter believes an inequity exists. Although the fire officer likely would not have the ability to change the organization's pay structure, he or she may be able to provide other outcomes that are desired by the fire fighters. This might include a more flexible time trade policy or recognition.

Expectancy Theory

Another motivational theory is based on the premise that people act in a manner that they believe will lead to an outcome they value. According to expectancy theory, the fire officer must address three considerations to motivate the individual:

- The employee's belief that his or her effort will achieve the goal
- The employee's belief that meeting the goal will lead to the reward
- The employee's desire for the reward or the reward's value to the employee

Using this method, the fire officer might indicate to the fire fighter that if he or she has a turnout time of less than 80 seconds for the next month, the officer will make sure that this performance is reflected on the fire fighter's evaluation, which is used for the upcoming promotional process in which the fire fighter is interested.

You Are the Fire Officer Conclusion

Captain Davis and Lieutenant Williams brainstorm on how they can resolve the issue of poor engine company response. Davis acknowledges that they had a near-miss experience and wants to avoid another close call. Williams tactfully points out that the crew mimics the captain's attitude of ignoring most of the departmental policies. Maybe it is time to "sweat" some of the fire-ground SOP details.

Wrap-Up

Chief Concepts

- Although the position of fire officer gives you the authority to accomplish certain tasks and objectives in the organization, this power does not make you a leader—it simply makes you the supervisor. Leadership attributes include beliefs, values, ethics, character, knowledge, and skills.
- Effective leaders are also good followers, supporting the fire department leadership.
- Three situational leadership styles are typically used:
  - Autocratic: An iron-hand approach used to maintain high personal control
  - Democratic: A consultative approach that recognizes the resources of the group
  - Laissez-faire: A free-rein approach that leaves decision making to the fire fighter rather than the fire officer
- Types of power may be categorized using several different methods. French and Raven refer to legitimate, reward, expert, referent, and coercive power. Yukl refers to personal and positional power.
- Most fire officer leadership activity is directed toward accomplishing routine organizational goals and objectives in nonemergency conditions.
- A core responsibility of a fire officer is to handle emergencies effectively. A fire officer always has direct leadership responsibility for the company that he or she is commanding. The first-arriving fire officer has additional important responsibilities—namely, to establish command of the incident and to provide direction to the rest of the responding units.
- The fire officer's primary responsibility is to the team of fire fighters under his or her direct supervision.
- Dangerous situations may develop suddenly during incident operations. A fire officer must use the autocratic leadership style when immediate action is required.
- The first-arriving company at an incident needs to describe conditions to the dispatch center concisely. A fire officer who provides a calm and complete description of the situation on arrival demonstrates leadership and ability to control the action that will occur.
- The first-arriving fire officer must demonstrate the ability to take control of the situation and provide specific direction to all of the units that are operating and arriving.
- The fire department workplace environment produces a special type of bonding among fire fighters, but it can easily produce a variety of productivity problems, as well as behavioral and personality traits that are often associated with a dysfunctional family.
- A fire officer must balance the expectations of the employer with the realities of a fire station work environment and the desire to create an effective team.
- Fire officers leading volunteer fire companies must rely on their leadership skills to an even greater extent than their municipal counterparts do. Pride, group identity, and personal commitment are factors that keep volunteers active and loyal to the organization.

Wrap-Up, continued

- One of the key components of leading is the ability to motivate. A variety of motivational theories have been developed:
  - Reinforcement theory: Behavior is a function of its consequences.
  - Motivation–hygiene theory: Hygiene factors are external to the individual and motivation factors are internal to the individual.
  - Goal-setting theory: The key to motivation is for the officer to set specific goals that will increase performance.
  - Equity theory: Employees evaluate the outcomes they receive for their inputs and compare them with the outcomes others receive for their inputs.
  - Expectancy theory: People act in a manner that they believe will lead to an outcome they value.

Hot Terms

<u>Equity theory</u> Motivational theory in which people evaluate the outcomes they receive for their inputs and compare them with the outcomes others receive for their inputs.

<u>Expectancy theory</u> Motivational theory in which people act in a manner that they believe will lead to an outcome they value.

<u>Followership</u> The characteristic that leaders can be effective only to the extent that followers are willing to accept their leadership.

<u>Hygiene factors</u> Conditions external to the individual, such as pay and work conditions.

<u>Leadership</u> A complex process by which a person influences others to accomplish a mission, task, or objective and directs the organization in a way that makes it more cohesive and coherent.

<u>Motivation factors</u> An individual's internal desire for recognition, achievement, responsibility, and advancement.

<u>Power</u> The capacity of one party to influence another party.

<u>Reinforcement theory</u> Motivational theory in which behavior is a function of its consequences.

References and Additional Resources

NFPA reprinted material is not the complete and official position of the NFPA on the referenced subject, which is represented only by the standard in its entirety.

Bennis, W., and W. G. Bennis. (2003). *On Becoming a Leader: The Leadership Classic—Updated and Expanded.* New York, NY: Perseus.

Benoit, J., and K. B. Perkins. (2001). *Leading Career and Volunteer Firefighters: Searching for Buried Treasure.* Halifax, NS, Canada: Henson College, Dalhousie University.

Brunacini, A. V. (1996). *Essentials of Fire Department Customer Service.* Stillwater, OK: Fire Protection Publications.

Campbell, D. S. (2012). *Determining the Difference in Effectiveness Between Lieutenants and Acting Lieutenants in the Delhi Township Fire Department.* Emmitsburg, MD: U.S. Fire Administration, Executive Fire Officer Program.

Caulfield, H. J., and D. Benzaia. (1985). *Winning the Fire Service Leadership Game.* New York, NY: Fire Engineering.

Compton, D. (2010). *Progressive Leadership Principles, Concepts and Tools.* Stillwater, OK: Fire Protection Publications.

Drucker, P. F. (1990). *From Mission to Performance: Managing the Nonprofit Organization.* New York, NY: HarperCollins, 53–106.

Gormley, W. T., and S. J. Balla. (2012). *Bureaucracy and Democracy: Accountability and Performance,* 3rd ed. Washington, DC: CQ Press.

Halberstam, D. (2002). *Firehouse.* New York, NY: Hyperion.

Hashagen, P. (1995). *A Distant Fire: History of FDNY Heroes.* Dover, NH: dmc associates.

Kiechel, W. K. III. (2012). "The Management Century." *Harvard Business Review* 90(11): 13.

Maslow, A. H. (1998). *Maslow on Management.* New York, NY: John Wiley & Sons.

McAniff, E. P., and J. J. Cunningham. (1974). *Leadership in the Fire Service.* Bayside, NY: McAniff Associates.

McCarl, R. (1985). *The District of Columbia Fire Fighter's Project: A Case Study in Occupational Folklife.* Washington, DC: Smithsonian Institution Press.

McGregor, D. (1944). "Conditions of Effective Leadership in Industrial Organizations." *Journal of Consulting Psychology* 8(2): 55–63.

Miller, K. (2010). *We Don't Make Widgets: Overcoming Myths That Keep Government from Radically Improving.* Washington, DC: Governing Books.

Myers, B. (2004). *Mentoring: Preparing Company Officers.* Emmitsburg, MD: U.S. Fire Administration Executive Fire Officer Program.

Page, J. O. (1973). *Effective Company Command.* Alhambra, CA: Borden.

Pfeifer, J. W. (2012). *Adapting to Novelty: Recognizing the Need for Innovation and Leadership.* New York, NY: With New York Firefighters, New York City Fire Department, 46.

Phoenix Fire Department. (2002). *The Phoenix Fire Department Way.* Phoenix, AZ: Phoenix Fire Department.

Raven, B. H. (2008). "The Bases of Power and the Power/Interaction Model of Interpersonal Influence." *Analysis of Social Issues and Public Policy* 8(1): 1–22.

Sadler, P., ed. (1998). *Management Consultancy: A Handbook of Best Practice.* London, UK: Kogan Page.

Singer, S. (2012). *Retention of Volunteer Membership at the Stafford Volunteer Fire Department.* Emmitsburg, MD: U.S. Fire Administration, Executive Fire Officer Program.

Snook, J. W., et al. (2011). *A Leadership Guide for Volunteer Fire Departments.* Sudbury, MA: Jones & Bartlett Learning.

The Washington Post. (2002 June 03) 'Firehouse" with David Halberstam. WashingtonPost.com Live Online. Washington, DC: The Washington Post Company. http://www.washingtonpost.com/wp-srv/liveonline/01/author/author_halberstam060302.htm.

Wilson, W. (1887). "The Study of Administration." *Political Science Quarterly* (2): 197–222.

Yukl, G. A. (2013). *Leadership in Organizations,* 8th ed. New York, NY: Prentice-Hall.

FIRE OFFICER in action

A factor in Quint 100's substandard performance at the industrial park fire was the inexperience of the crew in operating within a dynamic working fire situation. During the postincident review, the fire fighters assigned to the quint said that they did not believe their tasks included handling attack hose lines; they were doing search, forcible entry, and ventilation.

Lieutenant Williams points out that the quint is expected to perform both engine and truck duties. In the industrial park scenario, getting a 2.5-inch attack line in service was the highest priority after the fire roll-over.

1. What is the preferred way to deal with the emotions that may arise after a stressful incident?
   A. Not bringing them up is the best way; talking about things that have obviously bothered the crew is just another way to add tension.
   B. Everyone has different ways of dealing with stress, and the crew should respect that some individuals will want to talk about the incident while others will not.
   C. Being forced to talk about the issues in a roundtable-like discussion.
   D. Ignoring the issues if they are brought up.

2. On the next workday Lieutenant Williams has a meeting with the crew to see what can be done to improve quint company fire hose confidence. This is an example of _____ leadership style.
   A. democratic
   B. reinforcement
   C. expectancy
   D. laissez-faire

3. During the meeting, a senior fire fighter complains about the loss of tradition and the poor job the academy does teaching new recruits. When he is done, Lieutenant Williams makes no response. This is an example of _____ in reinforcement theory.
   A. positive reinforcement
   B. extinction
   C. punishment
   D. negative reinforcement

4. The crew agrees that they will practice a standardized fire-ground evolution every day, with the goal of completing the tasks more quickly than the target time identified in the drill manual. This is an example of _____ motivational theory.
   A. motivation–hygiene
   B. goal-setting
   C. equity
   D. expectancy

Fire Captain Activity

Chief Johnson is concerned about Station 100's response to activated fire alarms in commercial buildings after an early-morning near-miss event.

NFPA Fire Officer II Job Performance Requirement 5.2.1
Initiate actions to maximize member performance and/or to correct unacceptable performance, given human resource policies and procedures, so that member and/or unit performance improves or the issue is referred to the next level of supervision.

Application of 5.2.1
1. Using policies and procedures from a department familiar to you, prepare a proposed work improvement program for a fire company in responding to activated fire alarms in commercial occupancies.
   - Include specific observable behaviors and activities.
   - Reference the work improvement plan to an existing SOP on handling commercial fire incidents.
2. The substandard performance of Station 100 is due in part to the "Don't sweat the small stuff" approach taken by Captain Davis. What would an effective leader do about this issue?

Training and Coaching

Fire Officer I

Knowledge Objectives

After studying this chapter, you will be able to:

- Discuss the role of training in the fire service. (pp 142–143)
- Discuss the fire officer's role in training fire service personnel. (pp 143–147)
- Describe the four-step method of instruction. (pp 143–146)
- Describe on-the-job training and the order in which skills must be taught. (pp 147–148)
- Discuss the requirements for conducting live fire training (NFPA 4.2.3). (pp 148, 150–151)
- Describe how to develop a specific training program. (pp 151–152)

Skills Objectives

After studying this chapter, you will be able to:

- Direct fire company members in proper completion of a prepared training evolution (NFPA 4.2.3). (pp 148, 150–151)
- Develop a training program. (pp 151–152)

Fire Officer II

Knowledge Objectives

After studying this chapter, you will be able to:

- Discuss the role of professional development in the fire service. (pp 152–155)
- List and describe the components of a professional development plan (NFPA 5.2.3). (pp 155–156)

Skills Objectives

After studying this chapter, you will be able to:

- Create a professional development plan (NFPA 5.2.3). (pp 155–156)

CHAPTER 8

Additional NFPA Standards

- NFPA 1000: *Standard for Fire Service Professional Qualifications Accreditation and Certification Systems*
- NFPA 1001: *Standard for Fire Fighter Professional Qualifications*
- NFPA 1041: *Standard for Fire Service Instructor Professional Qualifications*
- NFPA 1142: *Standard on Water Supplies for Suburban and Rural Fire Fighting*
- NFPA 1403: *Standard on Live Fire Training Evolutions*

Principles of Fire and Emergency Service Administration (FESHE) Course Objectives

1. Acknowledge career development opportunities and strategies for success. (pp 152–156)
2. Recognize the need for effective communication skills both written and verbal. (pp 143–146, 150–152)
6. Discuss the various levels of leadership, roles, and responsibilities within the organization. (pp 143–148, 150–152)

You Are the Fire Officer

Battalion Chief Johnson conducts a quarterly review of fire fighter and fire officer skills and knowledge progress at every fire station. Recruit fire fighters are required to complete a field manual during their probationary period that concludes with an end-of-probation knowledge and skills examination. Fire officers and incumbent fire fighters are measured against the goals and objectives established during their annual performance review.

A probationary fire fighter was transferred to Quint 100 last week. Chief Johnson explains to Captain Davis and Lieutenant Williams that, after 9 months at another work location, the recruit has completed just one-third of the required field manual activity. Failing to complete the field manual within a year, or failing the end-of-probation test, is grounds for immediate dismissal.

According to Chief Johnson, the recruit wants the job. The problem with the recruit's progress was discovered after a captain and a battalion chief retired last month. Assistant Chief James Arrow had the recruit transferred to a new shift in a different battalion and wants the field manual completed within 3 months. Arrow is not sure if the issue is with the recruit or the retired officers. Two overdue accomplishments in this recruit's manual are demonstrating proficiency in operating a pumper and knowledge of the first-due district.

1. What should you do when a fire fighter has a performance problem?
2. How much time should you spend with individual fire fighter performance issues?
3. How can you develop a work improvement plan?

Introduction

Training and coaching have been core fire officer tasks since the establishment of the first organized fire departments. Training is defined as the process of achieving proficiency through instruction and hands-on practice in the operation of equipment and systems that are expected to be used in the performance of assigned duties. Fire service training has evolved in complexity and sophistication at a rapid pace as new areas of expertise have been added to the list of services performed by fire departments. Coaching is a method of directing, instructing, and training a person or group of people with the aim of achieving some goal or developing specific skills.

NFPA 1041, *Standard for Fire Service Instructor Professional Qualifications*, defines the standard and describes the requirements for three levels of instructor. Certification as an Instructor I is a prerequisite for Fire Officer I candidates.

Fire Officer I

Overview of Training

Training ensures that every fire fighter can perform competently and every fire company can operate as a high-performance team. Fire service training must anticipate high-risk situations, urgent time frames, and difficult circumstances. A wide variety of methods and practices are included in the overall category of training.

Initial training leads to basic skill certifications, such as NFPA Fire Fighter I and Emergency Medical Responder. These certifications are often required before a fire fighter is authorized to participate in field operations. Certification usually involves a formal training program, conducted by a training academy or equivalent organization, that includes both classroom and skills practice FIGURE 8-1. The trainee must pass both skills and knowledge evaluations to be certified.

After initial certification, most fire fighters work toward achieving progressively higher levels of certification and additional specialty qualifications. Often, a fire fighter is required to achieve additional qualifications, such as Fire Fighter II, driver/operator, or Emergency Medical Technician (EMT), within a specified time period. Some departments require higher-level certifications, such as Fire Officer I or II, for promotion or advancement. Many certifications require periodic refresher or update training, and some expire after a set period unless the fire fighter completes a refresher training requirement.

Several important components of training occur at the fire station or company level under the supervision of fire officers. This type of training is directed toward practicing basic skills and improving both individual and team performance. Many departments have a standard set of evolutions

CHAPTER 8 Training and Coaching

FIGURE 8-1 A formal training program includes both classroom and skills practice.

Assessment Center Tips

Four-Step Fire Fighter Development
The four-step method remains the method of choice if a candidate must describe how to develop a work improvement plan or explain a new procedure or device. When developing a written or oral response to such an issue, the candidate should identify each of the four steps in the training process.

that each company is expected to be able to perform without flaw, such as advancing a fire attack line over a ground ladder and into a third-floor window. Additional training at the company level often includes learning how to use new tools and equipment as well as refreshing, reinforcing, or updating knowledge and skills that are related to different aspects of the firefighting craft.

Additional company-level training often includes preincident planning and familiarization visits to different locations in a company's response area. These activities sometimes include a group of companies that would normally respond together to that location. Conducting these activities as a group is an excellent method of improving coordination between companies. Multicompany drills should be conducted periodically for the same reason.

Fire Officer Training Responsibilities

A basic responsibility of every fire officer is to provide training for subordinate fire department members. The specific training responsibilities assigned to fire officers vary, depending on the organization and the available resources. At a minimum, a fire officer must be prepared to conduct company-level training exercises and evolutions to ensure that the company is prepared to perform its basic responsibilities effectively and efficiently.

One of the enduring concepts of the 20th century is the four-step method of skill training. This prepare–present–apply–evaluate method originated during World War I, when the armed services were teaching farmers how to fly biplanes, drive tanks, and operate ships. The process was updated and renamed job instruction training when more than 1 million men and women received technical skill training during World War II.

Today, the four-step method is the foundation of the work performed at the Fire Instructor I level. Most fire officers use standardized curricula and training packages that are commercially prepared or developed by the local fire academy. Occasionally a company officer needs to start from scratch to develop a training program.

■ Review of the Four-Step Method
Step 1: Preparation
The fire officer conducts training to maintain proficiency of core competencies. Crews should be able to catch a hydrant, raise a ground ladder, and deploy attack lines so well that the task is automatic. In other words, they should be unconsciously competent with these tasks.

Sometimes the fire officer or the department determines the need for focused instruction. Three indicators that training is needed would be a near miss, a fire-ground problem, or an observed performance deficiency. For example, during a structure fire, the fire officer might observe that the fire fighters are having difficulty placing a 35-foot (11-meter) extension ladder at an upper-floor window, or perhaps the deployment of an attack line from a standpipe connection created a tangle of hose in the stairwell. These problems need to be addressed through additional training and practice.

Des Plaines Fire Department
Division of Training

Lesson Title

Training Date:

FireHouse Code:

Location of Training:

Safety Plan Required? Y ☐   N ☐

Topic:	Instructor(s):
Teaching Method(s):	Time allotted:
Handouts:	
AV needs:	Teaching resources:
Level of Instruction:	Evaluation Method:
NFPA JPR's:	Equipment needed:

LEARNING OBJECTIVES: *Upon completion of the class and study questions, each participant will independently do the following with a degree of accuracy that meets or exceeds the standards established for their scope of practice:*

General class activities for student application

| Safety Briefing | SAFETY RED FLAGS |
| 1. | ALL STOPS: |

FIGURE 8-2 Fire officers often write lesson plans in preparation for training activities.

The fire officer begins by obtaining the necessary material and teaching aids. If needed, the fire officer writes a lesson plan **FIGURE 8-2**. Components of a lesson plan include the following:

1. Break the topic down into simple units.
2. Show what to teach, in which order to teach it, and exactly which procedures to follow.
3. Use a guide to help accomplish the teaching objective.

If this is a new lecture or topic, the fire officer should practice delivering the lesson to make sure that the important

FIGURE 8-3 Check the physical environment to make sure it is conducive to adult learning.

items are covered in a timely fashion. In addition, the fire officer should preview any audiovisual items and check all of the equipment that will be used during the presentation.

The final preparation activity is to check the physical environment. Make sure you have an environment that is conducive to adult learning. This includes taking every reasonable effort to reduce distractions and student discomfort FIGURE 8-3 .

Step 2: Presentation

The presentation is the lecture or instructional portion of the training. The objective with Step 2 is to introduce the students to the subject matter, explain the importance of the topic, and create an interest in the presentation. During this step, the fire officer could be demonstrating or showing a skill or explaining a concept.

The instructor should present a skill one step at a time, delivering a perfect demonstration of how it should be performed FIGURE 8-4 . When presenting a concept or idea, recommended lecture practices should be followed. The objective of a lecture is to provide knowledge and develop understanding so that the fire fighter will be able to perform the skills properly. The overall goal is to increase fire company efficiency.

During a lecture, the instructor should use simple but appropriate language. Begin with simple concepts and move progressively to more complex information, relating the new material to old ideas. Lecture only on what is important at this time to achieve the teaching objective, and do not teach alternative methods. Teach in positive terms and avoid telling the fire fighter what *not* to do.

A lesson plan allows the fire officer to stay on topic and emphasizes the important points to be addressed. Increasing the number of senses engaged in the training session helps the fire fighter to retain more of the material. Audiovisual aids and training props should be used to enrich the presentation. Fire fighters retain information more effectively if they actually perform the skill in the process of learning it.

Step 3: Application

The fire fighter should demonstrate the task or skill under the fire officer's supervision. The objective is a correct demonstration of the task, safely performed. A good reinforcing technique is to have the fire fighter explain the task while demonstrating the skill. The fire officer should provide immediate feedback, identifying omissions and correcting errors FIGURE 8-5 . Success is achieved when the student can perform the task safely without input from the supervisor.

Step 4: Evaluation

At the end of the lesson or program, there should be an evaluation of the student's progress. Training that is related to a certification program always includes an end-of-class evaluation. Depending on the skill and knowledge sets involved, the evaluation may be a written or practical examination.

The fire officer can be certain that training has occurred only when there is an observable change in the fire fighter's performance when responding to a real situation where that task or skill is applied. For example, Fire Fighter Jones

FIGURE 8-4 Fire officers should deliver a perfect demonstration of how skills are performed.

FIGURE 8-5 The fire officer should provide immediate feedback to the fire fighter.

Fire Marks

Conscious Competence Learning Matrix

Dr. Thomas Gordon developed the Conscious Competence Learning Matrix in the 1970s. He described four levels of understanding of a skill in his book *Teacher Effectiveness Training*:

Level 1: *Unconscious Incompetence*. The fire fighter is not aware of the existence or relevance of a skill area. He or she is not aware of having a deficiency in this area and might deny the relevance or usefulness of the skill.

Level 2: *Conscious Incompetence*. The fire fighter becomes aware of the existence and relevance of a skill. In attempting to perform the skill, the fire fighter is aware of his or her deficiency. Training and practicing on the skill, however, will improve performance and move the fire fighter into Conscious Competence.

Level 3: *Conscious Competence*. The fire fighter can perform the skill reliably at will and without assistance, but the fire fighter still needs to concentrate and think to perform the skill. Repeated practice in the skill will get the fire fighter to Unconscious Competence.

Level 4: *Unconscious Competence*. The skill becomes so practiced that it enters the unconscious parts of the brain—it becomes second nature. The fire fighter will be able to perform other tasks while performing this skill, such as conducting a search and rescue while operating a fire attack line inside a burning structure. The tasks involved in wearing self-contained breathing apparatus (SCBA) and operating a fire attack line are second nature at this level, for example. Consequently, the fire fighter can simultaneously focus attention on performing the search and locating victims.

---

has not been coming to a complete stop at red traffic lights or stop signs when driving fire apparatus to emergency incidents. The fire officer provides a training session to address this behavior. If, after the training session, Fire Fighter Jones always comes to a complete stop at red traffic lights and stop signs, there has been an observable change of behavior. If the behavior does not change, the training objective has not been accomplished.

■ Ensure Proficiency of Existing Skill Sets

The fire officer needs to ensure that every fire fighter is proficient in skill performance. This is both an individual and a company-level requirement. Some departments have a standard set of evolutions that must be performed proficiently in which both task and time requirements must be met. The fire officer must invest some of the available training time practicing and reviewing these standard evolutions.

For best performance, many of these practice sessions should be performed while the fire fighters are wearing full personal protective clothing, operating within a realistic fire-ground situation. This may require construction of training props, such as an assembly that allows the fire fighters to practice opening a roof using a power saw. The fire officer should always be on the alert for opportunities to acquire abandoned structures where realistic fire-ground skills can be practiced.

■ Mentoring

Mentoring is a developmental relationship in which a more experienced person, or mentor, helps a less experienced person, referred to as a protégé. The unique organization of fire departments, where company officers function as working supervisors within a high-hazard environment, makes mentoring easy.

Mentoring is a one-on-one process in which the more experienced person provides a deliberate learning environment through instructing, coaching, providing experiences, modeling, and advising. Mentoring may be provided through feedback while picking up after an emergency incident. Failures as well as successes make up the mentoring experience.

The mentoring process extends beyond a particular fire company assignment and rank, usually lasting for a long period of time. Boston Fire Commissioner Leo Stapleton, writing in *Thirty Years on the Line*, described how he would get feedback from a senior fire fighter decades after Stapleton was his rookie.

The following qualities make for an effective mentor:

- A desire to help
- Current knowledge
- Effective coaching, counseling, facilitating, and networking skills

■ Provide New or Revised Skill Sets

On occasion, the fire officer is required to provide the initial training for a new or revised skill set. This is often related to a new device that has been acquired, such as a new type of breathing apparatus or a thermal imager. The fire fighters need to become familiar with the new equipment and proficient in its use. In the case of a thermal imager, the fire fighters have to learn how it works and how to maintain it, as well as how to incorporate its use into fire-ground operations. This type of training could also be required to introduce a change in standard operating procedures. Sometimes procedures are changed or additional training is mandated in response to a near-miss incident.

Teaching new skills takes more time than maintaining proficiency of existing skills. The fire officer should obtain as much information as possible about the device or procedure, especially identifying any fire fighter safety issues. The emphasis of fire station–based training should be on the safe and effective use of the device or procedure. The fire officer should plan to spend a couple of training periods developing competency and showing how the new skill relates to existing procedures. Active learning research recommends spending no more than 10 minutes presenting a formal lecture or video presentation; the student loses focus after 15 to 20 minutes of passive learning (Kiewra 2002). Practicing new skills is facilitated by encouraging adventure, challenge, and competition. For example, using a thermal imager to complete a scavenger hunt for heat sources around the fire station is one way to reinforce the capability of the device.

■ Ensure Competence and Confidence

The fire officer works as a coach when providing training to an individual or a team. After team members have learned the basic skills and can appropriately demonstrate them, the coach has to work with them to build competence and confidence. The coach must provide the guidance that advances them from being capable of performing the basic required skills to being able to perform those skills effectively, efficiently, and consistently.

Many fire fighter tasks involve psychomotor skill sets. Psychomotor skill levels can be classified into four categories. These levels may be described using the following example of a new driver/operator who is required to know, from memory, every address in the fire district.

- *Initial:* The driver/operator knows the main streets and has a basic understanding of how the street grid and numbering system work. The fire officer has to help by reading the map when responding in subdivisions and office parks.
- *Plateau:* The driver/operator can drive to more than 85 percent of the streets in the company's response area without assistance from the officer. At this level, the driver/operator is competent.
- *Latency:* The driver/operator can remember the route to an area of the district where the engine company has not had a response for several months.
- *Mastery:* The driver/operator can easily drive to any address in the district and knows at least two or three alternative routes to each area. The driver knows all of the subdivisions, the layout of every office park, and the locations of all hydrants and fire protection connections. At this level, the driver/operator is confident in his or her knowledge of the district.

To bring fire fighters up to the mastery level, the fire officer must work every day to reinforce their skills. Many fire fighter skill sets are used only infrequently, yet when that task is needed, the fire fighter must deliver a near-perfect performance under urgent or critical conditions. It is dangerous to ask a fire fighter to demonstrate a rusty skill set in a critical emergency situation.

The fire officer needs to provide enough repetition and simulations to maintain fire fighter confidence in seldom-used skill sets **FIGURE 8-6**. The continuing expansion of fire fighters' responsibilities as all-hazard mitigation specialists increases the fire officer challenge.

FIGURE 8-6 Skills practice may require training props.

When New Member Training Is On-the-Job

Many departments require that a new member obtain NFPA Fire Fighter I certification before responding in the field. Fire officers have special responsibilities when operating with inexperienced fire fighters and fire fighters in training. At the very first meeting with the new fire fighter trainee, the fire officer should explain the procedures in the fire station when the company receives an alarm, assign a senior fire fighter to function as a mentor, and describe any restrictions that are placed on fire fighters in training.

■ Skills That Must Be Learned Immediately

Federal regulations mandate that four topics are covered as part of any emergency service training program:

- *Bloodborne pathogens.* Even "suppression only" fire fighters are at risk of being exposed to blood and other bodily fluids. The Occupational Safety and Health Administration (OSHA) has issued regulation 29 CFR 1910.1030, *Occupational Exposure to Bloodborne Pathogens*, which requires all fire fighters to be trained about the department's exposure control plan, the personal protective equipment used by the fire fighter, and the reporting requirements if an exposure occurs. Usually, this training takes about 4 hours.

- *Hazardous materials awareness and operations.* Every public safety member is required to have training at the hazardous materials awareness and operations levels. OSHA regulation 29 CFR 1910.20, *Hazardous Waste Operations and Emergency Response (HAZWOPER) Training*, describes these requirements. At the awareness level, first responders are able to recognize a potential hazardous materials emergency, isolate the area, and call for assistance. At the operations level, first responders are able to recognize a potential hazardous materials incident, isolate and deny entry to other responders and the public, evacuate persons in danger, and take defensive actions such as shutting off valves and protecting drains without having contact with the product.
- *SCBA fit testing.* OSHA regulation 29 CFR 1910.134, *Respiratory Protection Training*, requires that anyone who uses respiratory protection during job tasks must be provided with appropriate training, be fit-tested for a mask, and be subject to a health monitoring program. This training is usually part of the Fire Fighter I training program.
- *National Incident Management System.* Homeland Security Presidential Directive 5 (HSPD-5), *Managing Domestic Incidents*, requires incident management training. The National Incident Management System (NIMS) provides a consistent framework that operates at all jurisdictional levels regardless of the cause, size, or complexity of the incident. NIMS creates an all-hazard template for fire fighters to operate within a multiple-jurisdiction domestic incident within the federal National Response Plan.

In addition to the federally required training, the fire department needs to provide emergency scene awareness training to reduce the risk that a probationary fire fighter will be injured at the emergency scene. This training would include such items as how to avoid being struck by a car when operating on an interstate highway. The fire officer should spell out the expected behavior of the trainee when operating at emergencies. This effort may include assigning the trainee to shadow, or work with, a senior fire fighter on the team.

The expected behavior falls into three areas: responding to alarms, on-scene activity, and emergency procedures. Responding to alarms includes the trainee behavior that should occur when an alarm is received at the fire station. Be specific about the appropriate actions necessary when a fire fighter is preparing to respond to an alarm. Emphasize the importance of not running in the station, donning protective clothing correctly, and always wearing a seat belt when riding in a fire vehicle.

On-scene activity includes clearly defining the expected location, activities, and behavior of the trainee at an emergency incident. Make sure that the trainee knows the safe locations when working at an incident on a roadway. Identify off-limits activity. The fire officer should know all the restrictions that apply to each individual and ensure that the trainee is equally aware of those limitations. A trainee who has not completed SCBA training and fit testing must not wear breathing apparatus or enter an immediately dangerous to life and health (IDLH) environment. If the trainee is a teenager, additional restrictions might apply; allowing a 15-year-old to operate inside a burning building is prohibited in many jurisdictions.

Finally, the officer should clearly identify the trainee's expected behavior when operating during an emergency situation, such as a mayday (fire fighter down or lost) or a potentially violent situation. The fire officer should ensure that the trainee will be in a safe location and stay out of the way until the emergency situation is controlled.

Skills Necessary for Staying Alive

Once the skills that must be learned immediately are covered, the fire officer should concentrate on the skills the fire fighter in training needs to know to stay alive. Most states and commonwealths require the trainee to pass a knowledge exam and a skills test for Fire Fighter I before engaging in interior structural firefighting. Some communities start with the trainee limited to "outside only" activities at fire scenes. Examples of skills and topics that should be taught first in a training program include the following:

- Fire-ground tasks that emphasize teamwork and require mastering the location of all of the equipment on the apparatus:
  - Supply line evolutions
  - Ropes and knots
  - Laddering the fire building
  - Lights, fan, and power deployment
- Motor vehicle crashes and medical emergencies
  - CPR and AED training
  - Outside-circle activities on a crash extrication
  - Helicopter landing zone procedure
  - Assisting the paramedics on a medical emergency

The trainee should have received enough training and demonstrated adequate stay-alive skills to begin responding to emergencies. This allows the student to gain experience by functioning in an "outside only" role while completing the Fire Fighter I certification training. Some fire departments do not allow trainees to respond to any emergency incidents until they have achieved the initial Fire Fighter I certification level.

Digital video technology is reshaping the way emergency service collects information and will influence how a fire officer provides training. Through this technology, you can download pictures and videos of incidents as they are occurring. A variety of Web sites are available from which to download training materials, including Firefighter Close Calls, Everyone Goes Home, commercial, and individual Web sites.

Live Fire Training

Fire fighters must receive appropriate training before participating in any live fire evolutions. NFPA 1403, *Standard on Live Fire Training Evolutions*, provides essential information for any type of live fire training session.

Student Prerequisites

Before participating in any live fire training evolutions, the student must have received training to meet the Fire Fighter I performance objectives from the following sections of NFPA 1001:

- Safety
- Fire behavior
- Portable extinguishers

VOICES
OF EXPERIENCE

In the fire service, *coaching* is a word used to connote a variety of circumstances. Some use it to describe mentoring new members, developing personnel for promotion, or even developing newly promoted officers.

In my time as an instructor with our high school programs, I have observed many opportunities to use coaching. Leading a recruit class is very similar to leading a company, or even building a sports team. In a recent program, I chose to implement some different coaching strategies and found them to be successful.

First, I spent time getting to know each of the students—who they were, what their goals were while in the program and beyond, and what their strengths and weaknesses were from their perspective. Then, I grouped them together in crews to build on their strengths in forming teams.

As the class progressed, I encouraged the students to get to know one another, evaluate one another, and, most importantly, develop into a functioning crew that could identify strengths and weaknesses and work together to overcome new challenges. A premium was placed on the success of the team, even while simultaneously recognizing individual accomplishments. Success of the group led to an appreciation for individual contributions as members practiced a concept known as introspection—looking at themselves to achieve things they might not have thought possible. This class, and the companies within it, began to study together away from class and travel together to classes, resulting in a closer-knit group. They knew where the other members were almost all of the time. In some cases, they knew more about each other than their own families.

> *Once I provided them with the tools to be successful, it was incumbent upon them to perform—and they did, to tremendous effect.*

I was amazed at the outcome. I watched a class of 13 individuals walk into a room, and 9 months later they had developed into four unique and highly effective companies that functioned with more efficiency, teamwork, communication, and determination than some of our more seasoned crews. The changes were clear to me early in the program; however, when the class entered live fire training, the crews' teamwork and abilities also became evident to our senior staff. On several occasions, our command officers observed these burn scenarios and compared the students' performance with that of experienced crews within the department. While these young men and women were certainly new and made some errors, they owned them, recovered, and pulled together as a team focused on their safety and the accomplishment of the mission.

It really was like building a championship team. Students learned quickly that while mastering basic skills and academics were individual events, scenario-based challenges offer an opportunity to fulfill a set of tasks beyond what any one individual could achieve. I was able to stand aside with incredible pride as they functioned effectively, surpassing my personal expectations of them. As a coach, I could not perform the tasks for them; the students had to accomplish them on their own. My job was to set the bar of expectations high enough to challenge them while making it possible for them succeed. Once I provided them with the tools to be successful, it was incumbent upon them to perform—and they did, to tremendous effect.

These same strategies can be used in the academy setting, in the station, or at the battalion level. No matter where we are in the fire service, we are a team. We are a group of individuals with different strengths, weaknesses, and abilities. However, when we move as one, with a common goal in mind, and capitalize on those individual dynamics, we will overcome any challenge we face.

Jason W. Burrow
Firefighter Medic III—Training Officer
Hanover Fire EMS
Hanover, Virginia

> ## Fire Marks
>
> **Fire Officer Convicted of Criminally Negligent Manslaughter After Live Fire Training**
> Alan G. Baird III was the instructor-in-charge of a live fire training session in an acquired structure that led to the death of a 19-year-old rookie and severely burned two other fire fighters. The National Institute for Occupational Safety and Health (NIOSH) report identified eight major mistakes made by Baird in setting up and running the evolution.
>
> The Oneida County district attorney filed charges of second-degree manslaughter. Part of Baird's defense was that he did not know about NFPA 1403. He was convicted of criminally negligent manslaughter. While sentencing Baird to 75 days in jail and 5 years' probation, Judge Michael Dwyer said that the incident was not an accident, but rather a series of bad decisions—decisions that never should have been made.

- Personal protective equipment
- Ladders
- Fire hose, appliances, and streams
- Overhaul
- Water supply
- Ventilation
- Forcible entry
- Building construction

In addition to the prerequisite training, the trainee must be equipped with full protective clothing, a personal alerting safety system (PASS) alarm device, and self-contained breathing apparatus that is compliant with the relevant NFPA standard.

Fire Officer Preparation Responsibilities

NFPA 1403 provides detailed instructions on how to conduct a safe live fire training evolution under five different scenarios: acquired structures, gas-fired training center buildings, non–gas-fired training buildings, exterior props, and exterior Class B fires. The fire officer should read the standard and, if available, consult with a training officer who has received additional instruction in the delivery of live fire training evolutions.

Several general points apply to all live fire evolutions. The fire officer who is the instructor-in-charge develops a training action plan that includes the following elements:

- Develop a preburn plan, including communications, personnel accountability, rehabilitation, and building evacuation procedures. The instructor-in-charge must remain mindful of severe weather conditions that might potentially increase the risk of injury or illness.
- Calculate the needed water supply. NFPA 1142, *Standard on Water Supplies for Suburban and Rural Fire Fighting*, provides a reference for this subject. Each attack and backup hose line should be capable of flowing a minimum of 95 gallons (360 liters) per minute. The water supply resources should be capable of delivering 150 percent of the minimum needed water supply at the training site from two separate sources.
- Provide emergency medical services (EMS) for all training. EMS must be transport capable if the training involves an acquired structure.
- Establish a designated and appropriately equipped rest and rehabilitation area.
- Inspect the structure to ensure it is safe for use in live fire evolutions. Determine the required Class A materials to be used to provide the fire load. Watch out for excessive fire load and for materials that could lead to early and unwanted flashover.
- Assign the dedicated positions of safety officer and ignition officer.
- Assign instructors to each functional crew. The maximum size of a functional crew is five trainees. An additional instructor is assigned to each backup hose line and functional assignment. All instructors must wear full protective clothing and SCBA. The weather, duration of the evolution, and size of the trainee group determine whether additional instructors are required.
- Conduct a preburn briefing session and walk-through for the instructors, safety crew, and trainees. Identify evacuation routes during the walk-through. All facets of each evolution should be discussed and assignments made before the fire is started.
- Conduct a postburn review, assuring all personnel are accounted for, any remaining fires are overhauled, and the training structure is inspected for stability and hazards. Conduct a critique of the training event, and assure all required records are prepared. Only then should you release the building and property to the owner.

> ## Safety Zone
>
> **Training Injury Patterns**
> From 1997 to 2010, deaths during training activities represented 77.8 percent of the nonemergency annual line-of-duty death (LODD) total. This is an unacceptable situation. Deaths and serious injuries that occur during live fire training are caused by the following factors:
> - Missing, incomplete, or inadequate preparation of the drill site.
> - Inadequate training of instructors and assistant instructors.
> - Inadequate orientation or training of recruits. Live fire evolutions are not the place to conduct or complete initial SCBA training. The students should be competent in SCBA use and emergency procedures before they participate in a live fire training exercise.
> - Inadequate planning of the training evolutions by the lead instructor.
> - A nonexistent incident management system, or no designated safety officer.
> - No provision of rehabilitation or emergency medical services at the drill site.

Getting It Done

Performance in Context

Whole-skill training or training in context is an effective tool allowing experienced fire fighters to improve their group performance at standard evolutions. Athletes use visualization and sequential task mastery to develop consistent and high-level performances. Members who prepare for the Fire Fighter Combat Challenge, Transportation Emergency Rescue Committee (TERC) Extrication Challenge, or other timed skill events also use training in context to improve performance.

■ Prohibited Live Fire Training Activities

NFPA 1403 identifies activities that are absolutely prohibited during live fire training. These prohibitions are the direct result of investigations of prior live fire evolutions that resulted in fire fighter deaths or serious injuries.

- No live "victims" may be used in live fire training evolutions.
- Flammable or combustible liquids cannot be used as fuel.
- Acquired structures have been the source of many line-of-duty deaths. Conduct only one fire evolution in an acquired structure at a time.

Developing a Specific Training Program

On occasion, the fire officer may need to develop a specific training program that is not covered by an existing certification training program or prepared lesson plan. This may be a work improvement plan or training related to a new device or procedure. NFPA Instructor II covers the development of a training program in more detail. Here is an overview of the five steps for developing a training program.

■ Assess Needs

The fire officer must first confirm that there is a need for a training program. Some performance problems may be better solved through an engineering solution. For example, the effective use of PASS devices was a problem in the past. Many fire fighters were operating in an IDLH environment while their PASS device was not armed. Despite training programs and, in some departments, progressive discipline, the devices remained unarmed. An engineering solution has resolved this performance issue. The PASS device is now integrated into the SCBA and is armed every time the high-pressure hose is charged.

■ Establish Objectives

Training has occurred when there is an observable change in behavior. Prior to undertaking the training program, you should identify the specific behavior you want the fire fighters to exhibit after the training. The desired behavior could be

Near-Miss REPORT

Report Number: 12-0000235

Event Description: During a pump operation evolution, a 5-inch supply line suffered a severe water hammer. The attack engine was flowing lines, with the supply engine supporting the operation via a humat valve. The attack operator had to ask for more pressure due to the size of the evolution. The attack operator finished the evolution and began to shut everything down on his end. He did not communicate with the supply engine that he was completing his evolution. Instead of communicating with the supply operator regarding decreasing the supply pressure, he went ahead and closed the piston intake.

When this happened, the supply operator was boosting the pressure at more than 120 psi. This is not normal supply pressure; however, due to the size of the evolution, it was needed. When the pressure reached the closed piston intake, the water hammer could be seen and heard traveling down the 5-inch hose line until it reached the weakest point, where it burst. Three fire fighters were close to the line and could have been injured. Also, when the supply line burst, it moved a steel structure that is attached to a six-story drill tower, indicating the seriousness of the water hammer.

Lessons Learned: Communication is just as important on the training ground as it is on the fire ground. To prevent another occurrence of this type, training officers must ensure students understand the seriousness of water hammer and how easily it can happen if the pump operator does not pay attention to detail.

that fire fighters will demonstrate the proper procedures for deploying a ground ladder. The description must include the conditions under which the behavior will be demonstrated. For example, if an expectation is that a ground ladder will be deployed while fire fighters are wearing full protective clothing and SCBA, then that condition must be part of the expected fire fighter behavior.

The final part of the objective is the measure of performance. Fire fighter evolutions are often timed events, so the measure of performance is usually described in terms of the time required to complete a task properly.

The completed behavioral objective could look like this:

> Given a fire department pumper and a building two or more stories high, a crew of two fire fighters will exit the pumper in full protective personal clothing and SCBA, select and remove the appropriate ground ladder from the pumper, and properly deploy the ground ladder to the assigned window within 2 minutes.

■ Develop the Training Program

Various methods exist for developing the training program. If training is needed for a new device, the manufacturer or vendor may have a training package available. If training is needed for a new procedure or company evolution, the department may have a template the fire officer can use. Other departments may have developed training programs, props, or resources that they will share. Organizations such as Fire Fighter Near Miss, Responder Safety, and special-interest Web sites may post resources available for downloading. A few minutes of research may provide a rich response.

The fire officer also needs to consider how the training will be delivered. Will it require a skill drill using multiple companies, or can the skill be practiced by an individual fire fighter?

■ Deliver the Training

New programs should be subjected to a pilot class or trial run before finalization. This sort of testing provides the opportunity to tweak the program, identify any problems, and correct any unforeseen issues.

Although the use of a lesson plan is part of the four-step method, it is important that the fire officer develop lesson plans for every training program. A good lesson plan satisfies four criteria:

- Organizes the lesson
- Identifies key points
- Can be reused
- Allows others to teach the program

■ Evaluate the Impact

Have you accomplished the change of behavior? Was the training program worth the fire fighters' or instructor's time? Was the instructional method appropriate for the learning objective? What can the instructor-developer do to improve the training program?

Many successful national fire service training programs, such as the "Saving Our Own" fire fighter rescue class and the "Car Busters" extrication seminars, have been developed by fire officers and fire fighters. They have the perspective, the understanding, and the need to develop and share vital emergency activity training.

Fire Officer II

Professional Development

Professional development refers to skills and knowledge attained for both personal development and career advancement. Professional development encompasses all types of facilitated learning, ranging from college degrees to certification training, continuing education, skill acquisition, and informal learning opportunities in the field. It has been described as intensive and collaborative, ideally incorporating an evaluative stage (Speck & Knipe 2005).

Each fire officer should develop a personal professional development plan. Obtaining professional designations and developing experience and skill sets makes the officer more valuable to the department and the fire service. More opportunities are available to those personnel with a vibrant professional development plan.

■ Training Versus Education

The relationship between training and education is confusing. Education is the process of imparting knowledge or skill through systematic instruction. Education programs are conducted through academic institutions and are primarily directed toward an individual's comprehension of the subject matter. Training, by comparison, is directed toward the practical application of education to produce an action, which can be an individual or a group activity. Thus there is an important distinction between these two types of learning.

Training is an essential fire service activity. The First Wingspread Conference on Fire Service Administration, Education and Research was sponsored by the Johnson Foundation and held in Racine, Wisconsin, in 1966. This conference brought together a group of leaders from the fire service to identify needs and priorities. They agreed that a broad knowledge base was needed and that an educational program was necessary to deliver that knowledge base. This agreement became the blueprint for the development of community college fire science and fire administration programs, as well as the bachelor's-level Degrees at a Distance program.

In 1998, the U.S. Fire Administration (USFA) hosted the first Fire and Emergency Services Higher Education (FESHE) conference. That conference produced *The Fire Service and Education: A Blueprint for the 21st Century*, which started a national effort to update the academic needs of the fire service. FESHE participants developed a model fire science curriculum that would go from associate's degree through master's degree. By the 2008 conference, the FESHE model undergraduate fire science curriculum had been adopted by most academic institutions, with each course supported with textbooks available through two or more publishers.

The direction of USFA changed in 2012 with the emergence of the National Professional Development Matrix (NPDM), which combined the FESHE and Training Resources and Data Exchange (TRADE) initiatives. The focus of the NPDM is on career enhancement for emergency responders, professional development matrix discussions, succession planning, and national professional development standards recommendations.

■ Cultural Change Issues in Education

In the 2013 *Fire Service Image Task Force Report: Taking Responsibility for a Positive Public Perception*, the International Association of Fire Chiefs (IAFC) described the issue of education:

> The fire service is looking to enhance its professional image by seeking higher education to support its work. Clearly, this is a good thing. However, it can create a dangerous shift in the traditional trade or paramilitary structure of the fire service.
>
> It's increasingly likely today that those entering the profession have some post-secondary education, and many in our ranks are seeking higher education either for their own enrichment or due to the growing number of departments requiring advanced degrees for promotion. The fire service is clearly benefiting from advanced education of its members both operationally and in regard to its image. So where's the harm? There are a number of possible pitfalls:
> - Failing to balance the values of classroom and experiential education. Mechanical and manual labor are still the necessary base skills for fire fighters, and only experience can prepare you for the high-pressure demands of the job.
> - The breakdown (real or perceived) of command and control, particularly in administrative issues, when this balance is not achieved.
> - Disenfranchising valuable and talented responders who don't have degrees.
> - Failing to balance the desire to attract professionals with degrees with building professional opportunities for young people within the community.
>
> On the one hand, higher education is consistent with public expectations around the education levels of those who are perceived as professionals. On the other hand, we run the risk of being perceived as elitist or condescending when we assert our position over community members who are less educated or cut off traditional avenues for non-college-bound community members to professionally excel and thrive.

■ Academic Accreditation

A prestigious accomplishment for a fire officer is to attend the 3-week *Senior Executives in State and Local Government* program at the John F. Kennedy School of Government. This on-campus program is one of 34 Executive Education lifelong learning programs offered at Harvard University. A handful of fire service members are awarded an annual scholarship through the Harvard Fire Executive Fellowship Program sponsored through a partnership among the IAFC, International Fire Service Training Association/Fire Protection Publications (IFSTA/FPP), the NFPA, and the USFA. This outstanding program carries no academic credit but does come with a completion letter.

Three types of academic accreditation are available:

1. Programs that meet a specific profession or vocation. For example, for paramedic programs, the Commission on Accreditation of Allied Health Education Programs (CAAHEP), administered through the Committee on Accreditation of Educational Programs for the Emergency Medical Services Professions (CoAEMSP), determines program accreditation.
2. Educational and training organizations that meet federal requirements for tuition reimbursement. This covers truck driver schools, computer training institutions, and colleges and universities.
3. Educational institutions that meet voluntary accrediting requirements to issue degrees and academic transcripts that are acceptable to other educational institutions.

<u>Accreditation</u> in higher education is defined as a collegial process based on self- and peer assessment for public accountability and improvement of academic quality. Peers assess the quality of an institution or academic program and assist the faculty and staff in improvement.

Voluntary accreditation has been in place for more than a hundred years. It was linked with federal tuition benefits in the 1952 Korean War GI Bill after issues arose with "diploma mills" after enactment of the original GI Bill in 1944.

The Council for Higher Education Accreditation (CHEA) was established in 1996 as a national advocate and institutional voice for self-regulation of academic quality through accreditation. CHEA is an association of 3000 degree-granting colleges and universities that recognizes 60 institutional and programmatic accrediting organizations. Accreditation through a regional organization is how most universities and colleges participate in the process. Accreditation through a regional organization means that the degrees awarded and courses completed are recognized for transfer or credit.

Universities and schools may participate in a voluntary regional accrediting process to issue degrees and academic transcripts that are acceptable to other educational institutions.

Under CHEA, six regional organizations provide academic accreditation:

- Middle States Commission on Higher Education
- New England Association of Schools and Colleges Commission on Institutions of Higher Education
- North Central Association of Colleges and Schools Higher Learning Commission
- Southern Association of Colleges and Schools Commission on Colleges
- Western Association of Schools and Colleges Accrediting Commission for Community and Junior Colleges
- Western Association of Schools and Colleges Accrediting Commission for Senior Colleges and Universities

A number of universities offer degrees entirely through distance education, delivering academic coursework either from an interactive Web-based service or through weekly assignments delivered by mail. Some have achieved regional academic accreditation, including those that have no bricks-and-mortar campus.

American Public University System (APUS) started as American Military University in 1991. APUS has regional and Distance Education and Training Council (DETC) accreditation. Despite its regional accreditation, APUS provides this guidance to students in its 2012 *Undergraduate Catalog*: "APUS cannot guarantee that its credit will be accepted as transfer credit into another university. Accreditation does not provide automatic acceptance by an institution of credit earned at another institution, as acceptance of credit is always the prerogative of the receiving institution."

■ Fire Fighter Certification Programs

Chief Engineer Ralph J. Scott of the Los Angeles Fire Department (LAFD) is one of the fathers of fire fighter certification training, creating a fire college in 1925. The LAFD training staff researched and documented every task a fire fighter might be required to perform. The list of almost 2000 entries evolved into a document that became known as *The Trade Analysis of Fire Fighting*. While serving as president of the IAFC in 1928, Scott convinced the U.S. Department of Vocational Education to accept this list as an official definition of fire fighter tasks.

The NFPA helped standardize fire fighter training by publishing the inaugural edition of NFPA 1001, *Standard for Fire Fighter Professional Qualifications*, in 1974. For the first time, there was a national consensus on the knowledge and skills a fire fighter should possess. The development of this standard started a trend toward developing national consensus standards on a wide range of fire service occupations TABLE 8-1.

■ Accreditation of Certification Programs

Every state or commonwealth in the United States has some type of fire service professional certification system in place. The particular programs range from local to national, depending on local history, government structure, legislation or regulation, and funding. Most of these systems are based on the NFPA professional qualifications standards.

Table 8-1 Standards for Fire Service Professional Qualification

- NFPA 472, *Standard for Competence of Responders to Hazardous Materials/Weapons of Mass Destruction Incidents*
- NFPA 1001, *Standard for Fire Fighter Professional Qualifications*
- NFPA 1002, *Standard for Fire Apparatus Driver/Operator Professional Qualifications*
- NFPA 1003, *Standard for Airport Fire Fighter Professional Qualifications*
- NFPA 1005, *Standard for Professional Qualifications for Marine Fire Fighting for Land-Based Fire Fighters*
- NFPA 1006, *Standard for Technical Rescuer Professional Qualifications*
- NFPA 1021, *Standard for Fire Officer Professional Qualifications*
- NFPA 1031, *Standard for Professional Qualifications for Fire Inspector and Plan Examiner*
- NFPA 1033, *Standard for Professional Qualifications for Fire Investigator*
- NFPA 1035, *Standard for Professional Qualifications for Fire and Life Safety Educator, Public Information Officer, and Juvenile Firesetter Intervention*
- NFPA 1037, *Standard for Professional Qualifications for Fire Marshal*
- NFPA 1041, *Standard for Fire Service Instructor Professional Qualifications*
- NFPA 1051, *Standard for Wildland Fire Fighter Professional Qualifications*
- NFPA 1061, *Standard for Professional Qualifications for Public Safety Telecommunicator*
- NFPA 1071, *Standard for Emergency Vehicle Technician Professional Qualifications*
- NFPA 1072, *Standard for Hazardous Materials/Weapons of Mass Destruction Emergency Response Personnel Professional Qualifications*
- NFPA 1081, *Standard for Industrial Fire Brigade Member Professional Qualifications*
- NFPA 1521, *Standard for Fire Department Safety Officer*

Development of the NFPA professional qualifications standards during the 1970s created a need to validate the many different certification systems that were already in use and to establish criteria for new systems. The NFPA standards define the minimum qualifications that an individual must demonstrate to be certified at a given level, such as Fire Officer I and

Fire Marks

Start of the FDNY Fire College

New York City began the first organized instruction of fire officers in 1869, when Commissioner Alexander Shaler established an Officer's School. That school published the *Manual of Instructions for Commanding Officers of Engine and Hook and Ladder Companies*. Formalized instruction of Fire Department of New York (FDNY) fire fighters began when the department purchased the first scaling ladders in 1882. The inventor of the Hoell Lifesaving Appliance came to New York to teach scaling ladder methods and techniques. The School of Instruction was established in 1883 and continued to evolve, becoming the FDNY Fire College in 1911. Four levels of instruction were conducted at the college: Officer, Engineer, Company, and Probationary Fireman Schools.

Getting It Done

DETC Accreditation Not Always Accepted

The Distance Education and Training Council provides national academic accreditation of colleges and universities that offer degree programs delivered through distance education methods. The DETC accreditation process uses the same CHEA standard as the six regional agencies. Nevertheless, fewer than half of the regionally accredited academic institutions will recognize the student's achievement if the college or university has only DETC accreditation. As a consequence, coursework from a college that just has DETC accreditation may not count toward a degree from a college or university with regional academic accreditation.

Fire Officer II. Accreditation establishes the qualifications of the system to award certificates that are based on the standards. A certification that is awarded by an accredited agency or institution is generally recognized by other accredited agencies and organizations.

Accreditation is a system whereby a certification organization determines that a school or program meets the requirements of the fire service. A group of impartial experts is assigned to thoroughly review a given program and determine whether it is worthy of accreditation. Two organizations provide accreditation to fire service professional certification systems: the National Board on Fire Service Professional Qualifications and the International Fire Service Accreditation Congress.

National Board on Fire Service Professional Qualifications

The Joint Council of National Fire Service Organizations created the National Professional Qualifications System (also known as the "Pro Board") in 1972. When the Joint Council dissolved in 1990, the Pro Board evolved into the independent National Board on Fire Service Professional Qualifications (NBFSPQ). The board of directors consists of representatives from national fire service organizations that have an interest in training and certification. The NBFSPQ has accredited the certification programs that are operated by 49 states, provinces, and other agencies, covering 85 levels in 18 different standards.

International Fire Service Accreditation Congress

Established by the National Association of State Directors of Fire Training, the International Fire Service Accreditation Congress (IFSAC) provides accreditation to certificate-issuing entities. IFSAC also accredits fire-related degree programs at the college and university levels. The IFSAC process follows the CHEA format and operates through Oklahoma State University. In 2012, there were 55 IFSAC-accredited fire service training programs and 23 fire science degree programs.

Building a Professional Development Plan

The IAFC's *Officer Development Handbook* defines professional development as "the planned, progressive, life-long process of education, training, self-development and experience." The handbook provides a guide for progressing from Supervising Fire Officer to Executive Fire Officer, building on the work done by the U.S. Fire Administration/National Fire Academy's Fire and Emergency Services Higher Education (FESHE) network. This integrated National Professional Development Model shows how education and training fit into a path that starts at Fire Fighter I and can end at Chief Fire Officer **FIGURE 8-7**.

■ Supervising Fire Officer Preparation

Training to become a supervising fire officer includes achieving the requirements laid out for NFPA Fire Officer I (NFPA 1021), Incident Safety Officer (NFPA 1521), Hazardous Material at Operations Level (NFPA 472), and the equivalent of NFPA Instructor I (NFPA 1041) and Inspector I (NFPA 1031).

Required lower-level college courses to become a supervising fire officer include English composition, public speaking, business communications, biology, chemistry, psychology, sociology, finite math (or algebra), business computer systems, health/wellness, American government, and human resources management. Specialized lower-level college courses include fire behavior, building construction, and fire administration.

The supervising fire officer candidate needs 3 to 5 years of experience in agency operations and 200 hours of experience as an acting unit officer handling emergency responses and nonemergency activities. He or she must develop and deliver training classes, participate in the planning process, participate in mass-casualty training exercises, and be capable of operating as a single-unit supervisor within an Incident Management System.

At this level, the supervising fire officer candidate has performed a personal and professional inventory. The goal of the inventory is to establish a career map that identifies personal traits, strengths, and areas for improvement.

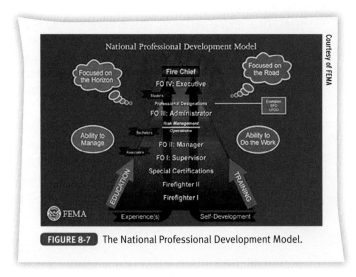

FIGURE 8-7 The National Professional Development Model.

Managing Fire Officer

Training to become a managing fire officer includes becoming an NFPA Fire Officer II, completing the National Fire Academy Leadership Development and Initial Company Operations series, and training for Unified Command for Multi-Agency and Catastrophic Incidents, Public Information Officer, and the equivalent of NFPA Investigator I (NFPA 1033) and Fire and Life Safety Educator I (NFPA 1035).

Lower-level college courses needed include statistics, interpersonal communication, philosophy, critical reasoning, professional ethics, professional report writing, accounting analysis, introduction to law, and introduction to planning. Specialized lower-level college courses are needed in fire service management, prevention, and education; fire protection systems; and fire protection hydraulics.

The managing fire officer candidate needs 2 to 4 years of experience as a qualified supervising fire officer. He or she must function as an acting officer for multicompany operations that include emergency and nonemergency activities. He or she must also be the supervisor or an aide to the incident commander of a multicompany operation, deliver performance appraisals and discipline, and participate in the development or updating of local emergency plans.

The managing fire officer candidate is expanding the career map to explore areas of special interest and to seek a mentor.

CPSE Designations

The Center for Public Safety Excellence (CPSE) provides the terminal focal point in fire officer professional development. It builds upon the NFPA Fire Officer Professional Qualification certification training and the IAFC *Officer Development Handbook* and, as described in the handbook, provides "personal guidance for career planning and development and recognizes lifelong career excellence and achievement."

Under CPSE, the Commission on Professional Credentialing awards the designation of Chief Fire Officer, Chief EMS Officer, Chief Training Officer, Fire Marshal, or Fire Officer to candidates who have completed an application and validation process. Successful candidates have demonstrated a strategy for continued career improvement and development. They also have met the education, certification, training, professional contributions, professional affiliation, and community involvement requirements. There were 851 Chief Fire Officers, 83 Chief EMS Officers, 23 Chief Training Officers, 52 Fire Marshals, and 124 Fire Officers in the United States as this book went to press.

You Are the Fire Officer Conclusion

Personnel regulations prohibit extending this recruit's probationary period, so Lieutenant Williams is tasked with developing an accelerated recruit training program. The plan is developed with the help of Apparatus Operators George Anders and Karen Rollo and senior Fire Fighter James Grynski.

The team meets with the recruit, with Lieutenant Williams describing how the recruit's workday will progress until the probationary field manual is completed. The recruit will be working with Apparatus Operator Anders to complete the pump operator proficiency knowledge and skills. Fire Fighter Grynski will work with the recruit to master knowledge of the first-due district by having the recruit use easel pad maps, driving drills, and Google Earth graphics. Apparatus Operator Rollo and other members of the shift will develop flash cards to test the specialized knowledge and procedures that are part of the probationary exam. Lieutenant Williams describes the recruit's on-duty work schedule for the next 10 weeks:

0700: Line-up, flash-card pop quiz
0730–0830: Apparatus and equipment check
0830–1030: Physical training, station maintenance, and pumping drill with Apparatus Operator Anders at the rear of the fire station
1030–1100: Street drill with easel pad maps (recruit and Fire Fighter Grynski)
1100–1200: Driving drill (find five locations in the district) and lunch run
1300–1700: Scheduled company training, productivity, and flash-card pop quiz
1700–1900: Dinner and station clean-up
1900–2100: Recruit works on drawing personal map of district, target hazard overview with Apparatus Operator Rollo, or making new knowledge flash cards

Lieutenant Williams will meet with the recruit weekly to assess progress and start providing practice tests to prepare for the end-of-probation exam. The recruit is encouraged to do additional work during off-duty days.

Chief Concepts

- Training and coaching have been core fire officer tasks since the establishment of the first organized fire departments.
- Fire service training must anticipate high-risk situations, urgent time frames, and difficult circumstances. A wide variety of methods and practices are included in the overall category of training.
- A fire officer must be prepared to conduct company-level training exercises and evolutions to ensure that the company is prepared to perform its basic responsibilities effectively and efficiently.
- The four-step method is a core part of most Fire Instructor I certification programs. It includes the following components:
  - Preparation
  - Presentation
  - Application
  - Evaluation
- For best performance, many skill practice sessions should be performed while fire fighters are wearing full personal protective clothing and operating within a realistic fire-ground situation. This may require construction of training props.
- Mentoring is a one-on-one process in which the more experienced person provides a deliberate learning environment through instructing, coaching, providing experiences, modeling, and advising.
- Teaching new skills takes more time than maintaining proficiency of existing skills. The fire officer should obtain as much information as possible about the device or procedure, especially identifying any fire fighter safety issues.
- Psychomotor skill levels are divided into four categories:
  - Initial
  - Plateau
  - Latency
  - Mastery
- At the very first meeting with a new fire fighter trainee, the fire officer should explain the procedures in the fire station when the company receives an alarm, assign a senior fire fighter to function as a mentor, and describe any restrictions that are placed on fire fighters in training.
- Four federal regulations govern fire fighter training:
  - OSHA regulation 29 CFR 1910.1030, *Occupational Exposure to Bloodborne Pathogens*
  - OSHA regulation 29 CFR 1910.20, *Hazardous Waste Operations and Emergency Response (HAZWOPER) Training*
  - OSHA regulation 29 CFR 1910.134, *Respiratory Protection Training*
  - Homeland Security Presidential Directive 5 (HSPD-5), *Managing Domestic Incidents*
- Once the skills that must be learned immediately are covered, the fire officer should concentrate on the skills the fire fighter in training needs to know to stay alive.
- Fire fighters must receive appropriate training before participating in any live fire evolutions. In addition to the prerequisite training, the trainee must be equipped with full protective clothing, a personal alert safety system device, and self-contained breathing apparatus that is compliant with the relevant NFPA standard.
- NFPA 1403 provides detailed instructions on how to conduct a safe live fire training evolution under five different scenarios:
  - Acquired structures
  - Gas-fired training center buildings
  - Non–gas-fired training buildings
  - Exterior props
  - Exterior Class B fires
- NFPA 1403 identifies activities prohibited during live fire training:
  - No live "victims"
  - No flammable or combustible liquids as fuel
  - One fire evolution can occur in an acquired structure at a time
- The fire officer may need to develop a specific training program that is not covered by an existing certification training program or prepared lesson plan. Five steps can help accomplish this goal:
  - Assess needs
  - Establish objectives
  - Develop the training program
  - Deliver the training
  - Evaluate the impact
- Education is the process of imparting knowledge or skill through systematic instruction, whereas training is directed toward the practical application of education to produce an action.
- The fire service benefits from advanced education of its members both operationally and in regard to its image.
- Three types of academic accreditation are available:
  - Programs that meet a specific profession or vocation
  - Educational and training organizations that meet federal requirements for tuition reimbursement
  - Educational institutions that meet voluntary accrediting requirements to issue degrees and academic transcripts acceptable to other educational institutions
- Chief Engineer Ralph J. Scott of the Los Angeles Fire Department, one of the fathers of fire fighter certification training, created a fire college in 1925.

Wrap-Up, continued

- The NFPA helped standardize fire fighter training when it published the inaugural edition of NFPA 1001, *Standard for Fire Fighter Professional Qualifications*, in 1974.
- Accreditation is a system whereby a certification organization determines that a school or program meets with the requirements of the fire service.
- The National Professional Development Model shows how education and training fit into a path that starts at Fire Fighter I and can end at Chief Fire Officer.
- The supervising fire officer must develop and deliver training classes, participate in the planning process, participate in mass-casualty training exercises, and be capable of operating as a single-unit supervisor within an Incident Management System.
- The managing fire officer must function as an acting officer for multicompany operations that include emergency and nonemergency activities.
- Under CPSE, the Commission on Professional Credentialing awards the designation of Chief Fire Officer, Chief EMS Officer, Chief Training Officer, Fire Marshal, or Fire Officer to candidates who have completed an application and validation process.

Hot Terms

<u>Accreditation</u> A system whereby a certification organization determines that a school or program meets with the requirements of the fire service.

<u>Coaching</u> A method of directing, instructing, and training a person or group of people with the aim to achieve some goal or develop specific skills.

<u>Education</u> The process of imparting knowledge or skill through systematic instruction.

<u>Job instruction training</u> A systematic four-step approach to training fire fighters in a basic job skill: (1) prepare the fire fighters to learn, (2) demonstrate how the job is done, (3) try them out by letting them do the job, and (4) gradually put them on their own.

<u>Mentoring</u> A developmental relationship between a more experienced person (a mentor) and a less experienced person (a protégé).

<u>Professional development</u> Skills and knowledge attained for both personal development and career advancement.

<u>Training</u> The process of achieving proficiency through instruction and hands-on practice in the operation of equipment and systems that are expected to be used in the performance of assigned duties.

<u>Unconsciously competent</u> The highest level of the Conscious Competence Learning Matrix developed by Dr. Thomas Gordon in the 1970s. At the Unconsciously Competent level, the skill becomes so practiced that it enters the unconscious parts of the brain—it becomes second nature.

References and Additional Resources

NFPA reprinted material is not the complete and official position of the NFPA on the referenced subject, which is represented only by the standard in its entirety.

American Public University Systems. (2012). *2012–2013 Undergraduate Catalog*. Charles Town, WV: American Public University Systems.

Anderson, L. W., et al. (2000). *A Taxonomy for Learning, Teaching, and Assessing: A Revision of Bloom's Taxonomy of Educational Objectives*. Old Tappan, NJ: Pearson/Allyn & Bacon.

Bear, M. P., with T. Nixon. (2006). *Bear's Guide to Earning Degrees by Distance Learning* (16th ed.). Berkley, CA: 10 Speed Press.

Bligh, Donald A. (2000). *What's the Use of Lectures?* San Francisco, CA: Jossey-Bass.

Bloom, B. (1956). *Taxonomy of Educational Objectives, Handbook I: The Cognitive Domain*. New York, NY: David McKay Company.

Coleman, R. J. (2000, November). "Professional Chief Ralph Scott." *Fire Chief*. Vol. 44, Issue 11.

Council for Higher Education Accreditation. (2013). *Directory of CHEA-Recognized Organizations 2012–2013*. Washington, DC: Council for Higher Education Accreditation.

Fahy, R. F. (2012). *U.S. Firefighter Deaths Related to Training, 2001–2010*. Quincy, MA: National Fire Protection Association.

Farley, T. (2007). *August—Training Activities*. Fairfax, VA: National Fire Fighter Near-Miss Reporting System.

Fire and Emergency Service Image Task Force. (2013). *Taking Responsibility for a Positive Public Perception*. Fairfax, VA: International Association of Fire Chiefs.

Gillen, A., D. L., Bennett, and R. Vedder. (2010, October). *The Inmates Running the Asylum? An Analysis of Higher Education Accreditation*. Washington, DC: Center for College Affordability and Productivity.

Gordon, T. (1975). *Teacher Effectiveness Training*. New York, NY: Crown.

International Association of Fire Chiefs and National Fire Protection Association. (2012). *Live Fire Training: Principles and Practice*. Burlington, MA: Jones & Bartlett Learning.

International Association of Fire Chiefs and National Fire Protection Association. (2014). *Fire Service Instructor: Principles and Practice* (2nd ed.). Burlington, MA: Jones & Bartlett Learning.

Kiewra, K. A. (2002). "How Classroom Teachers Can Help Students Learn and Teach Them How to Learn." *Theory into Practice* 41(2): 71–80.

Kimball, W. Y. (1966). *Fire Attack 1: Command Decisions and Company Operations*. Boston, MA: National Fire Protection Association.

Knowles, M. S., et al. (2005). *The Adult Learner: The Definitive Classic in Adult Education and Human Resource Development*. Burlington, MA: Elsevier, Butterworth, Heinemann.

Kuchinke, K. P. (1999). "Adult Development Toward What End? A Philosophical Analysis of the Concept as Reflected in the Research, Theory, and Practice of Human Resource Development." *Adult Education Quarterly* 49(4): 148.

McClincy, W. D. (2011). *Instructional Methods for Public Safety*. Sudbury, MA: Jones and Bartlett.

Merriam, S. B., and R. S. Caffarella, eds. (1999). *Learning in Adulthood*. San Francisco, CA: Jossey-Bass.

Metz, E. J. (2013). "New Venues for Discovering Fire and Emergency Services Literature." *Fire Technology* 49(2): 185–194.

Milan, K. O. (2005). *Evaluation of Electronic Student Response Technology in an Introductory National Incident Management System Training Course*. Emmitsburg, MD: Executive Fire Officer Program, U.S. Fire Academy.

National Association of EMS Educators. (2013). *Foundations of EMS Educators: An EMS Approach*. Independence, KY: Delmar/Cengage Learning.

National Volunteer Fire Council and National Fire Protection Association. (2012). *Understanding and Implementing Standards: NFPA 1500, 1720, and 1851*. Greenbelt, MD: National Volunteer Fire Council.

Page, J. O. (1973). *Effective Company Command*. Alhambra, CA: Borden.

Patrick, R. (2006). *Training Techniques and Professional Development: Managing Volunteer and Combination Emergency Service Organizations*. York, PA: VFIS, 133–156.

Professional Development Committee. (2010). *IAFC Officer Development Handbook* (2nd ed.). Fairfax, VA: International Association of Fire Chiefs.

Prouty, D. A. (2004). *Assessing the Impact of the Los Angeles County Fire Department Management Development Program*. Emmitsburg, MD: Executive Fire Officer Program, U.S. Fire Academy.

Secolsky, C., and Denison, D. B., eds. (2012). *Handbook on Measurement, Assessment, and Evaluation in Higher Education*. New York, NY: Routledge.

Smeby, L. C. Jr. (2014). *Fire and Emergency Services Administration: Management and Leadership Practices* (2nd ed.). Burlington, MA: Jones & Bartlett Learning.

Soptich, L. A. (2008). *Determining Core Emergency Scene Operational Standards for Eastside Fire & Rescue*. Emmitsburg, MD: Executive Fire Officer Program, U.S. Fire Administration.

Speck, M., and C. Knipe. (2005). *Why Can't We Get It Right? Designing High-Quality Professional Development for Standards-Based Schools* (2nd ed.). Thousand Oaks, CA: Corwin Press.

Stapleton, L. (1983). *Thirty Years on the Line*. Dover, NH: dmc associates.

Thiel, A. K., and R. Jennings. (2012). *Professional Development. Managing Fire and Rescue Services*. Washington, DC: International City/County Management Association, 247–272.

Zukoski, E. E. (1970). *Training and Education in the Fire Service*. Washington, DC: National Academy of Science.

FIRE OFFICER in action

Station 100 provides a wide range of services: fire suppression, hazardous materials, technical rescue, public education, and emergency medical services. There are dozens of highly critical activities that are infrequently used. Lieutenant Williams considers a recommendation made by public safety expert Gordon Graham: Spend 5 minutes each day covering one high-risk/low-frequency knowledge or skill item.

Lieutenant Williams asks the crew to start a list of items that need to be covered in a 5-minute drill that will be delivered as part of the morning line-up. Some items are part of the monthly training program; some topics come from the Fire Fighter Near Miss, Everyone Goes Home, and Firefighter Close Calls programs; and other topics come from standard operating procedures and regulations.

Fire fighter training and development has been a core fire officer task since the Officer's School was established in New York City in 1869. While the apparatus, tools, and educational techniques may have changed, the special responsibility of training and coaching fire fighters remains the same.

1. When can a person be utilized as a victim in a live fire training session?
   A. Never
   B. Only in gas-fired training center buildings
   C. Only when there is a safety officer assigned to the victim division
   D. Only in non–gas-fired training center buildings

2. Which of the following areas represents a federal regulation that affects every fire fighter?
   A. Fire Fighter I
   B. Rescue Systems I
   C. SCBA fit testing
   D. WMD awareness

3. Using a video to show an incident or situation is an example of the _____ step in the four-step process.
   A. preparation
   B. presentation
   C. application
   D. evaluation

4. "The apparatus operator can remember the route to an area of the district where the engine company has not had a response for several months" is a description of the _____ level of psychomotor skill.
   A. initial
   B. plateau
   C. latency
   D. mastery

FIRE OFFICER *in action* (continued)

Fire Captain *Activity*

Battalion Chief Johnson meets with Captain Davis: "We need to formalize our professional development program for company and command officers. I need a model plan for a Captain position in 2 weeks."

NFPA Fire Officer II Job Performance Requirement 5.2.3

Create a professional development plan for a member of the organization, given the requirements for promotion, so that the individual acquires the necessary knowledge, skills, and abilities to be eligible for the examination for the position.

Application of 5.2.3

1. **Survey peers.** Using a fire department with which you are familiar, survey 5 to 12 command-level officers to determine the three most important competencies a new captain must possess. Compare their answers to NFPA 1021 and identify gaps.
2. **Value of CPSE Fire Officer Designation.** The Center for Public Safety Excellence (CPSE), originally established to credential and designate Chief Fire Officers, has expanded to include Fire Officer designations. Determine whether the CPSE Fire Officer designation is a valuable element in a Captain professional development program within a fire department with which you are familiar.
3. **College credentials.** The U.S. Fire Administration has expanded its acceptance of academic work to include coursework from programs with only DETC national accreditation. Determine whether this is an issue in developing a Captain professional development program within a fire department with which you are familiar.
4. **Write a Captain professional development plan.** Write a professional development plan for a Captain position that includes a description of the education, training, self-development, and experience elements. Address any gaps discovered when completing Task 1. Include specific resources such as academic institutions, training programs/certifications, and fire department work assignments.

Evaluation and Discipline

Fire Officer I

Knowledge Objectives

After studying this chapter, you will be able to:

- Discuss the purpose of regular fire fighter evaluations (NFPA 4.2) (NFPA 4.2.5). (pp 164–166)
- Describe methods of positive discipline (NFPA 4.2.4) (NFPA 4.2.5). (pp 166–169)
- Discuss the role of documentation and record keeping for evaluations and discipline (NFPA 4.2). (p 171)

Skills Objectives

After studying this chapter, you will be able to:

- Demonstrate positive discipline of a fire fighter (NFPA 4.2.4) (NFPA 4.2.5). (pp 166–169)

Fire Officer II

Knowledge Objectives

After studying this chapter, you will be able to:

- List and describe the components of formal evaluation and discipline (NFPA 5.2) (NFPA 5.2.1) (NFPA 5.2.2). (pp 171–177)
- Discuss the purpose of employee assistance programs. (p 177)

Skills Objectives

After studying this chapter, you will be able to:

- Conduct formal employee evaluations (NFPA 5.2) (NFPA 5.2.1) (NFPA 5.2.2). (pp 171–174)
- Develop a work improvement plan (NFPA 5.4.1). (pp 172–173)

CHAPTER 9

Principles of Fire and Emergency Service Administration (FESHE) Course Objectives

4. Select and implement the appropriate disciplinary action based upon an employee's conduct. (pp 166–169, 171–177)
6. Discuss the various levels of leadership, roles, and responsibilities within the organization. (pp 164–165, 169, 171–173, 174–177)
7. Describe the traits of effective versus ineffective management styles. (pp 165–168, 173–174)
8. Identify the importance of ethics as it relates to fire and emergency services. (pp 167–168, 173–174, 177)

You Are the Fire Officer

It is March 18, and some of the incoming A-shift members are showing aftereffects from their vigorous Feast of Saint Patrick celebration. At 6:50 A.M., Lieutenant Williams notices that Fire Fighter/EMT Arthur Wirth has not arrived. Municipal City requires fire fighters to notify their supervisor at least 1 hour before the reporting time if they cannot get to work due to illness or unplanned events.

Williams notifies Captain Davis of the missing employee. An off-going B-shift fire fighter is placed on involuntary holdover, and Davis calls Wirth. This is the third time in 7 months that Wirth has failed to show up for work. The last occasion generated a written reprimand. Wirth says he will be at the station by 8 A.M.

1. How should you handle a fire fighter who has an existing disciplinary issue?
2. What should you do about this morning's late reporting to work?
3. Should you be involved in resolving off-duty problems that affect on-duty fire fighter performance?

Introduction

Evaluation and discipline are essential components of a fire fighter's development. In turn, the fire officer plays a key role in developing fire fighter success. For probationary fire fighters, the actions of their first fire officer set the stage for a 20- to 40-year career.

Supervision of fire fighters requires that the fire officer conduct regular evaluations to provide feedback on job performance, on-duty behavior, and problem resolution. Regular evaluations of employees are required and should be approached in a standard, professional manner.

The officer is also responsible, to a certain extent, for supervising off-the-job behavior when it reflects on the fire department. Performance and behavior are two different issues, but they are linked in the sense that the officer must monitor, evaluate, and deal with both types of problems.

Positive discipline is intended to help the employee recognize problems and make corrections to improve performance or behavior. Negative discipline is intended as punishment for unsatisfactory performance, unacceptable behavior, or failure to respond appropriately to positive discipline.

Discipline should be progressive, moving from positive to negative and from minor to major. The nature and the seriousness of the problem and the employee's efforts to correct reoccurring problems influence the starting point for this process. Progressive discipline might begin with verbal counseling; then move on to a verbal reprimand with a written note to be put in the employee's file; then to a formal written reprimand; and finally to suspensions, reductions in pay, or demotion. Dismissal is considered the ultimate level of negative discipline for an employee.

Some problems are so serious that the earlier steps in progressive discipline might be bypassed in favor of more severe punishment. Most fire departments or municipalities maintain a description of offenses that are considered to be so serious that they can lead to immediate suspension. Although these steps and actions are spelled out in written policies, a fire officer still uses judgment and discretion when performing evaluation and disciplinary functions.

Fire Officer I

Evaluation

Regular evaluations are performed to ensure that each fire fighter knows which type of performance is expected while on the job and where he or she stands in relation to those expectations. This process helps the fire fighter set personal goals for professional development and performance improvement and provides positive motivation to perform at the highest possible level.

Most career fire departments require a supervisor to conduct an annual performance evaluation for each assigned employee. The annual performance evaluation is a formal written documentation of the fire fighter's performance during the rating period. This permanent personnel record is important to both management and the fire fighter, but it cannot replace the ongoing supervisory actions that maintain and improve employee performance throughout the year.

Volunteer fire departments provide an equivalent form of periodic evaluation for each member by a higher-ranking individual. Although these documentation procedures are often less structured, every member deserves evaluation. The responsibilities of a supervisor are just as important in a volunteer fire department as in a full-time career organization.

Starting the Evaluation Process with a New Fire Fighter

Fire officers have a special responsibility when starting an evaluation process with a probationary fire fighter, because the fire officer is helping to shape that individual's fire department career. The officer will be shepherding the probationary fire fighter through his or her first real-life emergency experiences, providing feedback and guidance. Most fire fighters have vivid memories of their first supervisor and continue to respond to the expectations established by that officer.

New fire fighters start with wide variations in the range and depth of their skills. In some departments, probationary members can start to ride the apparatus after receiving a few hours of orientation. Many fire departments require new members to achieve NFPA 1001 Fire Fighter and EMT certification before they are authorized to respond to any alarms. It is the fire officer's responsibility to determine each individual's skills, knowledge, aptitudes, strengths, and weaknesses and then to set specific expectations for each new fire fighter.

Recruit Probationary Period

Most fire departments include structured in-station training as part of the recruit fire fighter's probationary period. Regardless of the new fire fighter's level of certification or pre-employment experience, the fire officer is responsible for evaluating each individual during the probationary period. Often a classified job description specifies all of the required knowledge, skills, and abilities that a fire fighter is expected to master within a specified time period to complete the probationary requirements. The fire fighter should be provided with a copy of these specific requirements to use as a checklist FIGURE 9-1.

The recruit fire fighter is expected to obtain experiences and demonstrate competencies related to the classified job description during the probationary period. This period often includes assignments on different types of companies and specialty units operated by the department, such as an engine company, a ladder company, an ambulance, and the telecommunications center. Competencies may include demonstrating all of the skill sets required for NFPA 1001 Fire Fighter II certification.

Structured probationary programs require the fire officer to complete a monthly evaluation of each probationary fire fighter. This evaluation typically assesses the probationary fire fighter's progress in four areas:

- Fire fighter skill competency, including proficiency as an apparatus operator
- Progress in learning job-specific information not covered in basic training, such as the local fire prevention code and the department's rules, regulations, and standard operating procedures
- Progress in learning the fire company district, including streets and target hazards
- Performance of other job tasks, such as recording deliveries, performing housework, conducting in-station tours, and completing reports

In volunteer fire departments, it is common for different individuals to certify that a probationary fire fighter has met the requirements for different components of the program. One officer should be specifically assigned to oversee the progress of each individual probationary member, thereby ensuring that the overall program requirements are being accomplished successfully.

The fire fighter probationary period lays the foundation for a long career. The fire officer has a special opportunity to prepare future departmental leaders by providing a comprehensive and effective probationary period.

Providing Feedback After an Incident or Activity

Performance evaluation should be a continual supervisory process, not a special event that is performed only when a scheduled rating has to be submitted. Frequent feedback from the fire officer should keep fire fighters aware of how they are doing, particularly after incidents or activities that present a special challenge.

Feedback on individual performance is most effective when delivered as soon as possible after an action or incident. That means providing essential feedback before the ashes get cold after a structure fire. At this point, the fire fighter is intensely aware of the event and wants to know how well he or she performed. The fire officer needs to be ready to provide specific information to recognize or improve fire fighter performance FIGURE 9-2.

Fire Marks

Dodd J. Miller Training Academy
Graduates from the Dodd J. Miller Training Academy at the Dallas, Texas, Fire-Rescue Department have completed a 15-month training program that includes fire suppression and paramedic certifications. After completing a 6-month EMT and fire fighter training program, the recruits are temporarily assigned to a fire station while they complete their paramedic training.

Safety Zone

Immediately Correct Unsafe Conditions
Although negative feedback should be issued to an individual in private, the fire officer must correct unsafe conditions as soon as they are noticed. There is no excuse for allowing an unsafe action or situation to occur without taking corrective action, even if that means shouting an order at a crowded fire ground or issuing a direct instruction over the radio. Once the incident is under control, the fire officer needs to follow up with a private face-to-face meeting with the fire fighter or fire fighters who created or ignored an unsafe condition.

City of Charlottesville Fire Department
Fire Fighter Job Description

General Definition of Work
The fire fighter performs responsible service work in fire suppression and prevention; does related work as required. Work is performed under the regular supervision of a company and/or shift commander.

Typical Tasks
Responds to alarms, drives and operates equipment and related apparatus, and assists in the suppression of fires, including rescue, advancing lines, entry, ventilation and salvage work, extrication, and emergency medical care of victims.
Performs cleanup and overhaul work, establishes temporary utility services.
Assists in maintaining and repairing fire apparatus and equipment, and cleaning fire stations and grounds.
Checks fire hydrant flows.
Makes fire code inspections of business establishments and prepares pre-fire plans.
Responds to emergency and nonemergency calls, pumps out basements, inspects for gas leaks, secures vehicle accidents, inspects chimneys, etc.
Participates in continuing training and instruction programs by individual study of technical material and attendance at scheduled drills and classes.
Conducts station tours for the public, school, and community demonstrations and programs.
Backs up for dispatching personnel, monitors alarm boards, receives and transmits radio and telephone messages.
Performs related tasks as necessary.

Knowledge, Skills, and Abilities
General knowledge of elementary physics, chemistry, and mechanics; general knowledge of technical firefighting principles and techniques, and principles of hydraulics applied to fire suppression; general knowledge of the street system and physical layout of the city; general knowledge of emergency care methods, techniques, and equipment; ability to understand and follow written and oral instructions; ability to establish and maintain cooperative relationships with fellow employees and the public; ability to keep simple records and prepare reports; possess a strong mechanical aptitude; ability to perform heavy manual labor; skill in operation of heavy emergency equipment.

Education and Experience
Any combination of experience and training equivalent to graduation from high school.

Special Requirements
Possession of a valid driver's permit issued by the Commonwealth of Virginia.

Future Requirements at Three Years of Service
NFPA 1001   Fire Fighter Level II
NFPA 1002   Driver/Operator Certification
Commonwealth of Virginia Emergency Medical Technician or greater

FIGURE 9-1 Sample fire fighter job description.

Discipline

Discipline is a moral, mental, and physical state in which all ranks respond to the will of the leader. The fire officer builds discipline by training to meet performance standards, using rewards and punishments judiciously, instilling confidence in and building trust among team leaders, and creating a knowledgeable collective will.

Within the fire department, discipline is divided into positive and negative sides. Positive discipline is based on encouraging and reinforcing appropriate behavior and desirable performance. Negative discipline is based on punishing inappropriate behavior or unacceptable performance. Both positive and negative discipline can be applied to a full range of activities, including emergency incidents and administrative functions.

Positive discipline should be used before negative discipline is applied. Progressive discipline refers to starting out to correct a problem with positive discipline and then increasing the intensity of the discipline if the individual fails to respond to the positive form, perhaps by using mild negative discipline. Negative discipline might have to be used to an increasingly greater extent in a situation in which an individual fails to respond in an appropriate manner to correct the problem. There are exceptions to this rule: Some actions or behaviors are so unacceptable that they must result in immediate negative discipline.

Near-Miss REPORT

Report Number: 12-0000088

Event Description: Our department's aerial apparatus was requested in a neighboring jurisdiction for a second alarm at a church fire. The apparatus responded with a crew of four, including the on-duty shift commander (captain), lieutenant, and two fire fighters. The first problem incurred was that the fire fighter assigned to the jump seat told the fire fighter assigned to drive that he would drive instead. This action was not known at the time of the response by the captain, who ordinarily responds in a separate car, but instead responded on the apparatus due to the second alarm request.

Upon arrival at the scene, the aerial apparatus was directed to the B/C corner of the pavement, to ventilate the roof. The vehicle weighs 72,000 pounds, and the captain advised the driver/operator to stay on a paved walkway as much as possible. The vehicle was positioned as ordered. Because of the changing fire conditions, the captain went to do a face-to-face meeting with the incident commander as the vehicle was set up. The lieutenant and a fire fighter readied the needed equipment to prepare to open the roof for ventilation. As the captain discussed operations with the incident commander, orders were changed to ventilate the A/B area instead.

With the aerial in position and fire fighters ascending the ladder upon return to the truck, there was no operator on the turntable. The driver/operator took it upon himself to "hand-jack" a 4-inch LDH supply line some 200 feet away from the vehicle. Upon noticing this, the captain immediately assigned a department member from a dispatched ambulance to take the turntable position and the initial driver/operator was reassigned to EMS.

Following the incident, the captain and lieutenant met with the driver/operator about his actions. The concerns were the safety aspects of leaving fire fighters in a position exposed to danger. The driver/operator felt that water supply was more critical and he felt that he "didn't have to babysit" the turntable. These actions violated department standard operating guidelines, which state that the operator shall remain with the turntable anytime fire fighters are working on the aerial ladder. It also violated the department's SOGs regarding chain of command, in that the driver/operator had both a captain and a lieutenant whom he could have consulted before initiating his actions. It was stressed to the fire fighter what his responsibilities are as the driver/operator and which concerns arise regarding the safety of the fire fighters operating on the aerial.

Lesson Learned: Following the incident, the captain and lieutenant met with the fire fighter, who continued to defend his actions based on his perception of the safety concerns for water supply. Situational awareness and risk were discussed and, given previous concerns and disciplinary actions, the fire fighter was disciplined by the captain. Information on the incident was sent through the chain of command, and the department is investigating the fire fighter's actions and considering further disciplinary measures. However, he has already been given copies of department SOGs covering these actions and had already received remedial training on operating the aerial, and more remedial training is now scheduled.

The problem is not so much with the physical operation of the truck but rather with the situational awareness and the risk involved. As part of the remedial training, safety, risk versus benefit, and the importance of situational awareness will be addressed. As a small department, our fire fighters have to perform multiple duties, ranging from EMS to driving many types of equipment. The department is concerned about this fire fighter's actions in similar situations, such as being the pump operator, or other positions of responsibility in which others are dependent on his actions. The department is in no position to single him out and assign him to only certain positions. One immediate action is to ensure that he is always assigned to a crew with an officer.

■ Positive Discipline: Reinforcing Positive Performance

Positive discipline is directed toward motivating individuals and groups to meet or exceed expectations. The key to positive discipline is to convince these parties that they want to do better and are capable of and willing to make the effort. A fire officer provides positive discipline by identifying weaknesses, setting goals and objectives to improve performance, and providing the capability to meet those targets.

The starting point for positive discipline is to establish a set of expectations for behavior and performance. Once these expectations are known, there must be a consistent and conscientious effort to meet them. The expectations have to apply to the entire team as well as to each individual team member.

FIGURE 9-2 Performance feedback should be delivered as soon as possible after an action or incident.

Positive discipline is reinforced by recognizing improved performance and rewarding excellent performance.

If an individual fire fighter's performance needs improvement in a particular area, the fire officer should coach that person, providing guidance and extra opportunities to correct the problem. Often, this step is needed when the fire fighter is unable to perform the task or skill because the task or skill was never learned. Sometimes, simply pointing the individual in the right direction and offering encouragement can achieve the objective. In many cases, the officer can arrange for another fire fighter to work with the individual to correct the problem.

Teamwork is a key factor in ensuring success for fire companies. In addition to each individual having the required knowledge and skills, the company must be able to work efficiently and effectively as a team. All of the company members have to work, learn, and practice together to become capable and confident. The fire officer has to provide the leadership to make this cohesiveness happen.

The officer sets the stage for positive discipline by setting clear expectations and by "walking the talk." Fire officers are working supervisors; that is, they supervise while directly participating in firefighting activities and performing nonemergency duties. The officer should demonstrate a personal commitment to the department's goals, objectives, programs, rules, and regulations by participating in all of the activities that are expected of fire fighters, such as physical fitness training and regular company drills. Fire fighters can gauge an officer's level of commitment by observing his or her own self-discipline **FIGURE 9-3**.

Competitiveness can also be used as a stimulant in positive discipline, particularly at the company level. Fire fighters are naturally competitive, and most fire companies work hard to prove that they can be better, faster, more skillful, or more impressive than a rival company. The officer has to point that competitive energy in a positive direction, making sure that the ultimate objective is high performance.

Empowerment

Empowerment is one of the most effective strategies within the realm of positive discipline. Fire fighters often complain that

FIGURE 9-3 Fire officers set the stage for positive discipline by "walking the talk."

they have little control over their work environment; they are told where to go, what to do, and when to do it. Fire officers can help make fire fighters feel stronger by learning how to control their own destiny. An officer who identifies an area where improvement is needed can often empower fire fighters to correct the problem on their own. It is important for the officer to identify the target and provide the resources, but doing the work on their own and demonstrating their capabilities can be a very positive motivator for the fire fighters.

Providing information to help fire fighters learn more about the job can support the empowerment process. The first component could be described as "Local Government 101." This information helps fire fighters learn more about how the fire department and local government work. It could include reviewing the approved budget for the next fiscal year or discussing which functions are performed by each part of the organization. The more fire fighters understand about how the organization and local government work, the more they can feel a sense of participation.

The second component could be described as "Success 102." This information identifies the tools that others have used to achieve success within the fire department. If some companies are consistently recognized for positive performance, what are they doing right? When a few individuals achieve high scores on every promotional examination, what is their secret method? Do fire fighter self-study groups improve promotional examination results? Answering these questions may take some research by the fire officer, such as identifying the individuals or groups that have performed well in the promotion process and asking them how they prepared. An officer can often help fire fighters succeed by identifying successful practices.

Company Tips

Hands-On Skill Drills
A commitment to in-station training produces competent and confident fire fighters and fire companies. A fire officer should provide frequent hands-on opportunities for fire fighters to practice their emergency service delivery skills, such as laying out hose, throwing ladders, forcing doors, and cutting roofs. Some type of activity should be scheduled every day, as skills have been shown to deteriorate if not used within 90 days of training.

Many fire departments have developed training props for practicing with hand and power tools. As a confidence builder, fire fighters should practice working with hose lines and tools in simulated fire-ground scenarios while wearing full personal protective clothing **FIGURE 9-4**. Fire officers should also keep an eye out for opportunities to conduct realistic exercises in buildings scheduled for demolition.

Assessment Center Tips

Evaluation and Discipline Procedures
Evaluation and discipline procedures are part of almost every promotional exam. The candidate must be completely familiar with the local procedures in these areas. For example, preparing for a captain exam will require knowledge of evaluation and discipline activities at the battalion chief level. Frequently, the candidate is asked to write a document pertaining to these issues for a superior fire officer's signature as part of the promotional process.

One group of captain candidates struggled with a promotional writing exercise that required them to document why one of their subordinates should receive a "Fire Fighter of the Year" award from a service club. All of their promotional preparation had focused on issuing negative discipline.

Oral Reprimand, Warning, or Admonishment
An oral reprimand, warning, or admonishment is the first level of negative discipline, considered "informal" by many organizations. An "informal" discipline action stays with the fire officer and does not become part of the employee's official record.

For example, a new fire officer observes that the apparatus driver does not comply with the department regulation requiring that units responding to an emergency come to a complete stop when encountering a red traffic light. When this occurs, the fire officer should have a private face-to-face meeting with the apparatus driver. In this meeting, the fire officer would determine why the driver did not stop at a red traffic light and would clearly state the policy and expectation for all subsequent responses. If the fire officer determines the reason was willful noncompliance with the department regulation, he or she would take the following steps:

1. Tell the driver that the officer expects compliance with the regulation requiring that rigs stop at red lights before proceeding through an intersection. Provide the driver with a printed copy of the regulation.
2. Inform the driver that this is a verbal reprimand and that the fire officer will maintain a written record of the reprimand for whatever time period is required by the authority having jurisdiction. (A common practice is to maintain the written record for 366 days. At the end of that period, if there have been no further problems related to the same individual or issue, the paper record is removed and destroyed.)
3. Inform the driver that continued failure to comply with the regulation will result in more severe discipline.

In most situations, these actions will suffice to correct the fire fighter's behavior. If the fire fighter continues to have difficulty complying with the regulation, however, this is the first step in progressive negative discipline. As part of this process, determine whether the fire fighter is unable to meet the required performance, which will require additional focused training, or unwilling to perform, which will start a fact-finding process.

Informal Written Reprimand
Some fire departments require the fire officer to use a standard form when issuing an oral reprimand, warning, or admonishment. Such a form covers the three items described earlier and provides a space for the employee to respond to the reprimand. The form ensures that the fire officer covers all of the requirements of an informal reprimand. In addition, the form allows the fire fighter to understand clearly that this is a disciplinary issue and to have an opportunity to respond. This type of contemporaneous record is valued by the personnel office if the issue becomes a grievance or results in a civil service hearing.

FIGURE 9-4 Fire fighters should practice working with hose lines and tools in simulated fire-ground scenarios while wearing full personal protective clothing.

VOICES OF EXPERIENCE

Fire service leaders must set clear expectations of their crew members from the start. Many company officers have a list of rules and expectations they present to crew members. My first crew was a group of folks with many years on the job, and all were older than me. I came into that crew with what I call a *conversation of expectations*. I sat down with the crew and had a brief discussion about my expectations. I assumed that this would be a great starting point for them to understand my expectations. Over the next year and a half I learned that many people need tangible written guidelines. This helps you, as a company officer, reinforce your expectations when the time comes—and it will.

One example I have is of a certain individual who had a small company on the side. From the beginning, I explained to him that our organization had a mission and that he was not to make money elsewhere while making money at the fire house. On the surface, this seemed to me a pretty cut-and-dried explanation of my expectations. The client phone calls stopped, and I thought all was good. However, one day I found him doing invoices on a fire department computer. His justification was that it was after hours; he stated, "What's the problem? I could just be sleeping or watching TV." I again reinforced my expectation that he would not run his business while on duty.

> **I had attempted to correct actions verbally over the course of a year, assuming my employees understood my expectations and shared the same morals and values.**

We had a few other corrective moments, and in his mind most of his actions had a good justification. He would bring up folks surfing the Net for hunting gear or vacation information. He would imply that I was singling him out. In another incident, he was found doing payroll and lining out his employees while the rest of the crew was doing morning checks. His justification once again was vague: "Last weekend Fire Fighter A had his kids at the station for an hour while we were doing chores. What is the big deal?" After this encounter I sat him down and documented what we call a PIP, or performance improvement plan. I had attempted to correct actions verbally over the course of a year, *assuming* my employees understood my expectations and shared the same morals and values. This was a big mistake. My approach made it harder to manage my other crew members, and it reduced my effectiveness as a leader. I once heard that good soldiers know who the bad soldiers are; pay attention.

Since this learning experience I have made my expectations clear and clearly followed our organization's disciplinary and corrective action policy. This practice is critical to ensuring the good folks we have—who are the majority—are confident in the organization's ability. Company officers are on the front lines of ensuring this credibility. In the end, our human resources will test us much more than a large commercial fire ever will.

Ed Mezulis
Battalion Chief, Sedona Fire District
Yavapai College
Clarkdale, Arizona

Documentation and Record Keeping

Municipal personnel rules usually require all of the official records of an employee's work history to be in a secured central repository. This repository includes all of the official documents accumulated during employment, including the following items:

- Hiring packet (application, Candidate Physical Aptitude Test score, and medical examination results)
- Tax withholding, I-9 status, and insurance forms
- Personnel actions (changes in rank and pay)
- Evaluation reports (probationary, annual, and special)
- Grievances
- Formal discipline

Some fire departments maintain a second personnel file at fire headquarters. That file may include letters of commendation, transfer requests, protective clothing record, work history, copies of certifications, and other fire department–specific information that does not need to be kept in the secured personnel file.

Fire Officer II

Formal Evaluation and Discipline

NFPA 1021 identifies Fire Officer II as the level of officer who issues formal evaluations and discipline. The captain, or managing fire officer, may delegate that task to a lieutenant or supervising fire officer. Every agency has established an evaluation and discipline policy and procedure. The fire officer should know and comply with these agency requirements.

■ Annual Evaluations

The personnel regulations in most career fire departments require that every employee who has completed the initial probationary period receive an annual written evaluation from his or her immediate supervisor. These annual evaluations become a formal part of the employee's work history.

The fire officer may receive an annual evaluation form from the human resources or personnel division a month or two before the employee's annual evaluation is due to be submitted. Some organizations expect the officer to keep track of the dates and submit a completed form when the evaluation for each employee is due.

The annual evaluation process consists of four steps. First, the supervisor fills out a standardized evaluation form. This form asks the supervisor to evaluate the subordinate on a number of knowledge areas, skills, and abilities appropriate for the subordinate's rank and classified job description. There is usually an opportunity for the supervisor to add comments in a narrative section. The supervisor's manager is required to review the evaluation before it is issued to the subordinate.

In the second step, the subordinate is allowed to review and comment on the officer's evaluation. The subordinate has an opportunity to provide feedback or additional information that would affect the performance rating.

The third step is a face-to-face feedback interview between the supervisor and the subordinate to discuss the evaluation. This meeting is an opportunity to clarify points and review the performance over the last rating period. A goal of the feedback interview is that both the supervisor and the subordinate understand the results of the evaluation.

The final step, usually completed at the end of the feedback interview, is to establish goals for the subordinate to accomplish during the next evaluation period. These are usually specific, measurable goals that allow the subordinate to improve on the knowledge, skills, and abilities of the current job or prepare for the next higher job level.

■ Conducting the Annual Evaluation

The completion of the annual evaluation forms should be viewed as a formality. The fire officer should be providing continual evaluation and feedback to the fire fighter throughout the year, so the information that goes onto the form should not come as a surprise. Unless there is a problem with the employee's performance, the scheduled evaluation should be used primarily as an opportunity to discuss future goals and objectives.

The methods outlined in the following section are designed to track an employee's performance and progress throughout the year. Using the information gathered from a year of entries in a performance log or T-account, the fire officer can provide an objective and well-documented annual performance evaluation. This documentation is essential, regardless of whether the fire fighter receives an "outstanding" rating or an "unsatisfactory" rating.

Keeping Track of Every Fire Fighter's Activity

In a performance log, the fire officer maintains a list of the fire fighter's activities by date, along with a brief description of performance observations FIGURE 9-5.

The T-account is a slightly more sophisticated documentation system, similar to an accounting balance sheet listing credits and debits. A single-sheet form is used to list the assets on the left side and the liabilities on the right side, with the result resembling the letter "T" FIGURE 9-6.

Using either method, the fire officer compiles an extemporaneous "when, what, and how" record of each fire fighter's work history throughout the evaluation period. This record can be a powerful evaluation and motivational tool.

Establishing Annual Fire Fighter Goals

The fire officer should require all fire fighters who have completed probation to identify three work-related goals they want

> March 11, Incident 3467. Trouble positioning apparatus to hook up soft suction and hydrant. Charged wrong attack line. First structural fire as engine driver.

FIGURE 9-5 Sample notation in a performance log.

to achieve during the next evaluation period. This provides focus beyond the minimum day-to-day activities and helps fire fighters prepare for future promotional examinations or assignments. Examples of individual fire fighter goals might include:

- Enrolling in and completing a building construction course at the local community college
- Becoming qualified as an aerial operator
- Learning how to use the computer-aided design software to develop tactical preincident plans

The fire officer should track the progress toward these goals made during the evaluation period, recording that progress in a T-account or a performance log.

Informal Work Performance Reviews

Most civil service and personnel regulations require an annual evaluation, but many departments also encourage fire officers to conduct informal performance reviews with each fire fighter throughout the year. During these informal sessions, the fire officer can review the T-account or performance log with the fire fighter and see what the officer can do to assist the fire fighter in meeting the established goals. Together they can identify situations or work conditions that impede progress. For example, a fire fighter who is spending 70 percent of the time detailed to cover positions at other stations will have difficulty qualifying as a backup aerial operator. The fire fighter and officer could discuss changing the goal or adjusting assignments so that the fire fighter would have more time at the station to work on meeting the goal.

Mid-Year Review

The mid-year review is another informal performance review that requires a higher level of documentation. The officer should have each fire fighter write a self-evaluation in preparation for the session. This task helps the fire fighter focus on the job description and the personal goals that were set at the beginning of the year. The officer and the fire fighter should review this document together and discuss how well expectations are being met.

If necessary, the officer can identify resources or assistance that would help the fire fighter meet the mutually established goals. For example, the officer might suggest meeting with another fire fighter who is experienced at working with the preincident planning software or point out that only one semester remains to enroll in the building construction course at the college during the evaluation period.

Also at this time, the personal goals can be adjusted because of changes in the work environment. It would be unrealistic to maintain the goal to attend a building construction class at the community college, for example, if the department has selected the fire fighter to attend paramedic school for the next 9 months.

Advance Notice of a Substandard Employee Evaluation

A fire fighter who is not meeting expectations should know there is a problem long before the annual evaluation is undertaken. The subordinate should be given adequate time to change the behavior or improve the skill, particularly if it is jeopardizing a scheduled pay raise or continuing employment.

Municipalities typically require a supervisor to provide a formal notification to an employee who is likely to be evaluated below the level necessary to obtain a pay increase or a "satisfactory" rating. The notice must identify the aspect of the job performance that is substandard and indicate what the employee must do to avoid receiving a substandard annual evaluation. A common practice is to require such a notification as soon as an issue arises, or at least 10 weeks before the annual evaluation is due.

If the employee receives a substandard annual evaluation, then the municipality might require a work improvement plan. This plan should cover a specific span of time, such as 120 calendar days, that would be designated as a special evaluation period. The employee is expected to participate fully in a work improvement plan during this special evaluation period. He or she is given an opportunity to demonstrate the desired workplace behavior or performance no later than the end of the special evaluation period.

+	−
March 11, Incident 3467. Rescue of elderly female from hallway outside burning bedroom in a one-story, single-family dwelling.	March 11, Incident 3467. Had to return to Engine 4 to retrieve forcible entry tools and firefighting gloves during initial actions.

FIGURE 9-6 Sample notation in a T-account.

The following is an example of the human resources division requirements that could be applied to a work improvement plan:

- The work improvement plan shall be in writing, stating the performance deficiencies and listing the improvements in performance or changes in behavior required to obtain a "satisfactory" evaluation.
- The work improvement plan has been reviewed and approved by a senior management officer.
- During the special evaluation period, the employee shall receive regular progress reports.
- If, at the end of the special evaluation period, the employee's performance rating is "satisfactory" or better, the time-in-rank pay increase will start at the first pay period after the work improvement period.
- If, at the end of the special evaluation period, the employee rating remains unsatisfactory, then no time-in-grade pay increase will be issued. In addition, the supervisor will determine whether additional corrective action is appropriate.

Regardless of the outcome, a special evaluation period performance evaluation is filled out and submitted as a permanent record in the employee's official file.

The employee's performance is officially evaluated during the annual review. Conducting informal reviews enables a fire officer to identify any underperforming fire fighter early in the annual evaluation period. The advance-notice procedure should provide the fire fighter with enough time to change the behavior or improve the necessary skills before the formal evaluation is scheduled to occur. If the employee has still not improved, a substandard evaluation would not come as a surprise to either the fire fighter or the fire officer's supervisor.

Six Weeks Before the End of the Annual Evaluation Period

The fire fighter should conduct another self-evaluation approximately 6 weeks before the official annual evaluation is due. The goal of this self-evaluation is to identify how well the fire fighter has met the organizational expectations and personal goals for the year. The fire officer should review this document and provide feedback. Together, the fire officer and fire fighter should develop the final, formal evaluation report for this rating period, which includes developing three new work-related personal goals for the next evaluation period.

Completing annual evaluation reports is an important fire officer responsibility. It offers an opportunity to identify and evaluate the work performed by subordinates. The fire officer should recognize outstanding accomplishments, encourage improvement, and, in some cases, identify those personnel who should be considering another career option.

The fire officer who maintains a performance log or T-account, conducts bimonthly informal reviews, conducts a mid-year review, and has the fire fighter complete a self-evaluation 6 weeks before the annual evaluation will have a wealth of information with which to present an accurate, well-documented, and comprehensive annual evaluation.

■ Evaluation Errors

Using a T-account or performance log to assemble a detailed list of work behavior observations over the evaluation period provides excellent background when the fire officer is preparing a fire fighter's evaluation. Evaluation is a largely subjective process that is vulnerable to unintentional biases and errors.

Leniency or Severity

Some fire officers tend to rate all of their fire fighters either higher or lower than their actual work performance. This practice is called leniency or severity, respectively. Leniency reduces conflict, because a positive evaluation is likely to make the evaluation a more pleasant experience and avoids confrontation. Leniency is common when the fire officer is required to conduct a face-to-face meeting with the fire fighter to review the evaluation.

Some fire officers lean in the opposite direction and rate all fire fighters with "needs improvement" or "unsatisfactory." Some officers think that low ratings will cause fire fighters to be motivated to work harder. Newer fire officers sometimes make this mistake when they have not received adequate training in preparing performance evaluations.

Personal Bias

Personal bias is an evaluation error that occurs when the evaluator's perspective skews the evaluation such that the classified job knowledge, skills, and abilities are not appropriately evaluated. Fire officers must not allow an evaluation to be slanted by factors such as race, religion, gender, disability, or age. This is the area most likely to be referenced when an employee files hostile workplace or discrimination charges.

Recency

Recency is an evaluation error in which the fire fighter is evaluated only on incidents that occurred in the last few weeks, rather than on all of the events that occurred throughout the evaluation period. Fire fighters who are aware of this tendency are on their best behavior in the weeks leading up to the fire officer's evaluation.

Central Tendency

Most evaluation systems involve some type of rated scale, ranging from unsatisfactory to outstanding. A fire officer demonstrates a central tendency when a fire fighter is rated in the middle of the range for all dimensions of work performance. A central tendency evaluation holds little value for the fire fighter or the evaluation process. Being rated as "OK in all areas" is not very informative or helpful.

Frame of Reference

In a frame of reference evaluation error, the fire fighter is evaluated on the basis of the fire officer's personal ideals instead of the classified job standards. For example, a fire officer who spent hours every day fixing, improving, and polishing the engine when he was an apparatus driver/operator might issue "unsatisfactory" or "needs improvement" ratings to apparatus operators who are meeting all of the departmental standards but are not meeting his personal ideals.

Halo and Horn Effect

Like life, fire fighters' performances are not just black and white. Sometimes, however, the fire officer may concentrate on only one aspect of the fire fighter's performance, which is either exceptionally good or bad, and apply that perception across the board to all aspects of the individual's work performance. This type of evaluation error is called the halo and horn effect.

Contrast Effect

Contrast effect is an evaluation error that can occur when the fire officer compares the performance of one subordinate with the performance of another subordinate instead of against the classified job standards.

■ Negative Discipline: Correcting Unacceptable Behavior

Whereas positive discipline is directed toward encouraging desirable behavior and high performance, negative discipline is aimed at discouraging unacceptable behavior and poor performance. Negative discipline is a stronger force than positive discipline.

Sometimes, an effort that begins with positive discipline (e.g., motivation, training, and coaching) evolves into negative discipline because of the continuing inability or unwillingness of a fire fighter to meet the required performance or behavioral expectations. If an individual does not respond to positive efforts to correct a problem, the next logical step is to punish continuing unsatisfactory performance. Progressive negative discipline moves from mild to more severe punishments if the problem is not corrected. The ultimate goal remains fire fighter performance improvement. In an extreme case in which progressive discipline fails to solve the problem, it may become necessary for the organization to terminate an employee who is ineffective and unwilling to improve.

The disciplinary process is designed to be consistent and well documented. Typical steps in a progressive negative discipline system include the following:

- Counsel the fire fighter about poor performance and ensure that he or she understands the requirements. Ascertain whether there are any issues contributing to the poor performance that are not immediately obvious to the supervisor. Resolve these issues, if possible.
- Verbally reprimand the fire fighter for poor performance.
- Issue a written reprimand and place a copy in the fire fighter's file.
- Suspend the fire fighter from work for an escalating number of days.
- Terminate the employment of a fire fighter who refuses to improve.

Some employee behaviors require the fire officer to implement negative discipline immediately. This is a common policy when there is willful misconduct, as opposed to inadequate performance. Personnel regulations usually provide a list of behaviors that will lead to immediate negative discipline, such as these acts:

- Knowingly providing false information affecting an employee's pay or benefits or in the course of an administrative investigation
- Willfully violating an established policy or procedure
- Being convicted of a criminal offense that affects the ability of the employee to perform his or her job
- Displaying insubordination
- Behaving in a careless or negligent way that leads to personnel injury, property damage, or liability to the municipality
- Reporting to work when under the influence of alcohol or a controlled substance
- Misappropriating fire department property or funds

As the penalty increases, negative discipline requires the participation of higher levels of supervision. A fire officer is often required to consult with his or her supervisor before issuing formal negative discipline. Many jurisdictions provide detailed descriptions of the "due process" and "just culture" principles that guide the administration of progressive negative discipline under a labor-management contract. Here is an example of how the negative discipline process might proceed:

- *Informal oral or written reprimand*: Reprimand is issued by a supervising or managing officer. Reprimand remains at the fire station level and expires after a time period no longer than 1 year.
- *Formal written reprimand*: Document initiated by a fire officer (usually after consulting with his or her supervisor). Some organizations require a battalion chief or other command officer to issue a written reprimand. A copy of the letter goes into the individual's official personnel file. It expires after a set time period, usually 1 year.
- *Suspension*: May be initiated or recommended by a fire officer. The suspension notice is usually issued by a battalion chief or a higher-level command officer after consultation with the fire chief or designe. The record of a suspension remains permanently in the employee's official file. Occasionally, the record is removed after a grievance, arbitration, or civil service hearing, or in the wake of a lawsuit.
- *Termination*: Although a termination is usually recommended by a lower-level command officer, the fire chief issues the formal termination notice after consulting with the personnel office.

Formal Written Reprimand

A formal written reprimand represents an official negative supervisory action at the lowest level of the progressive discipline process. Even when the department requires a command-level officer to issue a formal written reprimand, this document is often prepared by a fire officer. The fire officer is the closest person to the issue and will be involved in the remediation effort. He or she should consult the personnel regulations and departmental guidelines when preparing a written reprimand. In departments in which a fire officer can issue a written reprimand, the fire officer's supervisor should review and approve the document before it is issued to the fire fighter.

The written reprimand should contain the following information:

- Statement of charges in sufficient detail to enable the fire fighter to understand the violation, infraction, conduct, or offense that generated the reprimand

Metro County Fire Department
Memorandum

March 4, 2014

To: Fire Fighter Arthur Johnson
Engine 6

From: Captain William Schwartz
Engine 6

Subject: Written Reprimand: Driving through red light intersection

On March 2, 2014, you were driving the pumper in response to a second alarm commercial fire at 16171 Enterprise Avenue, incident #0164. While driving north on Waterway Boulevard with emergency lights on and siren sounding the traffic light at Edgerly Street turned red for Waterway. While you slowed down, you did not come to a complete stop for the red light signal as required in Standard Operating Procedure 3.4.1.b: **Emergency Vehicle Response.**

This is the fourth time you have failed to come to a complete stop when encountering a red light intersection when responding to a working structural fire.

Jan 17: Verbal reprimand by Captain Schwartz for incident #1009
Dec 22: Counseling by Lieutenant Scopes for incident #0789
Sept 03: Counseling by Captain Schwartz for incident #2084

Continued failure to come to a complete stop at red light intersections may result in more severe proposed discipline, including suspension or termination. This reprimand will remain in your permanent personnel folder for 366 days.

Chapter 8 of the Metro County Personnel Regulations explains your rights and options if you wish to appeal this discipline.

I have read and received a copy of this written reprimand:

_____   _____
Fire Fighter Arthur Johnson   Date

cc: Personnel file
Battalion Chief Andrea Walters, 3rd Battalion

FIGURE 9-7 Sample written reprimand.

- Statement that this is an official letter of reprimand and will be placed in the employee's official personnel folder
- List of previous offenses in cases in which the letter is considered a continuation of progressive discipline
- Statement that similar occurrences could result in more severe disciplinary action, up to and including termination

A written reprimand starts the formal paper trail of a progressive disciplinary process **FIGURE 9-7**. If the behavior is not repeated, most written reprimands are removed from the employee's file after a designated time period, usually 1 year after the reprimand is issued.

The written reprimand may potentially be seen by a wide audience. If it is appealed or becomes part of a larger disciplinary action, the fire chief, labor representatives, civil service commissioners, and attorneys will read the reprimand. Some departments require that an impartial third-party panel review formal discipline recommendations. Fire officers must focus on the work-related behaviors and clearly explain the behavior or action that generated the reprimand.

Suspension

Suspensions are the next step of a progressive negative discipline path but have many different forms. A <u>suspension</u> is a negative disciplinary action that removes a fire fighter from the work location and prohibits him or her from performing any fire department duties. A disciplinary suspension usually results from a willful violation of a policy or procedure or another specific act of misconduct. For career fire fighters, suspension is a personnel action that places an employee on a leave-without-pay (LWOP) status for a specified period. For volunteer fire fighters, a suspension means they are not allowed to respond to emergencies and, in some cases, are prohibited from entering the fire station or participating in other fire department activities. Suspensions usually run from 1 to 30 days.

A fire officer usually recommends a suspension to a higher-level officer, providing the required documentation. In some organizations, a fire officer has the authority to suspend a fire fighter immediately for the balance of a work period, pending a formal disciplinary action. Depending on the nature and the severity of the offense, career fire fighters can also be suspended with pay or placed on restrictive duty while an administrative investigation is being conducted. Restrictive duty is usually a work assignment that isolates the fire fighter from the public, often an administrative assignment away from the fire station environment.

In a few situations, the municipality can suspend a fire fighter before an investigation is completed. Such a suspension might occur when the employee is being investigated by the fire department or a law enforcement agency for an offense that is reasonably related to fire department employment, or when an employee is waiting to be tried for an offense that is job related or a felony.

Termination

Termination means that the organization has determined that the employee is unsuitable for continued employment. In general, only the top municipal official, such as the mayor, county executive, city manager, or civil service commission, can terminate an employee. Many senior municipal officials delegate this task to the agency or department heads, such as the fire chief. Terminations are high-stress events, with labor representatives, the personnel office, and the fire chief's office all involved in the process.

Predetermined Disciplinary Policies

With some common disciplinary issues, a predetermined policy for specific offenses may have already been developed. For example, here is how one department deals with fire fighters who are late reporting to work, without mitigating circumstances:

- *First offense*: Oral admonition and LWOP covering the period of time the employee was late. Admonition remains active for 366 days with the immediate supervisor.
- *Second offense (within 366 days of first offense)*: Written reprimand and LWOP for the period of time the employee was late.
- *Third offense (within 366 days of second offense)*: Suspension for one workday (8 hours for day-work staff or 12 hours for shift workers).
- *Fourth offense (within 366 days of third offense)*: Suspension for four workdays (32 hours for day-work staff, 48 hours for shift workers).
- *Fifth offense (within 366 days of fourth offense)*: Suspension for 10 workdays (80 hours for day-work staff, 120 hours for shift workers).
- *Sixth offense (within 366 days of fifth offense)*: Proposed termination.

Alternative Disciplinary Actions

Depending on the nature and the severity of the offense, a variety of penalties may be imposed:

- *Extension of a probationary period*: If the fire fighter is in a probationary period, as occurs with a recruit or after a promotion, the probationary period can be extended until the work performance issue is resolved.
- *Establish a special evaluation period*: An incumbent fire fighter might be given a special evaluation period to resolve a work performance/behavioral issue. For example, a fire fighter who fails a required recertification examination could be placed in a special evaluation period until he or she is recertified.
- *Involuntary transfer or detail*: In an involuntary transfer or detail, a fire fighter is transferred or detailed to a different or a less desirable work location or assignment.
- *Make financial restitution*: For example, the fire fighter might be required to pay the insurance deductible after a property damage incident. One department has a $2500 deductible insurance policy on fire and EMS vehicles and equipment. If the fire fighter is judged to be responsible for a department vehicle crash, he or she is required to repay the deductible amount. The restitution payment period can take up to a year and the payment made through payroll deductions.
- *Loss of leave*: A fire fighter might lose annual or compensatory leave. This is equivalent to the practice of paying a cash fine.
- *Demotion*: A demotion occurs when an individual is reduced in rank, with a corresponding reduction in pay. Demotions are more common in the supervisory ranks.

Predisciplinary Conference

In most cases, a predisciplinary conference or hearing must be conducted before a suspension, demotion, or involuntary termination can be invoked. The degree of investigative effort and the opportunities for fire fighter response before these punishments are issued are set higher than for less severe levels of discipline. Most fire departments require a formal disciplinary hearing before a suspension is issued to provide an opportunity for the fire fighter to formally respond to the charges.

Known as a Loudermill hearing, this process resulted from the 1985 U.S. Supreme Court decision in *Cleveland Board of Education v. Loudermill*. The Cleveland Board of Education hired James Loudermill as a security guard in 1979. Loudermill received a letter from the Board on November 3, 1980, dismissing him because of dishonesty in filling out his employment application. He had not listed a 1968 felony conviction for grand larceny on his application. The U.S. Supreme Court combined the Loudermill case with a similar appeal filed by Richard Donnelly, a Parma Board of Education bus mechanic fired in August 1997 after failing an eye examination.

The court ruling indicated that a pretermination hearing, including a written or oral notice, in which the employee has an opportunity to present his or her side of the case, and an explanation of adverse evidence were essential to protect the worker's due process rights. Such a hearing serves as a check against a possible mistaken decision, as it seeks to determine whether there are reasonable grounds to believe that the charges against the employee are true and whether they support the proposed action.

Not all disciplinary actions require a Loudermill hearing before suspension; the rules vary from state to state. In 1997 in *Gilbert v. Homar*, the U.S. Supreme Court ruled that East Stroudsburg University did not violate the constitutional due process rights of a university police officer when it immediately suspended him without pay after learning he had been charged with a drug felony.

A predisciplinary hearing can be conducted by a disciplinary board, by the fire chief or another ranking officer, or by a hearing officer. The designated individual or board reviews the case and makes a recommendation to the fire chief. Before the hearing, the fire fighter receives a letter that outlines the offense and the results of the investigation. The proposed duration of the suspension without pay could also be stated in the letter.

Here is a typical example of the process:

- The fire officer investigates the alleged employee offenses promptly and obtains all pertinent facts in the case (time, place, events, and circumstances) including, but not limited to, making contact with persons involved or having knowledge of the incident.
- The fire officer prepares a detailed report outlining the offense, the circumstances, the individual's related prior disciplinary history, and recommended disciplinary action. The report is submitted to a higher-ranking officer in the chain of command.
- The fire department representative consults with the human resources director or his or her designee, if necessary, when suspensions are contemplated.
- The disciplinary board hearing is scheduled, and an advance notice letter is prepared.
- The disciplinary board considers the charges and hears the employee's response.
- The disciplinary board makes a recommendation to the fire chief, who issues the final decision.

At the hearing, the accused individual has an opportunity to refute the charge or present additional information about mitigating circumstances. In most career fire departments, a union representative is present at the hearing to advise the individual. An alternative or lower level of discipline might be proposed, based on past department practices. The disciplinary board makes a final recommendation to the fire chief.

The fire fighter might or might not have the ability to appeal the disciplinary action through a grievance procedure, depending on the personnel rules of the organization. The final resolution of a disciplinary action usually resides with the civil service commission or the city personnel director. Some municipalities use arbitration to resolve these issues.

Employee Assistance Program

An employee assistance program (EAP) is designed to deal with issues such as substance abuse, emotional or mental health issues, marital and family difficulties, or other difficulties that affect job performance. Put simply, EAPs help the employees cope with underlying issues that might be affecting workplace performance. Fire department EAPs are comprehensive programs that deal with a wide range of issues that can affect fire fighters. When an EAP is available, it is a resource that fire fighters can turn to when in crisis.

One important characteristic of an EAP is its ability to maintain the value to the organization of highly trained emergency service professionals. Earlier in this chapter, a six-step progressive discipline process was discussed, followed by a description of how one department handles chronic tardiness. If the fire fighter is unable to correct this behavioral problem, and the process is followed, then termination is inevitable. The underlying problem, however, could be an off-the-job issue that the individual cannot solve without assistance. Termination of an otherwise good employee would be a tremendous waste in resources because it could cost the department as much as $100,000 to find and train a replacement. If EAP involvement can help solve the problem, it is worth the effort.

Consider some of the reasons a fire fighter might continue to report late for work:

- Child or elder care issues; the employee is late due to unanticipated coverage problems
- A family crisis, such as a divorce proceeding or a dying family member
- Alcoholism or substance abuse
- Coming from a second job that is needed to handle a financial crisis, such as a child with a significant health problem not covered by insurance, a crushing debt, or gambling losses
- A psychological condition or chemical imbalance

Often, these situations cause stress for the employee. A referral to the EAP may assist in addressing the issues. For an EAP to be successful, however, the fire officer must be able to recognize stress in an employee. Signs that may be noticed at work could include absenteeism, unexplained fatigue, memory problems, irritability, insomnia, increased use of products with caffeine/nicotine, withdrawal from the crew, resentment toward management/co-workers, stress-related illnesses, moodiness, or weight gain or loss. Although any of these points individually may not indicate the employee is stressed, multiple signs may indicate that the fire officer should ensure that the employee is aware of the EAP.

If the stressful situation goes unresolved, it can seriously affect performance on the job. Obviously, an employee who is absent reduces the efficiency of a fire company. More subtle effects are produced by symptoms of stress such as physical exhaustion from stress-related fatigue or lack of sleep. The fatigue or indecision caused by this condition can have serious safety consequences for every person on scene. It may also affect the group dynamics of the crew. A crew member who is stressed may not intend on taking it out on the crew at work, but the tension created is often evident to the rest of the company.

The goal of the EAP is to provide counseling and rehabilitation services to get the employee back to full productive duty as soon as possible. Fire department EAPs have been successful in lowering employee turnover and reducing absenteeism, tardiness, accidents, and injuries. In addition, fewer employee grievances and severe disciplinary actions are encountered when an EAP is in place.

Successful EAPs place a high value on confidentiality and require that fire fighters enter the program voluntarily. Although a fire officer can recommend or suggest that a fire fighter consider seeking assistance from an EAP, the fire officer cannot know the details of any fire fighter/EAP interaction. The fire officer's focus is on the fire fighter's job performance.

You Are the Fire Officer Conclusion

Captain Davis and Lieutenant Williams conduct a private meeting with Fire Fighter/EMT Wirth. Captain Davis reviews the Work Hours SOP, pointing out that this incident will generate a proposed 12-hour suspension without pay. Davis notes the continuing issue of running red lights when responding to emergencies. The captain also points out Wirth's excessive use of annual/sick leave and work exchanges. There appears to be an increasing pattern of work performance issues, any of which could lead to termination. What can the department do to help Wirth?

Wirth explains that he recently separated from his wife and is a single parent. This morning the babysitter did not show up. Davis acknowledges that the situation is difficult and reviews the discussion from the January mid-year informal review, in which Wirth rejected reaching out to the employee assistance program or requesting a transfer to day work. Davis concludes the session by urging Wirth to use EAP and to consider alternative ways of resolving the outside issues that make him late, burn up his leave, and distract him during his emergency driving.

Wrap-Up

Chief Concepts

- Supervision of fire fighters requires that the fire officer conduct regular evaluations to provide feedback on job performance, on-duty behavior, and problem resolution.
- Fire fighter evaluations should be an ongoing process throughout the year. The annual performance evaluation is a formal written documentation of the fire fighter's performance during the rating period.
- The fire officer must determine each new fire fighter's skills, knowledge, aptitudes, strengths, and weaknesses and then set specific expectations for each new fire fighter.
- Regular feedback from the fire officer should keep fire fighters aware of how they are doing, particularly after incidents or activities that present a special challenge.
- The fire officer builds discipline by training to meet performance standards, using rewards and punishments judiciously, instilling confidence in and building trust among team leaders, and creating a knowledgeable collective will.
- Positive discipline is directed toward motivating individuals and groups to meet or exceed expectations.
- Municipal personnel rules usually require all of the official records of an employee's work history to be in a secured central repository.
- NFPA 1021 identifies Fire Officer II as the level of officer who issues formal evaluation and discipline.
- The personnel regulations in most career fire departments require that every employee who has completed the initial probationary period receive an annual written evaluation from his or her immediate supervisor.
- The fire officer should provide continual evaluation and feedback to the fire fighter throughout the year, so the information that goes onto the annual evaluation form does not come as a surprise.
- Evaluation is a largely subjective process that is vulnerable to unintentional biases and errors.
- Negative discipline is aimed at discouraging unacceptable behavior and poor performance.
- An employee assistance program is designed to deal with issues such as substance abuse, emotional or mental health issues, marital and family difficulties, or other difficulties that affect job performance.

Hot Terms

<u>Central tendency</u> An evaluation error that occurs when a fire fighter is rated in the middle of the range for all dimensions of work performance.

<u>Contrast effect</u> An evaluation error in which a fire fighter is rated on the basis of the performance of another fire fighter and not on the classified job standards.

<u>Demotion</u> A reduction in rank, with a corresponding reduction in pay.

<u>Discipline</u> A moral, mental, and physical state in which all ranks respond to the will of the leader.

<u>Employee assistance program (EAP)</u> An employee benefit that covers all or part of the cost for employees to receive counseling, referrals, and advice in dealing with stressful issues in their lives. These problems may include substance abuse, bereavement, marital problems, weight issues, or general wellness issues.

<u>Formal written reprimand</u> An official negative supervisory action at the lowest level of the progressive disciplinary process.

Frame of reference An evaluation error in which the fire fighter is evaluated on the basis of the fire officer's personal standards instead of the classified job description standards.

Halo and horn effect An evaluation error in which the fire officer takes one aspect of a fire fighter's job task and applies it to all aspects of work performance.

Involuntary transfer or detail A disciplinary action in which a fire fighter is transferred or assigned to a less desirable or different work location or assignment.

Loudermill hearing A predisciplinary conference that occurs before a suspension, demotion, or involuntary termination is issued. The term "Loudermill" refers to a U.S. Supreme Court decision.

Oral reprimand, warning, or admonishment The first level of negative discipline. Considered informal, this discipline action remains with the fire officer and is not part of the fire fighter's official record.

Performance log An informal record maintained by the fire officer that lists fire fighter activities by date and includes a brief description; it is used to provide documentation for annual evaluations and special recognitions.

Personal bias An evaluation error that occurs when the evaluator's perspective skews the evaluation such that the classified job knowledge, skills, and abilities are not appropriately evaluated.

Pretermination hearing An initial check to determine if there are reasonable grounds to believe that charges against an employee are true and support the proposed termination.

Progressive negative discipline A process for dealing with job-related behavior that does not meet expected and communicated performance standards. The level of discipline increases from mild to more severe punishments if the problem is not corrected.

Recency An evaluation error in which the fire fighter is evaluated only on incidents that occurred over the past few weeks rather than on the entire evaluation period.

Restrictive duty A temporary work assignment during an administrative investigation that isolates the fire fighter from the public and usually is an administrative assignment away from the fire station.

Special evaluation period A designated period of time when an employee is provided additional training to resolve a work performance/behavioral issue. The supervisor issues an evaluation at the end of the special evaluation period.

Suspension A negative disciplinary action that removes a fire fighter from the work location; he or she is generally not allowed to perform any fire department duties.

T-account A documentation system similar to an accounting balance sheet listing credits and debits, in which a single-sheet form is used to list the employee's assets on the left side and liabilities on the right side, so that the result resembles the letter "T."

Termination A situation in which the organization ends an individual's employment against his or her will.

Work improvement plan A written document that is part of a special evaluation period. The plan identifies performance deficiencies and lists the improvements in performance or changes in behavior required to obtain a "satisfactory" evaluation.

References and Additional Resources

NFPA reprinted material is not the complete and official position of the NFPA on the referenced subject, which is represented only by the standard in its entirety.

Abbott, S. (2011). *Improving the Impact of Performance Evaluations upon Positive Behavioral Change*. Emmitsburg, MD: Executive Fire Officer Program, U.S. Fire Academy.

Abrams, R. L., and D. R. Nolan. (1985). "Toward a Theory of 'Just Cause' in Employee Discipline Cases." *Duke Law Journal* 1985: 594.

Barnett, C. C. (2012). *Performance Evaluations: Are We Using the Right Metrics for Success?* Emmitsburg, MD: Executive Fire Officer Program, U.S. Fire Academy.

Drucker, P. F. (1990). *Managing the Nonprofit Organization*. New York, NY: HarperCollins.

Drucker, P. F. (1995, February). "Really Reinventing Government." *Atlantic Monthly*. 50, 52.

Edwards, S. T. (2009). *Fire Service Personnel Management*. Upper Saddle River, NJ: Brady/Prentice Hall Health.

Fire and Emergency Service Image Task Force. (2013). *Taking Responsibility for a Positive Public Perception*. Fairfax, VA: International Association of Fire Chiefs.

Gormley, W. T., and S. J. Balla. (2012). *Bureaucracy and Democracy: Accountability and Performance*, 3rd ed. Washington, DC: CQ Press.

Horton, J. (2010). *Comparative Analysis of Estero Fire Rescue Company Officer and Firefighter Performance Evaluations Relative to Reward, Promotion, Corrective and Disciplinary Action*. Emmitsburg, MD: Executive Fire Officer Program, U.S. Fire Academy.

International Association of Fire Chiefs and National Fire Protection Association. (2014). *Fire Service Instructor: Principles and Practice*, 2nd ed. Burlington, MA: Jones & Bartlett Learning.

IOCAD Emergency Services Group. (1999). *Guide to Developing Effective Standard Operating Procedures for Fire and EMS Departments*. Emmitsburg, MD: U.S. Fire Administration.

LoSasso, C. (2012). *Mentoring Volunteer Officers Pre- and Post-Promotion*. Emmitsburg, MD: Executive Fire Officer Program.

Mintzberg, H. (2011). *Managing*. San Francisco, CA: Berrett-Koehler.

Scott, G., and W. F. Jenaway. (2006). *Human Resources Management: Managing Volunteer and Combination Emergency Service Organizations*. York, PA: VFIS, 95–113.

Smeby, L. C. Jr. (2014). *Fire and Emergency Services Administration: Management and Leadership Practices* (2nd ed.). Burlington, MA: Jones & Bartlett Learning.

Smith, K. B., and A. Greenblatt. (2013). *Governing States and Localities*. Washington, DC: CQ Press.

Standish, T. A. (2012). *Improving the Successfulness of Sioux City Fire Rescue's Company Officers*. Emmitsburg, MD: Executive Fire Officer Program, U.S. Fire Academy.

Stefano, D. A. (2010). *Formal Performance Evaluation for Reserve Firefighters*. Emmitsburg, MD: Executive Fire Officer, U.S. Fire Administration.

Walters, J. (2007). *Measuring Up 2.0: Governing's New, Improved Guide to Performance Measurement for Geniuses (and Other Public Managers)*. Washington, DC: Governing Books.

Weider, M. (2010). *Fire Service Reputation Management* [White paper]. Berkeley Springs, WV: Cumberland Valley Volunteer Firemen's Association.

FIRE OFFICER in action

Captain Davis and Lieutenant Williams implemented a new informal review process using quarterly and mid-year informal reviews to support the formal annual review. The last captain used a one-size-fits-all template that met the department's requirements but said nothing specifically positive or negative about each fire fighter or the unique issues that arise as part of a quint company. Davis wants to paint an accurate picture of each individual.

The fire officer is the starting point for fire fighter evaluation or disciplinary actions. Striving to provide an accurate and complete picture of each person under supervision provides an excellent foundation for building confident and competent fire fighters. The required tools include a detailed understanding of the agency's personnel regulations, labor agreements, and departmental regulations.

1. The proposed 12 hours of suspension without pay that Wirth will receive for a third incident of violating the work hours regulation is an example of:
   A. informal discipline.
   B. compliance reinforcement.
   C. formal discipline.
   D. an event not covered by progressive discipline.

2. Lieutenant Williams conducted a private counseling session with a rookie fire fighter about standing in the traffic lane while operating at a motor vehicle crash. The fire fighter is to remain within the lane blocked by fire apparatus or police to avoid getting struck by a vehicle. While assisting on a patient extrication on the interstate, Williams observes the rookie backing into an active traffic lane. What is Williams's next action?
   A. Issue a written reprimand after returning to the fire station
   B. Make a notation to bring it up at next informal performance review
   C. Digitally record the fire fighter's behavior
   D. Immediately order the fire fighter to get out of the traffic lane

3. Fire Fighter/EMT Neal Kinders fell in love 2 months ago. As a result, Kinders spends hours on the phone and is not completely participating in station cleaning or training activities. To focus on that issue as the main theme of an annual evaluation is an example of:
   A. central tendency.
   B. the halo and horn effect.
   C. the contrast effect.
   D. recency.

4. When should positive or negative feedback be delivered?
   A. At the next beginning-of-shift line-up
   B. At the next informal work performance review session
   C. As soon as possible after the incident or event
   D. As part of the annual performance review

Fire Captain Activity

It is April 15, and Fire Fighter/EMT Arthur Wirth failed to report to work. When called at 7:15, Wirth said that he is sick today and cannot come to work. Captain Davis reviews the supervisory notes on Fire Fighter/EMT Arthur Wirth.

August 10: Verbal reprimand, 40 minutes late reporting to duty and did not call fire station. First violation of MCPR Standard Operating Procedure 1.5.1.d: Work Hours, tardy.

September 6: Counseling by Captain Davis, Engine 100 ran a red light while responding to a first-due house fire, Incident #2084.

October 11: Written reprimand, 30 minutes late reporting to duty and did not call fire station. Second violation of MCPR Standard Operating Procedure 1.5.1.d: Work Hours, tardy.

December 26: Counseling by Lieutenant Schorr (fill-in), Engine 100 ran a red light while responding to a child struck, Incident #0789.

FIRE OFFICER *in action* (continued)

January 8: Mid-Year review.
- Completed Building Construction college course
- Authorized Quint 100 backup driver
- Continuing concern—stopping at red light intersections
- In Step 3 of Tardy procedure
- Taking leave/getting exchanges for 2 to 4 days per month since August
- Annual leave is 36.0 hours
- Using sick leave twice as fast as earning it—balance is at 46.6 hours
- Suggested EAP if Wirth needs assistance with issues that are consuming leave

January 17: Verbal reprimand by Lieutenant Taylor, Quint 100 ran a red light while responding to a working commercial fire, Incident #1009.

February 7: Delivered in-station drill on SOP 3.4.1 (work improvement plan—red light).

March 4: Written reprimand—failing to stop at a red light intersection.

March 11: Counseling by Captain Davis and Lieutenant Williams. FF/EMT Wirth's sick and annual leave balances are at zero. Urged to utilize employee assistance program (EAP) resources. Wirth says will consider EAP, does not want a transfer to a day-work position.

March 18: Arrives 55 minutes late, did not call fire station. Proposed 1-day suspension without pay for third violation of MCPR Standard Operating Procedure 1.5.1.d: Work Hours, tardy.

April 3: Chief Johnson meets with Davis and Wirth. Formally notifies Davis of a 12-hour suspension without pay on June 5.

Chief Johnson reviews the document with Wirth, noting that another violation of the Work Hours SOP could result in a second, longer suspension or a recommended termination. The chief also recommends utilization of the EAP and that Wirth review the "Fitness for Duty" SOP.

NFPA Fire Officer II Job Performance Requirement 5.2.1

Initiate actions to maximize member performance and/or to correct unacceptable performance, given human resource policies and procedures, so that member and/or unit performance improves or the issue is referred to the next level of supervision.

Application of 5.2.1

1. Using policies and procedures from a department with which you are familiar, prepare a proposed "Fitness for Duty," "Disciplinary Diversion," or "Termination" notification to Fire Fighter/EMT Wirth.

NFPA Fire Officer II Job Performance Requirement 5.2.2

Evaluate the job performance of assigned members, given personnel records and evaluation forms, so each member's performance is evaluated accurately and reported according to human resource policies and procedures.

Application of 5.2.2

1. Using policies and procedures from an agency with which you are familiar, describe how that organization could conduct a mid-year informal review of fire fighter performance.

Organized Labor and the Fire Officer

Fire Officer I

Knowledge Objectives

After studying this chapter, you will be able to:

- Discuss the impact of the International Association of Fire Fighters on fire fighters and emergency medical service personnel. (p 185)
- Describe how to establish a strong supervisor–employee relationship (NFPA 4.1.1) (NFPA 4.2.5). (pp 185–186)
- Discuss the value of positive labor–management relations (NFPA 4.1.1). (p 186)
- Describe the fire officer's role as a supervisor (NFPA 4.1.1) (NFPA 4.2.5). (pp 186–187, 189–190)

Skills Objectives

After studying this chapter, you will be able to:

- Demonstrate the initial handling of an employee grievance (NFPA 4.2.5). (pp 186–187, 189–190)

Fire Officer II

Knowledge Objectives

After studying this chapter, you will be able to:

- Describe the legislative framework for collective bargaining (NFPA 5.1.1). (pp 190–192)
- Discuss the fire service's relationship with labor unions through history (NFPA 5.1.1). (pp 192–193)
- List and describe labor actions in the fire service (NFPA 5.1.1). (pp 193–195)
- Discuss the emergence of labor organizations as a means of exerting political influence (NFPA 5.1.1). (p 195)
- Discuss the role of labor–management alliances and list four national examples (NFPA 5.1.1). (pp 195–197)

Skills Objectives

There are no Fire Officer II skills objectives for this chapter.

CHAPTER 10

Principles of Fire and Emergency Service Administration (FESHE) Course Outcomes

4. Select and implement the appropriate disciplinary action based upon an employee's conduct. (pp 186–187, 189–190)
5. Explain the history of management and supervision methods and procedures. (pp 184–185, 190–197)
7. Describe the traits of effective versus ineffective management styles. (pp 185–186, 191–197)
8. Identify the importance of ethics as it relates to fire and emergency services. (pp 189–190, 195–197)

You Are the Fire Officer

It is Saturday afternoon when Chief Johnson notifies Lieutenant Walters that a spot is available in next week's pump operator certification course. This NFPA 1002–compliant, week-long course is held at the regional fire academy, and this is the only time the course is available. After checking his records, Walters tells Fire Fighter Kinders to report to the academy Monday morning at 0800.

"Sorry, but I can't go," states Kinders. "Our shift is on scheduled days off Monday, Tuesday, and Wednesday next week, and I have plans." Walters points out that Kinders needs the class before he can become a backup engine driver.

"Are you really sure about this?" Kinders asks. "I think the new labor contract requires 72-hour notice for any work schedule changes."

Frustrated, Lieutenant Williams states, "This is a direct order. You are to report to the academy on Monday at 8 A.M."

"Okay," says Kinders, "this is an unfair labor practice and a violation of the labor agreement between the IAFF local and the city fire department. This is your notice of my step 1 grievance. I am officially requesting that you comply with the 72-hour advance notice of a work schedule change. I will be contacting my battalion representative."

1. Does a labor contract override personnel regulations?
2. What is the role of a supervising fire officer in a labor dispute?
3. How can this situation be resolved fairly and appropriately?

Introduction

Wages, working conditions, and many other aspects of the work environment are directly influenced by labor–management relations. It is important for the fire officer, particularly in a career fire department, to understand some of the history of labor relations to function better as a first-line supervisor. The range, scope, and tasks of a fire officer's supervisory activities are defined by three primary components:

- The local labor contract
- The municipality's personnel regulations
- The fire department's rules, regulations, and procedures

In many cases, a fire officer is both a supervisor representing management and a member of the bargaining unit represented by the union. This is an unusual situation when compared with most work environments, in which a very clear distinction is made between labor and management. It is especially important for a fire officer to understand how this dual role applies to the specific organization and the position he or she is occupying.

The U.S. fire service has been significantly influenced by organized labor activities, including several laws and regulations that affect the working conditions of fire fighters. Some results benefit all fire departments, from the largest all-career department to the smallest all-volunteer fire company. Although a career fire officer is much more likely to be involved in an organized labor situation than a volunteer officer is, most aspects of the relationship between the workers and the organization are similar within the volunteer organization.

At the fire company level, most career fire fighters work under a labor contract or some form of written agreement, such as a memorandum of understanding (MOU), between labor and management. The contract or MOU covers various working conditions, promotion/assignment practices, and problem-solving procedures. A labor contract is a negotiated legal agreement between the labor organization and the local jurisdiction. An MOU is a less powerful form of written agreement that is often used in jurisdictions where government employees do not have formal collective bargaining rights. Collective bargaining is a method whereby representatives of employees (unions) and employers determine the conditions of employment through direct negotiation; such negotiation normally results in a written contract setting forth the wages, hours, and other conditions to be observed for a stipulated period (e.g., 3 years).

Organized labor focuses on the working conditions and benefits of the union's members. In departments with a contract, collective agreement, or MOU, each fire station or work location will have a shop steward—that is, a union member who is appointed or elected to be the first line of labor representation at the workplace. The shop steward is a member of the workforce who has received additional training in labor relations. He or she enforces the contract or labor agreement and represents the union members at that fire station or work location. The steward may serve as the initial labor representative when handling issues of discipline, policy, or procedures.

The relationship between the employer and the labor organization is determined by labor laws and regulations at the federal, state, and local levels. During the 20th century, the balance of power between labor (representing the employees) and management (representing the fire chief/local government) swung back and forth, with each party having the upper hand during some eras. This balance of power depends on legislation and policies enacted by the federal government and decisions made by the Supreme Court. In the early part of the 20th century (about 1915 to 1930), the labor movement had a political advantage and was able to make considerable progress. With the lingering effect of the 2007–2009 recession, however, management has appeared to enjoy an advantage in recent years.

Fire Officer I

The International Association of Fire Fighters

The largest fire service labor organization in the United States, the International Association of Fire Fighters (IAFF), represents 300,000 fire fighters and emergency medical services (EMS) personnel in the United States and Canada **FIGURE 10-1**. The IAFF has provided almost 100 years of support and advocacy for career fire fighters, and its accomplishments influence many aspects of a fire fighter's job. It is a very powerful organization, both politically and within the fire service at the national level.

The International Association of Fire Fighters was established on February 28, 1918. The three principal objectives of the IAFF at that time were to obtain pay raises, to establish the two-platoon or 12-hour workday schedule, and to ensure that appointments and promotions were based on individual merit, not political affiliation.

The IAFF is unique among labor organizations in its dominance in representing a profession. In other public service professions, such as law enforcement and education, two or more national labor organizations compete to represent the employees. Although the International Brotherhood of Teamsters and the American Federation of State, County, and Municipal Employees also organize fire fighters, they represent relatively few fire service personnel and have little influence over the firefighting profession as a whole.

All fire fighters, including volunteers, have benefited from the efforts of the IAFF and its local and state affiliates. Labor advocacy has improved the quality of protective clothing, the safety of firefighting equipment, the content of training programs, and advanced techniques of emergency incident operations. The generalized process for handling a grievance, which is covered in this chapter, has also emerged from the labor–management relationship.

Establishing a Strong Supervisor/Employee Relationship

The basis for a strong, positive, and effective supervisor/employee relationship is open, honest, and constant communications between the fire officer and the fire fighter. Once those communication paths are opened and trusted, they will prove very beneficial through good times and tough times alike. It is also through that open, honest communication that some form of agreement, compromise, or answer can be found.

Key recommendations that form the foundation of any strong supervisor/employee relationship between a fire officer and a fire fighter include the following:

1. Schedule regular one-on-one meetings between you as the fire officer and each member of your company. Such routine contact establishes a personal connection and trust between the fire officer and each fire fighter. During the meeting, discuss job performance and expectations on the part of both people involved. Give guidance and coaching where necessary.
2. Schedule regular meetings with the company as a whole. Use this time to discuss new policies and procedures, any concerns about station procedures (e.g., checking out apparatus, housekeeping, kitchen duty), and upcoming personnel or policy changes. This is also a good opportunity to obtain feedback and input from the company members. Keeping an open line of communications is critical to an effective supervisor/employee relationship.
3. If a disagreement arises, work together to articulate the concern and to develop possible solutions. When both parties decide together what they want, they are usually very successful in attaining a mutually acceptable goal. When only one side decides what they want, the success rate drops off sharply.
4. If the personal and professional relationship between you and a fire fighter is rocky from the beginning and both of you have decided to work to improve it, start by listing the areas in which you can succeed together. Set goals and deadlines. Start with goals that are easily attainable ("low-hanging fruit") and build on those successes. Ask yourselves if the goal you are defining will have a positive impact on the company and the other station personnel.

Maintaining a good relationship does not mean that you will always agree on everything; rather, it means that you "agree to disagree" when you have discussed the issue and

FIGURE 10-1 The International Association of Fire Fighters (IAFF) is the largest fire service labor organization.

cannot find middle ground. In some cases, a third party will need to be brought in to help mediate the discussions to arrive at an acceptable solution.

To trust each other's intentions, the company officer and the fire fighter must be honest and up front. Sometimes the supervisor/employee relationship can take a detour when either side holds back information, exaggerates, or deceives. Unfortunately, suspicion usually leads to acrimony, and acrimony tends to lead to even more suspicion. In any long-term relationship, bluffing or threats can prevail only so long and are generally counterproductive. Whenever a company officer feels that communications are not going well, the officer should focus on bringing them back into alignment by discussing feelings and concerns in an open exchange.

The progression of the firefighting profession is just that: professionals working together for the common good and common goals. A cooperative, collaborative supervisor/employee relationship is the profession at its best. The chapter *Handling Problems, Conflicts, and Mistakes* provides specific tools and techniques to support a collaborative fire officer/fire fighter relationship. With hard work and open, honest communication, fire officers and fire fighters can continue to make the relationship better and raise the level of all fire service personnel's professionalism, trust, and stature with those whom they serve.

Positive Labor–Management Relations

The value of a positive and productive labor–management relationship is widely recognized. A healthy relationship is essential to produce positive outcomes and avoid the strife and consequences of a confrontational climate. Successful relationships are built on trust, respect, and open lines of communication. Each side must be willing and able to focus on the mutual benefits of a positive relationship or face the consequences of a negative outcome.

The root cause of almost every labor disturbance is a failure to manage the relationship between labor and management properly. The traditional way of thinking was based on the premise that either labor or management must score a victory over the other side to settle every point. Today, however, a philosophical shift in labor–management relationships is moving away from confrontational strategies and toward cooperative relationships, often through mediation. Mediation is the intervention of a neutral third party in an industrial dispute. As with many traditions in the fire service, the ability and the willingness to change were produced by necessity.

Tremendous amounts of time, energy, and money can be wasted in the process of two sides trying to overwhelm each other instead of working together. A poor labor–management relationship usually produces casualties on both sides. Both fire chiefs and union presidents can lose their positions of power and influence in the aftermath of such conflicts. Positive relationships based on mutual respect and understanding are much more likely to produce positive results.

Some fire departments exist merely to meet an internal determination of what makes up the basic requirements for public safety. Others are committed to excellence and continual progress. Most observers agree that the most successful and progressive fire departments put significant effort into managing their labor–management relationships instead of engaging in continual confrontations and power struggles.

In some instances, management uses the moral ethos of public service as leverage against labor, and vice versa. Public support is usually viewed as vital by both sides because elected officials, who represent the public, have ultimate control over the economic and policy issues.

The Fire Officer's Role as a Supervisor

One duty of a fire officer is to supervise the activities of subordinate fire fighters. The basic authority of a supervisor and the duties of subordinates are defined by the personnel rules of the city or governmental organization, as well as the specific rules, regulations, and procedures of the fire department. In addition, when a collective bargaining agreement is in effect, other details of the relationship are spelled out in the contract or MOU. Supervisors are expected to follow all of the established rules and procedures in assigning duties and all other aspects of the relationship with their subordinates. In many cases, it is a significant challenge for a newly promoted fire officer to learn which rules and regulations apply to which situation and how they are interpreted and applied.

In most organizational structures, there is a clear distinction between labor (the workers) and management (the managers and supervisors). The managers and supervisors represent the organization, and the union represents the workers. If any doubt or disagreement about the application or interpretation of the contract arises, a process should be in place for labor and management representatives to meet and resolve the problem.

This line between labor and management is more complicated in fire departments because the first-level supervisors are often members of the same collective bargaining unit as the fire fighters they supervise. As a consequence, the fire officer's relationship to the organization is often covered by the same contract that the officer has to follow and enforce. The formal line between labor and management is often established at a higher level, such as administrative fire officer. In some cases, the supervisory and managing fire officers are members of a bargaining unit that is separate from that of the fire fighters.

As a supervisor, a fire officer is generally the first point of contact between the workers and the fire department organization **FIGURE 10-2**. If a disagreement occurs about the interpretation or application of a work rule that is covered by the contract, the first-level supervisor is the individual who should have the first awareness of the problem and the first opportunity to resolve it. An officer who is a member of the same bargaining unit as an individual who is dissatisfied must clearly understand the established problem-solving processes.

■ Grievance Procedure

A grievance is a dispute, claim, or complaint that any employee or a group of employees may have about the interpretation, application, or alleged violation of some provision of the labor agreement or personnel regulations. A grievance

Near-Miss REPORT

Report Number: 11-0000241

Event Description: At 2106 hours our engine company was dispatched to the [de-identified by reviewer] restaurant for a fall injury. The engine company found the patient, who was a restaurant employee, at the top of a stairwell that leads to the basement storage area. Crew members began the regular line of questioning and treatment for what seemed to be a standard medical call. The patient was a 24-year-old female, who was pregnant. The patient stated that she was going into the basement to check on something and became lightheaded and fell. One of the other employees heard the fall and went to the stairwell to help the patient. Both employees exited the stairwell and called 911 to report the "fall injury". As the captain from the engine was questioning the patient and one fire fighter was checking vitals, the other fire fighter and the engineer went into the basement to see if the patient had tripped or slipped on something.

Shortly after entering the basement, both crew members became lightheaded and exited the basement. The crew had no information from anyone that would even give us the slightest thought that something else might be wrong. Upon exiting the basement, the engineer fell and both members reported dizziness and a bitter taste in their mouths. The captain immediately called for a hazardous materials assignment and evacuated everyone out of the building. The hazmat team made an entry into the building in bunker gear and SCBA. The goal of the entry was to meter the basement for what was suspected to be a carbon dioxide leak.

The manager of the restaurant told the crews that they had just had the carbon dioxide tank filled a couple of hours prior to the call. The crews made entry with two combustible gas-indicating meters and two flammability-detection meters. As the crews descended the basement stairwell they started to get decreased oxygen readings and slightly increased volatile organic compounds readings on the combustible gas-indicating meters. As the crews continued into the basement, the oxygen readings continued to decrease (the lowest reading was 17.5%). One of the many interesting things about this call was the readings the crews were getting on the flammability-detection meters, which was reading 100% LEL. When switched to "percentage of gas," the readings dropped to 25%. The readings were obtained at both ground level and at ceiling level. These reading prompted the hazmat team to exit the building and start to mitigate the potential hazards. Due to the elevated LEL readings, the hazmat team shut off the natural gas at the meter and attempted to shut down the power from the exterior. It has since been determined that the over-saturation of carbon dioxide caused the flammability detection meter to false-positive.

Notes:
- This call quickly changed from a routine medical call to a life-threatening hazmat call.
- The entire first six minutes of the EMS call were conducted at the top of the stairs with no indication of any hazard.
- The patient was crying hysterically the entire time and would/could not communicate with crew or give any information.
- Little information was given to the crew from the restaurant employees, who continued to work normally until the hazard was identified.
- The patient may have been scared to get in trouble with the boss if she thought she did something wrong.

Lessons Learned: This restaurant chain has recently started using large CO_2 tanks to provide propellant to the fountain beverages, and the tanks are not regulated by any code. Do not rely on signage, as we found no warning signs, placards, or identifying signs indicating that bulk compressed gas was on site. The bulk CO_2 is typically stored in the basement, and if there is a leak it will displace any ambient oxygen, causing a dangerous atmosphere. We could not find any CO_2 alarm system in place or functioning. Non-hazmat crews are routinely dispatched to "check odor" calls. Should they all be hazmat?

---

procedure is a formal structured process that is employed within an organization to resolve a grievance. In most cases, the grievance procedure is incorporated in the personnel rules or the labor agreement and specifies a series of steps that must be followed in a particular order. If the problem cannot be resolved in a mutually acceptable manner at one level, it can be taken to the next level, and so on up to some ultimate level, where an individual or body has the final authority to impose a binding decision.

The grievance procedure should specify a sequential process and a timeline to move through the steps. The grievance can be resolved at any point by management accepting the

VOICES
OF EXPERIENCE

In the fire service, an officer has a responsibility to ensure his or her men and women perform their duties in the fire station, during routine responsibilities such as inspections and educational opportunities outside the fire station, and during emergency situations. Sometimes this requires patience as a younger fire fighter learns the ropes, but there are times when it will call for the officer to initiate disciplinary action. All of this occurs each day, but what also occurs each day is that same officer sitting at the lunch and supper table and breaking bread with the men and women they are responsible for supervising.

In many areas of business this is an unworkable model, but in the fire service we find it works well for us in most occasions. Many experts in the management of personnel will insist this model is doomed to failure and there will be too much familiarity, a lack of discipline or adherence to policies, and a breakdown in job performance. The key is to empower your officer and identify individuals who can walk the line between being a buddy to their crew and providing teamwork and leadership.

Similarly, many people believe there must be a separation between a union member and the officer corps of the fire department. However, many of our union representatives demonstrate the same traits as fire officers, including the ability to balance the responsibility to the organization and the responsibility to the individual.

I was fortunate to have had one of these officers when I was a new fire fighter. He was a solid fireground leader who took his profession as seriously as anyone I have ever met. He protected his crew on and off the fireground. If someone had an issue with a crew member, they had to go through him and explain the circumstances of the issue. This officer was fair, and when a situation warranted it, the member was called on the carpet.

This officer was also the vice president of the IAFF Local, representing the membership. Just as he was protective of his crew, he was protective of the men and women of the division. He demonstrated a talent for defending members yet still holding them accountable to the rules and regulations of the division. He believed in accountability, but he also believed in consistency and fairness in how that accountability was handled. For example, there was an occasion where an officer was not completing his duties as required. The battalion chief side preferred departmental charges as warranted. The union side ensured the charged officer proceedings were fair and proceeded according to the Collective Bargaining Agreement. Accountability was addressed fairly and appropriately.

After working for this officer for a number of years and serving with him on the Executive Board, I have come to believe that some of our best fire officers can also be the best union representatives. A solid officer believes in protecting his crew, he believes in crew performance, he believes in accountability for mistakes and rewards when merited, and most of all he knows how to handle his crew fairly and appropriately in all situations. Sounds a lot like my captain and vice president.

Sean DeCrane
Battalion Chief/Director of Training
Cleveland Division of Fire
Cleveland, Ohio

> *I have come to believe that some of our best fire officers can also be the best union representatives.*

established to resolve disputes in fire departments where there is no formal labor contract, even in all-volunteer organizations. The most important responsibility for a fire officer is to know and follow the procedures that apply in his or her organization.

Sample Step 1

The grievant presents his or her complaint verbally to a supervisor, shortly after the occurrence of the action that gave rise to the grievance. In some organizations, this nondocumented verbal notification is called an "informal grievance" or step zero. Even this informal step requires the grievant to provide three important pieces of information:

- The article and section of the labor agreement or personnel regulation alleged to have been violated
- A full statement of the grievance, giving facts, dates, and times of events, as well as specific violations
- A statement of the desired remedy or adjustment

Sample Step 2

The second step initiates the formal part of a grievance procedure. If the problem is not resolved at step 1, the employee may prepare and submit a written grievance **FIGURE 10-3**. This is usually submitted on a specified grievance form document. The employee, the employee's supervisor, and the personnel office all receive a copy of the grievance.

The supervisor has 10 calendar days to reach a decision and provide a written reply to the grievant. Failure to respond to the grievance within 10 days means that the supervisor has denied the grievance and the grievant can immediately go to step 3.

Sample Step 3

A step 3 grievance is written out on another specific grievance form and again specifies the article and section of the contract or personnel regulation alleged to have been violated; the dates, times, and specific violations that are alleged to have taken place; and the desired remedy or adjustment. Copies of the step 2 grievance form and the supervisor's response are attached to this document.

A step 3 grievance is submitted to a second-level supervisor, typically an administrative fire officer, who has 10 calendar days to respond. If the grievance is denied or the administrative fire officer does not respond within the specified time, the grievant can move to step 4 and present his or her grievance to the fire chief.

Sample Step 4

If the grievance remains unresolved, the grievant can present it to the fire chief or designee as the fourth step. The same written information must be submitted, along with all of the documentation from the previous steps. The fire chief has 10 days to respond to a step 4 grievance.

If the fire chief does not respond within the time frame, or if the grievance remains unsettled, the process moves out of the fire department to a mediator, personnel board, or civil service board for resolution. In this example, the grievance goes to the county administrator and, if not resolved at that level, goes to federal or state arbitration.

FIGURE 10-2 A fire officer is generally the first point of contact between the workers and the organization.

complaint and the corrective action requested by the grievant or by both sides reaching a negotiated settlement that is acceptable to each party. If management rejects the claim, the grievance can be taken to the next level. The timeline ensures that a grievance will not remain stalled at any level for an excessive time period.

An employee can contact a union representative at any time to discuss a situation, including how the union interprets the rule in question and whether a grievance should be submitted. The employee's union representative usually becomes formally involved at either the first or second step of the grievance process. The union representative acts as an advocate for the individual or group that submitted the grievance. The union becomes more deeply involved as the process moves through the various steps of the grievance procedure, particularly in cases in which the problem has broad impact within the organization.

The objective should always be to resolve the problem at the lowest possible level and in the shortest possible time. Grievances that must be processed through multiple steps are disruptive, time consuming, and costly to both sides. The ability to resolve problems at a low level is an indication of a healthy organization with a good labor–management relationship, whereas a steady stream of grievances moving up to the highest levels is a symptom of major relationship problems.

The following section outlines one version of the grievance procedure. The details and sequences described here may differ in other fire departments. A similar process can be

SAMPLE GRIEVANCE FORM

STEP TWO: TO BE COMPLETED BY UNION OR EMPLOYEE

Date of filing: _____

From: _____ _____ _____
     Employee   Rank   Assignment/shift

Grievance form must be submitted within 15 calendar days of the incident being grieved.

STATEMENT OF GRIEVANCE Must (1) contain a statement, as complete as possible under the circumstances, of the grievance and the facts upon which it is based, including dates, times, locations, names of witnesses, and other appropriate information; (2) identify the section(s) of the Contract Agreement that affect this grievance; (3) state requested remedy or corrective action.
Additional pages may be attached to the Sample Grievance Form.

(*this is where the employee enters the statement*)

Original copy of the completed Sample Grievance Form shall be delivered to the employee's immediate supervisor, with a copy to the Human Resources office and the Union representative.

_____   _____
Employee   Date

TO BE COMPLETED BY IMMEDIATE SUPERVISOR within 10 calendar days of receipt of Sample Grievance Form

_____ _____ _____ _____
Supervisor's Name Rank Work Location/Shift Date Grievance received

(*this is where the immediate supervisor enters a response*)

_____   _____
Supervisor Signature   Date

If employee is satisfied with Supervisor's answer, sign the original Sample Grievance Form acknowledging agreement and submit it to the Human Resources Director for placement in your employment records. If employee is NOT satisfied, shall sign the original Sample Grievance Form acknowledging disagreement and immediately notify the Union in writing. The original Sample Grievance Form shall then be submitted by the employee to the Deputy Chief within ten (10) calendar days of the decision of the immediate supervisor.

Agree_____   Do not agree:_____

FIGURE 10-3 Example of a grievance form.

Fire Officer II

Legislative Framework for Collective Bargaining

Collective bargaining is regulated by a complex system of federal and state legislation. Federal legislation establishes a basic framework that applies to all workers, and the states have discretionary powers to adopt labor laws and regulations that do not violate the federal requirements. The application of this basic model to governmental employees is much more complex. Some of the federal labor laws do not apply to federal government employees, state government employees, or employees of local government agencies within the states.

Four major pieces of federal legislation have established the groundwork for the rules and regulations of the present collective bargaining system. Before the adoption of these federal laws, each labor case was decided by a judge in a local court, who applied the broad concepts used in common-law decisions. The federal legislation provides a set of guidelines for how each state or commonwealth can regulate collective bargaining. Fire fighters employed by local government agencies are subject to state law that can require, permit, or prohibit collective bargaining for local public employees.

The four federal laws that regulate the collective bargaining system are the Norris-LaGuardia Act of 1932, the Wagner-Connery Act of 1935, the Taft-Hartley Labor Act of 1947, and the Landrum-Griffin Act of 1959. These federal laws, together with the Railway Labor Act of 1926 and some antitrust legislation, created the legal foundation for collective bargaining in the United States. Like most federal legislation, each act was designed to address a specific aspect of collective bargaining or to correct a problem. Each subsequent act built on the earlier legislation. Like a pendulum, each subsequent act may change the direction of the federal government in relation to collective bargaining issues.

■ Norris-LaGuardia Act of 1932

The Norris-LaGuardia Act of 1932 specified that an employee could not be forced into a contract by an employer as part of obtaining and keeping a job. Prior to this act, some employers required workers to sign a pledge that they would not join a union as long as the company employed them; these pledges were called yellow dog contracts.

The local courts generally sided with management in any labor dispute where a yellow dog contract was in effect. The judge would order an injunction that either prohibited striking or prohibited picketing during a strike. The police would enforce these injunctions.

In reaction to this practice, the Norris-LaGuardia Act of 1932 said that yellow dog contracts were not enforceable in any court of the United States. This act made it almost impossible for an employer to obtain an injunction to prevent a strike.

President Franklin Roosevelt took many steps to bolster economic growth during the Great Depression. One of his initiatives was the National Industrial Recovery Act (NIRA) of 1933. Section 7a of the NIRA guaranteed unions the right to collective bargaining as part of an effort to keep wages at a level that would maintain the purchasing power of the worker. After the NIRA passed, workers flocked to join both the American Federation of Labor (AFL) and the new Congress of Industrial Organizations (CIO). The Supreme Court struck down the NIRA as unconstitutional in 1935.

■ Wagner-Connery Act of 1935

When the NIRA was overturned, employers were free to use unfair labor practices in the management of their employees. The ensuing abuses led to the Great Strike Wave of 1933–1934. During this period, labor organized against management and conducted citywide strikes and factory takeovers in numerous industrial sectors. In response to the ongoing labor unrest, Senator Robert Wagner of New York introduced the Wagner-Connery Act, which was quickly passed by Congress in 1935, to mitigate the revolutionary labor climate and avert further economic disruption. A 1936 strike in the automobile industry quickly brought the Wagner-Connery Act before the U.S. Supreme Court, where it was upheld as constitutional. The Wagner-Connery Act established the procedures that are commonly called collective bargaining.

The Wagner-Connery Act forms the basis of formal labor relations in the United States. It grants workers the right to decide, by majority vote, which organization will represent them at the labor–management bargaining table. The act also requires management to bargain with duly elected union representatives and outlaws yellow dog contracts.

Provisions of the Wagner-Connery Act also established the National Labor Relations Board (NLRB), which has the power to hold hearings, investigate labor practices, and issue orders and decisions concerning unfair labor practices. The act defined five types of unfair labor practices and declared them illegal:

1. Interfering with employees in a union
2. Stopping a union from forming and collecting money
3. Not hiring union members
4. Firing union members
5. Refusing to bargain with the union

The Wagner-Connery Act and favorable court decisions resulted in some unions becoming extremely powerful. Union membership swelled from 4 million members in 1935 to 16 million members in 1948. The next two acts described here were adopted to balance the power between unions and management.

■ Taft-Hartley Labor Act of 1947

The Taft-Hartley Labor Act of 1947, which was passed by Congress over the veto of President Harry Truman, was formulated during the period that followed World War II. At that time, industry-wide strikes threatened to undermine a smooth return to civilian production. The Taft-Hartley Labor Act was designed to modify the Wagner-Connery Act and swing the pendulum back toward the middle by reducing the power of unions. It also spelled out specific penalties, including fines and imprisonment, for violation of the act.

Taft-Hartley gave workers the right to refrain from joining a union and applied the unfair labor practice provisions to unions as well as to employers. It specifically prohibited a union from forcing management to fire antiunion or nonunion workers. Unions were required to engage in good faith bargaining, and a 60-day "cooling off" period was created, which comes into effect when a labor agreement ends without a new contract. The act also regulated many of the unions' internal activities.

The most significant provision of the Taft-Hartley Labor Act that affects fire fighters is referred to as "strikes during a national emergency." In the event that an imminent strike could affect a major part of an industry and imperil the health and safety of the nation, the president is granted certain powers to help settle the dispute. The president can order employees back to work, compel arbitration, or provide economic, judicial, or political pressure to achieve a resolution of the dispute.

■ Landrum-Griffin Act of 1959

After the AFL and CIO merged in 1955 to become the AFL-CIO, Senator John McClellan conducted hearings that revealed evidence of crime and corruption in some of the older local unions. The Labor-Management Reporting and Disclosure Act, otherwise known as the Landrum-Griffin Act, passed in 1959, at the height of the furor.

Landrum-Griffin established a bill of rights for members of labor organizations. It required that unions file an annual report with the government listing the assets of the organization as well as the names and assets of every officer and employee. Minimum election requirements were mandated, as were the duties and responsibilities of union officials and officers. This act also amended portions of the Taft-Hartley Labor Act.

Collective Bargaining for Federal Employees

Collective bargaining rights for public employees have traditionally lagged behind those in the private sector. Federal legislation passed in 1912 prohibits federal employees from striking. There were fewer than 1 million unionized government employees in the United States in 1956. Federal legislation during the 1950s and 1960s, however, allowed unions to grow in the public sector. By 1970, the number of unionized government employees had grown to 4.5 million. Relations between management and unions had matured.

President John F. Kennedy issued Executive Order #10988 in 1963, which granted federal employees the right to bargain collectively under restricted rules. This step represented a significant milestone for public-sector unions. President Richard Nixon further expanded the rights of employee unions within the federal government and established the Federal Labor Relations Council, which is similar to the National Labor Relations Board for private-sector unions.

State Labor Laws

In addition to the federal laws, each state exercises legislative control over several aspects of collective bargaining. Each state determines whether it will engage in collective bargaining with state government employees and whether local government jurisdictions within the state may engage in collective bargaining with their employees. Some states require municipal governments to bargain collectively with fire fighters' unions, some permit collective bargaining, and others limit or prohibit collective bargaining.

Strikes by state employees are illegal in all but 10 states, and many states also prohibit local government employees from striking. Forty percent of local governments have forbidden employee strikes; even so, numerous municipal strikes occurred between 1967 and 1980. After this period, several states and municipalities adopted legislation to prohibit strikes or other adverse labor actions.

Fire fighters have the right to bargain collectively in half of the states. Those states authorize municipalities to recognize a local fire fighter labor organization as the bargaining unit and enter into a binding contract that covers fire fighter pay, benefits, working conditions, conflict resolution, promotions, and department practices. The remaining states place restrictions on bargaining with fire fighters.

Since the 2007 recession, labor organizations have vigorously challenged the dismantling of collective bargaining, pension, and benefit packages at the state and federal levels.

Right to Work

The collective bargaining rights of public employees in different states are related to the right-to-work issue. Under the Taft-Hartley Labor Act, workers have the right to refrain from joining a union. In 2013, right-to-work laws were in effect in 24 states TABLE 10-1. In those states, a worker cannot be compelled, as a condition of employment, to join or pay dues to a labor union. The same states tend to limit or restrict collective bargaining for local government employees.

The purpose of the original right-to-work legislation in 1947 was to prohibit the practice of closed shops, in which a worker must be a member of a particular union to work for the company. Open shops provide the worker with the option of remaining outside the union. Proponents of the open shop statute believe that the practice eliminates union collusion and exclusionary practices, which are now deemed illegal, and protects an individual's right to refrain from joining an organization.

Opponents of open shops express the opinion that right-to-work statutes reduce a union's bargaining power and place an unfair burden on the union members. They state that, in essence, the union has to bargain for the entire workforce, even for nonmembers. A worker who chooses not to join the union is afforded the same benefits as union members, but without paying dues to support union activities, pay for attorney fees, or conduct research studies. Under the federal regulations, however, an individual who chooses not to join a union can still be compelled to pay a share of the cost of the union's representation in collective bargaining.

Right-to-work laws remain a source of controversy among labor organizations. Although unions support overturning existing right-to-work statutes, advocacy groups defend the existing statutes and promote their adoption in additional states. Workers entering the labor market should be aware of the prevailing labor laws that pertain to them.

Table 10-1 Right-to-Work States

Alabama	Nebraska
Arizona	Nevada
Arkansas	North Carolina
Florida	North Dakota
Georgia	Oklahoma
Idaho	South Carolina
Indiana	South Dakota
Iowa	Tennessee
Kansas	Texas
Louisiana	Utah
Michigan	Virginia
Mississippi	Wyoming

Data from the National Right to Work Committee.

Organizing Fire Fighters into Labor Unions

Regardless of the historical period, statutory environment, or work culture, labor–management relationships have been present in every work environment that includes an employer and employees. The first paid fire department in the United States was established in 1853 in Cincinnati, Ohio.

The career fire fighter's work environment at the start of the 20th century was grim. New York City fire fighters worked "continuous duty," 151 hours per week, with just 3 hours off each day to go home for meals. San Francisco fire fighters got 1 day off after five consecutive 24-hour duty periods. Career

Fire Marks

Federal Labor–Management Conflicts

Two notable strikes involving federal government employees marked the beginning and the end of a significant era in the evolution of labor–management relations: the postal service workers strike of 1970 and the air traffic controllers strike of 1981.

Postal Workers' Strike: 1970

In 1970, U.S. postal workers initiated an illegal strike. At that time, the Postmaster General was legislatively restricted from negotiating with the striking workers; even so, negotiations were conducted and a settlement was reached. The striking postal workers were reinstated without penalty, and the negotiated wage increases were adopted. Subsequently, Congress recognized the postal workers' union for the purposes of collective bargaining. This strike set the tone for future civil service strikes, including several that involved fire fighters.

PATCO Air Traffic Controllers' Strike: 1981

One reason why public employees rarely strike today could be the results of the 1981 strike involving U.S. air traffic controllers. In that year, the Professional Air Traffic Controllers Organization (PATCO) went on strike after reaching an impasse in contract negotiations with the Federal Aviation Administration.

President Ronald Reagan cited the example of the 1919 Boston, Massachusetts, police strike when he described the actions he would take to end the walkout. When the PATCO workers failed to report to work as ordered, Reagan fired every striking member and decertified the union. Military air traffic controllers were brought in as temporary replacements. Air traffic volume was reduced for 18 months while the Federal Aviation Administration hired and trained replacement workers.

Unlike in the 1970 postal workers' strike, none of the PATCO strikers were hired back, and many of the union officials were subjected to years of aggressive litigation by the federal government. This action set the tone for employer–employee relations during the remainder of the Reagan administration. The frequency of public-sector strikes in the United States decreased significantly over the following two decades.

---

by the federal courts to prohibit union representation of multiple labor groups. Consequently, individual local unions could not organize to form regional or national labor organizations. This prohibition lasted until the Clayton Act was passed in 1914.

Labor Actions in the Fire Service

The fire service has experienced many adverse labor actions in response to poor labor–management relations **FIGURE 10-4**. A strike—the act of withholding labor for the purposes of effecting a change in wages, hours, or working conditions—is one of the most drastic labor actions. Although a strike is precipitated by a conflict between labor and management, its impact is often felt beyond the parties that are directly involved in the relationship. In the private sector, the consumer is often the ultimate victim of a strike, through a combination of inconvenience and economic impact. In the public sector, the safety and the welfare of the general public may be directly affected by major labor actions.

The potential impact of a strike on public safety is so severe that many states prohibit fire fighters from walking out. In many of the cases in which fire service strikes have occurred, the public and media have questioned labor's right to strike as a matter of ethics. The IAFF started with the premise that fire fighter strikes are inadvisable, and from 1930 to 1968 the IAFF charter included a no-strike clause. Nevertheless, in three periods during the 20th century, municipal fire fighters did go on strike in various U.S. cities.

FIGURE 10-4 Picketing is one of the most visible labor actions.

---

fire fighters spent most of their time in cramped, unhealthy, and unsafe fire stations, sharing living accommodations with the horses.

As paid fire departments were established, fraternal or benevolent fire fighter support organizations gradually became political advocates for the workers and began to define the labor–management relationship. The genesis of these organizations occurred entirely at the local level. Antitrust protection, legislated by the 1890 Sherman Act, was interpreted

Striking for Better Working Conditions: 1918–1921

The 1918–1921 strikes occurred during a period of economic instability following World War I. Most of the fire fighter strikes were efforts to establish a two-platoon system, obtain more pay, or simply gain the right to form a labor organization that would be recognized by the municipality. Organized labor reached a turning point in 1919. First, there was an industry-wide strike in the steel industry. Then, the country witnessed the first "general strike" that shut down Seattle, Washington, on February 6. A similar strike occurred in Winnipeg, Manitoba, Canada, on May 15, 1919.

Fire fighters in Memphis, Tennessee, threatened to strike in 1918 and succeeded in obtaining a two-platoon work schedule, allowing fire fighters more time at home. The old system in Memphis had allowed just one to three days off each month; the new system allowed fire fighters to spend every other day away from the fire station.

Many IAFF local unions were able to gain a two-platoon schedule and/or a pay raise by threatening to strike. Fire fighters on the two-platoon system could finally have a home away from the fire station, although they were still required to work 84 hours per week. In turn, they could marry and raise a family. Today, most municipal fire fighters work between 42 and 56 hours per week. The workweek for most federal and military fire fighters still exceeds 60 hours.

Striking to Preserve Wages and Staffing: 1931–1933

The Great Depression began with the stock market crash of 1929. For the next several years, local governments reduced wages, initiated furloughs (unpaid leave), and reduced workforce size. In Seattle, fire fighters had to take two 25-day furloughs in 1933. In addition to the furloughs, the city closed seven fire stations and eliminated eight engine companies, two hose companies, two squad companies, two fireboats, and one truck company to reduce expenditures.

Several fire fighter strikes between 1931 and 1933 occurred in reaction to wage reductions and elimination of fire fighter jobs. One local union went on strike because the fire fighters had not been paid by the city for 2 months.

Organized labor and fire fighter strikes during this time seem to have accomplished the IAFF's mission of improving working conditions. Davis Ziskind's 1940 book, *One Thousand Strikes of Government Employees*, examined 39 fire fighter strikes and lockouts from 1900 to 1937. Approximately half of the strikes resulted in partial or complete victory by labor. In contrast to the chaos associated with the Boston police riot, Ziskind could find only one large-loss fire that occurred during a labor action—a large industrial fire in Pittsburgh, Pennsylvania, during the first IAFF organized strike in 1918.

Staffing, Wages, and Contracts: 1973–1980

The most recent series of fire fighter strikes occurred between 1973 and 1980. The IAFF voted to eliminate the 50-year-old no-strike clause in its constitution at its 1968 convention, at a time when the United States was mired in political turmoil. Most of the ensuing strikes occurred after labor and management reached an impasse while negotiating a new contract. An impasse occurs when the parties have reached a deadlock in negotiations. None of the strikes resulted in a net gain for organized labor.

In the 1970s, local governments were facing another recession that hit hard at their finances. Several large cities were facing bankruptcy. In the depths of the recession, New York City laid off more than 40,000 city employees on July 2, 1975, including 1600 fire fighters. Although 700 of the fire fighters were hired back within 30 days, 900 others lost their permanent fire fighter jobs. It would take 2 years before the city could rehire all of the laid-off fire fighters.

Negative Impacts of Strikes

There have been fewer fire fighter strikes in the United States and Canada since the 1980s. The negative public reaction to such labor actions, the political response of changing legislation, and the lasting legacy of lost trust have caused fire fighter strikes to be viewed as counterproductive. Measures such as picketing and alternative pressure tactics that maintain emergency services have been used to influence public opinion in several fire departments. Fire service strikes, or job actions, have occurred more recently among other industrialized nations throughout the world, particularly in Europe, where military personnel have been called upon to fill in for fire fighters.

The Fair Labor Standards Act and Fire-Based EMS

The Fair Labor Standards Act (FLSA), originally passed in 1938 as part of Roosevelt's New Deal, is an example of federal legislation that significantly affects the fire service. Once passed, federal legislation remains in effect unless it is ruled unconstitutional (as with the NIRA), amended (as with Taft-Hartley), or repealed. The FLSA gradually evolved with the adoption of new provisions and regulations that expanded its application to include fire departments.

The primary purpose of FLSA was to establish minimum standards for wages and to spell out administrative procedures covering work time and compensation, including overtime entitlement. For most workers, FLSA requires overtime to be compensated at time-and-a-half pay—that is, 150 percent of an employee's regular hourly wages. This legislation was adopted during a period of profound underemployment and was designed to encourage employers to hire more workers instead of paying current employees to work additional hours.

At the time of its original passage, FLSA did not cover government employees; however, additional provisions were adopted in 1974 to make FLSA applicable to federal employees. In 1985, the U.S. Supreme Court ruled in *Garcia v. San Antonio MTA*, 469 US 528 (1985), that local union public employers must also abide by the FLSA regulations unless Congress specifically legislates otherwise.

Amended regulations, adopted in 1986, established a special overtime rule for fire fighters. Most employees are entitled to be paid overtime starting at the 41st hour of a 7-day workweek. The average workweek for fire fighters across the United States at that time was determined to be 53 hours. Under the revised federal regulations, public safety agencies do not have to pay fire fighters overtime until after they have worked 212 hours in a maximum 28-day cycle—which averages out to a 53-hour workweek. This so-called 207(k) exemption covers individuals whose work involves "the prevention, control, or extinguishment of fires" for 80 percent of their work time.

Anne Arundel County, Maryland, fire fighter/paramedics filed suit in 1990 claiming improperly calculated overtime payments. The FLSA 1986 amendment described fire fighter activities to include "housekeeping, equipment maintenance, lecturing, attending community fire drills, and inspecting homes and schools for fire hazards" as incidental functions but did not include EMS activity. Some of the employees were assigned to paramedic ambulances and did not routinely participate in firefighting activities. Other employees, who were assigned to fire suppression companies, were actually spending more time handling EMS first-responder calls than performing fire suppression activities.

The plaintiffs claimed that the 207(k) exemption did not apply to them because more than 20 percent of their time was spent on nonfirefighting activities. They argued that their overtime pay should start at the 41st hour and not the 54th hour of work in an average workweek. U.S. District Court Judge Walter Black ruled in favor of the employees, creating a $4 million back-payment obligation for the county.

The Anne Arundel case was one of many similar lawsuits and union grievances that were going on throughout the United States during the same time period. Dozens of cities were involved in FLSA lawsuits and were becoming obligated for huge amounts of retroactive back pay. Some of the settlements went back to the 1986 adoption of the amended FLSA regulations. The situation became even more complicated when the eighth U.S. Court of Appeals issued a ruling that was in conflict with the ruling of the fourth U.S. Court of Appeals in the Anne Arundel case. When the U.S. Supreme Court declined to review the Anne Arundel case, the conflicting legal opinions were left to stand within the regions subject to each circuit court's interpretation.

U.S. Representative Robert Ehlrich, a Maryland Republican, introduced House Resolution 1693, the Fire and Emergency Services Definition Act. This legislation expanded the FLSA definition of fire protection activities to include paramedics, emergency medical technicians, rescue workers, ambulance personnel, and hazardous materials workers. The Ehrlich resolution, which had the support of both the IAFF and the International Association of Fire Chiefs (IAFC), was passed on November 4, 1999. The U.S. Senate passed the same resolution a week later, and President Bill Clinton signed the Fire and Emergency Services Definition Act into law on December 9, 1999. This law amended a portion of the original FLSA that was passed in 1938 and expanded by the Supreme Court decision in 1985 to cover local government employees. It took almost a full decade to resolve the issue of FLSA coverage for fire fighters performing EMS duties.

Today there remain active court cases that represent millions of dollars of unpaid overtime. Some of these cases involve staffing practices that started after the 1999 revision of the FLSA legislation. For example, a 2011 decision provided $30 million to metro Louisville, Kentucky, fire fighters.

The Growth of IAFF as a Political Influence

The economic turmoil and wage reductions in the 1990s were as severe as those in the 1930s, but organized labor took a different path. Instead of striking, labor worked to get political candidates who were supportive of labor elected to local and national political offices. The attempted recall of elected officials in Memphis in 1979 may have been the beginning of this new era of labor influence.

Fire fighters are particularly respected in political circles for the efforts they can put behind a political candidate who supports the objectives of the IAFF or a particular local union. In addition to assisting with political campaigns, organized labor found that "money talks." The creation in 1978 of FIREPAC, the IAFF's political action committee, was a natural progression of the labor organization's activity as a major influence in improving fire fighters' work environment. A political action committee (PAC) is a special-interest group that can solicit funding and lobby local and national elected officials on behalf of its cause. Funded by donations from individual IAFF members, FIREPAC promotes the legislative and political interests of the IAFF. The money is used to educate members of Congress about issues important to fire fighters and EMS personnel and to elect candidates to office who support those issues. In the 2011–2012 election cycle, FIREPAC raised more than $5.8 million in voluntary donations. Its support was not completely one-sided in regard to political parties; almost 15 percent of the funds donated to candidates for federal elections went to Republicans.

Using money and off-duty fire fighters to assist political candidates or causes may appear unseemly. Hal Bruno, former director of political reporting for ABC television and radio networks, repeatedly pointed out in his *Firehouse* magazine columns that the fire service must be fully engaged in the political process.

Labor–Management Alliances

In addition to political action, the IAFF has developed strategic labor–management alliances with the IAFC to promote mutually agreeable goals. In some situations, both labor and management want the same goals and they work together to achieve those goals. This chapter examines four national examples.

■ Fire Fighter Safety and Deployment Study

Building on the work started by the IAFF/IAFC EMS Systems Performance Measurement program, a Department of Homeland Security (DHS)–funded research effort started in 2008 to gather data and conduct time-on-task experiments to develop a prospective deployment model. Joining the IAFF and IAFC in this effort were three national groups: the Center for Public

Safety Excellence Commission on Fire Accreditation, the Department of Fire Protection Engineering at Worcester Polytechnic Institute, and the National Institute of Standards and Technology (NIST).

Crews from Montgomery County, Maryland, and Fairfax County, Virginia, participated in field experiments to determine time-on-task baseline performance capability and compare different staffing schemes with outcomes. In total, 22 time-on-task fire suppression evolutions were evaluated in more than 60 live-fire field experiments. These capabilities were similar to the 1979 Los Angeles Fire Department (LAFD) Measure of Effectiveness task studies, which used three to six fire fighters on engine companies. *Report on Residential Fireground Field Experiments, NIST Technical Note 1661,* summarized the results: "The four-person crews operating on a low-hazard structure fire completed all the tasks on the fireground (on average) seven minutes faster—nearly 30%—than the two-person crews. The four-person crews completed the same number of fireground tasks (on average) 5.1 minutes faster—nearly 25%—than the three-person crews" (NIST 10).

A follow-up study looked at 16 high-rise time-on-task capabilities, conducting 48 field experiments. *Report on High-Rise Fireground Field Experiments, NIST Technical Note 1797,* included this summary point: "when responding to a medium growth rate fire on the 10th floor, 3-person crews ascending to the fire floor confronted an environment where the fire had released 60% more heat energy than the fire encountered by the 6-person crews doing the same work" (NIST 13).

■ EMS Systems Performance Measurement

Working with the IAFC, the EMS Systems Performance Measurement organization constructed an instrument that was field tested, validated, and released in 2002. This instrument consists of 15 EMS quality indicators, their definitions, and performance measures that are useful at the local level in reporting the value of a fire-based EMS system to local decision makers and the public.

In 2007, the two organizations, along with other fire service groups, released a white paper on fire-based EMS, *Prehospital 9-1-1 Emergency Medical Response: The Role of the United States Fire Service in Delivery and Coordination,* as well as an associated video presentation, *EMS: The Right Response.*

■ The IAFC/IAFF Labor–Management Initiative

The IAFC/IAFF Labor–Management Initiative (LMI), formerly known as the Fire Service Leadership Partnership, was developed in 1999 to assist fire chiefs and union presidents in fostering cooperative and collaborative labor–management relationships. The same fundamental concepts that were involved in developing the Wellness–Fitness Initiative (discussed in the following section) provided the foundation for the LMI.

Scores of fire chiefs and union presidents have attended LMI programs to learn how to enhance their labor–management relationships. Many issues that face the fire service require continued cooperation and problem solving from both labor and management. Without the combined talent of all members of the fire service, fire fighters and fire chiefs will always be behind in their reaction to emerging issues, such as fire-based EMS,

Fire Marks

Seattle Medic One

The Seattle Medic One paramedic ambulance service began as a university research project on March 7, 1970. The Seattle City Council declined to fund continuation of the Medic One project within the fire department budget in 1972. Members of IAFF Local 27 scrambled to fund the life-saving project through a special countywide tax levy. Two years later, in 1974, the CBS news magazine *60 Minutes* profiled Seattle as "the best place to have a heart attack." Organized labor had to scramble again in 1997 when voters defeated a plan to continue the levy, requiring a special referendum in February 1998 to keep 22 paramedic ambulances running in Seattle and King County. Both the jurisdiction and labor were aggressive in promoting the levy, which passed in 2007.

Company Tips

The Phoenix Way: Relations by Objectives

In 1984, the Phoenix Fire Department adopted a policy of relations by objectives (RBO). The RBO process had been used in other labor–management disputes in which the two sides held significantly different positions on values, goals, and trust for each other. The Phoenix Fire Department and the United Phoenix Fire Fighters union used the RBO process when addressing each major issue. This process includes a means of analysis, decision, education, implementation, revision, and review. Each step has specific parameters in which labor and management cooperatively work on issues.

With RBO, the Phoenix Fire Department has been able to collaborate on issues such as incumbent drug testing, apparatus specifications and deployment, safety, and employee wellness. Because both labor and management participated in the development of many programs and issue resolution, both sides of the employee/employer environment shared equal successes and failures.

This cooperative effort led to the culture of the Phoenix Fire Department changing into an organization where each member, including civilian staff, feels like a part owner of the department. Whether this cultural shift was intended or not, the result was the development of a work environment in which each member has the opportunity to make significant contributions.

"The Phoenix Way" was severely tested after the March 12, 2001, line-of-duty death of Fire Fighter Bret Tarver at a supermarket fire. The recovery process took more than a year and resulted in significant operational changes within the Phoenix Fire Department and a national awareness of the problem of applying residential firefighting assumptions when operating at a commercial fire.

urban–wildland interface, occupational safety, and—perhaps the greatest challenge to the fire service—response to terrorism. It is clear in this environment of new challenges that a good labor–management relationship is essential. Learning the skills to develop solid relationships should be a regular part of a fire officer's curriculum and not an ancillary elective.

Fire Service Joint Labor–Management Wellness–Fitness Task Force

The IAFF and the IAFC joined forces to develop a comprehensive and mutually beneficial wellness–fitness program. Released in 1997, the program is designed to serve fire fighters throughout their careers. Ten fire departments and their IAFF local unions participated in the task force to develop a comprehensive and nonpunitive program:

- Austin, Texas, and IAFF Local 975
- Calgary, Alberta, and IAFF Local 255
- Charlotte, North Carolina, and IAFF Local 660
- Fairfax County, Virginia, and IAFF Local 2068
- Indianapolis, Indiana, and IAFF Local 416
- Los Angeles County, California, and IAFF Local 1014
- Metro-Dade County, Florida, and IAFF Local 1403
- New York City, New York, and IAFF Locals 94 and 854
- Phoenix, Arizona, and IAFF Local 493
- Seattle, Washington, and IAFF Local 27

The completed program package was released at both the IAFF Redmond Symposium and the IAFC Conference in August 1997. The Wellness–Fitness Initiative (WFI) task force released the third edition of the Wellness–Fitness Initiative in 2008. In addition, the WFI task force developed the Candidate Physical Ability Test (CPAT), a standardized physical ability test that has been validated as an assessment of a fire fighter candidate's physical ability to perform the critical tasks of a fire fighter. Finally, the WFI task force developed the IAFF/IAFC/ACE Peer Fitness Trainer certification program and initiated changes to the traditional critical incident stress management (CISM) behavior health model.

You Are the Fire Officer Conclusion

Lieutenant Williams learns that the labor contract overrides city personnel regulations. The personnel regulations state that a supervisor may change an employee's work hours with 1 day's notice, but the labor contract specifically requires a minimum of 72 hours' notice. The labor contract also requires that employees be provided with the appropriate educational resources (books, practice tests, and study guides) 9 work days before they are scheduled to attend an employer-mandated certification class.

The situation is resolved at the first step of the grievance process after Williams calls Chief Johnson for guidance. Following his advice, Williams apologizes to Fire Fighter Kinders. He is not ordered to attend the pump operator certification course. The union representative calls Kinders later that day to confirm that the problem has been resolved.

Johnson also clarifies the situation. He wanted to know if any of the fire fighters would volunteer to attend the pump operator course next week because the academy had a last-minute vacancy. The labor contract allows employees to voluntarily attend an employer-mandated certification course without the required advance notice.

The labor contract was not one of the items covered in a promotional exam for fire officers. Quickly correcting the problem and apologizing for having been wrong creates a positive impression with the Fire Station 100 members.

Wrap-Up

Chief Concepts

- Wages, working conditions, and many other aspects of the work environment are directly influenced by labor–management relations.
- The largest fire service labor organization in the United States, the International Association of Fire Fighters (IAFF), represents 300,000 fire fighters and emergency medical services personnel in the United States and Canada.
- The following key recommendations form the foundation of any strong supervisor/employee relationship between a fire officer and a fire fighter:
    - Regular one-on-one meetings
    - Regular meetings with the company as a whole
    - If a disagreement arises, working together to articulate the concern and to develop possible solutions
    - If the relationship between an officer and a fire fighter is rocky, working together to improve it
- A healthy labor–management relationship is essential to produce positive outcomes and avoid the strife and consequences of a confrontational climate.
- The basic authority of a supervisor and the duties of subordinates are defined by the personnel rules of the city or governmental organization as well as by the specific rules, regulations, and procedures of the fire department.
- An employee can contact a union representative at any time to discuss a situation, including how the union interprets the rule in question and whether a grievance should be submitted.
- Four major pieces of federal legislation have established the groundwork for the rules and regulations of the present collective bargaining system:
    - The Norris-LaGuardia Act of 1932 made yellow dog contracts unenforceable.
    - The Wagner-Connery Act of 1935 established the National Labor Relations Board and the procedures commonly called collective bargaining.
    - The Taft-Hartley Labor Act of 1947 established good faith bargaining and a 60-day "cooling off" period.
    - The Landrum-Griffin Act established a bill of rights for members of labor organizations.
- Federal legislation passed in 1912 prohibits federal employees from striking.
- Each state determines whether it will engage in collective bargaining with state government employees and whether local government jurisdictions within the state may engage in collective bargaining with their employees.
- Regardless of time, statutory environment, or work culture, labor–management relationships have been present in every work environment that includes an employer and employees.
- Although a strike is precipitated by a conflict between labor and management, the impact of a strike is often felt beyond the parties that are directly involved in the relationship.
- From 1918 to 1921, fire fighter strikes focused on better working conditions.
- Between 1931 and 1933, fire fighter strikes focused on preserving wages and staffing.
- Fire fighter strikes between 1973 and 1980 focused on staffing, wages, and contracts.
- The negative public reaction to strikes, the political response of changing legislation, and the lasting legacy of lost trust have caused fire fighter strikes to be viewed as counterproductive.
- Fire fighters are particularly respected in political circles for the efforts they can put behind a political candidate who supports the objectives of the IAFF or a particular local union.
- In addition to political action, the IAFF has developed strategic labor–management alliances with the IAFC to promote mutually agreeable goals.
- Scores of fire chiefs and union presidents have attended Labor–Management Initiative (LMI) programs to learn how to enhance their labor–management relationships.

Hot Terms

Arbitration Resolution of a dispute by a mediator or a group rather than a court of law. Any civil matter may be settled in this way; some labor–management agreements include a binding arbitration clause.

Collective bargaining Method whereby representatives of employees (unions) and employers determine the conditions of employment through direct negotiation, normally resulting in a written contract setting forth the wages, hours, and other conditions to be observed for a stipulated period (e.g., 3 years). This term also applies to union–management dealings during the terms of the agreement.

Fair Labor Standards Act (FLSA) Federal legislation passed in 1938 that provides the minimum standards for both wages and overtime entitlement and spells out administrative procedures by which covered work time must be compensated. Public safety workers were added to FLSA coverage in 1986.

Good faith bargaining A legal requirement of both the union and the employer arising out of Section 8(d) of the National Labor Relations Act. Enforced by the National Labor Relations Board, the parties are required to meet regularly to bargain collectively for wages, hours, and other conditions of employment.

Grievance A dispute, claim, or complaint that any employee or group of employees may have in relation to the interpretation, application, and/or alleged violation of some provision of the labor agreement or personnel regulations.

Grievance procedure A formal structured process that is employed within an organization to resolve a grievance. In most cases, the grievance procedure is incorporated in the personnel rules or the labor agreement and specifies a series of steps that must be followed in a particular order.

Impasse A situation in which the parties in a dispute have reached a deadlock in negotiations; also described as the demarcation line between bargaining and negotiation. A declaration of an impasse in labor–management negotiations brings in a state or federal negotiator who will start a fact-finding process that will lead to a binding arbitration resolution.

Mediation The intervention of a neutral third party in an industrial dispute. The object is to enable the two sides to reach a compromise solution to their differences, which the mediator usually does by seeing representatives of both sides separately and then together.

Negotiation Mutual discussion and arrangement of the terms of an agreement.

Political action committee (PAC) An organization formed by corporations, unions, and other interest groups that solicits campaign contributions from private individuals and distributes these funds to political candidates.

Right to work A worker cannot be compelled, as a condition of employment, to join or not to join or to pay dues to a labor union.

Shop steward A union member appointed or elected to be the first line of labor representation at the workplace. The steward enforces the contract, collective agreement, or memorandum of understanding and represents the union members at that fire station or work location.

Strike A concerted act by a group of employees who withhold their labor for the purposes of effecting a change in wages, hours, or working conditions.

Unfair labor practices Employer or union practices forbidden by the National Labor Relations Board or state/local laws, subject to court appeal. It often involves the employer's efforts to avoid bargaining in good faith.

Yellow dog contracts Pledges that employers required workers to sign indicating that they would not join a union as long as the company employed them. Such contracts were declared unenforceable by the Norris-LaGuardia Act of 1932.

References and Additional Resources

NFPA reprinted material is not the complete and official position of the NFPA on the referenced subject, which is represented only by the standard in its entirety.

Aitchison, W. (2005). *The Rights of Firefighters*. Portland, OR: Labor Relations Information System.

Antonellis, P. J. Jr. (2012). *Labor Relations for the Fire Service*. Tulsa, OK: PennWell/Fire Engineering.

Brehm, D. P. (2012). "Human Resource Management." In: *Managing Fire and Emergency Services*, A. Thiel and C. R. Jennings, eds. Washington, DC: International City/County Management Association.

Brunacini, A. V. (1996). *Essentials of Fire Department Customer Service*. Stillwater, OK: Fire Protection Publications.

Bugbee, P. (1971). *Man Against Fire: The Story of the National Fire Protection Association, 1896–1971*. Boston, MA: National Fire Protection Association.

Carter, H. R., and E. Rausch. (2008). *Management in the Fire Service*. Quincy, MA: National Fire Protection Association.

Caulfield, H. J., and D. Benzaia. (1985). *Winning the Fire Service Leadership Game*. New York, NY: Fire Engineering.

Compton, D. (2010). *Progressive Leadership Principles, Concepts and Tools*. Stillwater, OK: Fire Protection Publications.

Edwards, S. T. (2009). *Fire Service Personnel Management*. Upper Saddle River, NJ: Brady/Prentice Hall Health.

England, R. E., et al. (2012). *Managing Urban America*. Washington, DC: CQ Press.

Epler, P. (1997, November 13). "Fire Truculence. Is It a Case of Rogue Firefighters or Meddling City Managers? Whatever It Is, the Phoenix Fire Department Isn't Used to the Scrutiny It's Getting from City Hall." *Phoenix New Times*. Phoenix, AZ: Phoenix New Times LLC. Accessed 7/29/13 at: http://www.phoenixnewtimes.com/1997-11-13/news/fire-truculence/.

Fire and Emergency Service Image Task Force. (2013). *Taking Responsibility for a Positive Public Perception*. Fairfax, VA: International Association of Fire Chiefs.

Flood, J. (2010). *The Fires: How a Computer Formula, Big Ideas and the Best of Intentions Burned Down New York City—and Determined the Future of Cities*. New York, NY: Riverhead Books.

Gormley, W. T., and S. J. Balla. (2012). *Bureaucracy and Democracy: Accountability and Performance.*, 3rd ed. Washington, DC: CQ Press.

Hashagen, P. (2000). *New York City Fire Department History*, Fire Department City of New York, J. Kimmerly, ed. Paducah, KY: Turner Publishing Company, 17–230.

Kreiger, A., and T. Masten. (2012). *Beyond the Consent Decree: Gender and Recruitment in the San Francisco Fire Department*. Emmitsburg, MD: U.S. Fire Academy, Executive Fire Officer Program.

Matejka, M. G. (2002). *Fiery Struggle: Illinois Fire Fighters Build a Union, 1901–1985*. Chicago, IL: Illinois Labor History Society.

McAniff, E. P., and J. J. Cunningham. (1974). *Leadership in the Fire Service*. Bayside, NY: McAniff Associates, Inc.

McCarl, R. (1985). *The District of Columbia Fire Fighter's Project: A Case Study in Occupational Folklife*. Washington, DC: Smithsonian Institution Press.

National Fallen Firefighters Foundation. (2011). *Understanding and Implementing the 16 Firefighter Life Safety Initiatives*. Stillwater, OK: Fire Protection Publications.

National Institute of Standards and Technology. (2010). *Report on Residential Fireground Field Experiments, NIST Technical Note 1661*. Washington, DC: U.S. Department of Commerce.

National Institute of Standards and Technology. (2013). *Report on High-Rise Fireground Field Experiments, NIST Technical Note 1797*. Washington, DC: U.S. Department of Commerce.

Wrap-Up, continued

O'Leary, R., and C. M. Gerard. (2013). "Collaborative Governance and Leadership: A 2012 Survey of Local Government Collaboration." In: *The Municipal Year Book 2013*. Washington, DC: International City/County Management Association. pp 57–70.

Pratt, F. D., et al. (2007). *Prehospital 9-1-1 Emergency Medical Response: The Role of the United States Fire Service in Delivery and Coordination*. Washington, DC: IAFF, IAFC, and NVFC.

Rainey, H. G. (1997). *Understanding and Managing Public Organizations*. San Francisco, CA: Jossey-Bass.

Schrag, Z. M. (1992). "Nineteen Nineteen: The Boston Police Strike in the Context of American Labor. In: *Social Studies*. Boston, MA: Harvard University Press, 122.

Smeby, L. C. Jr. (2014). *Fire and Emergency Services Administration: Management and Leadership Practices*, 2nd ed. Burlington, MA: Jones & Bartlett Learning.

Starling, G. (2001). *Managing the Public Sector*. Stamford, CT: Wadsworth.

Varone, J. C. (2011). *Legal Considerations for Fire and Emergency Services*. Clifton Park, NY: Delmar/Cengage Learning.

Wage and Hour Division. (2011, May). *The Fair Labor Standards Act of 1938, as Amended*. Washington, DC: U.S. Department of Labor.

Ward, M. J. (1992). *Labor Strikes in the American Fire Service: A Special Report for the Japanese Local Government Center*. Fairfax, VA: International Association of Fire Chiefs.

Ward, M. J. (2003). "Brother vs. Brother." *Fire Chief*, 46–48.

Yukl, G. A. (2013). *Leadership in Organizations*, 8th ed. New York, NY: Prentice-Hall.

Ziskind, D. (1940). *One Thousand Strikes of Government Employees*. New York, NY: Columbia University.

FIRE OFFICER in action

Failure of a city enterprise project has created a large fiscal liability. This morning, local television reports that the city plans to reduce the public safety budget by 45 percent. That would mean disbanding fire companies and losing fire fighter jobs. There are many angry voices at the firehouse kitchen table during morning line-up, some talking about strikes.

Captain Davis agrees to get more information. Battalion Chief Johnson is also surprised by the news item. Davis suggests to Fire Fighter Anders that he call the shop steward or one of the local labor leaders. At noon, Davis and Anders share what is known so far. The suggestion appears to be a trial balloon launched by the city auditor, one of many budget-reducing ideas announced today.

The fire officer is in the middle of the labor–management relationship. While a managing or supervising fire officer represents the management of the department at the fire station, the officer is living with the crew and often is affected by the same work conditions. Understanding how labor interacts with management improves the effectiveness of the fire officer.

1. A fire fighter is asked to identify the pro-union members of the fire company to the administrative fire officer. This is an example of:
   A. collective bargaining.
   B. yellow dog contracts.
   C. unfair labor practices.
   D. right to work.

2. What does "right to work" mean?
   A. Employees are required to be members of a union.
   B. Employees have the right to bargain their positions.
   C. Workers cannot be compelled to join a union as a condition of employment.
   D. Employees are not permitted to strike.

FIRE OFFICER *in action* (continued)

3. If the contract dispute continues, some states would require binding arbitration. What is this?
   A. Giving a specific time period to agree on contract
   B. Dispute resolution by a neutral third party
   C. Opening the ability to strike
   D. Mediation

4. What is included in the third step of the grievance procedure?
   A. The grievance is presented verbally to a review board.
   B. A statement is made of the desired remedy or adjustment.
   C. The economic impact of granting the relief is analyzed.
   D. The grievance is submitted to the personnel or human resources office.

Fire Captain *Activity*

Chief Johnson is meeting with Captain Davis and Lieutenant Williams, discussing the budget crisis and the training assignment issue. The chief points out that the environment that the fire department is working in requires better understanding of labor procedures.

The chief directs the fire officers to "Find ways that we can work better with labor to meet our departmental needs. Working together we can accomplish much more."

NFPA Fire Officer II Job Performance Requirement 5.1.1
The organization of local government; enabling and regulatory legislation and the law-making process at the local, state/provincial, and federal levels; and the functions of other bureaus, divisions, agencies, and organizations and their roles and responsibilities that relate to the fire service

Application of 5.1.1
1. Select one of the IAFC/IAFF joint initiatives. At an organization participating in the initiative, interview one administrator and one labor official to get a current description of the initiative from their perspective. List two positive and two "needs work" points from the interviews.
2. Because of a chronic funding shortfall, it is proposed that Municipal City Fire and Rescue be part of a statewide fire and rescue department. Identify one issue with this proposal from a labor perspective and suggest a solution.
3. Municipal City proposes entering into Chapter 9 bankruptcy. What will that mean for existing labor agreement and fire fighter pension plans?

Working in the Community

Fire Officer I

Knowledge Objectives

After studying this chapter, you will be able to:

- Discuss the role of demographics in fire department–community relations (NFPA 4.3.1). (pp 204–206)
- Discuss the role of fire safety education in risk reduction (NFPA 4.3)(NFPA 4.3.3). (pp 206–207)
- List and describe opportunities for public education (NFPA 4.3)(NFPA 4.3.1)(NFPA 4.3.3). (pp 207, 208–212)
- List and describe steps to develop public education programs at the local level (NFPA 4.3)(NFPA 4.3.1). (pp 212–213)

Skills Objectives

After studying this chapter, you will be able to:

- Develop a public education program using a five-step program (NFPA 4.3)(NFPA 4.3.1). (pp 212–213)

Fire Officer II

Knowledge Objectives

After studying this chapter, you will be able to:

- Describe the role of the media in getting fire department information to the community. (p 213)
- Describe the role of the public information officer in working with the media (NFPA 5.3)(NFPA 5.3.1)(NFPA 5.4)(NFPA 5.4.4). (pp 213–217)
- Discuss the role of social media in community relations. (p 217)

Skills Objectives

After studying this chapter, you will be able to:

- Prepare a press release that conforms to the locally preferred format (NFPA 5.4)(NFPA 5.4.4). (pp 215–216)
- Conduct a media interview as the fire department representative (NFPA 5.3)(NFPA 5.3.1). (pp 215–217)

CHAPTER 11

Principles of Fire and Emergency Service Administration (FESHE) Course Outcomes

1. Acknowledge career development opportunities and strategies for success. (pp 205–206, 212–217)
2. Recognize the need for effective communication skills both written and verbal. (pp 206–207, 209, 213–217)
6. Discuss the various levels of leadership, roles, and responsibilities within the organization. (pp 206–207, 213–217)
8. Identify the importance of ethics as it relates to fire and emergency services. (pp 206, 213)

You Are the Fire Officer

It is almost noon. Quint 100 is returning from a structure fire that was started when discarded fireplace ashes were placed in a plastic trash can inside a garage, but then expanded to necessitate a greater alarm. Hours after the homeowner left for work, a neighbor called 911 to report smoke coming from a garage. As the fire companies started to arrive, the fire ignited improperly stored gasoline, creating an impressive explosion and pushing the fire into the home.

Lieutenant Williams notices a private vehicle parked at the front apparatus door. A citizen approaches as soon as you climb down from the cab and asks if you can take care of her problem right away. She has been waiting for more than an hour for someone to help her install a child car seat. She made an appointment and is unhappy because no one was at the station when she arrived.

Lieutenant Williams asks Apparatus Operator Rollo, the designated car seat technician, to help the citizen. While talking to the frustrated mother, Williams sees a group of children from a local daycare center arrive for a scheduled station tour. The two adults and eight preschoolers are so excited to visit a fire station.

1. What are your options for dealing with the station tour at this point?
2. How do you respond to the citizen who is unhappy about having to wait more than an hour for your return?
3. What will you do with your visitors if you get another alarm within the next 30 minutes?
4. The departmental regulations require you to be clean and dressed in uniform when conducting public education activities. At this moment, the quint crew is wet and dirty. Engine 100 will be at the fire for another hour, and Medic 100 is on the scene of a cardiac arrest. What should you do?

Introduction

Many volunteer fire departments were initially established by community members after a local disaster occurred. During the 1800s, fire stations were frequently used for public meetings and assemblies, and membership in the local volunteer fire company was viewed as a stepping stone to greater power for politicians and power brokers.

Strong community ties still exist in rural and small town fire departments. Even in cities that transitioned to career firefighting, the bonds between local communities and their fire fighters remain strong. The apparatus doors of many fire stations are left open, and the outside bench remains a popular location for fire fighters to maintain close contact with the neighborhood.

The situation is different in most urban areas, where the benches have been moved inside and the apparatus doors are always closed. Decay and crime have changed the nature of the fire service–community relationship in many locales.

The fire station continues to be widely viewed as an important member of the community. Citizens tend to think of the fire department in relation to their local fire station. The financial crises of the 1980s showed the value of strong community ties, as residents in major urban centers, such as New York City, Baltimore, and Toronto, rallied to keep their local fire stations open in the face of budget cuts, measuring their personal perceptions of safety by the proximity of the closest fire station.

Today, the trend is moving toward more community-based local government. Next to schools, the fire department is the most decentralized and community-based function of local government. Some cities use their neighborhood fire stations as a primary point of contact for local government services.

The fire officer in each fire station is the official fire department representative and also ensures that the community's needs are being addressed by the department. The fire company is also a neighbor. This chapter provides the fire officer with information about how to learn more about the community, how to provide fire and life-safety education to meet local needs, and how to work with the news media to disseminate information clearly and appropriately.

Fire Officer I

Understanding the Community

Each community has special needs and different characteristics that should be considered in relation to every service and program provided by the fire department. The most significant information comes from understanding which types of people live and work in the community. The fire officer should develop a good understanding of the population and demographics of the particular areas where the company responds.

The federal government undertakes a nationwide census once every decade. The data gathered via the census are readily available and provide an excellent starting point to begin an analysis of the local community. The census collects and identifies a massive amount of information about the demographics of the nation. Demographic data describe the characteristics of human populations and population segments. The census data can be analyzed with sophisticated database software tools and digitized mapping to profile many characteristics of local populations down to the neighborhood level.

A variety of demographic analysis techniques may be applied. Demographic data are often used to identify and analyze consumer markets, to help retailers predict what consumers will buy, and to give advertisers insight into the messages that will be most effective in particular markets. Politicians make extensive use of demographic data to predict voters' response to policies and messages in each area. Both nonprofit and for-profit programs may be fine-tuned to reach certain segments of the population based on detailed analysis of what different types of people like, what they value, what they believe, and where they live. The same approach can be used to ensure that the fire department is delivering the appropriate services and information to the local community.

Understanding the cultural factors that influence particular behaviors will increase the effectiveness of fire department messages. As a fire officer, you can identify groups with special needs so as to improve the delivery of emergency services. An effective program should be designed to meet the needs of the particular community, and the message must be formulated to reach and be understood by the target audience. With each census, it has become apparent that the United States is growing increasingly more multiethnic. The country is experiencing a continual flow of new immigrants who bring a variety of languages, cultures, religions, traditions, and beliefs into their new communities.

One dimension of demographic analysis is the identification of communities where people share the same cultural background and language. Within a small neighborhood, there may be thousands of people who speak one dialect and share a culture that is totally different from that of the adjoining neighborhoods. This type of diversity often challenges the fire service's ability to meet the needs of different groups.

Safety information shared through public education campaigns should be applicable to the particular community and delivered in a format that the community can understand and act on. The format and the method of delivery should be different when fire fighters are meeting with a community of new immigrant families who understand very little English versus a community of university students living in fraternity houses.

Both emergency services and public education efforts must be fine-tuned to identify the needs and meet those needs in particular communities. The fast and loud aspects of the typical U.S. response to a fire/emergency medical services (EMS) incident could be overwhelming, terrifying, or embarrassing to a recent immigrant who called for assistance. The customs and traditions of another culture may clash with the aggressive "pit crew" response to a cardiac arrest. A shop owner may be angered by the arrival of a fire company to perform a fire safety inspection. Cultural sensitivity is required to help the customers appreciate the services the fire department is providing.

Company Tips

Child Safety Seats

Demographic analysis can assist the fire department in planning for new safety programs, such as child safety seat installations. In the late 1990s, public safety officials in many communities identified a need to provide public education and assistance regarding proper child car seat installation. The sudden popularity of this service overwhelmed fire departments that were not adequately prepared to handle the public demand for it.

The proper installation of a child car seat is complex. A specialized class is required for a public safety official to achieve certification as a child car seat technician. The demand for this type of service is predictably concentrated in areas where there are families with automobiles and young children.

The fire department can use demographic information to identify those fire stations that serve communities where there are concentrations of young families who would take advantage of the service. Fire officers can then survey those communities to identify the days and times when the demand for the service would the highest. To meet the public's need, specific days and times might be designated to install child safety seats. Dedicated staffing could be assigned on those days to ensure that a certified child car seat technician is available to perform the installation, regardless of the emergency workload. This information can be disseminated in press releases and posted on the department's website and on signs at every fire station **FIGURE 11-1**.

Municipal City Fire and Rescue Department

Child Safety Seat Installation

The Municipal City Fire and Rescue Department has a new program that offers free information and assistance on the proper installation of your child's car seat.

It is important to make an appointment with the installer on the day you would like to come in. Because of the nature of our work, we regret that we cannot accommodate "walk-in" requests, but we may be able to provide same-day service.

Appointments are available for weekday, evening, and Saturday on a first-come, first-served basis.

You can schedule an appointment or get more information at:
<department website> and clicking on "car seat installation," or calling (111) 555-SEAT.

Public Information Officer T. L. Green,
<email>
(111) 555-FIRE or (111) 555-3473

FIGURE 11-1 Description of a car seat installation program.

Getting It Done

Se Habla Espanol?

In communities where English is not the primary language, it is important to be able to communicate in the language of the neighborhood. Many departments encourage their members to learn additional languages, and some even provide special classes or offer pay incentives for multilingual fire fighters. A few departments have experimented with language immersion programs, making the primary language of the neighborhood (e.g., Spanish) the primary language at a designated fire station. The fire fighters assigned to that station answer the telephone, speak to one another, and handle emergencies while speaking that language.

Risk Reduction

The fire service has a rich history of striving to reduce the risk of fire in the community. This has been achieved through emergency response, fire safety education, adoption of fire codes, and enforcement of those codes. More recently, fire departments have recognized that their primary goal is to save lives and property from more than just fires. This broader vision has become evident as the fire service has responded to extrication, drowning, heart attack, stroke, hazardous materials release, trench and building collapse, high-angle rescue, swiftwater rescue, underwater and ice rescue, and weapons of mass destruction incidents.

The fire service has taken on the role of prevention and mitigation of these types of incidents. The best method of preventing fire injuries and deaths, of course, is to prevent the fire from ever starting or to reduce the severity of the fire if it does start. The same is true for the other types of incidents to which the fire department is called.

This shift in thinking often requires a transformation of the culture of the fire department. Members, after all, joined to fight fires—not to check car seats. The company officer's challenge is to promote the concept that the fire fighter role is to save lives and property from every type of incident to which department members might be called. A link must be made between the incidents to which the fire department responds and the prevention of those types of incidents.

The fire officer should use incidents that are encountered to reinforce the need for risk-reduction programs. For example, suppose the fire company responds to a child who is seriously injured from head injuries in a bicycle accident, and the head injuries could have been prevented by wearing a helmet. There are two levels of needs in such an incident: systemic and individual.

Systemic needs can be addressed through the development of programs to eliminate or reduce risk, such as a bicycle helmet safety program directed toward children. Often, these programs are a community-wide response and involve many organizations and agencies. Many of these needs are identified at the departmental or community level rather than at the fire company level. Fire officers, as part of their role, may identify systemic needs from the calls to which the fire company responds; those needs should, in turn, be reported to the fire administration for consideration.

The fire officer is also in a unique position to identify the needs of the individual. The fire department is one of the few community resources that regularly responds to private homes. The needs identified in this way can vary from reporting signs of child abuse to heating assistance programs. Typically, programs are already in place to address the need. The company officer's role then becomes to connect the citizen in need with the community resource.

To fill this role effectively, the fire officer must be proactive about learning which community programs are available. Many relevant programs are run by the health department and the department of social services. Often, these departments know which assistance programs are available for low-income families and the elderly. The area safety council and hospitals are valuable resources for information on the prevention of accidents and injuries. The police department and the highway department have a vast amount of information on vehicle accident and injury prevention. Faith-based and civic groups also have many programs geared toward meeting community needs.

A fire officer should seek out this information before he or she actually needs it. Often, brochures are a good method of having the information readily available when it is needed; this approach allows the fire officer to address the need on the scene. For example, during a call, the fire officer notices a pool with an unlocked gate. Discussing the issue with the resident and leaving a pool safety information packet may prevent the fire department from responding to a child drowning at the location. Other issues may not be able to be addressed on the scene. An elderly couple who lives alone and is unable to care for themselves, for example, will likely need to be referred to the appropriate community agency for a follow-up visit.

It is important to follow departmental policies for these situations. Sometimes, the fire officer is allowed to place a call directly to the agency that can provide the assistance. Fire officers are more likely to take action in situations calling for external assistance in fire departments that empower the fire officer to initiate action directly.

■ Responding to Public Inquiries

Sometimes, the public makes an inquiry to the fire officer or fire fighter for general information, such as a request for a description of the services the fire department or city provides or a request to remove a cat from a tree. Both of these situations have the potential to leave the citizen satisfied or for the citizen to go away with a negative view of the fire department.

The fire officer must treat all requests professionally and with respect. Even though you may find the request less than critical, the citizen believes that it is valid and important. Failure to approach the request with sincerity can have a lasting negative impact.

Every effort should be made to answer each question fully and accurately. The fire officer may not have all the information

> ### Getting It Done
>
> **Reducing Residential Fire Deaths and Injuries**
> Since 2004, the London Fire Brigade been engaged in strong community safety programs as part of an overall national strategy for improving fire safety. A unique aspect of this campaign is the home fire safety visit program.
>
> The British fire service visits high-risk homes, using a combination of line fire fighters and prevention specialists. The visits include installation and testing of smoke alarms, inspections for hazards, mitigation of hazards, and one-on-one education. Community safety specialists called "advocates" join fire fighters in visiting ethnic or high-risk households. Their specialties include foreign languages, problems of the elderly, problems of alcoholics, and problems of the hearing or mobility impaired.
>
> Home visits are scheduled via call centers established in the brigade, not by dispatchers. Home visits are often scheduled after referrals from social services or other agencies, or by households already visited who suggest others.
>
> From 2004 to 2006, residential fire death rates in England and Scotland dropped by 41 percent and 44 percent, respectively. The residential fire death rate dropped from 9.7 deaths per 1 million population in 1990 to 5.7 deaths per 1 million population in 2005–2006. A follow-up London Fire Brigade study, covering 2007 through 2009, shows this trend continuing (Schaenman, 2007):
>
> - Incidents attended decreased from 152,117 to 138,385.
> - Serious fires attended decreased from 15,093 to 13,605.
> - Home fire safety visits increased from 36,617 to 48,768.
> - Accidental fires in the home decreased from 6224 to 5781.
> - Injuries from accidental home fires decreased from 1046 to 737.
> - Residential fire deaths changed from 30 (2007), to 41 (2008), to 30 (2009).

> ### Getting It Done
>
> **"My Cat Is Stuck in the Tree. Can You Get It Down?"**
> Many citizens believe that the image of a fire fighter rescuing a cat from a tree is an accurate one. In reality, most fire departments do not provide this service. There is a trend to reconsider this position for both risk reduction and public relations reasons. First, if the cat does not come down, the owner experiences a loss. Second, if the resident attempts to retrieve the cat, he or she can be placed in grave danger.
>
> One method of dealing with such calls is to send a fire company to the scene to evaluate the situation. Most cats are perfectly able and willing to come down in a matter of time. An empathetic explanation to the owner, with a referral to animal control or a veterinarian, may allay concerns. If the cat is actually stuck in the tree, has remained there for several days, and is in jeopardy of dying, the fire officer may consider whether the fire crew can retrieve the cat in a reasonably safe manner. The fire officer must balance the safety of the fire crew against the seriousness of the situation, the risk to the citizen from the fire department's inaction, and the potential benefits of a positive result.

that the individual is seeking. For example, the fire officer may not know where the closest flu shot clinic is located. When this occurs, the fire officer should seek out the information immediately rather than just expressing that he or she does not know. A phone call or two can usually lead to the answer and leaves the citizen with a positive image of the fire department.

Some requests that citizens make may not be within the fire officer's authority. In these situations, the fire officer should provide a method of moving the request to the level where it can be resolved. If time allows, the best method is to get the citizen's contact information, write up a summary of the discussion, and forward it to the appropriate administrative fire officer at headquarters. The fire officer should also follow up to ensure that the citizen is contacted in a reasonable amount of time.

Some citizens may prefer to contact fire administration themselves. In these situations, the fire officer should give the citizen specific information on whom to contact. The fire officer should also contact the administrative or executive fire officer before the citizen does to provide information on the situation.

Many departments have specific policies that address how citizen inquiries are to be handled. The fire officer should understand and follow these policies. Failure to do so may lead the citizen to believe that he or she is being treated unfairly. It can also leave the fire officer open to disciplinary action.

Public Education

Goals, objectives, content, and delivery mechanisms for public education programs vary greatly among fire departments, depending on their resources and circumstances. Some large fire departments have staff bureaus that specialize in the development and delivery of public education programs. Other fire departments adopt national programs or adapt a program that was developed by some other department. In many cases, a public education program is developed at the local level to meet local needs. Public fire safety education programs include the following:

- Learn Not to Burn
- Risk Watch
- Stop, Drop, and Roll **FIGURE 11-2**
- Getting to Know Fire
- Change Your Clock—Change Your Battery
- Fire safety for babysitters
- Fire safety for seniors
- Wildland fire prevention programs

VOICES OF EXPERIENCE

Although the fire service does not produce a commodity, like the one some people think is necessary for determining a business model, we do provide a service just like any other service-based provider. I'd like to share an example of customer service that is far removed from any job description I have seen in the fire service.

Located directly across the street from my firehouse is a three-story, 74-unit apartment building reserved for occupants age 55 and older. This building is known for having an unreliable public elevator, and one morning this would come into play. As I reported for duty one sunny Sunday morning, I was informed by the off-going shift officer that the elevator was out of service. This notification became a courtesy among the crews so that if we received an EMS call to the building, additional personnel could be dispatched to help move patients via the stairwell from the upper floors to the ground floor.

> **Sometimes it's the little things that make the big differences. It's not about us; it's about the customers.**

At approximately 10 A.M., a call came in on our business line. Upon answering it, I noted that the voice on the other end was that of an elderly female, who was noticeably upset and sobbing. When asked what was wrong, she went on to tell me that she was upset because her "normal" Sunday morning routine of retrieving her newspaper from the lobby and going to church was impossible due to the elevator outage. She went on to explain that she lived on the third floor of this apartment building and, due to her medical conditions, was reliant on the elevator. While this inability to use the stairs has other, greater implications, especially during a fire or other disaster, this day it meant that her Sunday morning routine was interrupted.

Understanding the importance of church to many people, I quickly offered whatever assistance I could to arrange for her to get to her destination. She quickly stopped me and stated that it was not about church because she had already said her prayers, but rather it was about retrieving her newspaper from the lobby. I was amazed at how much this meant to her.

I decided that the best thing to do was to walk across the street and deliver her newspaper. Entering the lobby, I encountered at least a half-dozen newspapers for both the second and third floors that had not yet been retrieved, all nicely stacked with Post-it notes listing the respective apartment on each one. I scooped up the papers and made my way up the stairs to each of the floors, knocking on doors and delivering as I went along. The looks on the faces of the residents were priceless when I explained what I was doing, and the level of appreciation and outpouring of support were tremendous, giving more importance to the added value gained from this no-cost, low-effort expression of nontraditional customer service.

It just goes to show that sometimes it's the little things that make the big differences. Just remember, it's not about us, it's about them—the customers.

Joseph Knitter
South Milwaukee Fire Department
Milwaukee Area Technical College
South Milwaukee, Wisconsin

FIGURE 11-2 Stop, Drop, and Roll teaches young children what to do if their clothing catches fire.

FIGURE 11-3 The goal of public safety education programs is to prevent injury, death, or loss due to fire.

The delivery of public education programs often involves fire suppression companies and depends on fire officers to transmit the message to the intended audience. Fire officers can also be assigned to positions that involve specific responsibilities for public education program planning and development.

Company Tips

Not Everyone Is Wired

Consider the population you serve. If most of your community is elderly, a website survey or exclusive use of social media will be ineffective; instead, access to the department through the telephone or a postcard will likely be more effective. Having a fire department representative at community events or meetings is a powerful tool in serving senior citizens.

Use all methods possible to obtain feedback from residents and visitors. They are the ones who determine how well the fire department is serving the community. Consider providing a feedback postcard to those who received service.

Every fire officer should understand some of the basic principles of public education programs.

The fundamental goal of a public safety education program is to prevent injury, death, or loss due to fire or other types of incidents. A public safety education program can have four objectives:

- Educate target audiences in specific subjects so as to change their behavior.
- Instruct target audiences on how to perform specific tasks, such as Stop, Drop, and Roll or operating fire extinguishers.
- Inform large groups of people about fire safety issues.
- Distribute information on timely subjects to target audiences.

An educational presentation is successful when it causes an observable change of behavior. The fire officer should have a specific goal in mind when educating the public, just as there should be a goal when providing training for fire fighters **FIGURE 11-3**.

National and Regional Public Education Programs

The NFPA, the United States Fire Administration (USFA), and other specialized associations and industry groups have developed a variety of national and regional public education programs that can meet local community needs. Some of these programs are designed for general outreach, such as Change Your Clock—Change Your Battery and Fire Prevention Week themes. Programs targeted toward particular problems and population groups are also available.

Fire Prevention Week

The history of Fire Prevention Week has its roots in the Great Chicago Fire, which began on October 8, 1871, but continued into and did the most damage on October 9, 1871. In just 27 hours, this tragic conflagration killed 300 people, left 90,000 residents homeless, and destroyed more than 17,400 structures.

FIGURE 11-4 National Fire Prevention Week is observed the anniversary week of the Great Chicago Fire and is sponsored by the NFPA.

> ### Getting It Done
>
> **Public Safety Education Messages**
>
> In the "You Are the Fire Officer" scenario, the fire station tour for the daycare center provides an opportunity to deliver a public safety education message that is suitable for a preschool audience. The NFPA has developed age-appropriate fire prevention and safety messages that can be delivered as part of a fire station tour. For preschoolers, knowing how to Stop, Drop, and Roll if clothes are on fire is a very appropriate safety message. The tour should include the opportunity for the children to demonstrate how to Stop, Drop, and Roll.
>
> Fire officers should develop scripts to guide fire station visits for different age groups. These scripts can include the safety message or messages that are appropriate for the particular audience. The USFA Kids website provides general lesson plan examples. Audiovisual aids, handouts, or souvenirs should be available for planned visits. Do not forget the adults who accompany the children. In addition to the fire safety messages, these adults might be interested in a handout describing how the department serves the community, including a breakdown that shows what the department has purchased from tax revenues and community donations.

On the 40th anniversary of the Great Chicago Fire, the Fire Marshals Association of North America (now known as the International Fire Marshals Association) decided that the date of this occurrence should be observed in a way that would keep the public informed about the importance of fire prevention. In 1920, President Woodrow Wilson issued the first National Fire Prevention Day proclamation. Every year since 1922, National Fire Prevention Week has been observed on the Sunday-through-Saturday period in which October 8 falls. In addition, the President of the United States has signed a proclamation announcing the national observance of Fire Prevention Week every year since 1925. NFPA has officially sponsored Fire Prevention Week since the observance was first established **FIGURE 11-4**.

Risk Watch

Risk Watch is a comprehensive program directed at injury prevention. It was developed by the NFPA, with co-funding from the Lowe's Home Safety Council and in collaboration with a panel of respected safety and injury prevention experts. A school-based program that links teachers, safety experts, and parents, Risk Watch gives children and their families the skills and knowledge they need to create safer homes and communities.

The curriculum is divided into age-appropriate lessons, each of which addresses the following topics:

- Motor vehicle safety
- Fire and burn prevention
- Choking, suffocation, and strangulation prevention
- Poisoning prevention
- Firearms injury prevention
- Bike and pedestrian safety
- Water safety
- Natural disasters

Community Emergency Response Team

The Community Emergency Response Team (CERT) concept was developed and first implemented by the Los Angeles City Fire Department (LAFD) in 1985. The CERT program helps citizens understand their responsibilities in preparing for disaster and increases their ability to help themselves, their families, and their neighbors safely in many types of situations.

The concept underlying the program was to provide basic training to residents and employees in local communities, as well as to government workers, that would allow them to function effectively during the first 72 hours after a catastrophic event. Experience had shown that in the event of an earthquake or similar event, emergency services would be overwhelmed with serious incidents and unable to respond promptly to every problem. The CERT program was developed to train citizens to help themselves.

CERT groups can provide immediate assistance to victims in their area and collect disaster intelligence that assists professional responders with prioritization and allocation of resources after a disaster **FIGURE 11-5**. The training also teaches the CERT members how to organize spontaneous volunteers who have not undergone the training.

The Whittier Narrows earthquake in 1987 underscored the area-wide threat of a major disaster in California as well as the effectiveness of the CERT program. As predicted, during that event, so many individual incidents occurred over such a wide area that fire department responders could not quickly get to every location where assistance was needed. Where the CERT program had been implemented, the value of having trained groups of citizens was clearly demonstrated. As a result, the LAFD Disaster Preparedness Division was established, and the CERT program was expanded.

FIGURE 11-5 CERT groups can provide immediate assistance to victims and help professional responders with prioritization and allocation of resources.

- **Session 4, Disaster Medical Operations, Part II:** Covers evaluating patients by performing a head-to-toe assessment, establishing a medical treatment area, performing basic first aid, and demonstrating medical operations in a safe and sanitary manner.
- **Session 5, Light Search-and-Rescue Operations:** Participants learn about search-and-rescue planning, size-up, search techniques, rescue techniques, and—most important—rescuer safety.
- **Session 6, CERT Organization:** Addresses organization and management principles and the need for documentation.
- **Session 7, Disaster Psychology:** Covers signs and symptoms that both the disaster victim and the worker might experience.
- **Session 8, Terrorism and CERT:** Provides an overview of terrorism and weapons of mass destruction.
- **Session 9, Course Review and Disaster Simulation:** Participants review their answers from a take-home examination. Finally, they practice the skills they have learned during the previous eight sessions in disaster activity.

Since the CERT program was moved to the Citizen Corps, there has been an increased emphasis on assessing community needs and developing CERT response goals that address local needs. The appendix in the CERT training manual includes lesson plans for 13 hazardous situations.

The Emergency Management Institute and the National Fire Academy adopted and expanded the CERT materials. In 1993, this training was made available nationally by the Federal Emergency Management Agency (FEMA). CERT was moved to the Citizen Corps section of the federal government in 2004. In 2013, Citizen Corps listed 2200 CERT programs.

CERT Course Schedule

The CERT course is delivered in the community by a team of first responders. The instructors should complete a CERT Train-the-Trainer Program, conducted by their State Training Office for Emergency Management.

Training is usually delivered in nine 2.5-hour sessions, with the sessions taking place one evening per week. The training covers the following topics:

- **Session 1, Disaster Preparedness:** Addresses the different hazards to which people are vulnerable in their community. Materials cover actions that participants and their families can take before, during, and after a disaster. As the session progresses, the instructor begins to explore an expanded response role for civilians to become disaster workers. The CERT concept and organization are discussed, as are applicable laws governing volunteers in that jurisdiction.
- **Session 2, Fire Safety and Utility Control:** Briefly covers fire chemistry, hazardous materials, fire hazards, and fire suppression strategies. The thrust of this session is safe use of fire extinguishers, size-up of the situation, control of utilities, and extinguishment of a small fire.
- **Session 3, Disaster Medical Operations, Part I:** Participants practice diagnosing and treating airway obstruction, bleeding, and shock by using simple triage and rapid treatment techniques.

Maintaining CERT Involvement

CERT members should receive recognition for completing their training. Some communities issue identification cards, vests, and helmets to graduates.

Getting It Done

Catastrophe Planning

Some fire departments have developed media communication/public education plans to assist their communities in responding to natural and human-made catastrophes. For example, the San Francisco Fire Department accepted the services of a retired local television anchorperson to develop preassembled public safety messages and citizen instructions that may be issued in a wide variety of major emergencies.

Fire departments have to be able to respond to dynamically changing conditions. For example, national concerns, such as methicillin-resistant *Staphylococcus aureus* (MRSA) outbreaks, have required fire departments to develop specific public information messages quickly, using information updates from the Centers for Disease Control and Prevention. In 2012, Hurricane Sandy required more than a dozen federal, state, and local public safety agencies to coordinate their efforts to provide a consistent message to the general public.

Near-Miss REPORT

Report Number: 12-0000259

Event Description: Engine [1] was at a local high school to oversee a bonfire event. (Identifying information in squared brackets [ ] has been eliminated or changed.) The engine [1] captain was asked to ignite the fuel to start the bonfire. The captain was using a match to ignite the fire from about 10 feet from the fuel source. Once the match was struck, the vapors from the fuel ignited. [The captain was burned.] He was not wearing appropriate PPE. He was transported to an area hospital.

Lesson Learned: Department members are not permitted to ignite any substance for activities not related to department training. In addition, the captain should have had full PPE on during this event, and a fuel trailer of at least 25 feet is required for any ignition.

Courtesy of the National Fire Fighter Near-Miss Reporting Systems (firefighternearmiss.com)

It is important to keep the graduates involved and their skills sharp after they have completed the basic training course. Trainers should offer periodic refresher sessions to reinforce the basic training. In addition, CERT teams can sponsor events such as drills, picnics, neighborhood clean-ups, and disaster education fairs, which keep them involved and trained.

■ Locally Developed Programs

Like the CERT program, many regional and national public education programs started within local fire departments. Brush abatement programs for wildland fire areas and fireplace ash disposal programs in the northern states are two examples of regional programs that meet specific needs.

In large fire departments, the development of new public education programs is usually assigned to an individual or a work group specializing in this area. In a smaller department, a fire officer could be given this assignment as a special project.

The five-step planning process can help local fire departments create and develop programs. This process was created by the USFA in the 1970s and updated in 2003 by FEMA.

Identify the Problem

The first task is to identify a fire or life-safety problem. In many cases, a public education project is initiated to address a problem that demands attention, such as an increasing number of fire deaths in a community. The process sometimes works in the opposite direction, with the decision to undertake a public education effort being made first, before the objective is identified. In either case, the fire officer should attempt to identify the problem clearly before jumping into program development.

The fire officer should look at local emergency incident data to identify the types of events that generate deaths or serious injuries as well as the most frequent causes of fires. This analysis should consider both the frequency and the severity of different risks. For example, the fire data for a particular city might indicate an average of 100 food-on-the-stove fires and 25 bedroom fires per year. The food-on-the-stove fires result in three serious burn injuries each year and one fatality every 3 years, whereas the bedroom fires average three fatalities and six serious burn injuries annually. This analysis indicates that bedroom fires are the more critical life-safety problem in this community. There are more kitchen fires than bedroom fires, but the bedroom fires cause more deaths and serious injuries. (The *Safety and Risk Management* chapter discusses the concept of hazard and risk in more detail.)

Data from another city might point out that the total number of fire deaths is smaller than the number of young children who drown in backyard swimming pools in an average year. In this city, a public education project that is directed toward the drowning problem would be a higher priority than a program designed to prevent fire deaths.

Public education programs are, in many cases, developed in response to a single, high-profile incident. Sometimes the high-profile incident directs public attention toward a problem that has been ignored in the past but has now created an opportunity to provide valuable public education. Unfortunately, in other cases, this type of program is more reactive than effective and causes resources to be wasted on low-priority problems.

Select the Method

After identifying the problem, the fire officer should select the most effective method to address it. Public life-safety education is not necessarily the best or the only response for every type of problem. The fire officer should try to determine whether an engineering solution or an enforcement solution would be more efficient.

Design the Program

When designing a public education program, the fire officer should review the ABCDs of course preparation, just as in fire fighter training:

- Identify the *audience*.
- Explain the desired *behavior* the student should demonstrate after training.
- Under which *conditions* will the student perform the task?
- Which *degree* of proficiency is expected?

The fire officer must also identify the specific objectives of the educational program. The program format should be matched to the message, the audience, and the available resources. Demographic information assists in describing the audience. The development process should include delivery of a pilot class to a target audience to see how well it works. Most new programs will benefit from revision after the pilot is evaluated.

Implement the Program

Implementation occurs when the program is actually delivered to the target audience. Timing is often an important consideration regarding when a public education program is implemented. For example, a fire safety program for university students would be most effective during the first few weeks in the fall semester, so that it can reach all of the new residents. The objective could be to reach every on- and off-campus student housing location at the beginning of the academic year.

Evaluate the Program

Two types of evaluation may be undertaken: immediate feedback and long-term evaluation. Immediate feedback is obtained from the students at the conclusion of the training, usually via a short survey form. Immediate feedback can help the fire officer determine whether the specific objectives were met by the presentation. This type of immediate feedback is most effective in evaluating the mechanics of the class presentation, such as interaction between the speaker and the audiovisual media.

A longer-term evaluation should focus on the effectiveness of the program in relation to the desired effect. Success is accomplished when the students actually demonstrate the desired behavior in the anticipated situation. For example, the goal of a program at a university could be to reduce the number of malicious false alarms from pull stations. To evaluate the effectiveness of the program, the number of malicious false alarms in a time period before the training would be compared with the number that occur during an equivalent period after the training.

Fire Officer II

Media Relations

Most fire departments have frequent interactions with the local news media when reporters want to obtain information about a situation that is still occurring, as well as in newsworthy situations that are related to other fire department programs or activities. In addition, in many situations, the fire department has information or an important message to communicate to the public.

Emergency incidents are high-profile news events that capture public attention. Members of local print, radio, and television organizations can be expected to appear at major emergency incidents or to call the fire department seeking information. Other activities, ranging from inspection and public education programs to promotions, retirements, awards ceremonies, and delivery of a new fire truck, are also likely to generate media interest.

The news media have an important and legitimate mission to obtain and report information to the general public. The fire department is an organization that performs a public function, operates in the public view, and is usually supported by public funds. It is always in the best interest of the fire department to maintain a good relationship with the local media. Providing information to the news media should be a normal occurrence.

Although the news media have a legitimate purpose in seeking information, there are limitations on the information that can be released in many circumstances. The fire officer should know the applicable departmental procedures and guidelines and understand how to respond to a request for information that the fire officer is not authorized to release. The fire officer should recognize that the interaction that occurs between the fire department and the news media is likely to have a direct impact on the department's public image, reputation, and credibility. The fire department depends on public trust and confidence to be able to perform its mission.

■ Fire Department Public Information Officer

Some fire departments have a full-time or part-time public information officer (PIO) who functions as the media contact person and the source of official fire department information. Where this position exists, most interactions with the news media should go through this individual. Other officers then have limited responsibilities and opportunities to interact with the news media. On some occasions, the PIO may refer a reporter to another fire officer to obtain information about a particular subject or situation.

Many smaller departments rely on the local or a staff officer to function as the PIO and interact with the local newspaper, radio, and television media as the need arises. In these situations, the individual should be guided by the same basic principles as a regularly assigned PIO. NFPA 1035, *Standard for Professional Qualifications for Fire and Life Safety Educator, Public Information Officer, and Juvenile Firesetter Intervention Specialist*, covers the job requirements of a public information officer.

Getting It Done

Engineering, Education, or Enforcement?

- *Fireplace ashes:* Fire departments in cold climates often experience large-loss fires due to the improper disposal of fireplace ashes. In many cases, the ashes are placed in combustible containers and left in a garage or on a deck. The smoldering ashes may then eventually ignite the container and cause a fire. An engineering solution for this problem might be for the fire department to provide free metal ash cans to the community as part of a seasonal fire safety education program.
- *Dead smoke alarms:* Many fire departments encounter single-station smoke alarms with dead or missing batteries in residences. Some departments make it a practice to check the smoke alarm during every fire company response to a residential occupancy in the community. After stabilizing the initial incident, the fire officer may request permission to check the smoke alarms in the home. A simple engineering solution can quickly solve this problem. Every fire company carries a supply of 9-volt batteries that are used to replace any dead or missing batteries. Some fire departments even issue complete smoke alarms to their fire companies with instructions to install one in any home that does not have a functioning smoke alarm. The fact that the fire department takes this problem so seriously conveys an added public education lesson.
- *Brush abatement:* Failure to maintain a clear area around a building is a common cause of structure fire losses in wildland fires. Los Angeles County uses all three factors—education, engineering, and enforcement—in efforts to reduce the impact of brush fires on the built environment. Education starts with seasonal public safety messages conveyed through the media in multiple languages. Engineering includes a description of the size of the required clear area and a program to certify brush clearance contractors. Enforcement includes local government inspection of properties, as well as issuance of fines and work orders if the brush has not been cleared.

FIGURE 11-6 Reaching out to a community is easier when solid contacts and relationships with members of the media have been established.

This section presents basic recommended practices for a PIO who has to establish and maintain an ongoing relationship with the local media. This aspect of community relations is particularly important when the fire department has information it needs to release to the general public. Reaching out to a community is much easier for an organization that has already established solid contacts and relationships with the local media **FIGURE 11-6**.

The NFPA provides a "Get Your Message Out" media primer to help the fire service publicize community outreach campaigns and other events associated with public safety education. The basic concepts are valuable for any fire officer who has to be prepared to interact with media representatives. The NFPA publication recommends three steps when working with the media:

- Build a strong foundation.
- Use a proactive outreach.
- Use measured responsiveness.

Step 1: Build a Strong Foundation

Look at your relationship with the media as a business arrangement: the media have something you need (an audience and the means to communicate with them), and you have something they need (news and information). If you work collaboratively, both parties can achieve their objectives. If you expect too much or if you do not anticipate the media outlet's needs, someone will end up disappointed, which could prove counterproductive.

The most important media asset you will ever have is a good relationship with the media. Members of the local media need to be familiar with you and comfortable working with you; they must have confidence in the information you provide and, most importantly, need to trust you. Whatever else you do, always tell the truth to the media, even if the truth is "I do not know" or "I cannot release that information at this time" or "It is premature to answer that question." Do not guess, speculate, or lie.

Do not wait until you need a reporter to establish media contacts; lay the groundwork in advance to ensure a collaborative working relationship. Build a list of local media contacts and keep it up-to-date. Obtain the names, titles, and contact information for key individuals. Learn who has responsibility for covering fire and life-safety issues at the newspapers, radio stations, and television stations.

Find out the deadlines for different news outlets and know how to reach reporters with late-breaking information. Television reporters need to have the information before their broadcast time, and print reporters need it before the paper is printed.

Be a consumer of your local media. Read the local newspapers, watch television, surf the online news sites, visit appropriate blogs, and listen to the radio. Pay attention to the kinds of stories these media outlets tend to cover, the angles they use, and the personal style of individual reporters. If a reporter does a good job reporting a fire department story, make a thank-you call. If a story contains inaccurate information, make a call to provide the correct information, keeping the tone positive.

Make sure that the news representatives know how to contact you whenever they need information. If you will not be available, make sure that someone else is accessible. Your objective is to make sure that news media representatives know that your department is interested in working with them and in responding to their inquiries, that you are the person to contact, and that they can have confidence in any information you give them.

Step 2: Proactive Outreach
If you have sufficient resources to undertake one, a proactive media communications plan can be much more beneficial than a reactive approach. Simply responding to media inquiries as they come to you is a reactive strategy. You should be as helpful as possible, providing information within the resource capabilities of your department.

Being proactive means that in addition to being responsive, you actively look for opportunities to use the media to communicate your department's objectives and mission. The media need factual, timely stories to report, and the fire department generates broad public interest. You can help them by providing announcements of newsworthy issues, topics, and events.

Once the working relationship has been established, reporters will begin to look to you for story ideas, resources, and quotes. You can call a reporter or editor and suggest a story that could be interesting to him or her. Learn the types of stories that are most interesting to different reporters and call the appropriate person when you have something that fits his or her interests. Reporters will not automatically accept every story idea, but they will listen to what you have to suggest.

When you have information that you want to distribute to several news outlets, use a press release or hold a press conference. A press conference is a staged event to which you invite the news media to come and hear something important that you have to announce. Press conferences should be reserved for topics and situations that definitely have broad interest; on these occasions, you know that the media will want to attend and ask questions.

Step 3: Measured Responsiveness
When dealing with the news media, remember that one of your responsibilities is to present a positive image of the fire department. In certain situations, this is difficult, particularly if something negative has occurred. It is no use denying that something has happened when the facts are clearly evident. Sometimes, the best that you can do is to be honest and factual to maintain the credibility of the organization.

Be wary of situations in which someone could be trying to misrepresent a situation in a manner that reflects badly on the department. The reporter who comes to you may have already been given a highly inaccurate or slanted viewpoint on a story by some other source. Do not allow yourself to be placed in a situation in which you make a statement that is inaccurate, misleading, or damaging to the fire department. Make sure that your response is accurate and that the information is appropriate for release before you react.

■ Press Releases

A press release is used to make an official announcement to the news media from the fire department PIO. It could highlight a special event, a promotion or appointment, an award ceremony, a fire station opening, a retirement, or any similar occurrence that the department wants to make known to the public. A press release could also provide information about a new or ongoing program, such as the launch of a new juvenile fire-setter counseling program or the release of an annual report showing a 34 percent decrease in residential fires since a major public education program was implemented.

A press release should be dated and typed on department stationery, with the PIO contact information at the top. It should be as brief as possible, ideally one to two pages in length, using an accepted journalistic writing style. Make the lead paragraph powerful to entice the reporter to read on, and answer the basic "who, what, when, where, and why" questions **FIGURE 11-7**.

Depending on the preference of your local media contacts, you can mail, fax, or e-mail press releases. Reporters and editors are more likely to read your releases if they receive only those that pertain to their particular beat or to topics in which they are interested. Personalize the release to an individual at each destination by name and title. For example, a release announcing the official opening of a new fire station could go to the assignment editor at a television station and to the newspaper reporter who usually covers the fire department. A release on cooking fire safety could be directed to the food editor.

■ The Fire Officer as Spokesperson

Every fire officer should be prepared to act as a spokesperson for the fire department. Even if the department has a PIO who is the designated official spokesperson, there are likely to be occasions when other fire officers will have to use these skills.

The most common situation in which a fire officer would appear as an official spokesperson is an interview. The NFPA provides these guidelines for the fire officer conducting an interview with the media:
- Be prepared.
- Stay in control.
- Look and act the part.
- It is not over until it is over.

Be Prepared
Before agreeing to be interviewed, make sure you are authorized to speak on the subject as a departmental spokesperson and be sure you have the appropriate information. Determine the reporter's story angle and identify what he or she already knows about the topic. Ask with whom he or she has already spoken or plans to contact and what he or she hopes to learn from you.

Press release

Public Information Officer T. L. Green
www.fire.jbpub.com
(978) 443-5000

Civilians Rescued from Burning Home

Twenty-eight fire fighters responded to an early-morning house fire in the Grandview District. The 911 call was at 10:47 P.M. on Friday, December 16, 2013; the caller reported smoke coming from a townhome.

Jones & Bartlett fire fighters arrived at 11:07 P.M., encountering smoke and flames coming from the first-floor windows at 10934 Braniff Way. Two elderly females were found by fire fighters on the third floor of the townhome.

An aerial tower was needed to remove the residents from the building. After treatment by paramedic/fire fighters, the women were transported to University Hospital.

It took 22 minutes of aggressive fire suppression before Battalion Chief Frank Johnson declared the fire "under control." There was extensive damage to the first floor of the home.

The cause remains under investigation. No monetary loss has been calculated.

FIGURE 11-7 Example of a press release.

If you are uncomfortable with the situation angle or do not have the information, you do not have to agree to be interviewed. You should be polite but firm if you decline a request for an interview. If possible, refer the reporter to someone else who can provide the information.

Be on time for an agreed-on interview, but do not begin the interview until you are ready. The reporter probably has a deadline that you should try to respect, but not at the expense of your readiness.

When preparing for an interview in which you want to deliver a specific message, identify no more than three key message points in advance and practice saying them in varied ways. Learn to use wording that emphasizes what you are saying is important, such as "The number-one thing to remember is _____" or "More than anything else, people should realize that _____." Practice your talking points with a colleague or a tape recorder. Do not memorize your message points, but be very familiar with them. Check any statistics, trends, or other information you would not know on a casual basis.

Stay in Control

Be cooperative, but stay focused on your key message points. Listen carefully to the reporter's questions and ask for clarification if you do not fully understand a question. If you need a few seconds to think through your answer, take the time necessary to formulate your answer. If you do not know the answer, say so and offer to find the information and get back to the reporter. Avoid jargon and highly technical language that confuses or distracts your audience. When you have answered the question, stop talking.

Remember that the reporter came to you looking for information and expertise. Be confident and authoritative, steering the interview in the direction that you want it to go. Showcase your message points and your department's mission and work.

Look and Act the Part

When doing a television interview, make sure your appearance is clean, tidy, and professional. Wear your uniform properly, and show pride in the department that you are representing **FIGURE 11-8**. Adopt the appropriate demeanor for the subject. If you are on the scene of a fire fatality, a somber yet authoritative tone is appropriate. If you are conducting a television interview from a fire station open house, a more enthusiastic tone is acceptable.

FIGURE 11-8 Make sure your appearance is clean, tidy, and professional, and wear your uniform properly.

It Is Not Over Until It Is Over

When speaking to a reporter, do not assume that anything you say is "off the record." Instead, assume that everything you say could be quoted, including conversations before and after the actual interview. When you are wearing a microphone, anything you say can be overheard and recorded.

After the interview, double-check to ensure that the reporter has the correct name and spelling for you and your department and accurate information about any program or event that you are promoting. Leave your business card to ensure that the proper name, title, and organization attribution are recorded.

Social Media Outreach

At the NFPA 2012 Conference and Exposition, NFPA leadership provided 10 tips on using social media to expand safety message outreach. The goal of social media outreach is to reach people who are on the Internet (Carli, Hazell, & Backstrom, 2012).

1. *Plan a strategy.* Select a digital media and find your customers. Plan on two digital postings per day. NFPA's Web publisher Mike Hazell suggests a blog is a good way to interact with subscribers and readers.
2. *Commit.* You need to spend time every day getting the message out. NFPA social media manager Lauren Backstrom recommends spending 25 percent of the time listening to customers, 50 percent of the time interacting, and 25 percent of the time creating new material.
3. *Be authentic.* Social media use represents a big departure from earlier message delivery procedures, and creating a human and authentic digital presence is more effective in getting your message across than issuing a cold, stiff news release.
4. *Be current.* Backstrom recommends connecting fire prevention information with current events.
5. *Be social.* Expect responses to digital postings and provide prompt and authentic responses.
6. *Maintain quality.* Provide accurate information that is useful to the reader.
7. *Tailor content to each venue.* Instead of posting the same information on all social media venues, tailor the post to utilize each venue's capability best. Consider establishing focused blogs or websites to meet a particular group's need. NFPA vice president of communications Lorraine Carli noted large growth in the number of age 55 and older residents who are using social media.
8. *Be interesting and entertaining.* Hazell described a contest to get a new voice for Sparky the Fire Dog. The contest got 26 people to audition for the role.
9. *Consider timing.* Identify the usage patterns for each social media venue and time messages to have the largest impact on the target audience.
10. *Track results.* Determine readership, level of engagement, and any measurable change in reader behavior or impact on the identified community issue.

> **Company Tips**
>
> **Grammar Counts**
> Press releases are public documents that are read by reporters, the public, and local elected officials. The presentation and the content both reflect on the image of the fire department. Invest in a dictionary, thesaurus, and grammar book or style guide to improve your written presentation skills. Check the document's spelling, particularly names, and be sure that ranks, titles, and telephone numbers are correct.

■ Social Media Challenges

Social media describes forms of electronic communication through which users create online communities to share information, ideas, personal messages, and other content. Online communities include Facebook, Twitter, and Google+, as well as specific-interest groups.

The immediacy and far reach of social media can have the same effect on a reputation as a 2½-inch attack line has on a room-and-contents fire. Virtual flash mobs can quickly elevate a minor event into an international incident. Media expert Dave Statter coined the phrase "social media–assisted career suicide syndrome" (SMACSS) to describe the impact of social media videos, pictures, and postings that lead to a sudden (and often unfortunate) change of status for the author of the post.

> **Assessment Center Tips**
>
> **Television Interview Behavior Tips**
> The following recommendations from the NFPA will help you if part of the assessment center process includes conducting a television interview as a fire officer.
>
> Look into the reporter's eyes, not into the camera. Keep your gaze steady and avoid rolling your eyes, blinking excessively, or closing your eyes when you are thinking about your answer. If you are standing, do not sway. Plant your feet about 18 inches apart, and keep your hands down to your sides or clasped together loosely in front of you. Do not put your hands in your pockets. If you are sitting, ask for a stationary chair. If you must sit in a swivel chair, plant your feet to help you keep the chair still. Sit with your knees together and your feet flat on the floor or your ankles loosely crossed. Keep your hands folded comfortably in your lap. Do not clench your fists, crack your knuckles, pick your nails, or play with your earrings. Not only are these behaviors distracting, but they also signal to viewers that you are nervous, which can be interpreted as a lack of confidence about the subject. (Imagine a close-up camera angle of your clenched fists or fidgety fingers.)

You Are the Fire Officer Conclusion

Williams apologizes to the mother who waited more than an hour for the crew members to return to the fire station, explaining that the quint was handling a structure fire. The car seat technician will install the child safety seat as soon as the apparatus is ready for the next emergency. Offer to reschedule the installation if this arrangement is inconvenient for her.

Williams directs the fire fighters to clean up quickly. The lieutenant greets the children and teachers and starts the tour while the fire fighters put on clean uniforms. Williams explains that they just came back from a fire and that the other fire fighters are cleaning up. When the cleaned-up fire fighters return, they demonstrate Stop, Drop, and Roll. They get the children to practice while Williams goes to clean up. At the end of the demonstration, the fire fighters are in clean uniforms and the kids have practiced rolling on the floor. The parents express appreciation of the authentic experience.

Wrap-Up

Chief Concepts

- Today the trend is in favor of more community-based local government. Next to schools, the fire department is the most decentralized and community-based function of local government.
- Each community has special needs and different characteristics that should be considered in relation to every service and program provided by the fire department.
- The best method of preventing fire injuries and deaths is to prevent the fire if possible or at least reduce its severity. The same is true for the other types of incidents to which the fire department is called.
- Some requests that citizens make may not be within the fire officer's authority. In these situations, the fire officer should provide a method of moving the request to the level where it can be resolved.
- Goals, objectives, content, and delivery mechanisms for public education programs vary greatly among fire departments, depending on their resources and circumstances.
- The NFPA, the USFA, and other specialized associations and industry groups have developed national and regional public education programs geared toward meeting local community needs.
- The development of new public education programs is usually assigned to an individual or a work group specializing in this area (in large fire departments) or to a fire officer as a special project (in a smaller department).
- Most fire departments have frequent interactions with the local news media during which reporters want to obtain information about a situation that is still occurring, as well as in newsworthy situations that are related to other fire department programs or activities.
- Some fire departments have a full-time or part-time public information officer (PIO) who functions as the media contact person and the source of official fire department information.
- A press release is used to make an official announcement to the news media from the fire department PIO.
- Every fire officer should be prepared to act as a spokesperson for the fire department, even if the department has a PIO who is the designated official spokesperson.
- The immediacy and far reach of social media can have the same effect on a reputation as a 2½-inch attack line has on a room-and-contents fire.

Hot Terms

Community Emergency Response Team (CERT) A fire department training program to help citizens understand their responsibilities in preparing for disaster and increase their ability to safely help themselves, their families, and their neighbors in the first 72 hours of a catastrophe.

Demographics The characteristics of human populations and population segments, especially when used to identify consumer markets; generally includes age, race, sex, income, education, and family status.

Risk Watch A comprehensive NFPA school-based program focused on injury prevention.

Social media Forms of electronic outlets through which users create online communities to share information, ideas, personal messages, and other content.

References and Additional Resources

NFPA reprinted material is not the complete and official position of the NFPA on the referenced subject, which is represented only by the standard in its entirety.

Bramble, D. (2011). *Facebook and the Fire Department: Who Is Using It and How?* Emmitsburg, MD: Executive Fire Officer Program.

Brunacini, A. V. (1996). *Essentials of Fire Department Customer Service.* Stillwater, OK: Fire Protection Publications.

Bugbee, P. (1971). *Man Against Fire: The Story of the National Fire Protection Association, 1896–1971.* Boston, MA: National Fire Protection Association.

Carli, L., M. Hazell, and L. Backstrom. (2012, June 13). *Social Media: What's Your Strategy?* Presentation at NFPA Conference and Exhibition, Las Vegas, NV.

Carpenter, Q. (2002). "Kid Drownings: After Decades of Failure, The Well-Intentioned Still Don't Get It." *Phoenix (Arizona) New Times.* Available at: http://www.phoenixnewtimes.com/2002-12-26/news/kid-drownings/. Accessed August 12, 2013.

Chaffin, A. J., and S. D. Harlow. (2005). "Cognitive Learning Applied to Older Adult Learners and Technology." *Educational Gerontology* 31: 301–329.

Coombs, W. T. (2012). *Ongoing Crisis Communication: Planning, Managing, and Responding,* 3rd ed. Thousand Oaks, CA: Sage.

Crawford, J. (2013). "Comprehensive Prevention Program." In: *Managing Fire and Emergency Services,* A. Thiel and C. R. Jennings, eds. Washington, DC: International City/County Management Association.

England, R. E., et al. (2012). *Managing Urban America.* Washington, DC: CQ Press.

Fire and Emergency Service Image Task Force. (2013). *Taking Responsibility for a Positive Public Perception.* Fairfax, VA: International Association of Fire Chiefs.

Gamache, S. (2008). "Reaching High-Risk Groups." In: *Fire Protection Handbook,* 20th ed. Quincy, MA: National Fire Protection Association. Chapter 5.5.

Howard, A. (2011). "Social Media in a Time of Need: How the Red Cross and the Los Angeles Fire Department Integrate Social Tools into Crisis Response." *O'Reilly Radar.* Sebastopol, CA: O'Reilly Media, Inc.

International Association of Fire Chiefs and National Fire Protection Association. (2014). *Fire Service Instructor: Principles and Practice,* 2nd ed. Burlington, MA: Jones & Bartlett Learning.

Keith, B. D. (2012). *Improving Communication with Non-English Speaking Populations in the City of Dalton, Georgia.* Emmitsburg, MD: Executive Fire Officer Program.

Lesperance, A., et al. (2010). *Social Networking for Emergency Management and Public Safety.* Oak Ridge, TN: Pacific Northwest National Laboratory: 95.

London Fire Brigade. (2009). *Case Study: Keeping Your Community Safe and Sound.* London, UK: London Fire Brigade.

Metz, E. J. (2013). "New Venues for Discovering Fire and Emergency Services Literature." *Fire Technology* 49(2): 185–194.

Miller, K. (2010). *We Don't Make Widgets: Overcoming Myths That Keep Government from Radically Improving.* Washington, DC: Governing Books.

Murphy, M. (2013). "Social Media and the Fire Service." *Fire Technology* 49(1): 175–183.

O'Neil, W. (1998). *Engine Company Tours That Educate.* Salem, OR: Pathfinder Enterprises.

Schaenman, P. (2007). *Global Concepts in Residential Fire Safety: Part 1—Best Practices from England, Scotland, Sweden, and Norway.* Arlington, VA: Centers for Disease Control and Prevention: 101.

Snook, J. W., et al. (2011). *A Leadership Guide for Volunteer Fire Departments.* Sudbury, MA: Jones & Bartlett Learning/IAFC.

Watson, C. (2012). *Using Social Media to Communicate with the Occupants of Large Residential Buildings During Fire Emergencies.* Emmitsburg, MD: Executive Fire Officer Program.

FIRE OFFICER
in action

While reviewing the second-alarm garage fire, Apparatus Operator Rollo points out that many of the large-loss fires in the wintertime are due to improperly disposed fireplace ashes. Other members chime in, sharing stories of recent fires and close calls. Many of these incidents are occurring in large multiple-family dwellings in your first-due area. Many of the apartments have a fireplace, and there have been a half-dozen near misses and small fires.

This property contains 10 three- and four-story garden-style apartment buildings with combustible vinyl siding and lightweight wood truss frames. The buildings have residential sprinkler systems and underground parking. There is a moderate turnover of tenants. The fire company members remember fireplace ashes starting fires on balconies, in the parking garage, and in the apartments.

Last year, a fire on a balcony spread up the vinyl siding and got into the attic, requiring a third-alarm assignment to control the noontime fire. With 10 buildings, this is a significant target hazard.

1. What is your first goal in this scenario?
   A. Determine whether it is an engineering, education, or enforcement issue
   B. Prepare a press release on the hazards of discarded fireplace ashes
   C. Define the problem
   D. Design the program

2. Approximately 30 percent of the tenants are recent immigrants who may not be fluent in English. Your message should:
   A. be provided in English to the apartment resident manager for translation.
   B. use pictures and graphics.
   C. provide an opportunity for tenants to learn English.
   D. be in a format and use a method of delivery that is effective and culturally appropriate for this group.

3. How can you measure the effectiveness of a fireplace ash disposal educational effort?
   A. Determine whether there are fewer fireplace ash–initiated fires after delivery of the program
   B. Count the number of times the program is mentioned in the local media
   C. Estimate the market penetration of the fire department–provided ash cans
   D. Count the number of requests for presentation of the program to civic groups and organizations

4. When preparing for a media interview, provide no more than _____ key message point(s).
   A. one
   B. two
   C. three
   D. four

FIRE OFFICER *in action* (continued)

Fire Captain *Activity*

It was ugly and not well coordinated, but Quint 100 managed to provide a preschool station tour and install a car seat while making sure the rig was available for emergency duty. Lieutenant Williams and Captain Davis reviewed the controllable and uncontrollable elements of a busy day.

NFPA Fire Officer II Job Performance Requirement 5.3.1
Explain the benefits to the organization of cooperating with allied organizations, given a specific problem or issue in the community, so that the purpose for establishing external agency relationships is clearly explained.

Application of 5.3.1
1. Using policies and procedures from a department with which you are familiar, propose a way to assure that car seats are inspected and installed even when the local fire company is committed to emergency activity.
2. Using policies and procedures from a department with which you are familiar, propose a partnership program to reduce the number of winter fires caused by improperly discarded fireplace ashes.
3. Using policies and procedures from a department with which you are familiar, propose a "best practices" approach to reducing the number of children who drown in private residential pools.
4. Using policies and procedures from a department with which you are familiar, respond to a suggestion by a community advocate to replace one fire fighter position on each engine with a CERT volunteer.

NFPA Fire Officer II Job Performance Requirement 5.4.4
Prepare a news release, given an event or topic, so that the information is accurate and formatted correctly.

Application of 5.4.4
1. Based on a local or national fire incident, write up a press release that meets the local jurisdiction format.
2. Using a local jurisdiction format, describe a new community initiative, such as a child drowning prevention effort, residential fire safety visits, child safety seat inspection, or other program.

Handling Problems, Conflicts, and Mistakes

Fire Officer I

Knowledge Objectives

After studying this chapter, you will be able to:

- Describe the interrelationships among complaints, conflicts, and mistakes. (pp 224–225)
- Describe the general decision-making procedure. (pp 225–229)
- Discuss how to manage conflict within the department. (pp 230, 232–234)
- Discuss how to recommend and implement policy changes (NFPA 4.4) (NFPA 4.4.1). (pp 234–235)
- Describe how to field and resolve citizen complaints (NFPA 4.3) (NFPA 4.3.2) (NFPA 4.3.3). (p 235)
- Describe the difference between customer service and customer satisfaction (NFPA 4.3). (pp 235–236)

Skills Objectives

After studying this chapter, you will be able to:

- Manage conflict within the department. (pp 230, 232–234)
- Manage citizen complaints (NFPA 4.3.2) (NFPA 4.3.3). (p 235)
- Develop a policy or procedure (NFPA 4.4.1). (pp 234–235)

Fire Officer II

Knowledge Objectives

There are no Fire Officer II knowledge objectives for this chapter.

Skills Objectives

There are no Fire Officer II skills objectives for this chapter.

CHAPTER 12

Principles of Fire and Emergency Service Administration (FESHE) Course Objectives

2. Recognize the need for effective communication skills, both written and verbal. (pp 230, 232–236)
4. Select and implement the appropriate disciplinary actions based on an employee's conduct. (pp 224–225, 230, 232–236)
7. Describe the traits of effective versus ineffective management style. (pp 224–229, 232–236)
8. Identify the importance of ethics as it relates to fire and emergency services. (pp 232, 235–236)

You Are the Fire Officer

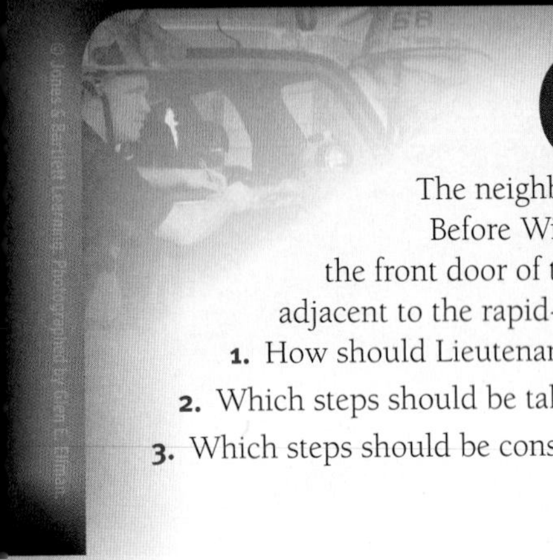

Quint 100 responds as a single unit to a late-night "smoke in the building" call during a severe thunderstorm. On arrival, Lieutenant Taylor Williams notes that the store is located at the end of a strip mall shopping center and there is a haze of white smoke in the store. The neighborhood is without electrical power.

Before Williams can complete a 360-degree size-up, Fire Fighter James Grynski attempts to forcibly open the front door of the business. Grynski breaks a large glass display window, right next to the fumigation notice and adjacent to the rapid-entry key safe.

1. How should Lieutenant Williams handle this situation?
2. Which steps should be taken immediately?
3. Which steps should be considered after the initial issues are addressed?

[handwritten note: problem is current vs desired]

Introduction

A problem is the difference between the current situation and the desired situation. If the way the apparatus driver parks at a highway incident places the fire crew at greater risk than is necessary, it is a problem. Forcing a commercial door and destroying a display window when a rapid-entry key safe is available is a problem. In each of these cases, there is a discrepancy between the current and the desired situations.

Fires and emergency incidents represent a unique category of problems that call for specialized problem-solving skills. Nonemergency situations require the application of conventional problem-solving skills and techniques. These situations include supervisory, management, and administrative activities in which the fire officer is directly responsible for solving the problem as well as initially processing situations that require resolution from an administrative or executive fire officer. These circumstances include situations that involve individuals or organizations outside the fire department.

Decision-making skills are used whenever the fire officer faces a problem or situation that requires a response. Promotional examinations evaluate the ability of fire officer candidates to exercise good judgment and make sound decisions. Decisions should always be guided by organizational values, guidelines, policies, and procedures. This is true even when the decision made does not agree with the fire officer's opinion or personal preferences.

Requiring the fire officer to act in the best interest of the department in solving problems and making decisions does not mean that other values and considerations are ignored, however. Several solutions to a problem may exist—some better than others and some more desirable to one set of interests than another. In most cases, there is a reasonable solution that serves multiple interests and concerns; in other cases, however, one concern must prevail over all others. Problem-solving techniques are designed to identify and evaluate the realistic potential solutions to a problem and determine the best decision.

Fire Officer I

Complaints, Conflicts, and Mistakes

Complaints, conflicts, and mistakes are special categories of problems. One of the key factors in decision making is how to deal with situations that involve conflicts or complaints.

These three terms are defined as follows:

- A complaint is an expression of grief, regret, pain, censure, or resentment; a lamentation; an accusation; or fault finding.
- A conflict is a state of opposition between two parties. A complaint is often a manifestation of a conflict.
- A mistake is an error or fault resulting from bad judgment, deficient knowledge, or carelessness. It can also be a misconception or misunderstanding. Mistakes happen; the issue is how to deal with a mistake, or the perception of a mistake, when someone complains to the fire officer about it.

Sometimes a fire officer has to make a decision or enforce a policy that is not popular with the crew members. A citizen could be frustrated with a fire department action or may be unhappy with a situation. People misbehave and make mistakes. Disagreements and differences of opinion occur. It is not possible to make everyone happy all of the time, but a fire officer must deal with all of these situations in a professional manner. Dealing appropriately with problems and conflicts requires maturity, patience, determination, and courage **FIGURE 12-1**.

CHAPTER 12 Handling Problems, Conflicts, and Mistakes

FIGURE 12-1 Dealing appropriately with problems and conflicts requires maturity, patience, determination, and courage.

The types of problems that a fire officer could be expected to encounter can be classified into four broad categories:

1. *In-house issues:* Situations or decisions occurring at the work location that are within the direct scope of supervisory responsibilities. An example might be a complaint about the assignment of duties to different individuals within a fire station. Most of these conflicts begin and end at the company officer level.
2. *Internal departmental issues:* Operational policies, decisions, or activities that go beyond the scope of the local fire station. An example is a conflict over where a reserve ladder truck will be housed and which company will be responsible for maintaining it. Another example is a dispute between companies over the tasks of a rapid intervention team in a high-rise fire. The resolution usually requires action by command officers at a higher level on the organizational chart.
3. *External issues:* Fire department activities that involve private citizens or another organization. An example is a citizen making a complaint about an inappropriate remark uttered during an EMS incident. External issues require the fire officer to perform one additional task early in the conflict resolution process: making sure that the fire officer's supervisor is not surprised. It is poor form for the battalion chief to learn about a fire department incident from the local media.
4. *High-profile incidents:* Any issues that are likely to become major events. An example might be a fire fighter who is arrested while on duty. The department must take immediate actions to respond to these events. Senior fire administrators often become directly involved in these situations or keep a close watch on how they are handled.

A problem should be solved at the lowest possible level within an organization. A fire officer is expected to manage problems within the level of authority for a supervising or managing fire officer. At the same time, the fire officer should recognize those problems that need to be handled at a higher level and make the appropriate notifications without delay. If there is any doubt, it is wise to discuss the situation with the officer at the next higher level in the chain of command.

General Decision-Making Procedures

A fire officer is called upon to make many different types of decisions about a wide variety of subjects. Most of the problems that arise are fairly uncomplicated, although they are not necessarily easy to solve. Moving up in the ranks means an exponential increase in decision-making situations. Often the problems become more complex at higher levels of the hierarchy, requiring the participation of multiple organizations.

The following systematic approach is recommended to ensure high-quality decision making:

1. Define the problem.
2. Generate alternative solutions.
3. Select a solution.
4. Implement the solution.
5. Evaluate the result.

Although this five-step technique appears to be designed for situations where plenty of time is available, the same basic approach is used for emergency incidents. A fire officer should be able to move quickly through these steps. Training and experience will prepare you as the fire officer to identify the pertinent problem, generate realistic solutions, and select the best option quickly.

■ Define the Problem

The first step in solving any problem is to examine the problem closely and to define it carefully. A well-defined problem is one that is half-solved. Poorly defined problems, in contrast, waste tremendous time and effort.

Pay Attention

Richard Gasaway is a retired fire chief who has researched fire officer decision making. Fire officers should know what is going on within the organization and address most difficult issues before they become major problems; Chief Gasaway calls this situational awareness, a topic covered in the *Managing Incidents* chapter. The best way to prevent major problems is to deal successfully with minor issues before they reach the crisis stage **FIGURE 12-2**.

Ask Basic Questions

Peter Drucker served as a Professor of Management at the Graduate Business School of New York University. His 1974 *Management: Tasks, Responsibilities, Practices* was a FDNY promotional exam reference for the Fire Department of New York. Drucker encouraged managers to question the value of each

FIGURE 12-2 Fire officers must pay attention to what is going on within the organization and deal with problems before they reach the crisis stage.

organizational activity once a year, because what may have been vital last year may have minimal importance this year.

Fire departments have a strong inclination to keep doing the same things in the same ways. The fire officer should identify which activities can be changed, improved, or updated. If there is a better way to do something, give it appropriate consideration. If we are doing something that is no longer worth doing, why not use the time to do something more productive?

How Quickly Do You Get Bad News?

Few things damage a new fire officer's reputation more quickly than not finding out about a situation that is going poorly until it is too late to fix it. Effective fire officers create a work environment that encourages subordinates to report bad news immediately. The sooner the officer receives the bad news, the faster he or she can implement corrective action. Fire fighters should be encouraged to tell their supervisor whenever an event or performance is out of balance with expectations **FIGURE 12-3**. This includes immediately reporting injuries or broken equipment, regardless of the time of day.

FIGURE 12-3 Fire fighters should be encouraged to report problems immediately.

The same principle applies to delivery of bad news during an emergency. Fire officers who create a barrier to receiving administrative bad news can suffer catastrophic results at an emergency scene. From the fire fighter's perspective, the fire officer who does not appreciate hearing bad news in the fire station will probably not want to hear it at the emergency scene, either. In some cases, a fire fighter might even be afraid to point out a problem that could save a life or prevent a serious injury.

Fear versus Trust

Every fire officer should foster a trusting relationship with his or her employees. Employees who do not trust their boss or one another are unlikely to make good decisions when faced with a problem. Employees who do not feel that their input is valuable will stop passing along vital information to the fire officer. Effective problem solving, however, requires good information.

Many fire officers believe that because they have been promoted, they can always identify the problem themselves. This is not true. The only way a fire officer can define the problem correctly is with the best information. Eliminating sources of valuable information because of fear and mistrust will cause the fire officer to make an inaccurate assessment.

■ Generate Alternative Solutions

Involve Anyone with Direct Knowledge of the Problem

The best people to solve a problem are usually those who are directly involved in the problem. The fire officer who is struggling with a new incident management clipboard in the rain has a valuable perspective on what might improve the clipboard's performance. The fire fighters who make roof ventilation openings have ideas about how those operations could be accomplished more efficiently. Company-level problems are most likely to be solved by involving the members of the company.

Brainstorming

Brainstorming is a method of shared problem solving in which all members of a group spontaneously contribute ideas. A typical fire company is a good size and a natural group for brainstorming. Fire fighters are usually very adept at solving problems.

The following eight steps will assist the fire officer when brainstorming alternative solutions:

1. Using a flip chart, whiteboard, or chalkboard, write out the problem statement. Everyone should agree with the words used to describe the problem.
2. Give the group a time limit to generate ideas. Although the scope of the problem is a factor, 15 to 25 minutes seems to work well for groups of 4 to 16.
3. The fire officer should function as the scribe. The scribe writes down the ideas and keeps the group on task.
4. Tell everyone to contribute alternative solutions. At this point, there is no commenting on ideas. All suggestions are welcome.
5. Once the brainstorming time is up, have the group select the five ideas they like the best.

6. Write out five criteria for judging which solution best solves the problem. Criteria statements include the word "should"—for example, "It should be legal" or "It should be possible to complete in 6 months."
7. Have every participant rate the five alternative solutions, using a 0 to 5 scale. The value 5 means the solution meets all of the criteria, whereas 0 means the solution does not meet any of the criteria.
8. Add up the scores for each idea. The idea with the highest score is the one that provides the best solution for your problem.

Some important constraints apply to brainstorming. Most notably, the process assumes that the problem statement is accurate and the criteria are valid. On occasion, the best solution that comes out of a fire station may crash at headquarters because of a criterion or restriction that is unknown to the fire officer, such as a new federal or state regulation.

Should the Fire Chief Participate?
Fire company group dynamics are different when the administrative or executive fire officer is in the station. The presence of a senior command officer significantly influences a group that is brainstorming alternative solutions. The verbal and nonverbal signals sent by the chief officer influence the group, and the participants are likely to self-censure their suggestions and limit the range of possible and plausible solutions because of how they think the chief will react to them. The best ideas emerge from the fire fighters directly affected by the problem or issue; each layer of hierarchy added to the group reduces the range of options.

Do Fire Fighters Feel Comfortable Sharing Ideas?
Fire fighters are naturally competitive. Even when the decision group is restricted to the fire fighters who are directly affected by a problem, such as in determining how fill-in apparatus drivers will be trained, they may be reluctant to share ideas. The fire officer can encourage participation by creating a positive and nonhostile work environment.

Is the Process Legitimate?
A legitimate problem-solving process has to be reasonable and based on logic and organizational values. Fire fighters should be able to anticipate that their decision will be implemented if it meets the criteria. Going through a process that results in no change or provides no feedback to the fire company members is the quickest way to destroy fire fighter participation in the decision-making process. A legitimate process reinforces the trust between the fire company and the fire administration.

■ Select a Solution
At this point in the decision-making process, the fire officer will have defined the problem, generated solutions, and ranked them based on criteria. Criteria are based on factors important to the department, the work group, and the fire officer. These may be found in the department's mission statement or core values. For example, if participation in local neighborhoods is one of the core values, a solution that increases the fire department's involvement in local neighborhoods would be preferred.

■ Implement the Solution
Once the decision has been made, the solution still has to be implemented. The implementation phase is often the most challenging aspect of problem solving, particularly if it requires the coordinated involvement of many different people. One reason for involving as many players as possible in making the decision is to capture their commitment to the plan when it is time for implementation.

Consider a plan for reducing the number of fire fatalities in residential occupancies. After careful analysis, the decision is made that the best strategy will be a community outreach program to check every smoke alarm in every multiple-family dwelling in a community over a 4-month period. This plan supports the fire department's mission statement and reflects the organization's values. To accomplish this objective, each fire company will have to invest 2 hours every evening for the next 4 months.

Before this plan can be implemented, buy-in is required. The plan has to be "sold" to the people who will perform the task, particularly to those who were not involved in the decision-making process. Rearranging schedules to do something different for 2 hours every evening involves a significant behavioral change, even if it is for a limited time and for a good cause. It would be possible simply to order everyone to "just do it"; however, this is not the best implementation strategy for a program that involves extensive public contact. Willing participation works better than involuntary compliance.

Who Does What When?
The fire officer must clearly assign tasks to individuals or teams. The work group benefits from the use of a project plan, which lists tasks, responsibilities, and due dates **FIGURE 12-4**. Most of the projects supervised by a company-level officer do not require a sophisticated planning system. A simple project control document can be used to divide a project into segments, with milestones established to identify progress. Complex and long-range tasks may require a formal project management plan and a designated coordinator, particularly if they require the coordinated activity of multiple agencies.

No Deadline Means No Implementation
An implementation plan must include a schedule to ensure that the goals are met. Deadlines focus effort and help prioritize activities. If the solution requires activity by other organizations, such as changing a local ordinance or submitting a budget request, then the fire officer must determine the time it will take to accomplish that task and incorporate that time into the schedule. A schedule is valuable only if it is followed and someone ensures that the deadlines are met.

Plan B
Many problems remain unsolved long after the problem has been clearly defined and a good solution has been selected. A problem is not truly solved unless its solution is implemented. There are many reasons why good solutions are not implemented—for example, the required approvals might not be obtained or the necessary resources might not be available. Quite simply, the organization might not have the capability to solve the problem or to implement the solution that was selected.

Fiscal Year 2018 Capital Improvement Training Room 12/15/2017

Goal		Due Date	
Repair and update fire station classroom		05/01/2018	
Task	**Assigned to**	**Due Date**	**Status**
Complete Learning Center design	Station Commander	08/16/2017	Done
Issue RFP for instructional technology	Logistics	09/04/2017	Issued 9/15
Review RFP responses	Resource Management/Station Commander	10/18/2017	Done
Award RFP	Procurement	11/06/2017	Award 11/3
Remove old furniture, tables, and carpet	B shift	11/13/2017	Done
Strip wall coverings and prepare to paint	A shift	11/20/2017	Done
Install fiber optics and intranet	Technical services	12/04/2017	Done 11/27
Order chairs and tables	Procurement	01/08/2017	
Install learning center system	Contractor	01/15/2018	Started 12/4
Paint and wallpaper party	C shift and volunteers	01/27/2018	
Install carpet	Facilities maintenance	02/05/2018	need to confirm
Wire projector, podium and speakers	Technical services	02/19/2018	
Move new furniture into training room	A shift	02/26/2018	

FIGURE 12-4 Example of a project control document for a fire station work project.

Courtesy of Mike Ward

Fire Marks

The Problem with Numbers

In the late 1960s, New York City hired the RAND Corporation to analyze city services. Using operations research methods, RAND provided recommendations to Mayor John Lindsay to increase the efficiency and effectiveness of city services. RAND published the research and New York City publicized the successes. Four decades later, an analysis of the outcome provided a different perspective.

In 2010, Joe Flood published *The Fires: How a Computer Formula, Big Ideas and the Best of Intentions Burned Down New York City*, a critical review of the RAND–New York City partnership. In an interview with *The Atlantic*, Flood described his experience:

> Interviewing the RAND researchers was a really interesting and sometimes frustrating experience. The short answer is that they're convinced that nothing was wrong with the studies . . . When I asked how a model could recommend closing one of the busiest fire companies in the city—in the whole world, really, because nowhere was burning like New York City—they'd cite a litany of equations and talk about how consistent the R-squares were, things like that. In effect, they were saying, "The Emperor is wearing clothes, because our calculations tell us he must be." One model actually says that traffic has absolutely no effect on how quickly a fire engine travels through New York City. Fire fighters just laugh and shake their heads when I mention that. But I couldn't get anyone from RAND to admit that there might have been something wrong with that model.

Ambinder, M. (2013 May 13) The Fires This Time: Joe Flood on Managing New York City. New York, NY: The Atlantic Monthly Group. Accessed 9/10/2013 from: http://www.theatlantic.com/politics/archive/2010/05/the-fires-this-time-joe-flood-on-managing-new-york-city/56682/

Fire officers should consider a "plan B" if the original solution cannot be implemented. Plan B could be an extended implementation schedule, a modified plan, or a completely different solution to the problem.

■ Evaluate the Results

After implementing the solution, the fire officer must assess whether it produced the desired results. Evaluation should be a standard part of the process of any problem-solving activity. The nature of the evaluation depends on the complexity of the problem and the solution; in most cases, an initial evaluation should be performed immediately after implementation, and then follow-up evaluations should be performed at regular intervals.

Determining whether the solution actually solved the problem requires some type of measurement that compares the original condition with the condition after implementation. For example, do the new hose loads really result in quicker deployment of attack lines? Answering this question requires data on how long the old way took versus how long the new way takes. The evaluation should also look for situations where the original problem is solved, but another unintended and equally bad situation is created. For example, if the hose is deployed more quickly with the new technique but comes out twisted and kinked, the negative impact could outweigh the positive.

Change the Plan If Necessary

If necessary, the fire officer needs to be prepared to adjust the plan or reevaluate the original decision. Many problems are solved in stages, with gradual progress being made toward a solution. In spite of the analysis, plan B may turn out to be a better choice. Changing a plan should not be viewed as a failure.

Feedback

Part of the evaluation process involves going back and listening to the people who identified the original problem and asking

CHAPTER 12 Handling Problems, Conflicts, and Mistakes

Near-Miss REPORT

Report Number: 09-0000878

Event Description: I was in charge of a group of three fire fighters. We were conducting a pump training evolution for a pump operator trainee. The pump operator trainee had begun his training more than a year prior to the incident but was forced to stop due to a work-related injury. His training restarted approximately 3 months prior to the incident. At the time of the incident we were trying to create a "real world" incident where multiple lines were being used and could not be shut down and master streams were needed to protect exposures. At the time of the incident, we had a smooth-bore nozzle attached to the rear discharge to simulate a primary attack line. We also had two 2½-inch lines attached to ground-mounted deck guns (permanently mounted at our training facility) to simulate one exposure line and one backup line.

The evolution was progressing smoothly until we added the engine-mounted deck gun. When the engine-mounted deck gun was charged, it began to leak water from the base. The deck gun was not sitting properly on the base (this had happened in the past). The pump operator trainee was told to shut down the deck gun. The trainee did so, but could not close the valve all the way. The trainee was asked to close the valve the rest of the way. As the trainee was trying to do so, the fire fighter operating the engine-mounted deck gun tried to push the deck gun down farther onto its mounting base.

Before I could stop the fire fighter on top of the apparatus, the pump trainee pulled the handle that controlled the deck gun valve with the engine still pumping the lines previously stated. The engine-mounted deck gun, while still being held by the fire fighter, lifted off its base, striking the fire fighter's helmet. The fire fighter then lost his balance, going face first into the stream from the deck gun base, then stumbled back across the engine cowling and fell on top of the cab, gasping for air. The evolution was shut down and everyone was evaluated for injuries. The fire fighter on top of the engine was transported to a local hospital for water inhalation and neck pain.

Lessons Learned: After the incident, when everyone had returned to work, the four of us sat down to discuss what we could do to keep this from happening again. The pump operator trainee stated he did not know that the fire fighter was attempting to push the deck gun onto its base (lack of communication and situational awareness). He also stated he did not think about what would have happened if he pulled the valve to the deck gun open and then slammed it shut (lack of training and situational awareness). The fire fighter on top of the engine stated he assumed that the pump operator trainee was closing the valve the rest of the way so he kept the twist lock handle to the unlocked position (lack of communication and situational awareness).

As for myself, I thought that the pump operator in training knew about the fire fighter on top of the engine, and I thought he knew what I wanted him to do (lack of communication). I should have used the fourth person in our crew to act as a safety officer (lack of decision making, task allocation, and situational awareness). Through this event, I discovered:

- The need to have better communication with the persons involved.
- To explain my objectives better and goals for the training session.
- When something happens that was not planned (the deck gun leaking) shut down the operation and start over after fixing the problem.
- I also need to have better situational awareness.

I never would have thought that the fire fighter on top of the engine would have tried to push the deck gun down while it was still flowing water. This is a person with 20-plus years of firefighting experience both as a volunteer and as a career fire fighter. I also should have had the fourth person in our crew act as a safety officer and oversee our whole operation. This way I could concentrate on the pump operator/trainee, and the safety officer could watch the rest of the operation. I think that with these steps, this incident would not have happened.

Courtesy of the National Fire Fighter Near-Miss Reporting Systems (firefighternearmiss.com)

them for feedback. A small city adopted a big city policy of sending a single fire company to a street fire alarm pull box. In the big city, the street pull box resulted in a false alarm for 99.9% of the activations. In the small city, fire companies were frequently encountering working structure fires when responding to a street pull box. The fire chief listened to the fire officers and dispatchers and learned that sending a single fire company to pulled fire alarm street boxes was providing a lower level of service and producing undesirable results. The policy was quickly changed.

Managing Conflict

One factor that distinguishes a fire officer from a fire fighter is the responsibility to act as an agent of the formal organization. A fire officer is the official first-level representative of the fire department administration when dealing with subordinates and enforcing policies and procedures. This responsibility places the fire officer in a position to be the initial contact in dealing with a wide variety of problems, including situations that potentially involve conflict, emotions, or serious differences of opinion.

Situations that involve conflicts and grievances require an additional set of skills that go beyond the general problem-solving model. The general model is designed to focus on solving the problem itself. In conflict situations, the issues are often much more complicated and sensitive. A relatively simple problem, for example, may become complicated by the ways that different individuals react to it or to one another. In some cases, the problem centers on the relationship between individuals or groups and plays out in relation to other issues.

■ Personnel Conflicts and Grievances

The close living relationships within a fire company can produce a variety of tensions, anxieties, and interpersonal conflicts. This friction can occur in addition to the types of conflicts commonly experienced in most workplaces. One of the most difficult situations for a fire officer is an interpersonal conflict or grievance within the company or directly involving a company member.

A fire officer may face four different types of internal conflict situations. A fire fighter might come to an officer with a complaint about:

- A co-worker (or co-workers)
- The work environment, including the fire station, apparatus, or equipment
- A fire department policy or procedure
- The fire officer's own behavior, decisions, or actions

The fire officer is the individual's first point of contact with the formal organization. The official response to the problem begins when the officer becomes aware that a problem exists. The relationship of the fire officer to the conflict and the complainant makes a significant difference in the role the officer can play in resolving the conflict.

Fire officers with staff assignments must also be prepared to deal with problems that involve conflict. Their relationship to the individuals involved is likely to be different, but their responsibility to officially represent the formal organization is the same.

> **Assessment Center Tips**
>
> **Conflict Resolution**
> You should anticipate handling some type of conflict issue during an assessment center. The four-step conflict resolution model provides an effective framework for managing internal and external conflicts. Active listening may provide the candidate with additional information that the role player may hold back if not questioned.

■ Conflict Resolution Model

The conflict resolution model is a basic approach that can be used in situations where interpersonal conflict is the primary problem or a complicating factor.

Listen and Take Detailed Notes

The first phase of the conflict management template is to obtain as much information as possible about the problem. The fire officer should encourage the complainant to explain the situation completely. If the details are even slightly complicated, the fire officer should take notes. The person who is making a complaint has a certain perspective on the situation. Whether you agree or disagree with that person's perspective, an important starting point is to find out what the complainant thinks about the situation.

Active Listening

When dealing with an individual who is expressing a concern or a problem, the fire officer should focus on active listening. Engaged or active listening is the conscious process of securing all kinds of information through a combination of listening and observing. The listener gives the speaker full attention, staying alert to any clues of unspoken meaning while also listening intently to every word that is spoken. The fire officer should be aware of nonverbal clues that may indicate agreement, dissatisfaction, anger, or other emotions. Often, these nonverbal clues provide great insight into the disposition of the speaker. The fire officer actively seeks to keep the conversation open and

> **Company Tips**
>
> **Fire Station Communication and Conversations**
> A fire officer has a different relationship with the company members than most supervisors have with their subordinates. Spending 10- to 48-hour shifts together in a fire station and engaging in emergency operations tend to create close relationships. As a consequence, the fire officer becomes aware of most internal problems early, when there is an opportunity to take preventive action. Unfortunately, the fire officer can become too close to some problems to deal with them effectively.
>
> The kitchen is the most important room in the fire station. Given enough coffee and time, the members will discuss and solve all of the department's problems. The kitchen is a great place for a fire officer to learn about the company's issues and concerns. It also provides a good opportunity for the fire officer to informally release information and measure reactions to different issues. Conversely, if the fire officer has important official information to share with the company, it should not be slipped into casual conversation.
>
> Although much information is exchanged in the kitchen, it should not substitute for official communication mechanisms. Although a fire officer is entitled to have a personal opinion, it is confusing to express mixed messages when presenting official communications.

VOICES OF EXPERIENCE

As a young company officer I was assigned to a multicompany, multiofficer firehouse. This placed 11 high-octane personalities in one building preparing for the next battle to fight. Right there I should have known what I was in for but it took me years to figure out why there seemed to be a persistent number of in-house squabbles, inter-shift battles and/or frustration with administration. In the nonemergent times there was never a shortage of battles to be waged.

As officers we teach our fire fighters to push through where others wouldn't go, to train to overcome obstacles, to solve problems through any means possible, and to fight against all odds. But in the nonemergent times we expect them to be compliant and cooperative. It is no wonder that the inter-shift battles and pettiness exists.

> **In the nonemergent times there was never a shortage of battles to be waged.**

So let me start with where this battle begins: "Who is the laziest shift?" and "Who is the whiniest shift?" The answers are that the laziest shift is the shift you follow because they never do what they are supposed to do, and the whiniest shift is the one that follows you because they are always whining about what you didn't do.

This problem is not unique to a multicompany house or to the fire service; it is an issue of perceived work load and that the other guy is always the problem. So how do you solve it? The answer is effective and persistent communication.

In my case, the fire administration had established what the expectation was via a policy outlining the daily apparatus and station duties. The department had established this expectation via training and foundation via policy, so what was missing? Communication and coordination of the six officers assigned to the fire house was the problem. As any fire fighter knows, policy is an organizational statement of expectation; however, for the policy to have value the company officers must communicate their support for the expectation and put it into practice for it to have the intended effect.

We worked to resolve the inter-shift battles by sitting down all six officers to discuss the frustrations and seek a long-term solution. As we talked it became apparent that we all had the same goals but saw the problem as the other guys. Once we came to the realization that our battles against each other and the lack of policy enforcement by us were the problems, we agreed that we would place value in the Department's Daily Apparatus and Station Duties policy through consistent compliance with policy and daily communication at shift change with the officers. This meant that we would support each other while holding ourselves accountable for the mission of the fire department.

This arrangement was successful as long as we stayed with the agreement of daily communication, because without it, when assignments were missed the perception of the other shift being lazy would bubble up as the battle of the day.

Your company's success is dependent on your ability to communicate, coordinate, and cooperate.

James P. Moore
Fire Rescue Chief
Crystal Lake Fire Rescue Department
Crystal Lake, Illinois

satisfying to the speaker, showing an interest in feelings and emotions as well as raw information FIGURE 12-5.

Paraphrase and Receive Feedback

The first objective should be to understand the issue and why the individual is complaining. After listening, the fire officer should be able to paraphrase the complaint and recite it back to the complainant. Paraphrasing the issue and receiving feedback from the complainant accomplishes two goals: The fire officer finishes this phase with a good understanding of the issue from the complainant's perspective, and the complainant feels that the fire officer really listened.

Do Not Explain or Excuse

In situations where the complaint is directly related to actions taken or policies enforced by the fire officer, it is understandable that the officer would want to respond immediately to the complaint. In this situation, it is important to listen and to process the information before deciding on an appropriate response. A reflexive explanation or excuse gives the individual an additional reason to complain. If the complainant feels strongly enough to complain about something the officer has done, that officer's explanation probably will not solve the problem.

■ Investigate

An <u>investigation</u> is a detailed inquiry or systematic examination. All complaints should be investigated, even if the foundation for the complaint appears to be weak or nonexistent. Fire department procedures should determine who will conduct the investigation, depending on the nature of the complaint and the relationship of the individuals who are involved. Sometimes the fire officer who received the complaint is assigned to conduct the investigation; however, a fire officer who is directly or personally involved in the problem should never be involved in conducting the investigation. Sensitive matters require an appropriate level of investigator.

The purpose of the investigation is to obtain additional information beyond the original complain. The investigator must be impartial in gathering and documenting information. The information could come from other individuals, reference documents, or incident-specific data. When investigating a human resources conflict, such as a payroll or work assignment issue, the fire officer might have to refer to specific departmental directives and regulations.

The product of an investigation is a report, which is provided in an appropriate format for the fire officer's immediate supervisor FIGURE 12-6. A complete investigative report has three objectives:

1. The report must first identify and clearly explain the issues.
2. The report should then provide a complete, impartial, and factual presentation of the background information and relevant facts.
3. The conclusion should be a recommended action plan, which is based on and supported by the information.

■ Take Action

Once the investigation is completed, the fire officer presents the findings and recommended action to a supervisor at a higher level. There are four possible responses:

1. *Take no further action.* The investigation may conclude that the complaint was unfounded or requires no further action. If the complaint was related to an earlier decision or action, that original decision is affirmed. The response should include the reasons why no further action is recommended.
2. *Recommend the action requested by the complainant.* The investigation may conclude that the complaint was justified and that the requested action is the best solution to the problem.
3. *Suggest an alternative solution.* The investigation may conclude that some alternative action or policy is the best solution to address the complainant's concerns. For example, a citizen might complain that the fire truck blocks several spaces in the parking lot at a local gym and does not want the fire fighters to go there. The fire officer meets with the citizen and proposes a more appropriate parking space for the fire truck. The compromise is acceptable to the citizen and to the department.
4. *Refer the issue to the office or person who can provide a remedy.* Other members of the fire department or some other municipal agency may be able to provide the relief the complainant seeks. Grievance procedures require that the employee start with the immediate supervisor for all complaints. If the employee is not satisfied with the response at that level, then he or she can take the grievance to a higher level. If the problem involves a paycheck deduction issue, it will probably have to be resolved by the payroll clerk or human resources office. The fire officer's duty is to refer the complaint to the appropriate person.

■ Follow Up

For many of the conflicts, the fire officer needs to follow up with the complainant to see whether the problem is resolved.

■ Emotions and Sensitivity

Fire fighters are passionate about their profession and are deeply concerned about issues that affect the job. They live,

FIGURE 12-5 Give the speaker your full attention.

Municipal City Fire and Rescue Department
Internal Memorandum

Date: May 08, 2017

To: Assistant Chief James Arrow, A Platoon Commander
Thru: Battalion Chief Frank Johnson, 3rd Battalion
Thru: Captain Jean Davis, Fire and Rescue Station 100
From: Lieutenant Taylor Williams, Quint 100

Subj: Civilian property damage
Ref: Incident #201705051473, 437 Western Way

On Friday, May 05 Quint 100 was dispatched for "smoke in the building" at Exotic Food Emporium. Municipal City was experiencing a severe thunderstorm and there was no electrical power in the neighborhood.

I observed a white haze in the store. While completing my size-up, Firefighter James Grynski started to force open the front door of the store. Swinging the flathead axe shattered the storefront window.

The white haze was from fumigation. The notice was posted on the front and back doors of the store. With the release of the fumigation gas, requested a haz-mat, EMS and chief response to our location.

There was no notation in the dispatch that the store was undergoing fumigation. Quint 100 staged at the A/B corner on arrival. I headed to Side C to complete the size-up while Grynski and Kinders headed to the Side A front door. My goal was to determine if there was a working fire in the store.

Investigation
Neither Grynski nor Kinders recall seeing the bright orange fumigation sign that was hanging on the front door.
After the window shattered Kinders located the rapid entry keybox that was located to the right of the door.

Action Taken
Once the event was stabilized, took statements from Grynski, Kinders and Rollo.
Took pictures of the damage.
Acquired contact information for the store owner and shopping center representative.
Placed plywood over the window and secured the front door.

Follow-up
Referred to Chief Johnson

FIGURE 12-6 Example of a fire officer investigation report.

breathe, eat, and dream about fire operations. They often are emotional when making a complaint. Michael Taigman, a consultant on emergency service performance issues, provides a conflict resolution model that is especially effective when emotions are high (Taigman & Dean, 1999). This model proved effective for Taigman when working with employees during the stressful creation of a paramedic ambulance service under a tight schedule. He continued to refine the model while working as a consultant. The model follows four steps:

1. Drain the emotional bubble.
2. Understand the complainant's viewpoint.
3. Help the complainant feel understood.
4. Identify the complainant's expectation for resolution.

Step 1: Drain the Emotional Bubble

The body reacts to emotional conflict or stress the same way it does when you are a member of the first-arriving company at a working structure fire. Both situations result in dumping of adrenaline to prepare the body to fight or run away. The same adrenaline-induced red haze that reduces fire fighter effectiveness at emergencies also happens to ordinary citizens. It tends to bring complaints to the surface and impedes resolution of conflicts. Adrenaline fills up the prefrontal lobes of the neocortex of the brain, creating an emotional bubble that interferes with the ability of the complainant to hear the fire officer or consider any response to the issue.

Taigman recommends listening deeply, actively, and empathetically to drain this emotional bubble. This is not an easy task for the fire officer, but writing detailed notes and not explaining or excusing can facilitate this process. The fire officer asks questions and encourages responses, draining the emotional bubble by allowing the complainant to express grief, regret, pain, censure, or resentment completely.

This discussion should be held in private and should adhere to a few ground rules. Most importantly, there should be no physical contact. If the discussion is between a fire fighter and a fire officer, avoid personal attacks and concentrate on the work issues.

Step 2: Understand the Complainant's Viewpoint

The initial complaint or behavior may be a sign or symptom of a larger problem. By draining the emotional bubble through active listening, the fire officer may identify the root cause or issue of the complaint.

Internal conflicts, grievances, or issues occasionally suffer from long memories. It may require some investigating to understand the current issue, which may be related to something that happened months ago or be wrapped up in history and tradition.

Step 3: Help the Complainant Feel Understood

The "listen and take detailed notes" part of the basic conflict management template includes the recommendation to paraphrase and feed back what you heard from the complainant. Some issues may be more readily resolved if the complainant feels that the fire officer understands the issue, conflict, or problem.

Step 4: Identify the Complainant's Expectation for Resolution

By this final step, the complainant has drained the emotional bubble, has been able to describe what is going on, and feels that the fire officer understands the issue, problem, conflict, or grievance. The fire officer should now ask what the complainant expects the department to do to resolve this issue. If the problem is an internal grievance, this is where the employee should be asked to describe the desired action.

Policy Recommendations and Implementation

Because the fire officer is in direct contact with fire fighters and citizens, the officer is in the best position to recommend new departmental policies. This change could be the result of a citizen or employee complaint or an identified problem. The problem could be anything that creates a disparity between the actual state and the desired state. For example, not having a standardized location for all equipment on the apparatus reduces efficiency on incident scenes; therefore, it is a problem.

■ Recommending Policies and Policy Changes

The fire officer must understand the procedure for adopting new policies within the department. Although many fire fighters may believe that this procedure simply consists of getting the chief to put his or her name on a piece of paper, usually it entails much more. Some departments have a policy on how to implement a new policy.

The most common method to get new policies adopted or to change existing policies is to outline the problem one or more fire fighters have identified to their supervisor with the recommendation that "someone ought to do something about this." This places all responsibility on the supervisor to come up with a solution and get the department to adopt it. The fact that this method is easy for the officer, however, is far outweighed by its ineffectiveness. Most of the time, nothing changes as a result of this process, and if it does, the change may create more problems than it solved.

To recommend a new policy or a change to an old policy successfully, the fire officer should carefully identify the problem and develop documentation to support the need for a change. This support could consist of statistical measures, anecdotal evidence, or both. The fire officer's opinion seldom carries enough weight to prompt a change because other officers may have different opinions. One effective technique is to discuss the situation with other fire officers initially to determine how widespread the perceived problem is.

> **Safety Zone**
>
> **Aggression, Anger, and Acting Out**
> Some people have trouble handling their anger. Some have very abrasive personalities, others launch vicious verbal attacks, and a few may physically act out their feelings. The fire officer needs to monitor the situation and the complainant. If, in attempting to drain the emotional bubble, there is a rise in rage, then take a time-out.
>
> If this situation involves a fire fighter, follow your organization's procedure on workplace violence or hostility. If the conflict is with a civilian at an emergency scene, ask for police assistance. Police officers have training in handling angry and potentially violent citizens.

Once evidence supporting the existence of a widespread problem has been obtained, the problem-solving techniques outlined earlier in the chapter should be used to develop and choose the best alternative. At this point, the fire officer is ready to write out a proposal to administration. The problem should be carefully outlined, along with the proposed policy change that will resolve it. Resources that will be needed for the solution must also be identified, including financial and time commitments. The benefits of the solution should be identified, as should any potential negative effects and the means by which they will be addressed.

Once a written proposal has been developed, the fire officer should approach his or her superior if that person has not already been part of the development. Along with the recommendation, the policy should be presented to the appropriate officer whose scope is to oversee the area affected by the policy. It is critical to get the agreement of this fire officer, because without this recommendation, the policy will likely not be implemented.

With this officer's recommendation, the proposal is usually forwarded to the chief of the department for review. Usually, a review committee composed of senior staff officers evaluates the proposal and makes changes and accepts or rejects the proposal. Once it is accepted, a draft policy is usually developed in the proper format. The draft policy is distributed to all personnel for review and comment. After a comment period, all comments are addressed, and a final policy is signed by the chief and distributed to the department.

■ Implementing Policies

Like the recommendation process, the process for implementing new policies varies widely. The fire officer should follow departmental procedures. In their absence, the officer may use the following methods to improve his or her communication and understanding.

Because policies are the backbone of order for the fire department, every individual should understand the policies that affect him or her. For this to occur, the fire officer must take responsibility for ensuring that the fire fighters are informed about the policy and fully understand it. The fire officer must also ensure that he or she follows all policies. Failure of the officer to follow all policies undermines their importance, and fire fighters will develop the attitude that they can choose which policies they will follow.

When a new or amended policy is distributed, the fire officer should make sure that it is communicated to the subordinates. At the beginning of each work shift, any new policies should be discussed. The fire officer should identify the points that are most relevant, particularly those areas that change the current practice. For example, if a revised driving policy requires every apparatus responding to an emergency to come to a complete stop at all red signals and stop signs rather than just slow to a speed necessary to avoid an accident, the fire officer should specifically point out this change.

The fire officer should require that all fire fighters read the policy and sign off that they understand it. This practice helps ensure accountability. Employees tend to follow policies more closely if they believe they will be held accountable for them. To make sure that employees understand the policy, the fire officer should provide fire fighters with situations covered by the policy and ask them to apply the policy. Many departments also require that a copy of the policy be posted on the bulletin board for a period of time. In addition, some require a notation in the station log book that the policy was distributed.

The fire officer should evaluate the employees' actions against the policy. If a policy is violated, the fire officer should review the policy with the employee involved. For repeated violations or a violation of a safety policy, disciplinary action should be considered. Employees should never get the feeling that it is acceptable to ignore policies.

On a long-term basis, a regular review of policies should occur. Selecting policies that are not routinely encountered and testing fire fighters' knowledge about them will help keep everyone up-to-date.

Citizen Complaints

The fire officer represents the department in dealing with citizens, public or private organizations, and other governmental agencies. A fire officer could be faced with three different types of citizen complaints. A citizen might complain about:

1. The conduct or behavior of a fire fighter (or a group of fire fighters)
2. The fire company's performance or service delivery
3. Fire department policy

Sometimes, a citizen may want to express an alternative viewpoint on an issue and try to see whether some resolution might be mutually agreeable to the department and the citizen. At other times, the citizen is making a formal complaint. When this occurs, the fire officer is functioning as the official recipient of the complaint. On other occasions, the fire officer may be the subject of the complaint. The first role of the fire officer is to respond to a complaint in a professional manner that effectively obtains the needed information and does not make the situation worse. The methods outlined for resolving conflict within the company also apply to a citizen complaint.

The fire officer must take notes and function as an active listener. By listening attentively and taking detailed notes, the fire officer demonstrates that the complaint is officially considered to be important and is receiving the fire officer's full attention. If the immediate response to a conflict is an explanation or excuse, the complainant will feel that the fire officer is not paying attention, does not care about the issue, or has something to hide. The person who had a complaint will likely stop providing information and feel even more strongly that the complaint was justified.

With some citizen complaints, the fire officer may be able to resolve the problem. With other issues, such as a complaint about using a siren at night, the fire officer should explain the rationale for the use of the siren as well as the laws that regulate its use. Be empathetic and listen to the complainant's statement of the problem without interrupting. Frequently, these kinds of complaints are a method of venting frustration by a citizen rather than a real expectation that the situation will change. Allowing the citizen to express the frustration is important to resolving the issue.

If the fire officer does not have the authority to make a decision on the issue or the citizen is dissatisfied with the officer's decision, the officer should ask whether the citizen would like the issue forwarded to the next level within the organization. If the citizen would like further action, the fire officer should determine the appropriate organizational level where the decision can be made. A preferred method is to consider the scope of the complaint. If the complaint deals with just an operational issue, it should be sent to the chief of operations. If it is an issue about codes, it would be sent to the fire marshal. If the issue affects all areas of the department, it would be forwarded to the chief of the department. Complaints about personnel should be forwarded to the supervisor of the individual who is involved for action.

All relevant facts should be identified and forwarded, along with the details of the complaint. In some departments, the proper procedure is to follow the chain of command. In other departments, fire officers are encouraged to forward the information directly to the decision maker to ensure prompt attention to the matter. Even in these circumstances, the fire officer should inform his or her supervisor about the situation. If any doubt arises about the appropriate response, discuss the issue with your supervisor before taking any action.

Customer Service versus Customer Satisfaction

Customer service is a term that public safety has borrowed from the retail business world. A focus on customer service fixes problems, straightens out procedural glitches, corrects errors of omission (or commission), and provides information.

Customer satisfaction focuses on meeting the customer's expectations. Fire departments often meet customers during one of the worst days of their life. They did not start the day planning to crash the car, melt the stove, or have trouble breathing.

FIGURE 12-7 Creating satisfied customers is one of the fire officer's most important activities.

Good customer service requires sensitivity on the part of every person involved. Although there is little competition for others to provide public safety services, having satisfied customers is important to the municipality **FIGURE 12-7**.

■ Complainant Expectations

When dealing with a civilian, the fire officer should ask what could resolve the situation. The fire officer needs to take the response from the complainant into account to resolve the problem.

In some situations, the resolution of the problem may be as simple as acknowledging that the complainant has had a bad experience with the department. In other cases, it may be as extreme as recommending that the department terminate the employee or employees involved. Unless it is within the scope of the fire officer's authority, the fire officer should make no promises or imply that certain actions will be taken in the discussion with the complainant. In many cases, the complainant's expectations must be relayed to the fire officer's supervisor as part of resolving the issue. If the proposed resolution involves discipline, the fire officer is also obligated to protect employee privacy and civil service due process **FIGURE 12-8**.

Keep the Complainant Informed

If the fire officer needs to do research, consult with others, or obtain direction from supervisors, he or she should keep the complainant informed during this process. Such communication demonstrates that the fire officer is not ignoring the issue, and it educates the complainant on the process.

Follow Up

Citizens frequently complain about local government unresponsiveness. By following up with the complainant, the fire officer reinforces the impression that the complainant's issue is important. This consideration is especially critical if the fire officer has referred the issue to another individual, agency, or organization.

Follow-up by the fire officer may be inappropriate, however, if the fire officer was the subject of the complaint or if the conflict was handled at a higher supervisory level. The fire officer may consider consulting with a supervisor before conducting a follow-up in some cases.

Company Tips

Use All of Your Tools to Solve the Problem at Your Level

You need to be proficient in solving problems. There is a natural tendency to push complaints and conflicts up the chain of command. In some departments, this practice can lead to the administrative fire officer handling an issue that should have remained at the supervising fire officer level. The farther up the chain of command you go with an issue or problem, the more likely the resolution will not work to your advantage.

If you wish to be promoted to higher positions within the department, developing mastery in handling problems, conflicts, and mistakes increases your value to the formal organization. Even if you just want to improve the situation where you work, the skills described in this chapter will help you meet that goal.

Municipal City Fire and Rescue Department Headquarters

May 24, 2017

Mrs. Caroline Marks
Exotic Food Emporium
437 Western Way
Municipal City

Mrs. Marks

We have completed our investigation into the fire and rescue department response to your store on May 5th. During an investigation for smoke in the building, the fire department damaged the front door, interrupted a fumigation service, and destroyed a storefront window.

The initial actions taken by our members were not consistent with departmental policies and procedures. The city will pay for the repairs to the storefront and a second fumigation. Cynthia Bowers from the Mayor's office will continue to be your point of contact during this process.

I apologize for the actions taken by our members. The members involved in this incident are receiving appropriate training and disciplinary actions. We are making sure that all of our fire fighters know how to properly respond to a building with a rapid entry key lock-box that is undergoing fumigation.

Sincerely,

Fire Chief
Municipal City Fire and Rescue Department

FIGURE 12-8 Example of a letter to a civilian in response to a complaint.

You Are the Fire Officer Conclusion

There are four problems Lieutenant Williams needs to solve after this hasty action.

1. *Mitigate the emergency.* Limit the exposure of your crew and anyone else to the fumigation chemical cloud. Call for a hazardous materials response team and the on-duty command officer. Read the fumigation notice posted on the building and call the contractor for decontamination recommendations. Contact the business owner and/or the building representative and ask for a representative to come to the scene.
2. *Document the damage.* Once the hazardous condition is resolved, start documenting the damage done to the storefront by taking pictures and notes. This includes obtaining names and contact information for everyone exposed at the scene. Make sure to meet with the owner or owner representative and walk through the damaged area.
3. *Make the appropriate notifications and documentation.* Although the administrative fire officer will complete the internal investigation, as the officer-in-charge Lieutenant Williams will be doing a lot of writing.
4. *Start the fire company problem-solving process.* Battalion Chief Johnson expressed concern about Quint 100 not knowing about the fumigation or that the business had a rapid-entry key safe. Issues may include both individual performance and system issues.

Wrap-Up

Chief Concepts

- Fires and emergency incidents present a unique category of problems that call for specialized problem-solving skills. Nonemergency situations require the application of conventional problem-solving skills and techniques.
- Complaints, conflicts, and mistakes are special categories of problems. One of the key factors in decision making is knowledge about how to deal with situations that involve conflicts or complaints.
- A systematic approach to high-quality decision making is recommended:
  - Define the problem.
  - Generate alternative solutions.
  - Select a solution.
  - Implement the solution.
  - Evaluate the result.
- The first step in solving any problem is to examine the problem closely and to define the problem carefully. A well-defined problem is one that is half-solved.
- The best people to solve a problem are usually those who are directly involved in the problem.
- One factor in deciding on the best solution is the core value system of the fire department.
- The implementation phase is often the most challenging aspect of problem solving, particularly if it requires the coordinated involvement of many different people.
- Determining whether the solution actually solved the problem requires some type of measurement that compares the original condition with the condition after implementation.
- As the official first-level representative of the fire department administration when dealing with subordinates and enforcing policies and procedures, the fire officer is in a position to be the initial contact in dealing with a wide variety of problems, including situations that potentially involve conflict, emotions, or serious differences of opinion.
- One of the most difficult situations for a fire officer is an interpersonal conflict or grievance within the company or directly involving a company member.
- The conflict resolution model is a basic approach that can be used in situations where interpersonal conflict is the primary problem or a complicating factor.
- All complaints should be investigated, even if the foundation for the complaint appears to be weak or nonexistent.
- The fire officer may take or recommend four actions after completing an investigation:
  - Take no further action.
  - Recommend the action requested by the complainant.

Wrap-Up, continued

- Suggest an alternative solution.
- Refer the issue to the office or person who can provide a remedy.
■ With many conflicts, the fire officer needs to follow up with the complainant to see whether the problem is resolved.
■ Fire fighters are passionate about their profession and are deeply concerned about issues that affect the job. They often are emotional when making a complaint.
■ A citizen might complain about the following issues:
  - The conduct or behavior of a fire fighter
  - The fire company's performance or service delivery
  - Fire department policy
■ Because the fire officer is in regular direct contact with both fire fighters and citizens, he or she is often in the best position to recommend new departmental policies.
■ A fire officer must understand the procedure for adopting new policies within the department.
■ The process for implementing new policies varies, and the fire officer should follow departmental procedures.
■ Customer service is an important part of customer satisfaction.
■ Unless it is within the scope of the fire officer's authority, the fire officer should make no promises or imply that certain actions will be taken in the discussion with the complainant.

Hot Terms

<u>Brainstorming</u> A method of shared problem solving in which all members of a group spontaneously contribute ideas.

<u>Complaint</u> Expression of grief, regret, pain, censure, or resentment; lamentation; accusation; or fault finding.

<u>Conflict</u> A state of opposition between two parties. A complaint is a manifestation of a conflict.

<u>Investigation</u> A systematic inquiry or examination.

<u>Mistake</u> An error or fault resulting from defective judgment, deficient knowledge, or carelessness; a misconception or misunderstanding.

<u>Problem</u> A condition in which the desired situation is different from the current situation.

References and Additional Resources

NFPA reprinted material is not the complete and official position of the NFPA on the referenced subject, which is represented only by the standard in its entirety.

Aitchison, W. (2005). *The Rights of Firefighters*. Portland, OR: Labor Relations Information System.

Ambinder, M. (2013, May 13). *The Fires This Time: Joe Flood on Managing New York City*. New York, NY: Atlantic Monthly Group. Retrieved September 10, 2013, from: http://www.theatlantic.com/politics/archive/2010/05/the-fires-this-time-joe-flood-on-managing-new-york-city/56682/.

Benoit, J., and K. B. Perkins. (2001). *Leading Career and Volunteer Firefighters: Searching for Buried Treasure*. Halifax, NS, Canada: Henson College, Dalhousie University.

Brehm, D. P. (2012). Human Resource Management. In: *Managing Fire and Emergency Services*, A. K. Thiel and R. Jennings. Washington, DC: International City/County Management Association, 219–246.

Brunacini, A. V. (1996). *Essentials of Fire Department Customer Service*. Stillwater, OK: Fire Protection Publications.

Carter, H. R., and E. Rausch. (2008). *Management in the Fire Service*. Quincy, MA: National Fire Protection Association.

Caulfield, H. J., and D. Benzaia. (1985). *Winning the Fire Service Leadership Game*. New York, NY: Fire Engineering.

Coombs, W. T. (2012). *Ongoing Crisis Communication: Planning, Managing, and Responding* (3rd ed.). Thousand Oaks, CA: Sage.

Drucker, P. F. (1974). *Management: Tasks, Responsibilities, Practices*. New York, NY: Harper & Row.

Drucker, P. F. (1990). *Managing the Nonprofit Organization*. New York, NY: HarperCollins.

Drucker, P. F. (1995, February). "Really Reinventing Government." *Atlantic Monthly*. Washington, DC: Atlantic Media Company. Accessed September 26, 2013 at http://www.theatlantic.com/past/politics/polibig/reallyre.htm.

Edwards, S. T. (2009). *Fire Service Personnel Management*. Upper Saddle River, NJ: Brady/Prentice Hall Health.

Fire and Emergency Service Image Task Force. (2013). *Taking Responsibility for a Positive Public Perception*. Fairfax, VA: International Association of Fire Chiefs.

Fleming, E. G. (2012). *Effects of Positional Authority on Communications, Leadership, and Followership*. Emmitsburg, MD: U.S. Fire Administration, Executive Fire Officer Program.

Flood, J. (2010). *The Fires: How a Computer Formula, Big Ideas and the Best of Intentions Burned Down New York City—and Determined the Future of Cities*. New York, NY: Riverhead Books.

Gasaway, R. B. (2013). *Situational Awareness for Emergency Response*. Tulsa, OK: Pennwell/Fire Engineering.

Gormley, W. T., and S. J. Balla. (2012). *Bureaucracy and Democracy: Accountability and Performance* (3rd ed.). Washington, DC: CQ Press.

Ignall, E. J., et al. (1982). *Analytical Approaches to Public Fire Protection*. Lexington, MA: Open Learning Fire Science Program.

LoSasso, C. (2012). *Mentoring Volunteer Officers Pre- and Post-Promotion*. Emmitsburg, MD: Executive Fire Officer Program.

Miller, K. (2010). *We Don't Make Widgets: Overcoming Myths That Keep Government from Radically Improving*. Washington, DC: Governing Books.

Mintzberg, H. (2011). *Managing*. San Francisco, CA: Berrett-Koehler.

National Fallen Firefighters Foundation. (2011). *Understanding and Implementing the 16 Firefighter Life Safety Initiatives*. Stillwater, OK: Fire Protection Publications.

Schilling, J. P. (2012). *Critical Decision Making: Improving Fire Officer Development*. Emmitsburg, MD: U.S. Fire Administration, Executive Fire Officer Program.

Smeby, L. C. Jr. (2014). *Fire and Emergency Services Administration: Management and Leadership Practices* (2nd ed.). Burlington, MA: Jones & Bartlett Learning.

Sprouse, C. B. (2012). *When Good People Make Bad Decisions: Assessing Decision Fatigue in Las Vegas Fire & Rescue*. Emmitsburg, MD: Executive Fire Officer Program.

Taigman, M., and S. Dean. (1999). "Complaints." In: *Secrets of Successful EMS Leaders: How to Get Results, Advance Your Career, and Improve Your Service*. Midlothian, VA: Sempai-Do.

Walker, W. E. (1978). *Changing Fire Company Locations: Five Implementation Studies*. Washington, DC: U.S. Department of Housing and Urban Development, Office of Policy Development and Research.

Walker, W. E., et al., eds. (1979). *Fire Department Deployment Analysis: A Public Policy Analysis Case Study*. Publications in Operations Research Series. New York, NY: Elsevier North Holland.

Weider, M. (2010). *Fire Service Reputation Management* [white paper]. Cumberland Valley Volunteer Firemen's Association.

Weinschenk, C., et al. (2008). "Analysis of Fireground Standard Operating Guidelines/Procedures Compliance for Austin Fire Department." *Fire Technology* 44(1): 39–64.

Yukl, G. A. (2013). *Leadership in Organizations* (8th ed.). New York, NY: Prentice Hall.

FIRE OFFICER in action

Investigation of the fumigation incident included the finding that there was an inappropriate procedure to ensure that the local fire companies knew when and where fumigations were occurring. Battalion Chief Johnson has tasked Lieutenant Williams with recommending a policy change to the notification procedure. In addition, Station 100 will prepare a field training drill on responding to fumigation incidents.

Williams assigned Fire Fighter Grynski to make a list of the address and, location of every rapid-entry safe location in Quint 100's response area. This information will be added to the property information contained in the computer-aided dispatch program.

The fire officer is the first supervisor or manager to deal with external and internal problems and complaints. Documenting and properly processing each issue is a required activity that frequently appears as a component in a promotional exam. Beyond complying with the internal process, a fire officer encounters unique opportunities to learn from each problem and complaint as part of a personal professional development journey.

1. Allowing a person with a complaint to express grief, anger, pain, or resentment completely is an example of:
   A. active listening.
   B. paraphrase and feedback.
   C. emphatic listening.
   D. draining the emotional bubble.

2. It has been 6 weeks since your department implemented a new procedure to reduce the time from dispatch to wheels rolling. The new policy has increased the time it takes to get wheels rolling by 30 percent. You should:
   A. continue the new procedure for another 6 weeks.
   B. re-interview the fire fighters to make them feel understood.
   C. change the plan.
   D. return to the original procedure.

3. While taking notes during a citizen's complaint, you realize that the resolution requires action by an executive fire officer. You should:
   A. complete active listening and explain that you will be forwarding the complaint to the appropriate administrative fire officer for investigation.
   B. stop active listening and provide the complainant with the contact information of the appropriate administrative fire officer.
   C. stop the process and ask to reschedule this session when the appropriate administrative fire officer is available.
   D. recommend that the complaint be filled out on a fire department form and sent to the fire chief's office.

4. Which of the following is *not* a response after completing an investigation?
   A. Take no further action.
   B. Recommend additional study.
   C. Suggest an alternative solution.
   D. Refer the issue to the administrative fire officer who can provide a remedy.

Preincident Planning and Code Enforcement

Fire Officer I

Knowledge Objectives

After studying this chapter, you will be able to:

- Discuss the fire officer's role in community fire safety. (pp 242–243)
- Discuss the purpose of preincident planning and list the relevant factors to consider (NFPA 4.5) (NFPA 4.6). (pp 243–247)
- Discuss the types of fire codes and their use. (pp 247–249)
- Discuss the purpose and function of built-in fire protection systems (NFPA 4.5.2). (pp 249–252)
- Discuss the purpose and process of a fire code compliance inspection. (p 252)
- Discuss the classification of buildings by occupancy (NFPA 4.5.1) (NFPA 4.5.2). (pp 252–256)
- Describe how to prepare for an inspection (NFPA 4.5) (NFPA 4.5.1). (pp 258–259)
- Describe how to conduct an inspection (NFPA 4.5). (pp 259–260)
- Discuss the creation and use of the written inspection/correction report (NFPA 4.6). (p 260)
- Identify general inspection requirements (NFPA 4.5) (NFPA 4.5.1). (pp 260–261)
- Identify use groups and their specific concerns (NFPA 4.5) (NFPA 4.5.1) (NFPA 4.5.2). (pp 261–264)

Skills Objectives

After studying this chapter, you will be able to:

- Prepare for an inspection (NFPA 4.5). (pp 258–259)
- Conduct an inspection (NFPA 4.5). (pp 259–260)
- Write an inspection/correction report (NFPA 4.6). (p 260)

Fire Officer II

Knowledge Objectives

After studying this chapter, you will be able to:

- Describe the preincident plan's role in a business continuity program (NFPA 5.6) (NFPA 5.3.1) (NFPA 5.6.1). (pp 264–265)

Skills Objectives

There are no Fire Officer II skills objectives for this chapter.

CHAPTER 13

Additional NFPA Standards

- *NFPA Fire and Life Safety Inspection Manual*
- NFPA 1, *Fire Code*
- NFPA 10, *Standard for Portable Fire Extinguishers*
- NFPA 12, *Standard on Carbon Dioxide Extinguishing Systems*
- NFPA 101, *Life Safety Code®*
- NFPA 220, *Standard on Types of Building Construction*
- NFPA 291, *Recommended Practice for Fire Flow Testing and Marking of Hydrants*
- NFPA 704, *Standard System for the Identification of the Hazards of Materials for Emergency Response*
- NFPA 1561, *Standard on Emergency Services Incident Management System*
- NFPA 1600, *Standard on Disaster/Emergency Management and Business Continuity Programs*
- NFPA 1620, *Recommended Practice for Pre-Incident Planning*
- NFPA 2001, *Standard on Clean Agent Fire Extinguishing Systems*

Principles of Fire and Emergency Service Administration (FESHE) Course Outcomes

2. Recognize the need for effective communication skills, both written and verbal. (pp 243–244, 247, 256, 260, 265)
6. Discuss the various levels of leadership, roles, and responsibilities within the organization. (pp 242–247, 251–252, 258–260, 264–265)

You Are the Fire Officer

A sign of a recovering economy is coming to Station 100's district. A long-empty big-box retail building is undergoing extensive reconstruction to become a personal manufacturing factory using 3D printing in an extremely environmentally friendly workspace. Captain Jean Davis and Battalion Chief Frank Johnson are attending a meeting at Fire Prevention headquarters with the "My Factory" team.

The architect proudly describes the new environmentally sound technology and techniques that will be components of the facility. The 3D engineers describe innovative technical processes and procedures that are not covered in NFPA standards. As the local fire company commander, the fire chief has assigned Chief Stuart Smith (prevention), Captain Jean Davis (Station 100), and Fire Protection Engineer Brenda Southerland (plans review) as fire department liaisons to the "My Factory" development team.

1. What is the best way to perform a preincident plan for a complex facility?
2. How can you prepare fire fighters for fire inspection duties?
3. How does the fire preincident plan interact with the emergency management and business continuity plan?

Introduction

The fire officer considers the built environment from several viewpoints. If it is burning, damaged, or expelling hazardous materials, the fire officer is expected to command the incident, rescue those in harm's way, mitigate the situation, and render the scene safe.

To accomplish those tasks, the fire officer looks at the building from two different perspectives. First, the fire department prepares to handle an emergency in the building by developing a preincident plan. Second, members of the fire department perform a fire and life safety inspection to ensure that the building meets the appropriate fire prevention code requirements. Although separate activities, preincident planning and code enforcement require similar skill sets, including an understanding of building construction and built-in fire protection systems.

Fire Officer I

The Fire Officer's Role in Community Fire Safety

Fire departments perform fire prevention, community risk reduction, preincident planning, and public education. Fire officers play multiple roles in relation to properties within their communities, including handling these critical tasks:

1. Identifying and correcting fire safety hazards through safety checks or code enforcement
2. Developing and maintaining preincident plans
3. Promoting fire safety through public education

In most areas, fire inspectors and fire officers working in staff assignments perform fire inspections and code enforcement duties **FIGURE 13-1**. Fire suppression companies are usually involved in preincident planning. In some areas, the local fire suppression company also conducts code enforcement inspections. Public education activities are often performed by a combination of staff personnel and fire companies.

Even where the role of fire companies does not include code enforcement, fire officers and fire fighters should conduct regular visits to properties to develop preincident plans. During

FIGURE 13-1 Fire officers often conduct preincident planning and fire inspection activities during the course of their workday.

Tactical Priorities

Address: 1500, 1510, 1520		
Occupancy Name:		

Preplan #: 02-N-01	Number Drawings: 1	Revised Date: 12/2002
District: E275A	Subzone: 60208	By: ACEVEDO

Rescue Considerations: Yes ( ) No (X)

Occupancy Load Day:	Occupancy Load Night:
Building Size:	Best Access:

Knox Box: NONE	Knox Switch: NONE	Opticom: NONE
Roof Type: X	Attic Space: Yes ( ) No ( )	Attic Height: X

Ventilation Horizontal:	Ventilation Vertical:

Sprinklers: Yes (X)	No ( )	Full (X)	Partial ( )
Standpipes: Yes (X)	No ( )	Wet ( )	Dry ( )
Gas: Yes ( )	No ( )	Lpg ( )	

Hazardous Materials: Yes (X) No ( )
DIESEL GENERATORS

1,000 GALLON TANKS

BATTERY ROOM

Firefighter Safety Considerations:
ELEVATOR PIT

Property Conservation And Special Considerations:
VENTILATION: AUTOMATIC SMOKE REMOVAL SYSTEM

3 OFFICE BLDGS; 2 PARKING STRUCTURES

6 FLRS - 1230 W. WASHINGTON ST.
4 FLRS - 1500 N. PRIEST DR.
4 FLRS - PARKING GARAGE

FIGURE 13-2 An example of a preincident plan.

preincident planning, they should also be on the lookout for fire and life safety hazards. A fire officer should always take proactive steps to reduce the impact of any potential fire emergency that could occur, including identifying and correcting conditions that could start or spread a fire, or restrict or prevent occupants from leaving the building, and that could ultimately increase the risk to citizens and fire fighters in the event of an emergency.

Preincident Planning

A preincident plan is described by NFPA 1620, *Recommended Practice for Pre-incident Planning,* as a document developed by gathering general and detailed data used by responding personnel to determine the resources and actions necessary to mitigate anticipated emergencies at a specific facility **FIGURE 13-2**.

The original purpose of a preincident plan was to provide information that would be useful in the event of a fire at a high-value or high-risk location. A high-value property contains equipment, materials, or items that have a high replacement value. Examples include properties containing agricultural equipment, electronic data processing equipment, or scientific equipment; fine arts centers; and storage or manufacturing sites.

A high-risk property has the potential to produce a catastrophic property or life loss in the event of a fire. Examples include nuclear power plants, bulk fuel storage facilities, hospitals, and jails. Preincident plans include information that could apply to a variety of situations that could potentially occur at the location covered in the plan.

Facilities that store or handle hazardous materials are required to submit information about those materials and the threats they pose to the fire department and the Local Emergency Planning Council (LEPC); this information is then incorporated into preincident plans for those facilities. Some plans include responses to natural or human-made catastrophic incidents.

Today, preincident plans are used for all types of buildings and occupancies within a fire department's response area. A preincident plan is meant to identify in advance the strategies, tactics, and actions that should be taken if a predictable situation occurs, and to make the fire fighters familiar with the building. Preincident plans can be particularly useful at the company level for practicing initial operations for buildings in the company's district.

■ A Systematic Approach

To ensure a systematic approach that collects all of the required data, the fire officer should use a standardized method for completing each preincident plan. NFPA 1620 outlines a six-step method:

1. Identify physical elements and site considerations.
2. Identify occupant considerations.
3. Identify fire protection systems and water supply.
4. Identify special hazards.
5. Identify emergency operation considerations.
6. Identify special or unusual characteristics of common occupancy.

Step 1: Identify Physical Elements and Site Considerations

The first step is to evaluate the physical elements and site considerations. Plot plans provide a representation of the exterior of a structure, identifying site access, doors, utilities access, and any special considerations or hazards. Floor plans are interior views of a building. Rooms, hallways, cabinets, and the like are drawn in the correct relationship to each other in such a plan.

The building's size and dimensions, including its overall height, number of stories, length, width, and square footage, should be determined and included in the plan. Connections between buildings and distances to exposures should be noted. Access routes and points of entry should be clearly indicated. In addition, the preincident plan should identify concealed spaces and windows that could be used for ventilation or rescue.

The preincident plan should also include detailed information about the construction of the building's roof, floor, and walls, as well as an assessment of their structural integrity. The focus is on those factors that could lead to collapse of building components or spread of fire or toxic gases. Any factor that might affect the ability of responding fire fighters to enter and safely perform interior operations should be highlighted.

Conditions that affect the safety of fire operations on the roof should be noted as well.

This part of the preincident plan should also document the location of utilities, including gas, electrical, and domestic water entry locations and shutoffs, and heating, ventilating, and air-conditioning controls. Note the presence of flammable liquids, compressed or liquefied gases, hazardous materials, and steam lines. Information about elevators should include their location in the building, floors served, type of elevator, and type of recall or override service.

Note any conditions that might potentially delay access, including weight-restricted bridges, low-overhead clearances, and roads subject to natural or human-made blockages. Security information, such as the presence of fences, 24-hour security forces, and guard dogs, should be part of the plan as well. The fire officer must also determine the environmental impact of any contaminants that could be released from the facility.

During a test of the preincident plan, the fire officer should document any interference or poor coverage of the two-way radio system. For example, when conducting a preincident planning survey of a high rise, be sure to test your portable radios to ensure that you can contact dispatch or the outside unit when in the inner core as well as in the basement. For many radio systems, these areas prove to be "dead" spots. Such a problem could be mitigated by having a fixed antenna system or repeaters installed in the building.

Step 2: Identify Occupant Considerations

The fire department is expected to provide a rapid and safe evacuation of a facility or plan for protection in place. Protecting in place would require the officer to determine which areas within the structure are resistant to the magnitude of fire that would be expected to occur.

If the preincident planners determine that the occupants should be removed, the plan should identify how that will be accomplished. It should describe both the process and the fire department and building resources that will be dedicated to occupant protection, including the occupant escape routes.

Document the number of occupants and their ages, their physical or mental conditions, and any need for assisting them to leave the building. Building-specific information should include hours of operation, occupant load, and location of occupants. Each of these points of information has a bearing on how many companies would be needed and what they can expect in terms of removing occupants.

The preincident plan should note the locations of exits and any special locking devices, such as a delayed release or stairwell unlocking system. The fire officer should also ensure that the preincident plan reflects coordination between the facility staff and the fire fighters. Past natural disasters have highlighted the need to be clear on the role of the facility and the role of the fire department. This might require a meeting between the fire officer and the staff of a nursing facility in which they predetermine the role of the facility and the responders when handling a fire or disaster. As mentioned earlier, the preincident plan should identify the locations of any occupants who need assistance to evacuate the building safely, along with the locations of stair chairs, stretchers, and lifts.

When large numbers of people are relocated, they need basic services, such as food, water, and sanitation facilities. In some communities, the public school system can assist the fire department by providing areas of shelter and transportation. Joint agreements and defined roles and responsibilities are part of the preplanning process.

As part of the Incident Command System, a tracking system for the occupants of a building should be established, particularly for locations where the number of occupants varies with the facility's operations and time of day. If evacuees will be relocated, their locations must be tracked. For larger evacuations during weather-related events, the American Red Cross has had considerable experience with tracking evacuees.

Some facilities, such as high-rise, healthcare, or detention facilities, cannot be completely evacuated in a short time. Consequently, the response plan could be to relocate occupants within other areas of the building. These facilities are usually designed to relocate occupants to certain areas until they can be evacuated or moved to another area or building.

During all of these operations, the products of combustion must be segregated from the occupants. Move those in the greatest peril to areas away from the fire. Close fire and smoke doors, and seal the individual rooms with wet towels to prevent smoke from filtering into the room. The heating, ventilation, and air-conditioning (HVAC) system is often designed to shut down automatically upon activation of a fire alarm. Sophisticated HVAC systems, such as those found in hospitals and high rises, are often divided into zones. With such systems, exhaust fans in unaffected zones can be redirected to eject smoke from the zone that is in alarm.

Step 3: Identify Fire Protection Systems and Water Supply

The required water flow is determined by evaluating the size of the building or buildings, their contents, construction type, occupancy, exposures, fire protection systems, and any other features that could affect the amount of water needed to control the fire. The adequacy of available water for sprinkler systems, inside and outside hose streams, and any other special requirements or needs should be considered.

Identify the available water supply. Document the locations of fire hydrants as well as their flow rates and the distribution system. The ideal hydrant would be fed from a large main that is part of a grid that allows water to flow from several directions. Water supply test data should be obtained and tests should be conducted in accordance with NFPA 291, *Recommended Practice for Fire Flow Testing and Marking of Hydrants*. When the demand exceeds the available supply, the preincident plan should identify an appropriate response to mitigate the deficiency. This might include a water shuttle operation or the initiation of relay operations. Some sites may have a private water system, including their own water tower.

The preincident plan should identify the location and details of every fire department connection, fire pump, standpipe

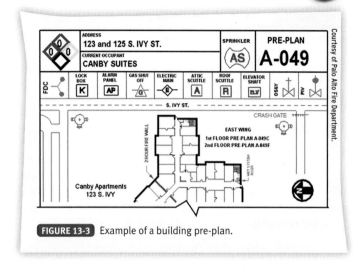

FIGURE 13-3 Example of a building pre-plan.

system, and automatic sprinkler system **FIGURE 13-3**. Smoke management and special hazard protection systems should also be detailed on the plan. Finally, the preincident plan should include data on the protective signaling system (fire alarm).

Step 4: Identify Special Hazards

Document special hazards in the facility and develop a plan to send the proper resources during an emergency. Special hazards might include flammable or combustible liquids, explosives, toxic or biological agents, radioactive materials, and reactive chemicals or materials. Request data on the maximum inventory of hazardous materials and highly combustible products found in the building. In addition, the preincident plan should note contact information for the facility hazardous materials coordinator and the location of material safety data sheets.

Some buildings contain specialized operations, processes, and hazards that can pose unique challenges during an emergency. The preincident plan should identify any area of the occupancy that contains gases or vapors that could present a hazard to emergency responders. This includes confined spaces, inert atmospheres, ripening facilities, and special equipment–treating atmospheres. Document emergency operating procedures and identify personnel who can provide technical assistance during emergency incident mitigation.

The preincident plan should include instructions for de-energizing electrical systems and isolating and securing mechanical systems. These systems should be shut down or monitored to protect emergency responders from electrocution, mechanical entrapment, or other perils. Some systems may be remotely controlled by an off-site service. Digital cameras are useful to document systems for training.

Step 5: Identify Emergency Operation Considerations

The incident action plan should be based on the priorities of life safety, incident stabilization, and property conservation, in that order. The preincident plan should address the appropriate and adequate departmental response to a working fire or emergency incident. The planned location of the incident

Near-Miss REPORT

Report Number: 11-0000265

Event Description: On [date omitted], I was assigned to a two-person tender company. At 0900 hours, we began inspecting fire hydrants for ISO. At 1000 hours, we were on our seventh hydrant inspection of the morning. The hydrant was close to the intersection of a two-lane road that was being widened for a right-turning lane. The crew positioned the apparatus with the tailboard just in front of the hydrant with lights on to provide traffic control and began our usual inspection process. The crew tagged the hydrant and flushed it.

The crew then attached a static pressure gauge and a gate valve with a flow gauge on the end of it. The gate valve was shut and my driver, who was standing in front of the large-diameter hose discharge, began to charge the hydrant. My driver had turned the hydrant on with approximately five turns when the entire large-diameter hose assembly, which weighed approximately 30 pounds, blew off and broke the retaining chain that connected it to the hydrant. The large piece traveled between his legs and landed 15 feet in the center of the roadway. The cap chain for the left 2½-inch ear was connected to the same chain and it landed on the other side of the roadway, 25 feet away. It was close, but no injuries were received.

The driver finished the shift. Upon inspection, none of the parts were broken or malfunctioning. The large-diameter hose discharge connection had sets of teeth that went in and then were one-fourth turned with a set screw to lock it into place. The set screw was found fully extended. This leads me to believe that the hydrant came from the manufacturer with the large-diameter hose connection not locked in place. The hydrant was an [brand name deleted by reviewer] and was manufactured in 2010.

Lesson Learned: Hydrants should always be charged with no personnel or equipment in front of any discharge.

---

command post and emergency operations center, if provided, should be noted.

The number of required fire companies is affected by fuel loading. Fuel load refers to the total quantity of all combustible products found within a room or space. The fuel load determines how much heat and smoke will be produced by a fire, assuming that all of the combustible fuel in that space is consumed. The size and the shape of interior objects and the types of materials used to create them have a tremendous impact on the objects' ability to burn and on their rate of combustion. The same factors influence the rate of fire spread to other objects.

Fires develop in four phases: incipient, free burning, flashover, and smoldering/decay. The incipient phase of a fire is the starting point of a fire. A fire in this phase usually involves only the object of origin. A preincident plan may identify what would be required to handle a fire at the incipient stage. For example, a loaded stream-type portable fire extinguisher may be required for the combustibles present in the building. A fire in the free-burning phase involves other objects in the fire. The preincident plan might indicate that a free-burning-stage fire would require a minimum of a 1¾-inch (44-mm) handline.

Once a fire has undergone flashover, depending on occupancy and construction factors, the preincident plan might recommend that a fire in this stage be fought defensively on arrival, or it might indicate that an attack with large handlines be initiated. When the fire has consumed all of the oxygen but has retained its heat and still has fuel available to it, it has entered the smoldering/decay phase. Plans should indicate how the building would be ventilated when this situation is encountered.

The preincident plan should identify anticipated areas of fire spread. Open pipe chases, elevator shafts, and balloon construction, for example, all contribute to vertical fire spread. Large open areas may conceal the magnitude of a fire that has only light smoke showing from the outside yet has the potential for significant horizontal fire spread. False ceilings and cocklofts may conceal horizontal fire spread.

The fire department's preincident plan should be fully coordinated with the internal evacuation or emergency operations plan. The facility should provide the fire department with an on-site liaison as soon as command is established.

Step 6: Identify Special or Unusual Characteristics of Common Occupancy

Particular hazards for each occupancy group should be identified. After gathering data on the specific facility, the fire officer should research the applicable codes, standards, and other information sources to assist in identifying the common and unusual hazards that are associated with that occupancy group. For example, multiple-fatality fires are the major risk factor that should be considered in places of assembly.

Effective preincident plans for simple sites or facilities can be developed with minimal amounts of data. Additional data may be required for more complex sites or facilities, such as locations with numerous hazards with the potential for high-risk incidents.

Time, owner resistance, and proprietary information are all factors that may potentially hinder the collection of data. For all these reasons, the fire officer might have to prioritize the data collection effort to obtain the most useful information. An incremental process can be used, in which very basic plans are developed and then updated at a later time as more information becomes available. The local building department and other agencies may have information that will assist you with developing the preincident plan. For example, they may have floor plans for specific buildings you can use.

■ Putting the Data to Use

The goal of preincident planning is to develop a written plan that would be valuable to both the owner of the building and the fire department if an incident occurs at that location. This plan should provide critical information that could be advantageous for responding personnel, in a format appropriate for emergency conditions. Many fire departments maintain preincident plans in electronic form instead of printing out hard copies. Data storage systems allow the needed information to be automatically retrieved when the dispatch system processes an alarm for a particular location. This information can be printed at the fire station, or on-board mobile data equipment can provide access to the information from a command post or while en route to the incident. Additional detailed information, such as building plans and fire alarm drawings, can be kept in a lock box or other secured area at the site.

The preincident plan includes a plot plan. The plot plan should show the relationship of the building to other buildings, streets, hydrants, utility controls, and other features. The plot plan visually represents these objects, allowing the officer to identify relevant facts quickly.

Some departments may also maintain drawings of the interior of the buildings. Like the plot plan, a floor plan allows the officer to identify considerations for a fire attack quickly. A floor plan is a drawing of the interior of the structure and is similar to an architect's blueprints. It notes stairwell locations, elevators, standpipe connections, hazardous material storage areas, fire alarm panel locations, and points of entry.

A written report also accompanies most preincident plot plans. This report identifies the address, type of occupancy, construction type, size of the occupancy, and required water flow for firefighting operations all in one section. In another section, the report identifies the resources available, including responding units and water sources. Another section includes any special hazards or considerations, such as a fire department connection, utility controls, hazardous material storage, and rapid-entry key box locations. It is important that written reports follow a standardized format and include standardized information to allow for quick access by the officer, whether the report is in an electronic form or a hard copy.

Understanding Fire Codes

Fire officers often perform inspections to enforce a fire code (or fire prevention code). A fire code establishes legally enforceable regulations that relate specifically to fire and life safety, including related subjects such as regulation of hazardous materials and process protection and operating features.

A variety of codes may be adopted by different jurisdictions. A state fire code applies everywhere in the state, whereas a locally adopted code can be enforced only within that particular jurisdiction. In many cases, a state or provincial fire code sets a minimum standard, and local jurisdictions have the option of adopting more stringent requirements. The local jurisdiction may not be able to exceed the state minimum (called a mini/max code).

Fire code requirements are often adopted or amended in reaction to fire disasters, an approach known as the catastrophic theory of reform. Many code requirements can be traced back to disasters, such as the 1903 Iroquois Theatre fire (Chicago: 602 dead), the 1911 Triangle Shirtwaist Company fire (New York: 146 dead), the 1930 Ohio State Penitentiary fire (Columbus, 320 dead), the 1942 Cocoanut Grove Nightclub fire (Boston: 491 dead), the 1944 Ringling Brothers–Barnum and Bailey Circus fire (Hartford, Connecticut: 168 dead), the 1977 Beverly Hills Supper Club fire (Southland, Kentucky: 164 dead), the 1990 Happy Land Social Club fire (Bronx, New York: 87 dead), and the 2003 Station nightclub fire (West Warwick, Rhode Island: 100 dead).

Authority having jurisdiction is a term used in NFPA documents to refer to "an organization, office, or individual responsible for enforcing the requirements of a code or standard, or for approving equipment, materials, an installation, or a procedure." The authority having jurisdiction for a state fire code is usually the state fire marshal. In the case of a provincial fire code, the authority having jurisdiction would be the provincial fire marshal or fire commissioner. The local fire chief, fire marshal, or code enforcement official would be the authority having jurisdiction for a local fire code.

The regulations contained in a fire code are enforced through code compliance inspections. These inspections could be conducted by the state fire marshal's office, by the local fire department, or by code enforcement officials who might or might not be a part of the fire department. The authority having jurisdiction delegates the power to enforce the code to the fire officers, fire inspectors, and other individuals who actually conduct inspections.

■ Building Code versus Fire Code

Both building codes and fire codes establish legally enforceable minimum safety standards within a state, province, or local jurisdiction. A building code contains regulations that apply to the construction of a new building or to an extension or major renovation of an existing building, whereas a fire code applies to existing buildings and to situations that involve a potential fire risk or hazard. For example, the building code might require the installation of automatic sprinklers, a fire alarm system, and a minimum number of exits in a new building; the fire code would require the building owner to maintain the sprinkler and alarm systems properly and to keep the exits unlocked and unobstructed at all times when the building is occupied. Sometimes, a fire code includes certain requirements that apply to new buildings that are beyond the scope of the building code, such as a regulation requiring the installation of fire lanes and hydrants.

State Fire Codes

Most U.S. states and Canadian provinces have adopted a set of safety regulations that apply to all properties, without regard to local codes and ordinances. Where a state or provincial fire code has been established, it is generally the minimum legal standard in all jurisdictions within that state or province. The state or provincial fire marshal usually delegates enforcement authority to local fire officials.

Most states allow local authorities the option of adopting additional regulations or a more restrictive code. A few states have adopted mini/max codes, which mean that local jurisdictions do not have the option of adopting more restrictive regulations.

The NFPA *Fire Protection Handbook* identifies seven different organizational patterns for state fire marshal organizations in the United States. The state fire marshal may work in any of these organizations:

- The department of insurance
- The department of public safety
- A separate government department
- A regulatory agency
- The state police
- A cabinet-level office
- The state fire commission

Local Fire Codes

At the local level, fire and safety codes are enacted by adopting an ordinance, which is a law enacted by an authorized subdivision of a state, such as a city, county, or town. The local jurisdiction adopts an ordinance that establishes the fire code as a set of legally enforceable regulations and empowers the fire chief to conduct inspections and take enforcement actions.

Fire Marks

Life Safety Code
NFPA 101, *Life Safety Code*®, is a model code document that contains requirements specifically related to protecting the lives of building occupants, covering detailed information on means of egress. When NFPA 101 is adopted by a jurisdiction, it can be applied to both new and existing buildings.

This authority can then be delegated to fire officers, fire inspectors, and other individuals.

Model Codes

Model codes are documents developed by a standards-developing organization, such as the NFPA, and made available for adoption by authorities having jurisdiction. In 1905, for example, the National Board of Fire Underwriters published the National Building Code. A model code is developed through a consensus process using a network of technical committees. Most jurisdictions use model codes developed by the NFPA and the International Code Council.

States and local jurisdictions may adopt a nationally recognized model code either with or without amendments, additions, and exclusions. A complete set of model codes includes a building code, electrical code, plumbing code, mechanical code, and fire code. The primary advantages of a model code are that the same regulations apply in many jurisdictions, and all of the requirements are coordinated to work together without conflicts.

The process in which a model code is adopted by a local jurisdiction may follow one of two paths. Adoption by reference occurs when the jurisdiction passes an ordinance that adopts a specific edition of the model code. For example, a local jurisdiction might adopt NFPA 1, *Fire Code* (2015), by reference. The requirements specified in NFPA 1 then become local requirements that can be enforced by designated local officials. A fire officer would need to obtain a copy of the 2015 edition of NFPA 1 to read the specific requirements.

Adoption by transcription occurs when the jurisdiction adopts the entire text of the model code and publishes it as part of the adopting ordinance. For example, a city might copy the language of the code and include it in its entirety within the ordinance. A fire officer would then need to read only the city's ordinance to identify the specific requirements rather than having to find the appropriate code book.

The fire officer must know which code and which annual edition are used by the local jurisdiction. Although the model code process updates the code every 3 to 5 years, the authority having jurisdiction must specifically adopt the new edition of a model code before it becomes legally enforceable. Different codes or different editions of the same code might apply to different occupancies. Some communities adopt only selected portions of one or more model codes, but defer to a state code or locally written ordinances to cover other issues. Some jurisdictions choose to maintain their own independent codes.

Retroactive Code Requirements

Regulations that applied to a particular building at the time it was built remain in effect as long as it is occupied for the same purpose. The fire officer may have to determine the specific code document, title, and year used when the building was built to determine whether a building is still in compliance with the applicable code requirements. Most codes include provisions that can be applied to buildings that were constructed before a code was adopted.

If a building is remodeled or extensively renovated, or if its occupancy use changes, most codes specify that all of the current requirements of the code must be met. New code requirements, meaning those adopted after a certificate of occupancy has been issued, do not apply unless specific language is included in the adopting ordinance. For example, Scottsdale, Arizona, established a mandatory requirement for sprinkler systems in all new residential construction in 1986. The ordinance did not require retroactive installation of sprinkler systems in residential structures built before that date.

On occasion, a state or local authority having jurisdiction passes a code revision that is specifically identified as applying retroactively to all affected occupancies. After the 1980 Las Vegas MGM Grand fire, which killed 84 people and injured 679, the state of Nevada required fire suppression systems to be installed in all existing casinos and hotels. Rhode Island and Massachusetts adopted retroactive sprinkler requirements for nightclubs in response to the 2003 fire in The Station nightclub, which killed 100 people and injured more than 200 patrons.

Understanding Built-in Fire Protection Systems

After clarifying means of access and egress, the status of the built-in fire protection features is the second reason for a fire company to perform inspections. Built-in fire protection systems are designed as tools to assist fire fighters in combating a fire. The officer should understand how systems work and how the codes are altered when fire protection systems are in place.

A jurisdiction's codes often allow more flexibility in the design of a building when built-in fire protection systems are included. A building with a sprinkler system can be larger and taller, the travel distance to exits can be longer, and the access for fire apparatus could be restricted. All of these trade-offs depend on a properly functioning sprinkler system.

If a fire occurs in such a building, you are depending on the built-in fire protection systems to assist you. NFPA's *U.S. Experience with Sprinklers* reports that sprinklers operated in 91 percent of all reported structure fires large enough to activate sprinklers, excluding buildings under construction and buildings without sprinklers in the fire area. When sprinklers operated, they were effective 96 percent of the time, resulting in a combined performance of operating effectively in 87 percent of all reported fires where sprinklers were present in the fire area and fire was large enough to activate them. The more widely used wet-pipe sprinklers operated effectively 89 percent of the time, while dry-pipe sprinklers operated effectively in 76 percent of cases (Hall, 2013).

The dependability of these systems is of prime importance to fire fighters' safety, and an inspection is the best method of ensuring the systems will work as intended. Los Angeles, Phoenix, and Fairfax County, Virginia, started testing existing fire protection systems in the late 1980s. In the first year, they encountered significant failures in built-in fire protection systems. In turn, a periodic retesting program was put in place to improve the performance of built-in fire protection services.

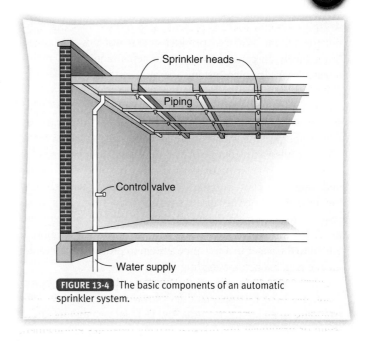

FIGURE 13-4 The basic components of an automatic sprinkler system.

Water-Based Fire Protection Systems

Automatic sprinkler systems, standpipe systems, and fire pumps are the three primary components of water-based fire protection systems. An automatic sprinkler system consists of a series of pipes with small discharge nozzles (sprinklers) located throughout a building. When a fire occurs, heat rising from the fire causes one or more sprinklers to open and release water onto the fire **FIGURE 13-4**. When the water starts flowing, a water flow alarm is activated. This alarm may be monitored by a central station alarm service or on-site safety/security service. Some systems are "local alarms" and sound a gong or bell only at the outside of the building **FIGURE 13-5**.

Depending on the usage and climate, automatic sprinkler systems may be wet pipe, dry pipe, deluge, or preaction. In a

FIGURE 13-5 Some systems are "local alarms" and sound a gong, horn/strobe device, or bell only at the outside of the building.

wet-pipe system, water is present in all of the pipes throughout the system. When a sprinkler opens, water is discharged immediately. In general, wet-pipe systems require less maintenance than dry-pipe systems. Because of the faster reaction, fewer sprinklers are activated to control most fires.

Dry-pipe systems are used in locations where a wet-pipe system would be likely to freeze, such as unheated storage facilities and parking garages. Instead of containing water, these pipes are filled with compressed air or nitrogen until a sprinkler opens. When the air pressure drops, the dry-pipe valve opens and water is released into the system. Dry-pipe systems require higher maintenance because activation of the sprinkler system requires the entire sprinkler system to be drained. Sometimes, instead of installing a dry-pipe system, an antifreeze solution is added to the water in a wet-pipe system to protect an unheated area (e.g., a freezer or a loading dock).

Deluge systems are special versions of wet- or dry-pipe systems in which large quantities of water are needed to control a fast-developing fire quickly. Deluge systems are most often found in ordnance plants, aircraft hangars, and occupancies with flammable liquid hazards. All of the sprinklers are open and ready to discharge water as soon as the control valve opens. Smoke, flame, or fire detectors are used to sense a fire and trigger the system.

Preaction sprinkler systems are similar to dry-pipe systems, but include a separate detection system that triggers the dry-pipe valve and fills the sprinkler pipes with water. At this point, it becomes equivalent to a wet-pipe system. A preaction system is designed to reduce the risk of water damage due to accidental sprinkler discharge or a broken pipe.

Some sprinkler systems are designed to discharge protein or aqueous film-forming foam as the extinguishing agent. They are used in high-hazard areas where the contents to be protected are flammable liquids, such as fuel storage and chemical process facilities. These complex systems require more frequent inspections and are custom designed for the storage area, processing plant, or fueling facility that is protected. Often there is an on-site or nearby technician who has been trained in the maintenance of these systems.

A standpipe system provides the ability to connect fire hoses within a building. Standpipes are an arrangement of piping, valves, hose connections, and allied equipment that allow water to be discharged through hoses and nozzles to reach all parts of the building. As with automatic sprinkler systems, dry standpipes may be installed in unheated areas. Standpipes are subdivided into three classes based on their expected use:

- Class I provides 2½-inch (64-mm) hose outlets, intended for use by fire department or fire brigade members trained in the use of large hose streams.
- Class II provides a 1½-inch (38-mm) hose coupling with a preconnected hose and nozzle in a hose station cabinet. The hose is designed for occupant use.
- Class III provides both 1½-inch (38-mm) and 2½-inch (64-mm) connections. The 1½-inch connection may have a preconnected hose line that can be used by the occupants until the fire department arrives.

FIGURE 13-6 A fire pump may be needed to maintain or increase water pressure in standpipe and automatic sprinkler systems.

During building surveys and inspections, you should pay particular attention to the condition of the standpipe system; it is one of the few built-in fire protection devices specifically designed to help fire fighters. Fire officers should identify fire hose access and ensure proper calibration of pressure-regulating devices. A pressure-regulating device limits the discharge pressures from standpipe hose outlets. During a 1991 high-rise office fire at One Meridian Plaza in Philadelphia, fire fighters discovered that the pressure-regulating devices were improperly set, and only a weak flow could be obtained from each outlet. Usually found on the lower floors of high-rise standpipes, the devices were installed on floors 26 through 30 of One Meridian Plaza and their poor performance impeded firefighting operations. Three fire fighters died while the fire consumed the 22nd through 30th floors in 19 hours.

Fire Pumps

Fire pumps increase the water pressure in standpipe and automatic sprinkler systems **FIGURE 13-6**. They are designed to start automatically when the water pressure drops in a system or a fire suppression system is activated. A field inspection of such a device generally is confined to a visual inspection to confirm that the pump appears to be in good condition and is free of physical damage. The fire officer should confirm that the fire pump has passed the annual performance test.

■ Special Extinguishing Systems

Four types of special extinguishing systems may be used in various structures: carbon dioxide, dry or wet chemical, Halon/clean agent, or foam. Note these specialized systems in the preincident plan. In general, the ongoing code compliance inspection consists of visual inspection, looking for physical damage, and confirming that the safety seals are intact, as well as verification of the documentation for required inspection, testing, and maintenance. A safety seal is placed on any handle that would activate the system. If this safety seal is broken, the system could have been discharged.

> ### Getting It Done
>
> **Fire Sprinkler Alarms**
> A fire alarm system is present in most buildings with automatic sprinklers. In many cases, smoke detectors and local fire alarm pull stations are also connected to the fire alarm system. Sophisticated alarm systems are often monitored by a central fire alarm service that calls in the fire alarm event to the local 911 center. Most of the notifications from fire alarm systems, however, are nonfire events. Some of these events are caused by situations that appropriately trigger an alarm, such as cigarette smoke or burnt food. Other activations may occur due to defective or dirty smoke detectors. The fire officer must remain vigilant because the same notification process is used when a water flow alarm occurs.
>
> A water flow alarm means that water has moved past a flow switch that is located inside the water supply pipe and set off the alarm. Nonfire events that may set off the water flow alarm often include public water supply surges, sprinklers that have been broken accidentally, and burst pipes. Many older buildings have a single water flow device for the entire structure, so the problem could be anywhere in the building.
>
> A trouble alarm in a monitored fire alarm system is like a warning light in a car: all it tells you is that something is not right. Some communities require that a log book be provided in buildings with large and sophisticated alarm systems, usually in the fire control room. Both the building maintenance department and the fire department make entries in this log book with every activated fire alarm. Fire officers who work in districts with many monitored buildings develop a detailed knowledge of the features and problems in each fire alarm system. Although the code may specify a certain type of system, these fire officers appreciate that every building has unique characteristics.
>
> Every activated alarm requires an inspection of the building. If the fire alarm system resets without going back into alarm, a surge in the municipal water supply may have tripped the water flow alarm.
>
> If you respond back to a second water flow activation at the same location within a couple of hours, invest the time to identify where the water is flowing. This effort may require a rigorous inspection of the fire suppression system and assistance from a building representative.

Carbon Dioxide

Carbon dioxide systems are fixed systems that discharge carbon dioxide from either low- or high-pressure tanks, through a system of piping and nozzles, either to protect a specific device or process (e.g., a printing press) or to flood an enclosed space. Carbon dioxide extinguishes fire by displacing oxygen. This gas is heavier than air, so it settles in low spaces. Fixed systems are generally required to comply with NFPA 12, *Standard on Carbon Dioxide Extinguishing Systems*. Your preincident plan should require the use of self-contained breathing apparatus (SCBA) if a fixed system has become activated.

Dry or Wet Chemical

Fixed chemical extinguishing systems discharge a chemical extinguishing agent through a system of piping and nozzles. These systems can be found protecting commercial cooking devices and industrial processes where flammable or combustible liquids are used. Observe the condition of the nozzles. Many contain a protective cap that is intended to protect the nozzle from becoming obstructed with cooking grease.

Wet chemical systems are preferred for protecting cooking equipment. The wet chemical agent reacts with hot grease to form a foam blanket, reducing the release of combustible vapors. The foam blanket cools the grill and reduces the possibility of a rekindle. Dry chemical systems leave a residue that is difficult to clean up.

Both types of systems may be activated in one of two ways: (1) a fusible link that melts on flame contact or (2) a manual pull station. Activation of the system turns off the cooking device by closing the cooking fuel valve or turning off the electricity.

Halon

Halon 1301 was the extinguishing agent of choice for fire protection in computer rooms and to protect electronic equipment from the 1960s to the 1990s. Based on the weight of the agent, Halon 1301 is about 250 percent more efficient than carbon dioxide for extinguishing fires. Unfortunately, it also depletes the ozone layer in the atmosphere. Since 2000, Halon has not been allowed to be manufactured or imported into the United States, but legacy systems may still be recharged.

NFPA 2001, *Standard on Clean Agent Fire Extinguishing Systems,* covers the use of alternative agents that have replaced Halon systems in protecting electrical and electronic telecommunications systems. Most of these systems are designed to flood a room or an enclosure and, like Halon, are toxic. The room or enclosure must be sealed, and the occupants must leave before the agent is discharged. The system can be automatically or manually fired; both methods include a prealert warning for the occupants to leave the room before the agent discharges. The preincident survey should identify the chemical used, the duration of the discharge (10 seconds to 1 minute), the enclosure protected, the automatic activation sequence, and the location of the manual activation station.

Foam Systems

A low-expansion foam system is used to protect hazards involving flammable or combustible liquids, such as gasoline storage tanks. These systems discharge foam bubbles over a liquid surface to create a smothering blanket that extinguishes the fire and suppresses vapor production. High-expansion foam is used in areas where the goal is to fill a large space with foam, thereby excluding air from the area and smothering the fire.

■ Fire Alarm and Detection Systems

A fire alarm system consists of devices that monitor for a fire and notify the appropriate personnel. Manual fire alarm boxes, smoke detectors, or heat detectors may activate the fire alarm system. The alarm may be activated by the water flow or pressure switch in a sprinkler system. The activated system notifies the appropriate personnel, including the building occupants, with audible and visual signals. These signals may be transmitted outside the building and to the fire department or a central station monitoring firm.

Understanding Fire Code Compliance Inspections

The objective of a fire code compliance inspection is to determine whether an existing property is in compliance with all of the applicable fire code requirements. Some codes call this a maintenance inspection. The goal for the inspector is to observe the housekeeping to ensure that no fire hazards exist and to confirm that all of the built-in fire protection features, such as fire exit doors and sprinkler systems, are in proper working order.

The fire department's authority and responsibilities for conducting code compliance inspections are usually included in the ordinance that adopts the local fire code. The responsibility for code enforcement is usually assigned to the fire chief or fire marshal. The fire chief can delegate the actual code enforcement activities to different individuals or units within the department.

In many fire departments, fire officers and fire inspectors who are assigned to a fire prevention bureau or code enforcement division conduct inspections of specific types of properties. In departments where fire companies conduct code compliance inspections, an individual or a group is usually assigned to coordinate these activities and provide technical assistance.

■ Fire Company Inspections

The purpose of conducting a fire inspection is to identify hazards and to ensure that the violations are corrected. This process helps the fire department be proactive in preventing fires. One method of maximizing the number of inspections carried out is for fire companies to conduct them.

Inspections conducted at the fire company level have been an activity advocated by the National Fire Protection Association since the first *NFPA Quarterly* was published more than 100 years ago. Through such inspections, local fire companies have the opportunity to become familiar with their response areas and take actions to prevent the start of fires and, if a fire starts, to minimize the fire spread and risk of destruction of the neighborhood.

Before conducting an inspection, it is important to understand the scope of code enforcement authority that is delegated to a fire officer in your jurisdiction. There is wide diversity in the scope and authority of inspections conducted by fire companies. In some areas, fire suppression companies are authorized to enforce all fire code regulations; one metropolitan city authorizes fire officers to issue corrective orders that would require the installation of automatic sprinklers in a business under renovation. In other fire departments, fire suppression companies are assigned to inspect only certain types of occupancies, while the fire marshal's office or fire prevention division inspects other types of occupancies. Some fire departments can provide only safety recommendations to citizens and property owners, whereas others can issue citations and compliance orders. The fire officer needs to determine the source and the scope of his or her code enforcement authority before conducting any fire safety inspection, as well as which particular codes are used by the jurisdiction.

Except in case of fire emergencies, a fire officer generally cannot enter private property without the permission of the owner or occupant. Most fire codes include a section that authorizes code enforcement officials to enter private properties at any reasonable time to conduct fire and life-safety inspections; however, the scope of this authority usually depends on the type of property involved. In most cases, the permission of the owner or occupant is required for entrance into a dwelling unit, whereas access to public areas is less restricted. The fire code often contains a section that, if necessary, allows for the issuance of a court order requiring the owner or occupant to allow the fire department agent to enter the occupancy to conduct an inspection.

Classifying by Building or Occupancy

Many of the code requirements that apply to a particular building or occupancy are based on its classification. The codes may classify a building by construction type, occupancy type, and use group. In turn, you as a fire officer need to know how to classify a building before you can determine the code requirements that apply to it. In addition to these three classifications, local zoning ordinances often regulate where certain types of occupancies or buildings are permitted.

■ Construction Type

The building itself is classified by construction type, which refers to the design and the materials used in construction. NFPA 220, *Standard on Types of Building Construction*, addresses building construction. The type of building construction has a significant influence on firefighting strategy. Construction type is a fundamental size-up consideration when firefighting strategy for a burning structure is being determined.

The most commonly used model codes classify construction into five basic types:

- **Type I: Fire resistive.** The construction elements are noncombustible and are protected from the effects of fire by encasement, using concrete, gypsum, or spray-on coatings **FIGURE 13-7**. Depending on the model code used, Type I construction is divided into subtypes, based on the level of fire protection provided. The level of protection is described by the number of hours a building element can resist the effect of fire.
  - Type I is the most durable and lasting structure. Extensive fire suppression operations may be carried out in such a building before it will collapse.
  - This type of construction often uses compartmentation instead of fire sprinklers to control fire spread, which in turn may create unsurvivable

FIGURE 13-7 A Type I building.

FIGURE 13-9 A Type III building.

interior furnaces when fire fighters are trapped on the fire floor. Metal structural elements may be failing due to age and rust.

- **Type II: Noncombustible.** The structural elements can be made from either noncombustible or limited-combustible materials **FIGURE 13-8**. Like Type I, Type II construction is classified into subdivisions, based on the level of fire resistance. Although the buildings are assembled from noncombustible components, the structural elements have limited or no fire resistance. A Type IIA structural frame is expected to resist fire for 1 hour, whereas the structural frame in a Type IIB building is not expected to resist the effects of fire. A strip shopping center with cinderblock walls, unprotected steel columns, and steel bar joists supporting a steel roof deck is an example of a Type IIB building.
  - Type II is a common 20th-century construction method and the type of building that Frances Brannigan referenced when describing the 20-minute interior firefighting rule of thumb in 1971.
  - Such a structure is durable, but is not a legacy building—it will require replacement in 30–40 years.

This type of building is frequently updated with Type V structural elements.

- **Type III: Limited combustible (ordinary).** The exterior load-bearing walls of the building are noncombustible masonry **FIGURE 13-9**. A masonry wall may consist of brick, stone, concrete block, terra cotta, tile, adobe, or concrete. The interior structural elements may be combustible or a combination of combustible and noncombustible. As with Types I and II, different levels of fire protection are possible in Type III buildings. The structural frame of a Type IIIA building is protected, which means it is encased in concrete, gypsum, or spray-on coatings and is expected to have a fire-resistive rating of 1 or 2 hours. A Type IIIB structural frame is unprotected and has no fire resistance rating.
  - Type III was used to build commercial, multiple-family, and mercantile types of buildings through the 1980s. Brannigan classifies these buildings as "Main Street USA."
  - This kind of building is usually no higher than four stories; it was designed to preserve the load-bearing walls if fire consumed the building. The connection between the floor and the load-bearing wall is designed to pull out without damaging the wall.
- **Type IV: Heavy timber.** The exterior walls are noncombustible (masonry), and the interior structural elements are unprotected wood beams and columns with large cross-sectional dimensions **FIGURE 13-10**. Mill construction, which was used in many New England textile buildings built in the 1800s, is an example of heavy timber construction. Mill construction features massive wood columns and wood floors.
  - As durable as Type I structures, most surviving Type IV buildings have been converted to residential, mercantile, or mixed-use spaces.
  - A well-seated fire in a nonsprinklered Type IV building may exceed the municipal water supply capability.
- **Type V: Wood frame.** The entire structure may be constructed of wood or any other approved material

FIGURE 13-8 A Type II building.

FIGURE 13-10 A Type IV building.

stone, or masonry veneer applied to the exterior walls. This type of wall covering does not enhance the strength or fire resistance of the building.

■ Occupancy and Use Group

The <u>occupancy type</u> refers to the purpose for which a building or portion of a building is used or is intended to be used. The code requirements are determined by the structure's <u>use group</u>. Occupancies are classified into use groups based on the characteristics of the occupants, the activities that are conducted, and the risk factors associated with the contents. Within each occupancy type are dozens of use groups.

Assembly

An assembly occupancy is used for the gathering of people for deliberation, worship, entertainment, eating, drinking, amusement, or awaiting transportation. This classification may be further divided into more specific types of assemblies. Examples of assembly-type occupancies include:

- Churches
- Taverns or bars
- Nightclubs
- Basketball arenas
- Restaurants
- Theaters

FIGURE 13-11. Sometimes, a masonry veneer is applied to the exterior, but the structural elements consist of wood frame.

- Type V buildings are the most common structures and include single-family, multiple-family, mercantile, and low-rise commercial buildings.
- Such a structure will maim or kill the first-arriving fire fighters if they enter the structure and fall through a fire-weakened floor. Underwriters Laboratory research in 2011 showed Type V residences achieving flashover in 3:30 to 4:45 minutes, compared to flashover times of 29:30 to 34:15 minutes for legacy residences (Kerber, 2012).
- Type V construction techniques and approach will be the foundation for future innovative building methods and materials.

It is often difficult to determine the type of construction from the exterior of a building. Many low-rise offices, apartment buildings, and residences appear to be ordinary construction, but are actually wood-frame buildings with a brick,

Business

A business occupancy is used for account and record keeping or transaction of business other than mercantile **FIGURE 13-12**. Examples of business occupancies include:

- Dental offices
- Banks
- Architects' offices
- Hair salons
- Colleges and universities
- Doctors' offices
- Investment offices
- Insurance offices
- Radio and television stations

FIGURE 13-11 A Type V building.

FIGURE 13-12 Example of a business occupancy.

FIGURE 13-13 Example of an industrial occupancy.

Educational
An educational occupancy is used for educational purposes through the 12th grade. Typically, this designation refers to schools. Educational occupancy may also cover some daycare centers for children older than 2½ years.

Industrial
In an industrial occupancy, either products are manufactured or processing, assembling, mixing, packaging, finishing, decorating, or repair operations are conducted FIGURE 13-13. This would include occupancies such as:

- Automobile assembly plants
- Clothing manufacturers
- Food processing plants
- Cement plants
- Furniture production facilities

Health Care
A healthcare occupancy is used for purposes of medical or other treatment or for care of four or more persons, where such occupants are mostly incapable of self-preservation due to age, physical or mental disability, or security measures not under the occupants' control. Such buildings would include hospitals and nursing homes.

Detention and Correctional
A detention and correctional occupancy is used to house four or more persons under varied degrees of restraint or security, where such occupants are mostly incapable of self-preservation because of security measures not under the occupants' control. This category could include prisons, jails, and detention facilities.

Mercantile
A mercantile occupancy is used for the display and sale of merchandise FIGURE 13-14. This group includes the following:

- Retail stores
- Convenience stores
- Department stores
- Drug stores
- Shops

FIGURE 13-14 Examples of mercantile occupancies.

Residential
A residential occupancy provides sleeping accommodations for purposes other than health care, detention, or corrections. This type of occupancy includes five subcategories:

- **One- and two-family dwelling units:** Buildings that contain no more than two dwelling units with independent cooking and bathroom facilities
- **Lodging or rooming houses:** Buildings that do not qualify as one- or two-family dwellings but that provide sleeping accommodations for a total of 16 or fewer people on a transient or permanent basis, without personal care services, with or without meals, but without separate cooking facilities for individual occupants
- **Hotels:** Buildings under the same management in which there are sleeping accommodations for more than 16 persons and that are primarily used by transients for lodging with or without meals
- **Dormitories:** Buildings in which group sleeping accommodations are provided for more than 16 persons who are not members of the same family in one room, or a series of closely associated rooms, under joint

occupancy and single management, with or without meals, but without individual cooking facilities
- **Apartment buildings:** Buildings containing three or more dwelling units with independent cooking and bathroom facilities

Storage

A storage occupancy is used primarily for storing or sheltering goods, merchandise, products, vehicles, or animals. Examples include:

- Cold storage plants
- Granaries
- Lumber yards
- Warehouses

Mixed

A mixed-use property has multiple types of occupancies within a single structure. An example would be an old commercial building that has been renovated to include multifamily residential loft apartments on the second and third floors and a bakery on the first floor.

Unusual

Some structures do not fit neatly into the other occupancy categories. These occupancies can be placed in a miscellaneous category that represents unusual structures, such as towers, water tanks, and barns.

You can find the use group classifications in the building code or NFPA 101 **TABLE 13-1**.

Once the use group classification is known, the allowable height, floor area, and construction type can be found in the building code. The building code (or NFPA 101) also includes the requirements for the number of exits, the maximum travel distance to an exit, and the minimum width of each exit. Requirements for the installation of fixed fire protection systems in new buildings appear in the building code (or NFPA 101) as well. Additional requirements for fire protection systems are sometimes included in the fire code, particularly for existing buildings.

Code enforcement starts by confirming that the occupancy is used for the approved purpose. For example, if a tenant space in a shopping center has been converted from a clothing store to a bar, the exit requirements and the built-in fire protection system requirements could be different.

NFPA 704 Marking System

Buildings with significant quantities of hazardous materials may be required to use a marking system. The most widely recognized standard is NFPA 704, *Standard System for the Identification of the Hazards of Materials for Emergency Response*. This marking system consists of a color-coded array of numbers or letters arranged in a diamond shape **FIGURE 13-15**.

Each color represents a specific type of hazard. Blue represents health hazards, red represents flammability hazards, and yellow represents the material's reactivity hazard. Within each color diamond, there is a number from 0 to 4 that represents the relative hazard of each. A 0 means that the material poses essentially no hazard, whereas a 4 indicates extreme danger. For example, on the exterior of a building, there might be a marking with a 2 inside the blue area, a 4 inside the red area, and a 0 inside the yellow area. This means that inside the building, there are materials that pose a moderate health hazard, a severe flammability hazard, and no real reactivity hazard. The last quadrant of the diamond is white and is used to indicate special hazards. This area could include letters or numbers. For example, it might include a placard that indicates to the responder that water-reactive material is present.

This marking system also requires labels to be affixed to containers inside the structure to indicate the hazards of the substance. Such labels might identify materials as "corrosive," "flammable," or "poison." In general, the system requires a 704 marker at each entrance to the building, on doorways to chemical storage areas, and on fixed storage tanks.

Table 13-1	Classification of Use Groups
Major Use Classification	**Occupancy Categories**
Assembly	Theaters, auditoriums, and churches Arenas and stadiums Convention centers and meeting halls Bars and restaurants
Health care	Hospitals and nursing homes
Detention and correctional	Prisons and penitentiaries
Mercantile	Retail stores
Business	Offices
Industrial	Factories
Storage	Warehouses Parking garages
Educational	Schools
Residential	Homes Apartments Dormitories Hotels

Reprinted with permission from NFPA 101®-2012, Life Safety Code®, Copyright © 2011, National Fire Protection Association, Quincy, MA.

FIGURE 13-15 Document any special hazard in the facility.

VOICES
OF EXPERIENCE

One Friday I was informed by a fire suppression contractor that the fire suppression system for a large warehouse in our district would be out of service for repairs due to a leak near one of its private fire hydrants on site. Periodically throughout the day, from the afternoon to the early evening, I received a couple of status reports on the progress of the repairs.

Late in the evening, I stopped by to see the actual repair work, as I had not received any confirmation that the project was completed and that the system had been placed back in service. Upon my arrival, the fire suppression contractor said the water line feeding the fire hydrant had been repaired and was ready for the water line to be charged with pressure. There was a large, excavated hole exposing only some of the plastic pipe and hardware on the line. The workers at the site stated that they were unsure/confused about this piping grid because another pipe was exposed and was actually connected to this water line in this same hole.

> **Even the most mundane, routine event can turn into an incident of injury.**

During the charging of the underground fire service water line, it ruptured, shooting a large, violent stream of water up in the air, flooding the hole immediately, and launching debris everywhere. Even the most mundane, routine event can turn into an incident of injury.

The previous work was now negated, and the entire fire suppression service for the warehouse was out of service for the next 12 hours while further repairs were undertaken. The confusion over the pipe was resolved through a set of plans produced by a supervisor who was called to the site. The pipe was part of the entire grid that provided additional needed water supply to the fire pump on site. The actual cause of the rupture was that the water line's valve connection had only locking lugs (e.g., Mega Lugs) and had not been secured with any additional thrust blocking and/or tie rods.

This incident illustrates why all relevant drawings should be reviewed prior to work being started on a repair. Third-party inspectors such as a private contractor, a fire official, a building official, and/or a utility official should be available for inspection of all repairs.

If the workers and I had not moved away from the hole during the charging of this water line, we would have certainly suffered significant injuries. We must understand and respect the power of water, especially under pressure, which may catch fire fighters and others off-guard. Charging handlines, opening hydrants, and doing hose testing are just some scenarios that can result (and have resulted) in unexpected failures and injuries.

Situational awareness is an attribute that is an absolute must in the fire service, whether you are a line fire fighter, a fire officer, or "just the fire prevention guy." Always prepare for the worst—maintain safe practices and wear your safety gear apparel.

Dave Belcher
President, Ohio Society of Fire Service Instructors
Lieutenant, Violet Township Fire Department, Fire Prevention Bureau
Pickerington, Ohio

Preparing for an Inspection

Preparation is required before conducting a code enforcement inspection. The assumption in this section is that the fire officer has the responsibility, authorization, and training necessary to conduct such an activity. The focus here is how to prepare to conduct such an inspection.

The fire officer should regard the owner or occupant as a professional partner. Their interaction begins with the fire officer preparing for the inspection by reviewing the applicable code provisions and any information on previous inspections that can be found in the occupancy file. For the business owner or occupant, a visit by the local fire company is a major and disruptive event. Some jurisdictions send a pre-inspection or self-survey form to the business or property a few weeks before an official fire department inspection will occur.

■ Reviewing the Fire Code

Before conducting an inspection, you should review the sections of the fire code that apply to the specific property. In some cases, you may discover unanticipated situations during an inspection that require you to research an additional section of the code. The NFPA's *Fire and Life Safety Inspection Manual* provides general information about a large variety of processes, equipment, and systems.

■ Review Prior Inspection Reports, Fire History, and Preincident Plans

Review the inspection results from past activities. You might detect a trend or a chronic problem. If a certain facility always requires two follow-up inspections to correct the violations that are noted during inspections, then that fact might indicate that the fire officer needs to turn up the salesmanship. For example, when calling to schedule the inspection, the fire officer might take the opportunity to review the violations noted during the previous inspection and encourage the owner to check those items.

Look at the fire history for the occupancy. If a barbeque restaurant has a history of a flue or hood fire every 6 to 9 months, you should plan to concentrate on the grill and hood system during your inspection. In such case, you should think about which measures you might recommend to reduce the frequency of accidental fires.

Bring a copy of the preincident plan with you. A code enforcement inspection is an excellent time for you to update contact information, such as names, e-mail addresses, and phone numbers. You can also use the preincident plan to identify any modifications and additions that have been made to the occupancy.

■ Coordinate Activity with the Fire Prevention Division

Coordination is needed in departments that implement both a fire prevention division and a company-level inspection program. Some jurisdictions assign a list of occupancies to be inspected by each fire suppression company on a monthly or quarterly basis. In buildings where a process, storage, or occupancy is required to have an annual fire prevention permit, the local fire company's ongoing compliance inspection could be scheduled for 6 months after the fire prevention division issues the permit.

The authority having jurisdiction usually determines the frequency of inspections for each type of occupancy. For example, daycare facilities might be required to have an annual inspection from the fire marshal's office, whereas business occupancies are inspected by a local fire company once every 2 or 3 years. The fire officer commanding the fire suppression company is expected to follow the inspection schedule and coordinate his or her actions with the fire marshal's office.

■ Arrange a Visit

It is good practice to contact the owner or business representative to schedule a day and a time for the fire safety inspection. Many businesses are cyclical, and some days or months are more difficult to accommodate a fire department inspection than others. For example, early April may not be the best time to inspect an accounting firm's office because it is the peak of accountants' annual work cycle.

In some cases, a time that is inconvenient for a business is an important time in terms of fire safety. Between October and December, retail stores are packed with extra stock for the holiday buying season. Some retail businesses earn 30 percent or more of their annual revenue during this period. A fire safety inspection of a store in an enclosed mall on December 1 may reveal boxes stored from floor to ceiling, obstructing the sprinklers. Empty cardboard boxes may be pushed up tight against the electrical panel, instead of maintaining a 30-inch clearance. The electrical panel may be hotter than usual because of all the extra power needed to run the holiday displays. Trash may have piled up in the storage room and blocked the rear emergency exit.

■ Assemble Tools and References

Once you have reviewed the information on the occupancy to be inspected and the applicable fire code, you are ready to perform the inspection. In addition to your knowledge, you will need to bring some equipment to assist you during the visit. At a minimum, you should carry the following tools:

- Pen, pencil, and eraser
- Inspection form
- Clipboard (Consider using the type of clipboard that law enforcement and EMS professionals use; it would allow you to store the completed forms in a safe place in case you need to respond to an emergency during a code enforcement period.)
- Graph paper and ruler
- Hand light
- Digital camera
- Coveralls
- Measuring device, tape measure, or walking meter
- Fire department business cards
- Reference code books

> **Company Tips**
>
> **Neatness Counts**
> Make sure that your uniform is neat and clean. Some areas will require wearing a safety helmet to complete a life-safety inspection or preincident plan. Many members will utilize their fire helmet to meet this safety requirement. A "salty" helmet with a heavy buildup of carbon and other products of combustion can create a poor impression. If your hand gets dirty when you touch your helmet, it is time to clean the helmet.

> **Company Tips**
>
> **Conducting Inspections and Surveys**
> A four-step system that meets the organization's need for fire company inspections and preincident surveys will help the fire company focus on what is important.
> **Step 1:** Schedule the inspections based on use group or occupancy. For example, inspections of all of the service stations and auto repair facilities in the district might be scheduled for November.
> **Step 2:** Hold an in-station drill to review the fire code sections applicable to the types of items that might be found during a typical code compliance inspection in that type of occupancy. Continuing the preceding example, in October, the fire officer schedules an in-station drill to review the fire code requirements for a service station or automotive repair shop and also delegates a senior fire fighter to schedule the inspections.
> **Step 3:** Break the fire company into two-person inspection teams. One team member will focus on the code enforcement, while the other will update the preincident survey form.
> **Step 4:** The company reviews their findings. This review is an ideal time to discuss any changes that are required on the incident action plans for the specific facilities, based on the inspections. For example, if an auto repair shop has added a vehicle spray booth, the plan for that occupancy should be revised.

Plan to wear your safety shoes and bring your fire helmet, eye protection, and protective gloves if you will be inspecting an area that is under renovation or an occupancy that requires such protective equipment.

Conducting the Inspection

Some departments require that the entire fire company perform the fire inspection together. Business owners, in turn, may complain about the disruption that occurs when four to six fire fighters come into a business. Portable fire radios are likely to be blaring, and sometimes the fire fighters seem to be more interested in sightseeing or shopping than identifying fire safety hazards. Other departments deploy the fire company in two- or three-person teams to conduct inspections in adjacent occupancies. Make sure you know which procedures and practices your organization follows.

The conduction of a fire inspection should be approached in a systematic fashion. The fire officer should begin every inspection in the same manner. First, circle the area as you park the apparatus to get a general overview of the property. Second, meet the property owner or manager to let him or her know that you have arrived and will begin the inspection. Third, begin the inspection at the exterior of the building and work systematically throughout the inside of the building, beginning at the lowest level and working up. Fourth, conduct an exit interview with the contact person. Finally, write a formal report on the inspection.

■ General Overview

Circle the property and observe all four sides of the building when you arrive. During this review, you are looking for any obvious access or storage problems and any new construction since your last visit. You should also confirm the locations of hydrants, sprinkler/standpipe hook-ups, and other outside features with your preincident survey sheet.

Park the fire apparatus in a location that does not disrupt the business and allows the fire company to respond if they are dispatched to an emergency. Some departments require that one fire fighter remain with the apparatus both to listen to the radio and to protect the rig from vandals or thieves. The apparatus should not be parked in a fire lane. It is hard to convince the business owner to comply with the fire code when the fire engine is violating the code by sitting in a fire lane when there is no emergency.

■ Meet with the Representative

Enter the business through the main door and make contact with the appropriate representative. If you have called ahead to schedule this visit, you already have a name and office location. Introduce your crew and briefly explain the goal of this visit. This is a great time to review and update all of the contact names, phone numbers, and information found in your preincident survey sheet. Ask to have a representative with the appropriate access cards and keys accompany the inspection team.

■ Inspecting from the Outside In, Bottom to Top

A fire company–level ongoing compliance inspection confirms the built-in fire protection systems are fully operational and makes sure the area is free of fire ignition sources. This type of assessment begins with a walk around the exterior of the premises. Confirm that the address is present and properly identified on both the front and the rear of the building. If the building has a fire department connection, is it free from foreign objects inside the piping, and are the threads in workable condition? Verify that all means of access and egress are clear and in proper operating order—there is no more important issue in this type of inspection. Exit problems require immediate correction.

The fire officer should ensure that the location of dumpsters does not present a fire hazard. Outside storage buildings should be checked for compliance, particularly with the storage of hazardous materials. Are required markings, like those specified in NFPA 704, present?

After the outside walk-around, go to the basement, where you will likely find utility rooms, fire pumps, fire protection system control valves, backup generators, and laundry rooms. These areas are susceptible to improper storage of combustibles near electrical panels, open junction boxes, improperly stored hazardous material, and blocked exits.

The fire officer should systematically work through the building, checking for fire safety issues. Like the exterior and the basement, the ability of occupants to exit a building quickly is a primary concern. Problems may include inventory stock that is blocking exits, locked doors, and exit and emergency lights that are not working.

Look for conditions that are prone to starting fires: open electrical wiring, use of extension cords, and storage of combustibles too close to heat sources. Watch for storage of flammable liquids, as well as the use of candles, portable heating units, and fireplaces or wood stoves.

Last, consider items that would assist in extinguishing a fire once it has started. These considerations would include the clearance between sprinklers and objects, the condition of smoke and heat detectors, and the operability of fire extinguishers. The fire officer should confirm that fire doors are not propped open and are in operable condition. If the structure is equipped with a commercial kitchen, the hood and duct system and fire suppression system should be checked for compliance.

■ Exit Interview

It is important to wrap up your ongoing compliance inspection/preincident survey by meeting with the owner or designated representative to review what was found and issue any required correction orders. Remember that one of your roles is to be a partner with the business. A written report needs to be completed, with one copy going to the occupant.

Writing the Inspection/Correction Report

Several different types of inspection/correction reports are possible. Many use a check-off system, in which the officer puts a check next to the corresponding deficiency. This allows the citation to have the appropriate code without the officer having to look up each violation. If a reinspection reveals that violations have not been corrected, a notice of hazard is described in a narrative rather than noted in a check-off form.

The report needs to describe clearly any needed corrections and to quote the appropriate sections of the code or ordinance. Some communities have developed report forms that list the most commonly occurring violations.

Once the report is complete, you should review it with the owner or representative. If violations are found during the inspection, you must explain to the responsible individual what needs to occur to correct each problem. You retain the original report for follow-up purposes. The owner or representative should sign the form and keep a copy of the report. The report should also be forwarded to the fire prevention division.

Life-threatening hazards, such as locked exits, must be corrected immediately. Less critical issues can be corrected within a reasonable time period, generally 30 to 90 days. You should arrange for any needed follow-up inspections to verify that the corrections have been made. Here is the procedure used by one fire department for non–life-threatening code enforcement issues:

- Fails first fire company–level inspection: Schedule follow-up inspection in 30 days.
- Fails second fire company–level inspection: Follow-up inspection in 15 to 30 days.
- Fails third fire company–level inspection: Issue sent to the fire prevention division for resolution. Depending on the nature of the issue and the history of this occupancy, the fire prevention division contacts the owner and makes an inspection within 2 to 30 days.
- Fails first fire prevention division inspection: The fire marshal issues a notice of violation or correction order. The time given to comply with the notice or order is determined by the situation and by local regulations. The owner can file for appeal or equivalency within 10 days of the first fire prevention division inspection.
- Fails second fire prevention division inspection: Second notice issued by the fire marshal. Generally, the time to comply is shorter. The owner is warned of more severe consequences if the issues remain unresolved.
- Fails third inspection: Follow-up inspection in days; the building representative may be subject to a misdemeanor charge punishable by fines or jail, or both. The occupancy may be required to cease operations until the matter is resolved.
- Fails fourth inspection: Fines and legal action are initiated.

General Inspection Requirements

The fire code includes general fire and life-safety requirements that apply to every type of occupancy. These items should be checked during every inspection. The general requirements include properly operating exit doors and unobstructed paths to egress travel. The general requirements also require built-in fire protection systems, such as automatic sprinklers and fire detection systems, as well as portable fire extinguishers, to be regularly inspected and properly maintained in operational condition. In addition, the fire code includes general precautions that are required with the use or storage of hazardous, flammable, or combustible goods. Some fire codes include topics related to emergency planning and conduct of fire drills, as well as features such as rapid-entry key boxes for fire department use.

■ Access and Egress

In every inspection, the fire department must confirm that there is sufficient means of egress for the occupants. These provisions of the fire code are often violated because improper storage causes the exits to be obstructed.

Proper storage practices mean that the goods for sale or the items used by the business or industrial process are safely stored in accordance with the fire code. In the example of a retail store during the holidays, the fire hazard is greatly increased when the store is jammed to the ceiling with stock. In some cases, a fire company may discover a very hazardous condition due to excessive, illegal, or dangerous storage practices. This situation requires immediate corrective action and, if available, assistance from the fire prevention division or the fire marshal's office.

Exit Signs and Emergency Lighting

Equally important considerations in occupant evacuation are exit signs and emergency lights. Exit signs indicate the direction of exit to occupants during a fire. Emergency lights light the path of the exit. These features are particularly important when the occupants may be unfamiliar with the location of exits other than the main entrance, such as at theaters and nightclubs. Often, officers find exit lights that are burned out or for which the backup battery no longer works. The same is true for emergency lights. The occupant is required to maintain documentation of monthly checks performed on these systems.

Portable Fire Extinguishers

Almost every building has portable fire extinguishers, which are provided for the occupants to control incipient fires. The fire code describes requirements for the size, type, and locations of extinguishers needed in various occupancies, usually referring to NFPA 10, *Standard for Portable Fire Extinguishers*, for the details. Verify extinguishers are of the appropriate size and type. In addition to visually inspecting each extinguisher for physical damage, confirm that the instructions are visible, the safety or tamper seal is present, there is current inspection and testing documentation, and the pressure gauge is in the normal range.

Built-in Fire Protection Systems

The officer should verify that all fire department connection caps are in place, and that they are unobstructed and accessible. For sprinkler systems, extra sprinklers and a changing wrench should be readily available. The officer should ensure that all control valves are in the open or correct position (some valves are normally closed) and are locked or have a supervisory alarm to protect the system from being accidentally shut down.

Only properly trained personnel can test fire protection systems. Los Angeles, Phoenix, and Fairfax County encountered embarrassing situations when they started their retesting programs. Using a pumper to charge a standpipe system blew one standpipe right out of the building wall. Another pressure test filled a mercantile basement with hundreds of gallons of water. Issues with liability and repair costs may arise when the local fire company is operating built-in fire protection systems in a non-fire-emergency situation. In some retesting programs, the fire prevention bureau requires a licensed fire protection contractor to perform the retest, and this contractor must submit a certified report of the findings.

Electrical

During the inspection, the fire officer should confirm that no combustibles are stored around the electrical panels and that the panel covers have not been removed. The use of extension cords is often abused. Extension cords are intended for temporary use, such as powering a drill or a fan; they are not designed to serve as permanent wiring to power a refrigerator or copy machine. Check for multiple extension cords being used in sequence, and for open electrical boxes for outlets, switches, or junctions.

Special Hazards

The nature of an industrial process by itself may be hazardous. For most processes with very high hazards, the occupancy is required to have a <u>fire prevention division or hazardous use permit</u>. Such a local government permit is renewed annually after the fire prevention division performs a code compliance inspection. The local fire company may be the first to discover a high-risk occupancy, like the Los Angeles fire company that discovered a fireworks warehouse during a routine inspection. Such surprises are especially likely if your district includes "on spec" or "spec built" industrial parks. In these developments, owners construct buildings that are generic spaces and rent out space to a variety of tenants. Such spaces have high occupant turnover, with each occupant being associated with different fire risk factors. A space that was used for storing building materials last year, for example, may be used for assembling computer hardware equipment this year and might be an auto repair shop next year.

Hazard Identification Signs

When required, visible hazard identification signs meeting NFPA 704 should be placed on stationary containers, on above-ground tanks, and at entrances to locations where hazardous materials are stored, dispensed, used, or handled. Individual containers, cartons, or packages must be conspicuously marked or labeled. Material safety data sheets must be readily available on the premises for regulated material.

Selected Use Group–Specific Concerns

Each use group classification involves special concerns and considerations in addition to the general concerns presented earlier in this chapter. Some of the more significant considerations are described in the following sections.

Public Assembly

The goal for a code compliance inspection in a public assembly occupancy is to ensure that all of the access and egress pathways are clear and in good order **FIGURE 13-16**. The number and the size of exits should have been approved before the notice of occupancy was given to the business.

A major problem noted in many assembly inspections is overcrowding. Assembly occupancies should have their occupant load posted, and during the inspection you should check for crowds that exceed this capacity.

FIGURE 13-16 All access and egress pathways must be clear.

> **Getting It Done**
>
> **LEEDing the Way: Saving the Environment One Building at a Time**
>
> Leadership in Energy & Environmental Design (LEED) is a "green" building tool that provides building owners and operators with a framework for identifying and implementing practical and measurable design, construction, operations, and maintenance solutions to reduce the environmental impact of the structure. LEED certification includes verification from an independent third party that a building, home, or community was designed and built using strategies that focus on environmental health—for example, water savings, energy efficiency, materials selection, and indoor environmental quality.

Confirm that the exits are not blocked or locked and are in good working order, and that the layout has not changed since the last inspection. The fire code will likely specify when panic hardware must be installed and the direction of swing of the doors. Like the number of exits, these features are not usually an issue unless remodeling has occurred. Storage of combustibles under exit stairs is not permitted.

Exit lighting and emergency lights are an important consideration for assembly-type occupancies. Typically, occupants are not familiar with exit locations, so methods to direct them to the exits in the event of an emergency are essential. Lighting requires battery backup or a backup generator. Evacuation plans may also be required.

Confirm use of flame-resistant material in the curtains, draperies, and other decorative materials. Watch for unacceptable storage of hazardous or flammable materials.

Commercial exhaust hoods and ducts also need to be inspected. The fire officer should check for unvented, fuel-fired heating equipment. These occupancies are also prone to occupants propping open fire doors and smoke barriers.

Business

Make sure that access and egress pathways are clear and in good order; this includes exits that are blocked with office furnishings. Exit emergency lights are problematic. Most occupants are familiar with building design, but fire protection equipment, such as fire extinguishers, often is not maintained properly.

These occupancies are notorious for inappropriate use of electrical cords. In older buildings, improper use of portable heaters, open electrical boxes, and spliced wiring tend to be problems. Business occupancies also tend to store small amounts of flammable liquids and other hazardous materials inappropriately.

Educational

A single school fire produced the swiftest example of the catastrophic theory of reform. The Our Lady of Angels School burned on December 1, 1958. Despite heroic efforts by the Chicago Fire Department, 95 people died, most of them children. NFPA's Percy Bugbee reported that within days of the disaster, state and local officials ordered the immediate inspection of schools. The NFPA partnered with Los Angeles Fire Marshal Raymond Hill to conduct 172 fire tests to develop new fire safety regulations for schools. Operation School Burning provided smoke and heat data that guided the technical committee in drafting new standards. It was reported that major improvements had been made in 16,500 schools throughout the United States within a year of the new standards' publication.

As in assembly-type occupancies, exit paths are essential for occupant safety in educational buildings. The fire officer should ensure that doors are not blocked or locked, thereby preventing a safe exit. In addition, the fire officer must verify that built-in fire protection systems are in good working order and are properly maintained. Pay attention to cooking facilities, science labs, and shop areas. Confirm that the evacuation plan is up-to-date and practiced. Staff should also be trained in fire emergency procedures.

Factory Industrial

Factory industrial occupancies include many of the same hazards as business occupancies. They also may utilize processes that are particular to that type of factory; these should be assessed for specific fire hazards. All too often, factories have improperly stored combustibles. They are also prone to having fire protection systems that have been shut off or are improperly maintained and fire doors that are propped open.

■ Hazardous

Special consideration should be given to preventing fires from starting when buildings house hazardous materials. Codes will specify which items and quantities can be stored in a facility, as well as which markings are required for hazardous substances. Signs should be posted that prohibit open flames and smoking. Fire doors and smoke barriers should be in proper working order. Fire protection systems should be regularly inspected and maintained. Fire emergency plans should be in place, along with evacuation plans, which should be practiced.

■ Health Care

Healthcare occupancies include hospitals, nursing homes, and similar occupancies where the occupants are likely to require special assistance to evacuate. Built-in fixed fire protection systems, including automatic sprinklers, smoke detection systems, and sophisticated alarm systems, are required to provide additional evacuation or relocation time. During an inspection of such an institutional occupancy, fire detection systems should be checked for proper maintenance and operation. These facilities should have a fire evacuation plan that should be up-to-date and practiced.

■ Mercantile

Mercantile occupancies include retail shops and stores selling stocks of retail goods. Stand-alone 24-hour convenience markets, department stores, and big-box home improvement and discount stores are all mercantile occupancies. Mercantile fires are responsible for a higher-than-average number of fire fighter line-of-duty deaths. As in the assembly classification, the people who are present in mercantile occupancies may not be aware of the locations of exits. Consequently, ensuring that exits are properly marked and lit is a prime concern during a fire inspection.

Housekeeping is another concern in these occupancies. Frequently, exits and aisles are blocked with merchandise, sometimes to the point of covering sprinklers, standpipe connections, and fire extinguishers. Exits may also be locked. The storage of flammable liquids might be noted as well.

■ Residential

When the fire officer is inspecting residential units, only common areas can be inspected unless otherwise requested by the occupant. Hallways, utility areas, entryways, common laundry rooms, and parking areas are all open for inspection. Residential concerns vary by specific occupancy type; however, in general, exits are a major concern. Fire doors are routinely propped open and may have items stored in their path. Exit and emergency lighting is often not maintained properly.

In residential occupancies, fire protection systems need to be checked for adequate maintenance and operation. Vandals often remove caps and place objects inside standpipes. Valves may be closed, preventing the system from operating. Smoke alarms may have been removed. Hose cabinets may have damaged threads.

Preventing fires from igniting is a prime concern in these buildings. Poor housekeeping, such as bags of trash in hallways, may provide inviting targets for fire setters. Managers of garden apartments often store mowers and gasoline under stairways both inside and outside the structure. Structures in high-crime areas may have bars on windows and doors preventing occupant egress.

■ Special Properties

The "special properties" class includes structures that hold a wide variety of hazards. Before inspecting a special-use occupancy, the officer needs to review applicable code requirements that are specific to the occupancy type. Such occupancies might include the following facilities:

- Covered shopping malls
- Atriums
- Motor vehicle-related occupancies
- Aircraft-related occupancies
- Motion picture projection rooms
- Stages and platforms
- Special amusement buildings
- Incidental use areas
- Hazardous materials

■ Detention

A working sprinkler system is essential in detention facilities because of the inability of the occupants to protect themselves from fires. As a result of the security measures implemented by the facility, the fire department is often delayed in accessing the seat of the fire. Verify that the fire department hook-up is accessible and in working order. Confirm that the standpipe threads match those of the hose used by the department.

Many of these facilities have standpipe systems because of the travel distance to the exterior. Often, they have "house lines," which are preconnected 1½-inch (38-mm) hose lines attached to the standpipe connection. These lines are used by workers on scene and may be located in a locked metal cabinet. Like the standpipe connections, fire extinguishers may be locked in a metal cabinet; however, they should be properly marked.

Typically, these facilities do not have a great deal of combustibles, and the housekeeping is excellent.

■ Storage

Storage facilities often have sprinkler and standpipe systems, which should be checked for proper inspection and maintenance records. Frequently, these records have not been completed as required. In addition, storage facilities regularly stock material too close to the sprinklers. Fire extinguishers may have been removed and set on the ground or may be damaged from forklifts.

These facilities may also store hazardous materials, albeit sometimes for brief periods. During the inspection, the fire officer should determine both the normal amount of on-site hazardous materials and the maximum amount that could be present.

Storage facilities often feature blocked access and egress paths. The need to use space for the business's primary purpose often causes the occupants to block exits with pallets of material. The side exit doors may also be locked, and the exit and emergency lights may not work.

Mixed

Each mixed-use building is considered individually in terms of its requirements, but the building as a whole must meet the most stringent requirement that applies to any of the occupancies inside. Consider a building that has a fireworks storage area in one unit and a barbershop in another unit. Although each occupancy in this building must meet the codes that apply to that specific occupancy, the building in general must meet the most stringent standard—in this case, the requirements that apply to fireworks storage.

Fire Officer II

Emergency Management and Business Continuity Plans

Natural and human-made catastrophes within the last decade have prompted significant changes in NFPA 1600, *Standard on Disaster/Emergency Management and Business Continuity Programs*. The standard focuses on planning, implementation, training and education, exercises and tests, and program maintenance and improvement in disaster/emergency scenarios.

The fire department preincident plan represents one component of a business continuity program. Emergency management and business continuity plans may be established for public, not-for-profit, and private entities, including risk management, security, and loss prevention practices. The fire department is a stakeholder in an organization's program. A business continuity program resembles an incident action plan within the National Response Framework.

Risk Assessment

Public, not-for-profit, and private entities conduct a risk assessment to identify hazards, monitor those hazards, determine the likelihood of their occurrence, and assess the vulnerability of people, property, the environment, and the entity itself to those hazards. These hazards are subdivided into natural hazards, such as an earthquake or hurricane; accidental and intentional human-caused events; and technologically caused events.

Business continuity planning (BCP) practices expanded with the migration from paper to electronic records, with the 1989 San Francisco earthquake requiring the first large-scale recovery of digital records after a disaster. Such planning can prove invaluable. For example, when 15 million ft^2 (1.4 million m^2) of office space was destroyed at the World Trade Center in the terrorist attacks in 2001, most of the affected financial organizations were able to operate from alternative centers minutes after the attack.

As stakeholders, local government agencies may help in BCP efforts by providing an analysis of the impact of such an event on the health and safety of persons in the affected area. The fire department will also be involved in the analysis of properties, facilities, and infrastructure. If the property handles reportable levels of hazardous materials, this analysis may have been done as part of a Local Emergency Planning Council activity.

As part of their BCP practices, senior officials of public, not-for-profit, and private entities will consider the regulatory and contractual obligations as well as the reputation or confidence of the organization. Each entity will evaluate the economic and financial conditions when developing the operations and delivery of services after a disaster or major emergency.

All parties will assess the impact of the potential event on the environment. Some incidents will meet the federal government's definition of "incidents of national consequence," whereas others may have regional or international implications.

Incident Prevention

Based on the risk assessment, a strategy is developed to prevent an incident that threatens people, property, and the environment. The public, not-for-profit, and private entities are expected to evaluate and update their risk assessments regularly, adjusting the level of preventive measures to be commensurate with the predicted risk.

Mitigation

Mitigation includes measures taken to limit or control the consequences, extent, or severity of an incident that cannot be reasonably prevented. The mitigation strategy is based on the results of hazard identification and risk assessment, impact analysis, program constraints, operational experience, and cost–benefit analysis. Such strategies include interim and long-term actions to reduce organizational vulnerability.

Resource Management and Logistics

The emergency/disaster management and business continuity plan document requires that the public, not-for-profit, and private entities provide the resources needed to run the proposed program. To do so, these organizations will need to locate, acquire, store, distribute, maintain, test, and account for services, personnel, resources, materials, and facilities used to support the emergency/disaster management program.

The fire department may provide subject-matter experts or specialized resources as part of the program. It is the responsibility of this organization to develop a detailed resource storage, deployment, and recovery plan. This may include providing reimbursements for fire department time or resources. For example, who will pay for personal protective clothing that has to be replaced after a hazardous materials incident?

■ Publishing the Plan

The emergency/disaster management and business continuity plan document is published in six parts: strategic plan, emergency operations/response plan, prevention plan, mitigation plan, recovery plan, and continuity plan. The emergency operations/response plan assigns responsibilities for carrying out specific actions in an emergency. This plan should link with the fire department incident action plan and be consistent with NFPA 1561, *Standard on Emergency Services Incident Management System*. The mitigation plan establishes interim and long-term actions to reduce the impact of hazards that cannot be eliminated. The recovery plan provides the short- and long-term priorities for restoration of functions, services, resources, facilities, and programs.

■ Training, Exercises, and Evaluation

The public, not-for-profit, and private entities are expected to develop a training program so that all members will obtain the skills to maintain and execute the emergency/disaster management and business continuity program. Periodic reviews, testing, and exercises should take place. In some cases, federal or industry requirements may mandate the frequency and size of a simulation exercise. Postincident analysis and reports, performance evaluation, and lessons learned can be used to review the plan. As a stakeholder, the fire officer may be participating in or evaluating these activities.

You Are the Fire Officer Conclusion

The "My Factory" team does a weekly internal update and provides a monthly report to the fire chief. Captain Davis is responsible for developing emergency response and business continuity plans. As this structure is the first factory of its type in the United States, the code process has not been developed. Neither is there a fire history with large-scale 3D manufacturing in an environmentally friendly converted retail store.

Wrap-Up

Chief Concepts

- The fire officer looks at a building from two different perspectives:
  - Handling an emergency in the building (developing a preincident plan)
  - Performing a fire and life-safety inspection to ensure that the building meets the appropriate fire code requirements
- Even where the role of fire companies does not include code enforcement, fire officers and fire fighters should conduct regular visits to properties to develop preincident plans and to look for correctable fire and life-safety hazards.
- A preincident plan is a document developed by gathering general and detailed data used by responding personnel to determine the resources and actions necessary to mitigate anticipated emergencies at a specific facility.
- The original purpose of a preincident plan was to provide information that would be useful in the event of a fire at a high-value or high-risk location.
- NFPA 1620 provides a six-step method of developing a preincident plan.
  - Step 1: Identify physical elements and site considerations.
  - Step 2: Identify occupant considerations.
  - Step 3: Identify fire protection systems and water supply.
  - Step 4: Identify special hazards.
  - Step 5: Identify emergency operation considerations.
  - Step 6: Identify special or unusual characteristics of common occupancies.
- The preincident plan should provide critical information that could be advantageous for responding personnel, in a format appropriate for emergency conditions.
- Fire code requirements are often adopted or amended in reaction to fire disasters, an approach known as the catastrophic theory of reform.
- A building code contains regulations that apply to the construction of a new building or to an extension or major renovation of an existing building, whereas a fire code applies to existing buildings and to situations that involve a potential fire risk or hazard.
- Most U.S. states and Canadian provinces have adopted a set of safety regulations that apply to all properties, without regard to local codes and ordinances.
- At the local level, fire and safety codes are enacted by adopting an ordinance, which is a law enacted by an authorized subdivision of a state, such as a city, county, or town.
- States and local jurisdictions may adopt a nationally recognized model code, with or without amendments, additions, and exclusions. A complete set of model codes includes a building code, electrical code, plumbing code, mechanical code, and fire code.
- Regulations that applied to a particular building at the time it was built remain in effect as long as it is occupied for the same purpose.
- Built-in fire protection systems are designed as tools to assist fire fighters in fighting a fire. The fire officer should understand how these systems work and recognize how the codes are altered when fire protection systems are in place.
- Automatic sprinkler systems, standpipe systems, and fire pumps are the three primary components of water-based fire protection systems.
- Several types of special extinguishing systems exist: carbon dioxide, dry or wet chemical, Halon, clean agent, or foam.
- A fire alarm system consists of devices that monitor for a fire and notify the building occupants and appropriate personnel when fire is detected.
- The objective of a fire code compliance inspection is to determine whether an existing property is in compliance with all of the applicable fire code requirements.
- The purpose of conducting fire inspections is to identify hazards and to ensure that any violations are corrected.
- Many of the code requirements that apply to a particular building or occupancy are based on its classification. The codes classify a building by construction type, occupancy type, and use group.
- The building itself is classified by construction type, which refers to the design and the materials used in construction.
- Occupancies are classified into use groups based on the characteristics of the occupants, the activities that are conducted, and the risk factors associated with the contents.
- Buildings with significant quantities of hazardous materials may be required to use a marking system.
- In preparing for an inspection, the fire officer should regard the owner or occupant as a professional partner.
- Before conducting an inspection, the fire officer should review the sections of the code that apply to the specific property.
- When preparing for an inspection, the fire officer should review the inspection results from past activities.
- Coordination is needed in departments that implement both a fire prevention division and a company-level inspection program.
- It is good practice to contact the owner or business representative to schedule a day and a time for the fire safety inspection.
- Once you have reviewed the information on the occupancy to be inspected and the applicable fire codes, you are ready to perform the inspection.

Wrap-Up, continued

- Conducting a fire inspection should be approached in a systematic manner.
  - Circle the property and observe all four sides of the building when you arrive. Note any obvious access or storage problems and any new construction since your last visit.
  - Enter the business through the main door and make contact with the appropriate representative.
  - A fire company–level ongoing compliance inspection confirms the built-in fire protection systems are fully operational and ensures that the area is free of fire ignition sources.
  - Wrap up the ongoing compliance inspection/preincident survey by meeting with the owner or designated representative to review the findings from the inspection and issue any required correction orders.
- The inspection/correction report should describe clearly any needed corrections and quote the appropriate sections of the code or ordinance.
- Some fire codes include topics relating to emergency planning and conduct of fire drills, as well as features such as rapid-entry key boxes for fire department use.
- General inspection requirements pertain to access and egress, exit signs and emergency lighting, portable fire extinguishers, built-in fire protection systems, electrical systems, special hazards, and hazard identification signs.
- Each use group classification involves special concerns and considerations in addition to the general concerns. Use groups include the following:
  - Public assembly
  - Business
  - Educational
  - Factory industrial
  - Hazardous
  - Health care
  - Mercantile
  - Residential
  - Special properties
  - Detention
  - Storage
  - Mixed
- The fire department preincident plan is a component of a business continuity program. Emergency management and business continuity plans are established for public, not-for-profit, and private entities, including risk management, security, and loss prevention practices.
- The fire department may provide subject-matter experts or specialized resources as part of the business continuity plan.
- The emergency/disaster management and business continuity plan document is published in six parts: strategic plan, emergency operations/response plan, prevention plan, mitigation plan, recovery plan, and continuity plan.

Hot Terms

Adoption by reference Method of code adoption in which the specific edition of a model code is referred to within the adopting ordinance or regulation.

Adoption by transcription Method of code adoption in which the entire text of the code is published within the adopting ordinance or regulation.

Authority having jurisdiction An organization, office, or individual responsible for enforcing the requirements of a code or standard, or for approving equipment, materials, an installation, or a procedure.

Automatic sprinkler system A system of pipes with water under pressure that allows water to be discharged immediately when a sprinkler head operates.

Business Continuity Planning An ongoing process to ensure that the necessary steps are taken to identify the impact of potential losses and maintain viable recovery strategies, recovery plans, and continuity of services.

Catastrophic theory of reform An approach in which fire prevention codes or firefighting procedures are changed in reaction to a fire disaster.

Construction type The combination of materials used in the construction of a building or structure, based on the varying degrees of fire resistance and combustibility.

Fire prevention division or hazardous use permit A local government permit that is renewed annually after the fire prevention division performs a code compliance inspection. A permit is required if the process, storage, or occupancy activity creates a life-safety hazard. Restaurants with more than 50 seats, flammable liquid storage, and printing shops that use ammonia are examples of occupancies that may require a permit.

Floor plans Views of a building's interior. Rooms, hallways, cabinets, and the like are drawn in the correct relationship to each other.

Fuel load The total quantity of all combustible products found within a room or space.

High-risk property Structure that has the potential for a catastrophic property or life loss in the event of a fire.

High-value property Structure that contains equipment, materials, or items that have a high replacement value.

Masonry wall A wall that consists of brick, stone, concrete block, terra cotta, tile, adobe, precast, or cast-in-place concrete.

Mini/max codes Codes developed and adopted at the state level for either mandatory or optional enforcement by local governments; these codes cannot be amended by local governments.

Mitigation Measures taken to limit or control the consequences, extent, or severity of an incident that cannot be reasonably prevented.

Wrap-Up, continued

Model codes Codes generally developed through the consensus process with the use of technical committees developed by a code-making organization.

Occupancy type The purpose for which a building or a portion thereof is used or intended to be used.

Ongoing compliance inspection Inspection of an existing occupancy to observe the housekeeping and confirm that the built-in fire protection features, such as fire exit doors and sprinkler systems, are in good working order.

Ordinance A law established by an authorized subdivision of a state, such as a city, county, or town.

Plot plan A representation of the exterior of a structure, identifying doors, utilities access, and any special considerations or hazards.

Preincident plan A written document resulting from the gathering of general and detailed data to be used by responding personnel for determining the resources and actions necessary to mitigate anticipated emergencies at a specific facility.

Regulations Orders written by a governmental agency in accordance with the statute or ordinance authorizing the agency to create the regulation. Regulations are not laws but have the force of law.

Risk assessment The process of identifying hazards, monitoring those hazards, determining the likelihood of their occurrence, and assessing the vulnerability of people, property, the environment, and the entity itself to those hazards.

Standpipe system An arrangement of piping, valves, hose connections, and allied equipment installed in a building or structure, with the hose connections located in such a manner that water can be discharged in streams or spray patterns through attached hose and nozzles, for the purpose of extinguishing a fire, thereby protecting a building or structure and its contents in addition to protecting the occupants. This is accomplished by means of connections to water supply systems or by means of pumps, tanks, and other equipment necessary to provide an adequate supply of water to the hose connections.

Use group A category in the building code classification system in which buildings and structures are grouped together by their use and by the characteristics of their occupants.

References and Additional Resources

NFPA reprinted material is not the complete and official position of the NFPA on the referenced subject, which is represented only by the standard in its entirety.

Brannigan, F. L. (2003). *Effect of Building Construction and Fire Protection Systems on Fire Fighter Safety.* Fire Protection Handbook. Quincy, MA: National Fire Protection Association: 7-169 to 167-185.

Bryan, J. L. (1997). *Automatic Sprinkler and Standpipe Systems.* Quincy, MA: National Fire Protection Association.

Bugbee, P. (1971). *Man Against Fire: The Story of the National Fire Protection Association, 1896–1971.* Boston: National Fire Protection Association.

Carson, W. G., and R. L. Klinker. (2012). *Fire Protection Systems: Inspection, Test and Maintenance Manual,* 4th ed. Quincy, MA: National Fire Protection Association.

Fitzgerald, C. P. (2012). *A Risk Assessment of the Meritex Lenexa Executive Park Underground Development.* Emmitsburg, MD: U.S. Fire Administration, Executive Fire Officer Program.

Fowler, D. E. (1990). *Report on the Operations of the San Francisco Fire Department Following the Earthquake and Fire of October 17, 1989.* San Francisco, CA: San Francisco Fire Department.

Hall, J. R. Jr. (2013, June). *U.S. Experience with Sprinklers.* Quincy, MA: National Fire Protection Association.

Hill, H. J. (2012). *Failure Point: How to Determine Burning Building Stability.* Tulsa, OK: Pennwell.

Karter, M. J. Jr., and J. L. Molis. (2012). *U.S. Firefighter Injuries—2011.* Quincy, MA: National Fire Protection Association.

Kerber, S. (2012). "Analysis of Changing Residential Fire Dynamics and Its Implications on Firefighter Operational Timeframes." *Fire Technology* 48(4): 865–891.

Kimball, W. Y. (1966). *Fire Attack 1: Command Decisions and Company Operations.* Boston, MA: National Fire Protection Association.

Metz, E. J. (2013). "New Venues for Discovering Fire and Emergency Services Literature." *Fire Technology* 49(2): 185–194.

Mills, S. E. (2012). *Effective Emergency Operations Preparedness for Recurring Planned Public Events.* Emmitsburg, MD: Executive Fire Officer Program.

National Fallen Firefighters Foundation. (2011). *Understanding and Implementing the 16 Firefighter Life Safety Initiatives.* Stillwater, OK: Fire Protection Publications.

National Fire Data Center. (2012). *National Estimates Methodology for Building Fires and Losses.* Washington, DC: U.S. Fire Administration.

National Fire Protection Association. (1959). *Operation School Burning: The Official Report on a Series of School Fire Tests Conducted by the Los Angeles Fire Department.* Boston, MA: National Fire Protection Association.

National Fire Protection Association. (1961). *Operation School Burning No. 2: Official Report on a Series of Fire Tests in an Open Stairway, Multistory School Conducted June 30, 1960 to July 30, 1960 and February 6, 1961 to February 14, 1961 by the Los Angeles Fire Department.* Boston, MA: National Fire Protection Association.

National Fire Protection Association. (2012). *NFPA Fire and Life Safety Inspection Manual.* Quincy, MA: National Fire Protection Association.

NFPA. Reprinted with permission from NFPA 1021-2014, Fire Officer Professional Qualifications, Copyright © 2013, National Fire Protection Association, Quincy, MA.

Spencer, R. A. (2012). *Putting CIKR Sites on the Map.* Emmitsburg, MD: U.S. Fire Administration, Executive Fire Officer Program.

TriData. (1998). *An NFIRS Analysis: Investigating City Characteristics and Residential Fire Rates.* Washington, DC: U.S. Fire Administration.

Writ, C. L. (1989). "Dillon's Rule." *Virginia Town and City* 24. Richmond, VA: Virginia Municipal League.

FIRE OFFICER in action

Every year, approximately one-third of the tenants in the industrial park and the strip-style shopping center change owners. Updating the preincident plans, Lieutenant Williams determines that some of the newer tenants are living in their strip shopping center stores. Some industrial park tenants are very busy between 9 P.M. and dawn, but are closed up tight during the day. Other industrial park tenants have built mezzanines or internal spaces without permits, inspections, or updates of the built-in fire protection systems. After checking with Captain Davis and Chief Johnson, Williams schedules a meeting with the fire prevention division to see what can be done to ensure occupant survival and fire fighter safety.

This chapter describes two different roles for the company officer. In preincident planning, the officer considers what might happen if the building burns, collapses, or expels hazardous materials. These considerations are developed into an emergency/disaster management and business continuity document that becomes part of an incident action plan in the event of an incident at that building.

The fire officer's second role is to perform fire and life-safety inspections to ensure that the building meets the appropriate fire code requirements. Such inspections also examine the status of built-in fire protection features and general housekeeping.

1. What is a Class III standpipe system?
   A. A system that provides both 1.5-inch (38-mm) and 2.5-inch (64-mm) fire hose connections
   B. A combination sprinkler system with 1.5-inch (38-mm) fire hose connections
   C. A system that provides a 2.5-inch (64-mm) connection with a reducer and 150 feet (48 m) of 1.5-inch (38-mm) fire hose
   D. A Class I system with pressure-reducing valves

2. A structure constructed entirely of wood is a _____ occupancy.
   A. Type II
   B. Type V
   C. Type W
   D. Type III

3. A nightclub is considered to be a(n) _____ occupancy.
   A. mixed-use
   B. business
   C. entertainment
   D. assembly

4. _____ is an emergency management task that includes measures taken to limit or control the consequences, extent, or severity of an incident that cannot be reasonably prevented.
   A. A business continuity plan
   B. Mitigation
   C. Risk assessment
   D. Incident prevention

Fire Captain *Activity*

Captain Davis researches response plans for other manufacturing occupancies, both in the captain's district and in others, and includes the relevant hazard information when writing the department's plan for My Factory.

NFPA Fire Officer II Job Performance Requirement 5.6.1
Produce operational plans, given an emergency incident requiring multi-unit operations, the current edition of NFPA 1600, and AHJ-approved safety procedures, so that required resources and their assignments are obtained and plans are carried out in compliance with NFPA 1600 and approved safety procedures resulting in the mitigation of the incident.

Application of 5.6.1
1. My Factory is placing grass and a garden on the roof of the original big-box store. Develop an operational preincident plan for handling a structure fire with a green roof. The roof is 600 feet by 900 feet.
2. In the new building, My Factory will install a photovoltaic roofing system. Develop an operational preincident plan that addresses the issues with a building-integrated photovoltaic roofing system.
3. Identify two fire suppression issues when operating within a LEED-certified commercial building.
4. My Factory is proposing an innovative flooring system that is not covered in ICC or NFPA code documents. Describe how to determine the relevant fire prevention/operations issues.

Budgeting

Fire Officer I

Knowledge Objectives

After studying this chapter, you will be able to:
- Describe the budget cycle (NFPA 4.1.1) (NFPA 4.4.3). (pp 272–273)
- Identify revenue sources. (pp 274–275)
- Discuss the impact of lower revenue on resources. (pp 275–276)
- Describe the purchasing process (NFPA 4.4.3). (pp 276–277, 279)

Skills Objectives

After studying this chapter, you will be able to:
- Prepare a budget (NFPA 4.4.3). (pp 272–273, 276–277, 279)

Fire Officer II

Knowledge Objectives

After studying this chapter, you will be able to:
- Identify revenue sources (NFPA 5.1.1) (NFPA 5.3.1) (NFPA 5.4) (NFPA 5.4.2). (pp 274–275, 280–281)
- Describe the purchasing process (NFPA 5.4.3). (pp 276–277, 279)
- Identify expenditures (NFPA 5.3.1) (NFPA 5.4) (NFPA 5.4.2). (pp 281–282)
- Discuss bond referendums and capital projects (NFPA 5.1.1) (NFPA 5.3) (NFPA 5.3.1) (NFPA 5.4) (NFPA 5.4.2). (pp 282–283)
- Describe the budgetary process (NFPA 5.1.1) (NFPA 5.3) (NFPA 5.4) (NFPA 5.4.2). (pp 283–286)

Skills Objectives

After studying this chapter, you will be able to:
- Prepare a budget (NFPA 5.4) (NFPA 5.4.2). (pp 274–275, 280–286)

CHAPTER 14

Principles of Fire and Emergency Service Administration (FESHE) Course Outcomes

2. Recognize the need for effective communication skills, both written and verbal. (pp 277, 279, 283–286)
6. Discuss the various levels of leadership, roles, and responsibilities within the organization. (pp 272–273, 275–276)
8. Identify the importance of ethics as it relates to fire and emergency services. (pp 275–276)

You Are the Fire Officer

Quint and Engine 100 were finishing a drill tower experiment using a device that projects a three-dimensional map of a fire building onto an SCBA face piece. This system also provides the incident commander with the location of every fire fighter in the building and data on the fire fighter's pulse, respiration rate, and oxygen saturation level. The transmitter/receiver is part of a PASS device. The project manager told Captain Davis that there was grant funding to provide a three-year demonstration project with the equipment in a fire department.

The crew at Station 100 is enthusiastic about the device and provides feedback to Battalion Chief Johnson. Chief Johnson later informs Captain Davis that the fire chief authorized applying for the grant. Davis and Williams are tasked with preparing the Fire Ground Location (FGL) grant documents.

1. How does a department procure funds?
2. What do you have to consider when preparing a budget?
3. How do you prepare a budget proposal?
4. Which procedures are required to purchase goods and services?

Introduction

A budget is an itemized summary of estimated or intended revenues and expenditures. Revenues are the income of an organization from all sources; expenditures are the money spent for goods or services. Every fire department has a budget that defines the funds that are available to operate the organization for one year. The budget process is a cycle consisting of six steps:

1. Identification of needs and required resources
2. Preparation of a budget request
3. Local government and public review of requested budget
4. Adoption of an approved budget
5. Administration of approved budget, with periodic review and revision
6. Close-out of budget year

Budget preparation is both a technical and a political process. The technical part relates to the calculation of funds required to achieve different objectives, whereas the political part is related to elected officials making the decisions about which programs should be funded among numerous alternatives.

The fire department represents one of the larger sections of a municipal budget, along with schools, police, and public works. The elected officials determine tax rates to generate the needed revenue. They then allocate the available funds among the different organizations and programs. Federal, state, and local regulations also influence the budget process.

Fire Officer I and II

The Budget Cycle

The budget document describes where the revenue comes from (input) and where it goes (output) in terms of personnel, operating, and capital expenditures. Annual budgets usually apply to a fiscal year, such as the year starting on July 1 and ending on June 30 of the following year. Using these dates, the budget for fiscal year 2018, referred to as FY18, would start on July 1, 2017, and end on June 30, 2018. The process for developing the FY18 budget would start in 2016, a full year before the beginning of the fiscal year.

Consider the timeline for a replacement pumper. If the initial request is submitted in August 2016 and it is approved at every stage of the process, the funding will not be authorized until July 2017. **TABLE 14-1** provides a description of the process.

■ Base and Supplemental Budgets

Most municipal governments use a base budget in the financial planning process. The base budget is the level of funding that would be required to maintain all services at the currently authorized levels, including adjustments for inflation and salary increases. A built-in process assumes that the current level of service has already been justified and that the starting point for any changes in the budget should be the cost of providing the same services next year.

Budgets can increase or decrease from the base level. Proposed increases in spending to provide additional services are classified as supplemental budget requests. When budgets have to be reduced, the change is calculated as a percentage of the base budget. A 15 percent budget cut, for example, means that only 85 percent of the base budget

Table 14-1	Fiscal Year 2018 Timeline
August–September 2016	Fire station commanders, section leaders, and program managers submit their FY18 requests to the fire chief. These are for proposed new programs, expensive new or replacement capital equipment, and physical plant repairs. This is the point at which a fire officer would request funds for a new storage shed or a replacement dishwasher. Documentation to support replacing a fire truck or purchasing new equipment, such as a thermal imager or a hydraulic rescue tool, would be submitted, and a proposal for a special training program would be prepared. These documents are prepared by fire officers throughout the organization.
September 2016	The fire chief reviews the requests and develops a prioritized budget, including new items and programs, for FY18. This is the point at which department-wide initiatives or new programs are developed. Purchasing a new radio system would be a new item.
October 2016	The budget office distributes FY18 preparation packages. Included are the forms for personnel, operating, and capital budgets. The senior local administrative official (mayor, city manager, county executive) provides guidelines. For example, if the economic indicators predict that tax revenues will not increase during FY18, the guidelines could restrict increases in the operating budget to 0.5 percent or freeze the number of full-time equivalent positions.
November 2016	All agency heads must submit their FY18 budget requests to the budget director.
December 2016	The budget director assembles all of the proposals. Any request that exceeds the amounts specified in the guidelines must be the result of a legally binding regulation, agreement, or settlement. Other submissions above the limit will need the support of elected officials.
January 2017	Local officials receive the preliminary budget from the budget director. There may be two or three alternative proposals that require a decision from the officials. Those decisions are made, and the proposal is completed. This is the "proposed fiscal year 2018 budget."
February or March 2017	The proposed budget is made available to the public for comment. Many cities make the proposed budget available on an Internet site, inviting public comment to the elected officials. In smaller communities, the proposed budget may be published in the local newspaper.
March or April 2017	Local officials hold a hearing or town meeting to receive input on the FY18 proposed budget. Based on the hearings and on additional information from staff and local government employees, the elected officials debate, revise, and amend the budget. During this process, the fire chief may be called on to make a presentation to explain the department's budget requests, particularly if large expenditures have been proposed. The department may be required to provide additional information or submit alternatives as a result of the public budget review process.
May 2017	Local leaders approve the amended budget. This becomes the "approved" or "adopted" FY 2018 budget for the municipality.
July 1, 2017	FY18 begins. Some fire departments immediately begin the process of ordering expensive and durable capital equipment, particularly items that have long lead times for delivery.
August–September 2017	The budget process for 2019 starts.
October 2017	Informal first-quarter budget review, looking for trends in expenditures to identify any problems. For example, if the cost of diesel fuel has increased and 60 percent of the motor fuel budget has been exhausted by the end of September, there will be no funds remaining to buy fuel in January. Amounts that have been approved for one purpose may have to be diverted to a higher priority.
January 2018	Formal mid-year budget review. The approved budget may be revised to cover unplanned expenses or shortages. In some cases, these changes come in response to a decrease in revenue, such as an unanticipated decrease in sales tax revenue. Expenditures for the remainder of the year have to be reduced.
April 2018	Informal fourth-quarter review. Final adjustments in the budget are considered after 9 months of experience. Year-end figures can be projected with a high degree of accuracy. Some projects and activities may have to stop if they have exceeded their budget, unless unexpended funds from another account can be reallocated.
May 2018	The finance office begins looking at the end-of-year budget projections. Departments with approved funding need to finalize spending funds allocated.
June 2018	All expenditures for 2018 are complete. No new items can be purchased after June 31 on the 2018 budget.
July 2018	The 2018 budget is closed out.

expenditures for the current year will be approved for the next fiscal year. The department would need to reduce its expenditures by more than 15 percent, however, due to inflation and already negotiated salary increases. The department may have to prioritize programs and activities to achieve the budget reductions. Instead of an across-the-board expenditure reduction, there may be elimination of an existing activity, such as the staffing of one engine, to accomplish the budget goal.

Increases in the fire department budget require early notification and support of elected officials. A new program that involves additional employees or a major capital expenditure would require funds that would have to be obtained by increasing revenues or decreasing expenditures from some other part of the budget. These decisions are made by the elected officials. As part of the effort to present a balanced proposed budget, the budget director usually has a good idea of what the elected officials are likely to approve.

Elected officials are both advocates and gatekeepers in developing the local budget. They want to keep taxes down and, at the same time, deliver the services that the voters expect. They do this through direct involvement in the budget preparation process and participation in public hearings. Elected officials often request private face-to-face meetings with fire department leadership to discuss issues that are expected to be controversial.

Revenue Sources

The revenue stream depends on the type of organization that operates the fire department and the formal relationship between the organization and the local community. There is a wide diversity both in organizations and in revenue stream sources. Each type of organization has a different process for obtaining revenue and authorizing expenditures.

Most municipal fire departments operate as components of local governments, such as towns, cities, and counties. A fire district is a separate local government unit that is specifically organized to collect taxes that support fire protection. In some areas, fire departments are operated as regional authorities or an equivalent structure.

■ Local Government Sources

The mix of revenues available to local governments varies considerably because state governments set the rules for local governments. The fire officer needs to know the rules that apply to his or her jurisdiction. General tax revenues can be spent for any purpose that is within the authorized powers of the local government.

Some funds are restricted and can be used only for certain purposes. A <u>fire tax district</u> is created to provide fire protection within a designated area. A special fire protection tax is charged to properties within that service area, in addition to any other county or municipal taxes. For example, the fire tax could be $0.07 per $100 of value for property within the defined area. All of the revenue from the fire tax is dedicated to pay for the provision of fire protection services in the fire tax district.

The use of other revenues can also be restricted. For example, a fuel tax can be used only to build or maintain roads. Likewise, taxes that are levied to repay capital improvement bonds can be used only for that purpose. As demonstrated during the Great Recession of 2007–2009, however, funding sources that are not restricted by federal legislation may be taken from one part of the budget to cover another part of the budget.

The United States Census summarizes the sources of state and local government tax revenues in nine general areas. The top three local tax revenue sources represent about half of the total revenues collected by local governments:

- General sales and gross receipts taxes
- Property taxes
- Individual income taxes

Other sources of revenues for local governments include federal support and charges levied on consumers, such as user fees.

Many fire departments also obtain revenue through direct fees for service. Fire departments that operate ambulances obtain revenue by charging for patient treatment and transportation to offset the operating costs of the service. Direct revenue might also come from administrative fees for fire prevention permits, special inspections, and other services that are provided to individual property owners or contractors. Some fire departments have tried to obtain reimbursement for routine activities, such as responding to automobile accidents or alerts generated by fire alarm systems.

> **Company Tips**
>
> **Direct Mail and Telecommunications**
> Some volunteer fire departments operate successful direct mail campaigns. Sophisticated marketing techniques can be used to customize the mail campaign to increase the size and percentage of returns they receive on their mail-outs. This is more efficient use of volunteer staff than running a pancake breakfast or door-to-door solicitation of funds.
>
> Another revenue source involves the installation of cellular telephone equipment on fire department property. Telecommunications companies rent space from some volunteer organizations and place their equipment on an existing fire department radio tower. Other volunteer departments rent a small piece of land on the fire station property and allow a telecommunications company to construct a monopole.

■ Volunteer Fire Departments

There are as many different ways to fund volunteer fire departments as there are paint schemes on fire apparatus.

Some volunteer departments are operated by municipal governments and are completely supported by local tax revenues, with all capital and operating expenses being handled by the jurisdiction. Such a budget process treats the volunteer fire department as a local government agency.

Other volunteer fire departments are organized as independent 501(c) nonprofit corporations. Some of these volunteer organizations raise their own operating funds through public donations or subscriptions and are entirely independent of local government. In other areas, local tax revenues are allocated to the volunteer corporations. Their relationship with a local jurisdiction may be defined through a memorandum of understanding, a contract for services, or a regional association of independent volunteer fire departments. Some jurisdictions pass local legislation or approve a charter that details the relationship, duties, services, and compensation that the volunteer fire department will provide to the community in return for the tax revenues.

Tax revenues support the entire operation of some volunteer fire departments, whereas others supplement their tax allocation with fund-raising activities. Sometimes the volunteer corporations own and operate the fire stations and apparatus, whereas other volunteer organizations occupy buildings and operate vehicles that are owned by the local government.

Volunteer organizations use a wide variety of fund-raising methods. Direct fund-raising often includes door-to-door solicitations or direct mail campaigns. Many volunteer corporations sponsor activities to generate revenue, such as dinners, bake sales, car washes, raffles, and bingo nights. Some volunteer departments generate revenue by renting out social halls or meeting rooms for private events.

Bingo and Other Gaming Activities

A traditional volunteer fire department fund-raising method is to sponsor bingo or other gaming activities. These activities are state or regionally regulated, with specific procedures for conducting the games and handling the money. Volunteer fire departments have a revenue advantage over other operators because they do not have to pay their members to run the games.

Some departments hold an annual carnival or other entertainment event that provides revenue to the department. Holding the event on fire department property and using unpaid staff allows the fire department to minimize operating costs. Many of these fund-raisers are major events in the community, creating goodwill and strong citizen awareness and support.

Real Estate and Portfolio Management

As a corporation, some fire departments have invested in real estate to generate revenue for fire department operations. Others have developed a financial portfolio that provides interest income to the department.

Lower Revenue Means Fewer Resources

Planned expenditures have to be balanced against anticipated revenues a year or more in advance. If the revenues do not meet expectations, adjustments must be made to reduce spending. If revenues exceed expectations, some extra funds could be available during the year. Local revenue has a history of cyclical behavior, driven by external and internal forces that affect the delivery of services.

Six of the seven recessions that have occurred in the United States since 1969 lasted an average of 10.8 months. The most recent, called "the Great Recession," spanned from December 2007 through June 2009. In the first six recessions, the economy sprang back within a year, restoring municipal revenue within two years. As the U.S. Bureau of Labor Statistics noted in its report *The Recession of 2007–2009*, however, three years after the most recent recession, many parts of the U.S. economy had not shown signs of recovery. Changes in the local economy often result in major changes in the amount and types of revenues collected by local government. If property values fall, property tax revenue decreases. Where local governments receive a percentage of sales taxes, a healthy economy provides increasing revenues, whereas a weak economy results in a revenue drop. A business that shuts down or reduces the number of employees can have a major impact on the local government budget.

Consider the impact of an empty office, retail, or manufacturing building. When the building is occupied and operating, it provides revenue to local government through property taxes and sales or gross receipts taxes. When the building is vacant, there are no sales or gross receipts to be taxed. Thus an empty building generates dramatically lower revenues to local government. For this reason, local officials closely monitor the vacancy rate for office, business, and mercantile space.

When a new business moves into a community, it brings additional revenue to the local government. When a business moves to another jurisdiction, its sales tax revenue also moves to the new jurisdiction. The local jurisdiction could also lose the income tax from people who move with the business. If a property owner goes into bankruptcy, the building could be abandoned and the property taxes may not be paid.

■ Lower Revenue Options

Budget reductions may drastically affect fire department operations. When faced with declining revenues, fire departments have to make difficult choices. Five such options, provided in increasing order of difficulty, are the following:

1. Defer scheduled expenditures, such as apparatus replacement and station maintenance.
2. Regionalize or consolidate services.
3. Privatize or contract out elements of the service provided by the department.
4. Reduce the workforce.
5. Reduce the size of the department.

Defer Scheduled Expenditures

Deferring scheduled expenditures means delaying the purchase of replacement fire apparatus or other expensive equipment. For example, a department that normally replaces pumpers on a 15-year schedule might decide to stretch the life of the rigs to 18 or 20 years. During the recession of the 1980s, Los Angeles and Chicago could not afford to replace aging ladder trucks. Instead, they obtained funding to rehabilitate existing rigs. The equipment was updated, rewired, repainted, and recertified as aerial apparatus. The Los Angeles Service Life Extension Program (SLEP) replaced gasoline motors and manual transmissions with diesel engines and automatic transmissions—a

> **Fire Marks**
>
> **The California Experience with Budget Calamities**
> In June 1978, California's Proposition 13 restricted the rate by which municipalities could raise property tax. This measure stopped the dramatic annual increases in property taxes that had been used to pay for municipal expenditures.
>
> Proposition 13 did not have the predicted impact of devastating local government services. It has, however, significantly restricted municipal growth. California's 4700 special service districts, which include many fire organizations, lack the flexibility that local municipalities had to find other sources of revenue.
>
> Orange County declared bankruptcy in 1994. During the two-year restructuring process, the Orange County Fire Authority was created. The new authority operates under finance rules that are different from those that apply to a county-operated fire department.
>
> In 2008, the City of Vallejo filed for bankruptcy. The bankruptcy voided four labor contracts that covered 400 employees and eliminated the city's obligation to fund retirement programs.
>
> *Governing* magazine reports that 33 municipalities in California had filed for bankruptcy by 2013, with Stockton being the largest and San Bernardino being the latest. Many fire protection districts are implementing extreme reductions in their service delivery due to dwindling funds.

rehabilitation program that squeezed an additional 5 to 7 years of operational service from the organization's rigs.

Privatize or Contract Out Elements of the Service Provided

Privatization means replacing municipal employees with contract employees. The concept is that the cost to the municipality to provide the service should be lower because a private company can operate more efficiently than a local government agency. Trash pickup, vehicle fleet maintenance, and school bus operations are three examples of services that are often contracted out by local government.

Some fire departments privatize or contract out services that are so specialized that a private company may be able to provide an economic advantage to the municipality. Paramedic training, special operations classes, hazardous materials site clean-up, ambulance billing, apparatus maintenance, and communications system maintenance are popular candidates for privatization.

Regionalize or Consolidate Services

Regional or consolidated fire departments are established to increase efficiency by reducing duplication in staff and services. Miami–Dade Fire Rescue Department in Florida started in 1935 as a fire patrol with one employee and one truck reporting to the Agriculture Department. Today this department provides fire suppression, emergency medical services, and specialized operations services in an 1883-square-mile area that includes the unincorporated portions of Dade County and 30 municipalities.

California provides many examples of consolidation. Los Angeles County provides fire protection and paramedic services to 58 cities and towns, as well as the Lifeguard Division. The Orange County Fire Authority contracts with 23 cities.

Regionalization of specialized and infrequently used resources is a way of reducing expenditures. The Mutual Aid Box Alarm System (MABAS) provides a resource response plan for Illinois, Wisconsin, and parts of Indiana, Iowa, and Missouri. MABAS encompasses 42 hazardous materials teams, 26 underwater rescue/recovery teams, and 41 technical rescue teams.

Metrofire is an association of 34 metropolitan Boston fire departments that includes a regional hazardous materials response team as well as evacuation/rehabilitation buses.

Safety Zone

Verify Your Sources

One of the best ways to ensure approval of a budget proposal is to identify a federal regulation, state law, or local ordinance that mandates equipment replacement or delivery of a program. Budget officials know that they must comply with all legally binding requirements, but they are not always eager to spend limited resources on recommended practices, non-binding consensus standards, or suggested best practices. Thus, if the state requires replacement of Level A chemical suits after 5 years, the budget request should identify the specific regulation to ensure funding for this item.

Fire Marks

The Bicentennial Layoff

New York started its fiscal year 1976 by laying off more than 40,000 city employees, including 1600 fire fighters. Although the city hired some workers back, 900 people lost their Fire Department of New York (FDNY) jobs. Some of the laid-off individuals became temporary employees under a federal Housing and Urban Development grant. They were assigned as the fourth or fifth member of an FDNY ladder company, with the job of boarding up roofs and windows of fire-damaged buildings. It would take 2 years before the city could rehire all the laid-off fire fighters.

Reduce the Workforce

Sometimes cities have to reduce their workforces. Fire departments try to protect staffing on emergency response vehicles by reducing positions in administrative and support areas. This approach has limited effectiveness, however, because these personnel represent only a small proportion of the workforce, and because they perform important tasks that support the emergency responders.

One option is to maintain the same number of companies but to reduce the staffing per vehicle. Another option is to limit the number of units in service. Neither option is popular or attractive; however, budget realities sometimes leave no other choices.

Some fire departments temporarily close fire companies for one day or work shift at a time, reassigning the on-duty fire fighters to fill vacancies in other fire stations. This practice may create huge political repercussions if a civilian dies or suffers serious injury in an incident where the nearest fire company was closed for the day.

Reduce the Size of the Fire Department

Departments have been asked to reduce their workforce in response to reduced local revenues. One fire department disbanded 23 engine and ladder companies between 1990 and 2012, reducing its fire fighter workforce by 26 percent. The seven fire companies that were closed in 2000 did not reduce the workforce, because the department instead shifted the staff from these companies to other positions, where they were used to reduce fire fighter overtime and staff additional ambulances.

Closing a fire station or eliminating a fire company often creates a significant and high-profile political issue that mobilizes the citizens as well as the fire fighters. No community wants to lose its neighborhood fire company. The Los Angeles Fire Department's 2011 deployment plan stopped three years of rotating closures and permanently closed 12 engine companies, 6 "light forces" (pumper and tiller truck), and command positions. Compared to 2007, there are 228 fewer fire fighters on duty each day in 2013 (Ward 2011).

The Purchasing Process

Most agencies have established a standardized method for making purchases. The fire officer must understand these policies and procedures of the organization—failure to do so can

Near-Miss REPORT

Report Number: 12-0000280

Event Description: Units responded to a working dwelling fire. The normal staffing for an on-duty platoon was three members. The normal response structure was two members on the engine and one member on the tanker. With a member being on paid time off (PTO), in this incident, one member responded with the engine and one member responded with the tanker.

On arrival, units were confronted with a well-involved single-family dwelling fire. There was also a small natural-cover fire going on at the time. The advanced stage of the fire was threatening the exposures on the D side of the property. The power lines feeding the dwelling came into the dwelling at the A/B corner. The fire was in the attic when the first units arrived.

The local police chief had also responded to the incident. He volunteered to take over the radio traffic, freeing up the chief officer present to assist with the deployment of a portable monitor and handlines. Before transferring the radios to the police chief, the chief officer issued a report over the tactical radio band advising all units to be aware of the hazardous conditions presented by the fire involvement and power lines. The chief also requested an emergency response from the power company.

The chief officer donned protective equipment and assisted with the deployment of the monitor. Once that operation was under way, the chief returned to the apparatus and deployed a 1¾-inch handline with a 15/16 smooth-bore nozzle. This line was deployed on the B side of the structure in an attempt to knock down a large body of fire, which was now present in the dwelling.

Approximately two minutes into the handline operation, the fire burned through the power lines, and the lines dropped to the ground. An energized power line fell within several feet of where the chief officer was operating the 1¾-inch handline.

Lesson Learned: A constant evaluation of a scene must be done to assess hazards. Staffing needs to be such that command officers are not forced to take actions at the tactical level. Safety officers should be put into place early at an incident to monitor hazardous situations. Clear hazards and risk of collapse must be identified and communicated to all individuals operating at an incident.

Courtesy of the National Fire Fighter Near-Miss Reporting Systems (firefighternearmiss.com)

---

have significant consequences for the organization. Because most fire departments are either political entities or nonprofit organizations, they are accountable to the public for the wise use of funds. Audits are used to monitor expenditures and identify potential or actual budget problems. Purchasing violations are sometimes found during the audit process.

■ Petty Cash

Policies and procedures for purchasing typically vary based on how much an item costs. For example, the fire department may allow any items costing less than $100 to be purchased directly in a noncompetitive manner. Often this is done through a petty cash system. In the petty cash system, a member of the fire department acts as the custodian of an amount of cash that is provided by the organization. As members of the fire department purchase small-dollar-value items, the petty cash custodian reimburses them for the expense in exchange for the sales receipt.

When the petty cash custodian is low on cash, the accumulated receipts are turned over to the finance department in exchange for a like amount of cash. In small organizations, the receipts are turned over to the governing body, such as the fire board, which approves the purchases and then writes a check to the petty cash custodian. This check is then cashed and the funds are placed back in the petty cash box. The petty cash custodian should always have the authorized amount on hand in the form of either cash or receipts. The petty cash account is regularly audited to ensure that it is being used properly.

Some personnel tend to think of the petty cash account as an endless fund, which is not true. On each receipt, the appropriate notation is made to identify which account, or part of the budget, is covering the expenditure. Petty cash allows for small purchases to be made without having to obtain purchase orders. Each time petty cash is spent, it lowers the account balances.

■ Purchase Orders

A purchase order is a method of ensuring that a budget account contains sufficient funds to cover a purchase. As with petty cash expenditures, most organizations place a limit on the maximum amount of a purchase order; for example, an organization may allow purchase orders to be used for purchases up to $2000.

The fire officer's role is to acquire the item at the most reasonable cost to the organization. To facilitate this outcome, most organizations allow phone bids for these items. For example, if a fire officer needs to purchase a pair of tires, he or she first determines which size is needed and then calls at least three places to get prices on the tires. The officer takes the lowest price and completes a purchase order (commonly called a "PO"). The purchase order typically requires an authorizing

VOICES OF EXPERIENCE

In the fire service we discuss many aspects of the job that impact us on a daily basis. Perhaps one of the most misunderstood ones, although it is at the core of the fire department's existence, is an understanding of the fire department's budget. Fire officers work with men and women who are truly enthused and energized with ideas, plans, and goals. One of the responsibilities we have as fire officers is to make sure that members within the organization have a basic understanding of budgeting and the components included within budgets.

The questions that are asked are where does funding it come from? How much money do we receive? What can we do with the funds, and how? Budgeting is a complete process that starts at understanding your jurisdiction, be it a municipal setting or a fire district. How do you develop a budget based upon revenues that include taxes, grants, EMS charges, inspections, and plan reviews?

Today's fire service leader must also understand investments and how they work to our advantage or perhaps disadvantage. How do we spend our funds? What are the components of budget preparation based on past practice, bidding for services, bidding for equipment, and economic restraints based upon what we can afford or what we can sustain?

The budgeting process must cover purchase orders based on competitive bids, competent vendors, and service agreements guaranteeing the necessities of operations.

Recently my focus with the budget process has been centered on balance. I have been dealing with increased pension contributions, increased insurance costs, maintaining the framework of a collective bargaining agreement, and assuring that the fire departments has an adequate reserve fund for future projects and purchases. Balance and sustainability without overburdening our tax payers has led to a period of looking at all sources of revenues where we can increase our budgets so that we can continue to provide essential service.

Over the past few years, I have taken a more hands-on approach with assistance from all officers when we review project requests and how they apply to our budget, mindful of our revenue flow. The best example I use is that of a wheel where the center hub holds spokes that attach to the wheel. The wheel is our fire department and the spokes are the functions and services we provide, along with the costs of providing these service. The hub is the budget, and without a solid budget we will be stressed to provide extended services, let along essential ones.

> **Budgeting is a complete process that starts at understanding your jurisdiction, be it a municipal setting or a fire district.**

Jim Grady III
Chief
Frankfort Fire Protection District
Illinois Fire Chiefs Association
Illinois Fire Service Institute
Frankfort, Illinois

signature by an official who has control over the budget area from which the funds are paid. In this example, it might be the chief. The purchase order is then entered into the purchasing system, which debits the maintenance account.

The fire officer then takes the purchase order to the location that has the tires and the sale is transacted. A copy of the purchase order is given to the vendor, who attaches the sales receipt to it and sends it to the fire department's finance department for reimbursement. In effect, the vendor allows the fire department to charge the merchandise until the purchase order is paid. For its part, by issuing the purchase order, the fire department indicates to the vendor that there are funds available for the purchase and the purchase has been approved. Not all vendors accept purchase orders, however.

■ Requisitions

For purchases that exceed a predetermined amount, such as $2000, a requisition is required, rather than a purchase order. This occurs for large purchases, such as a new hydraulic extrication tool. A requisition has even more stringent requirements than a purchase order.

The requisition differs from a purchase order in that the exact amount of the purchase is not known. Instead, the department requests that a specific amount of funds be encumbered, or set aside from the budget account, that will more than cover the cost of the purchase. In the case of the rescue tool, the department might set aside $15,000. A bidding process will be followed. Once it is complete, the requisition will allow for payment of the item.

■ The Bidding Process

Once funds have been encumbered, the fire department must go about the process of making the purchase. This is done in one of two ways. For smaller items, the fire department may develop specifications for bids. The second method, which is often used for very large or complex purchases, is a request for proposal (RFP).

To purchase an item based on specifications (commonly known as "specs"), the fire department writes up exactly what it desires in the product. For example, it might specify that a portable power unit for a fire truck must not exceed a size of 18" × 14" × 21" (46 × 36 × 53 cm). With a spec, every requirement must be met for the vendor to be considered. In this example, if a vendor wanted to bid but its power unit was 19" × 13" × 20" (48 × 33 × 51 cm) in size, its bid would be rejected. This type of restriction may be important if the compartment in which the power unit will be placed will accommodate only the specified size. Some fire departments use this method to ensure that only one vendor or product can meet the specifications; however, this is an illegal and unethical practice. Specs should be used to identify which criteria must be met for the purchase to be most effective and efficient for the fire department.

Once developed, a bid sheet listing the specifications is sent to vendors for pricing. Most organizations have a bidding list of vendors that wish to do business with them. The vendors place their names and addresses on the list so each time a bid comes up, a copy is sent to them for consideration. Local administrative rules may also require that a notice of bid be placed in the local paper. Other organizations place notices of bids in trade magazines and papers. The Internet also is becoming a method of getting information out to the public about upcoming bid processes.

Often, the use of specifications eliminates virtually every potential vendor or greatly increases the price because only more expensive items meet the specifications. Sometimes, a bid specification limits the bidders. For example, the purchase of a mobile data computer (MDC) system for use in the fire apparatus would likely use the RFP method rather than the bid specification method because of the complexity of the technology.

An RFP provides a formal process to solicit goods or services in a procedure that is determined by the authority having jurisdiction (AHJ). In an RFP, the fire department gives the general information about what is desired and allows the vendor to determine how it will meet the need. For example, the mobile data computer RFP might require that the system provide coverage to 99 percent of the city. One vendor might opt to place 10 antenna towers to deliver the required coverage, whereas another might opt for only 7 antenna towers but place them at a higher elevation. Both ideas would meet the need of the fire department. The fire department also determines how the proposals will be evaluated. For example, the price might be worth 50 percent of the overall consideration, whereas delivery time might be worth 5 percent. Once the RFP is developed, it is sent out to potential vendors for bidding.

Whether the process uses an RFP or bid specifications, vendors are given a specific amount of time to reply. Their replies are placed in a sealed envelope indicating the bid number on the outside of the envelope. At a time that was outlined in the bid specifications or RFP, all bids or proposals are opened in public view. With a bid process, the contract is then awarded to the lowest bidder and the amount is attached to the requisition for payment when delivery is made.

When an RFP is used, each proposal is evaluated to determine whether it best meets the needs of the department. The proposals are typically assessed based on performance as well as price. Once each performance requirement has been addressed and the price is considered, the RFP is awarded to the vendor with the highest score. The price of the winning proposal is added to the requisition for payment when the item is delivered.

In recent years, there has been a trend toward developing intergovernmental or interagency cooperative arrangements to streamline purchases and leverage greater discounts. The federal government, for example, issues many purchasing contracts for a wide range of equipment. Most vendors honor these rates for other governmental agencies, which in turn streamlines the process because the bidding process has already occurred. Local organizations can also band together to purchase common items. For example, a group of fire departments may jointly develop an RFP to purchase SCBA. This practice ensures a lower bid because the group is purchasing items in greater quantity, and it allows all agencies to use like equipment.

Fire Officer II

Revenue Sources

Grants

Municipal and volunteer fire departments can also obtain funds by applying for grants. One of the larger sources of grant funds is the Assistance to Firefighters Grant Program, which distributes federal funds to local jurisdictions.

Although the Assistance to Firefighters Grant Program is the largest fire service–oriented program in the United States, hundreds of smaller grant programs are offered by the federal government, state governments, private institutions, nonprofit organizations, and for-profit companies. Grants are competitive and require the local fire department to meet specific eligibility and documentation requirements. Many grant programs also require the local jurisdiction to provide a percentage of the total funding in the form of matching funds.

Some state agencies offer grants to local responders from specially designated funds, such as a fee on motor vehicle permits, a surcharge on traffic fines, or a tax on fire insurance premiums.

The U.S. Fire Administration provides a four-step method to develop a competitive grant proposal:

1. Conduct a community and fire department needs assessment.
2. Compare community needs to the priorities of the grant program.
3. Decide what to apply for.
4. Complete the application.

The process of following the four steps in developing a grant proposal requires compliance with the grant instructions. To qualify, applicants must follow a specific set of procedures and provide a detailed level of financial and operational information. These requirements may seem excessive and intrusive, but the grant administrators have to ensure that the money is spent wisely and within legislative guidelines.

The Assistance to Firefighters Grant Program applications are evaluated by a peer-based committee that ranks them on the basis of preestablished criteria. To be approved, the application must make a request that matches the goal of the grant program, show that the proposal is cost efficient, and demonstrate strong community and financial need for the requested goods or services.

When seeking grant funding, the first step is to determine whether a grant program exists that covers your problem area. Once you have identified the program, determine whether your organization is eligible under the rules of the grant program and which process is used to apply for the grant.

A grant application must describe how the department's needs fit the priorities of the grant program. The grant applicant needs to analyze the community, conduct a risk assessment, evaluate the existing capabilities of the fire department, and identify the department's needs.

There is strong competition for the limited amount of grant funds available, so the applicant must provide a sound justification for why the organization should receive the funding. Use the information gathered in the needs assessment in a narrative to describe the problem in a concise but detailed manner.

In addition to the needs assessment, the grant application must demonstrate the department's financial need and explain why it needs outside assistance to acquire these resources. The application should show that efforts have been made to obtain the funds from other sources. In describing the current financial situation, include factors such as an eroding tax base, expanding community growth, local/state legislation that restricts taxes, and any significant local economic problems, such as the loss of a major employer.

The application also needs to show that this solution provides a benefit at the lowest possible amount of funding. Factors include collaborating with other organizations to share expensive equipment. For the Fire Fighter Status and Location project described in the chapter-opening case study, the grant opportunity comes from a private source.

Nontraditional Revenue Sources

Fire departments have solicited donations from a targeted business or industry for a specific purpose. In one area, a resort hotel–motel association provided funding to purchase an additional aerial platform truck that serves the high-rise hotel district. Equipment for a hazardous materials team might be donated by a company in the community that produces hazardous materials. One fire department offered to sell advertising on its vehicles.

Cost Recovery

Local or state regulations may allow a fire department to recover the extraordinary expenses incurred in responding to a hazardous materials incident, particularly decontamination, spill clean-up, and recovery costs. The department can invoice the responsible party for extraordinary charges, such as overtime and replacement of expendable supplies (e.g., absorbents and neutralizing agents). In addition, the carrier may be responsible for the expense of repair or replacement of durable equipment, such

---

Getting It Done

Justifying New Gas Monitors

A fire department is preparing a grant request to purchase four-channel gas meters to replace flammable vapor detectors assigned to ladder and rescue companies. The old detectors can measure the concentration of a flammable vapor only in relation to the lower explosive limit, whereas the newer detectors also measure the percentage of oxygen and carbon monoxide in the atmosphere. As part of the needs assessment to justify a grant for these new monitors, the fire department would identify the increase in service calls to check on carbon monoxide alarm activations.

as contaminated monitoring equipment, chemical protective suits, fire fighter protective clothing, and fire suppression tools.

Cost recovery also applies to some special services supplied by fire departments. For example, if a movie production company is shooting a film that involves pyrotechnics, the fire department might provide an engine company to stand by for several hours. The cost of overtime to staff the unit and an hourly rate for the use of the apparatus could be charged to the production company.

Expenditures

The annual budget for the fire department describes how the funds available are spent during the year. In the case of a municipality, the fire department budget is usually a part of the overall budget for that governmental entity. The format of the budget should comply with recommendations of the Governmental Accounting Standards Board (GASB), which is the standards-setting agency for governmental accounting.

The local government budget typically includes a system of accounts that classifies all expenditures within certain categories and complies with the generally accepted accounting principles as outlined by the GASB. In most cases, the accounting system uses a database system to support a line-item budget. A line-item budget is a format in which expenditures are identified in a categorized line-by-line format. The accounting system may have a complex numbering system, such as 02-12301-876543, that includes a category for every type of expenditure. In this example, the line item describes the following:

- 02: The first two numbers identify the fund. In the example, the 02 indicates that this money is coming from the general fund.
- 12301: The second part (the five-digit number) identifies the department and the division or subdivision where the money is being spent. Called the "object code" in some systems, this part of the example says that the money is allocated to the Fire Department (12) and to the Suppression Division (301).
- 876543: The third part (the six-digit number) provides a detailed description of what the money is being used for. Called the "subobject code" in some systems, this one says that the money is to be used for replacement fire hose.

The accounting system allows budget analysts and financial managers to keep track of expenditures throughout a municipal government. For example, law enforcement, public works, schools, and fire departments all purchase work uniforms. The database system would allow a budget analyst to determine how much the municipality spent on all uniforms and to break down expenditures by department and type of uniform. This could be valuable information when one is considering the savings that could be obtained by purchasing all uniforms from one vendor.

Fire department expenditures are generally divided into three general areas: personnel costs, operating costs, and capital expenditures. The accounting system allows budget analysts and managers to keep track of how much money is spent on fire fighter salaries (personnel costs), fuel for vehicles (operating costs), and construction of new fire stations (capital expenditures).

■ Personnel Expenditures

More than 90 percent of a career fire department's budget is likely to be allocated to salaries and benefits. This includes fringe benefits, such as pension fund contributions, worker's compensation, and life insurance. Civil service regulations, local administrative decisions, and labor contracts often determine the cost of these benefits. For example, the worker's compensation rate is determined by the history of claims and payments. Fire fighters have a high worker's compensation rate when compared to other municipal employees.

Fringe benefits also reflect the estimated cost of providing sick and annual leave benefits to the employee. Many municipalities have a sliding scale, in which the amount of leave earned per pay period increases as the employee gains seniority. The base cost of this fringe benefit equals the amount of the employee's accrued leave. In many cases, the actual cost to the fire department also includes overtime to pay for another employee to fill the vacancy.

■ Operating Expenditures

The operating budget covers the basic expenditures that support the day-to-day delivery of municipal services. Uniforms, protective clothing, telephone charges, electricity for the fire stations, flashlight batteries, fire apparatus maintenance, and toilet paper are all examples of purchases that would be classified as operating expenses. These funds are allocated in categories that allow for some flexibility throughout the year. For example, a fire department might be authorized to spend $250,000 on protective clothing during the year. The actual numbers of coats, pants, helmets, boots, and gloves that would be purchased would depend on the needs and the cost per item at the time of purchase. When a budget needs to be trimmed, operating expenditures are usually the first area to be considered as a target for cost-cutting efforts. For example, employee training, travel, and consulting expenses are often eliminated when revenue falls below expectations.

The cost of operating a vehicle is calculated on a per-mile basis that includes fuel, scheduled maintenance, and anticipated repairs. Maintenance costs for fire department vehicles are high because they are loaded with additional equipment, such as emergency lighting and siren systems, two-way radios, and computers. Large fire apparatus, which is expensive to repair and accumulates low annual mileage, may have an operating cost of several dollars per mile. If the actual fleet management costs exceed the budget projections because of high fuel costs or unanticipated repairs, the additional amount is sometimes charged back to the fire department at the end of the fiscal year. That adjustment is included in the third-quarter adjustments.

As part of their operating expenditures, many municipalities include a per-mile vehicle replacement fee applied to automobiles, police cars, buses, ambulances, and light-duty vehicles. The precise amount is calculated by estimating the anticipated replacement cost and the projected number of years that the vehicle will be used. This accounting technique is also used by fire departments that have dozens of pumpers and aerial apparatus and a fire apparatus replacement schedule. In most cases, the replacement of a pumper, aerial apparatus, or specialized suppression rig will be a capital item.

Replacement uniforms and protective clothing are another expense included in the operating budget **FIGURE 14-1**. For

FIGURE 14-1 Replacement protective clothing is an operating expense calculated in the budget.

example, the fire department might budget a lump sum of $150 per fire fighter to cover the annual cost of cleaning and repairs to protective clothing and up to $400 per fire fighter for uniforms.

Mandated training, such as hazardous materials and cardiopulmonary resuscitation recertification classes, is also factored into the fire department's operating costs. That figure includes the cost of the training per fire fighter per class. The cost to pay another fire fighter overtime to cover the position during the mandated training could also be included.

■ Capital Expenditures

Capital expenditures comprise purchases of durable items that cost more than a predetermined amount and will last for more than one budget year. Local jurisdictions differ on the cost and service period used to define capital expenditures. Items such as computers, hydraulic rescue tools, washing machines, and

Assessment Center Tips

Remember Annual Expenses

When discussing or responding to a budget proposal, remember to focus on the annual expenses in addition to the initial purchase cost. The annual costs fall into two areas: continuity and personnel.

Consider the purchase of semi-automatic cardiac defibrillators for three fire companies. What happens when one of the devices breaks down? Continuity considers the need for a spare unit to be used when one of the new devices is broken or unavailable, as well as the cost of maintenance and repairs. Additional costs would include the initial training of fire fighters to use the devices, plus annual recertification classes.

Personnel costs are the largest expense for career fire departments. An additional ladder truck might cost $800,000 to purchase, but the annual cost of additional personnel to provide staffing for that vehicle could easily exceed $1 million. This cost could be difficult to justify for a company that will make fewer than 100 responses per year.

FIGURE 14-2 Two examples of equipment purchased as capital items. A. Computer. B. Tool.

self-contained breathing apparatus (SCBA) are examples of equipment purchased as capital items **FIGURE 14-2**. In general, the municipality establishes a unique inventory record that documents the purchase price, source, assignment, and disposal of each capital item.

Sedans, pick-up trucks, and sport-utility vehicles (SUVs) are usually part of the fire department's capital budget. Specialized vehicles, such as heavy fire apparatus, are also capital equipment, but they are often included in a special section of the capital budget because of their cost and complexity. In larger departments, replacement vehicles are purchased from a special set-aside fund, and only additional vehicles are included in the supplemental capital budget.

Capital improvement projects encompass the construction, renovation, or expansion of municipal buildings or infrastructure. These expensive projects are often funded through long-term loans or bonds issued by the municipality.

Bond Referendums and Capital Projects

Large capital improvement projects, such as new fire station construction or major renovations, are funded through bond programs. A <u>bond</u> is a certificate of debt issued by a government or corporation; the bond guarantees payment of the original investment plus interest by a specified future date. The same type of funding mechanism is also used to build roads, water distribution systems, sewer systems, parks, libraries, and similar public facilities. Voters must approve this type of expenditure through a local referendum. If the voters approve, the municipality is authorized to borrow money from investors by issuing bonds up to a set amount. The authorization to sell the bonds is usually valid over a period of 5 to 10 years.

The bonds are repaid over a period of 10 to 30 years and return a fixed interest rate to investors. Funds obtained by issuing bonds are classified as a special revenue source and require compliance with a different set of accounting controls and reporting procedures. Annual status reports are required, and specific restrictions are placed on which items can be purchased with the bond revenue. If a city is issuing $100 million in bonds to be repaid over 20 years, a fraction of a percentage difference in the interest rate can have a significant impact to the taxpayers.

Bond funding allows taxpayers to actually pay for the facility while it is being used, just as a mortgage allows a family to live in a home while they are paying off their loan. The repayment period should be at least equal to the anticipated life span of the facility. Bonds are repaid by collecting a special property tax or, if the facility will produce revenue, such as an airport or convention center, the bonds may be repaid from future income.

When bonds are used to build a new fire station, the authorized amount often includes the land, site improvements, building construction, permits, architectural and engineering fees, and furniture and equipment that go into the building, as well as the apparatus and equipment that will be assigned to the new station.

Occasionally, a bond program is used to fund a large fire department apparatus purchase. For example, the St. Louis Fire Department used a bond to cover the cost of an entire fleet of new apparatus in 1986. That large purchase allowed for the replacement of all existing pumpers and ladders with 30 new quints to establish the "Total Quint Concept." In fiscal year 2000, the city authorized another bond program so that it could purchase 30 replacement quints.

Navigating the Budgetary Process

Using the example presented in the chapter-opening scenario, let's navigate through the budgetary process. In the case study, the fire department is preparing a budget so that it can get funding from a private corporation to establish a Fire Fighter Safety and Location program.

■ Developing a Budget Proposal

Because this equipment is a new device providing a new service, the first step in the budget proposal process is to describe what the new unit will do and what the impact will be if the unit is not funded. The description should consider the audience that will review, authorize, and approve the application. Provide enough information in an understandable format to ensure understanding of the goals that the program will meet.

■ Overview of Fire Ground Location Program

The following is the narrative that is part of the budget submission. It describes what the <u>Fire Ground Location (FGL) program</u> will do, the impact of the FGL program on current operations, and the consequences if the new program is not funded.

The FGL is a two-person unit that responds to working incidents to track fire fighters operating in immediately dangerous to life and health (IDLH) environments and to monitor the medical status of members who rotate through the rehabilitation sector during an event. The FGL operates every hour of every day.

The FGL has two responsibilities at a working incident. Its first responsibility is to deploy the field antenna system to capture the three-dimensional global positioning system (GPS) data being transmitted from the integrated personal alert safety system (PASS) units that are activated on the self-contained breathing apparatus. The FGL operator will advise the Safety Sector of any individual who appears to be in distress or, upon activation of a PASS alarm, the location of the fire fighter within the incident.

The FGL's second responsibility is to track the vital signs of fire fighters who enter and exit the rehabilitation area during an incident. The primary goal is to identify those who may need immediate rehydration or medical attention. A secondary goal is to assure that the fire fighters are ready to leave the rehabilitation area by confirming that their vital signs conform with medical standards established by the operating medical director.

If the FGL program is not funded, the department will not reduce the identified risk factors that lead to fire fighter line-of-duty deaths and disability. The department will be unable to improve its ability to rescue a fire fighter in trouble within a burning building. The fire department will also be unable to provide early detection of life-threatening health or environmental conditions that may affect fire fighters.

■ FGL Annual Personnel and Operating Expenditures

This part of the budget submission shows the annual expenses to operate the FGL. Both personnel and operating expenses will be incurred in this example.

Personnel Expenses

The FGL will have a captain, three lieutenants, and four technicians. In this example, the personnel cost is described in two areas. The first area is the direct salary for each position. Most civil service pay classification systems use a pay level that includes steps tied to seniority. Budget protocol requires that new positions reflect the average seniority step that the incumbents occupy. That means that the captain is at seniority step 7; the lieutenant is at seniority step 5; and the technicians are at seniority step 3.

The second area is the calculation of the fringe benefits for each employee position. It is calculated as a percentage of the employee's salary. For the FGL example, fire fighter fringe benefits account for 28.7 percent of the salary. That means that every $100 of salary actually costs the city $128.70.

Operating Expenses

The cost of operating the FGL unit entails three general areas: vehicle cost, office operations, and personnel training. City fleet maintenance has determined the cost to operate and replace a small, all-wheel-drive emergency services hybrid SUV is $0.84 per mile. This amount covers the fuel and scheduled maintenance of the vehicle. It also includes a per-mile charge that is designed to cover the replacement of the vehicle when it reaches the end of its scheduled service life. Vehicle service life is determined in three ways: years of service, miles accumulated, and specific vehicle cost. It is estimated that the unit will accumulate 17,000 miles per year. The city replaces small emergency service vehicles at 80,000 miles or after 10 years. The FGL will need replacement after 4.7 years.

Office operations include rent, telecommunication charges, and equipment maintenance. In this example, the FGL is operating out of a fire station, so rent is zero. The office has two hard-wired lines to provide service for the phones, Internet access, and fax machine. The municipality's information technology (IT) office has an annual fee per hard-wired line that covers telephones and Internet broadband access. IT also handles computer maintenance and repairs. As with the vehicle, the annual maintenance and operating charge covers the anticipated annual costs of maintenance, software upgrades, and repairs. It also includes an amount that will allow replacement of the computers on a 5-year cycle. The FGL has two desktop computer workstations, one vehicle-based desktop workstation, and four laptop computers.

Operating expenses also include the expenses of mandated annual fire fighter training. The cost reflects the per-person cost of training, for both the tuition and the position replacement overtime to cover the FGL person who is receiving the training. This expense covers training required by federal or state regulations, such as annual bloodborne pathogen and hazardous materials refresher training.

■ FGL Capital Budget

The grant request would pay for the durable equipment needed to start the FGL program. Many of the required items would be purchased via federal, state, or regional purchasing contracts. The budget summary shows the entire first-year capital expenditures to start the program, including the response vehicle, the administrative office, the monitoring equipment, and the impact of eight new fire department positions.

FGL Vehicle

During the planning process with Chief Johnson, the team determines the appropriate type of vehicle. The state offers a range of vehicles for purchase. The following options are considered for the FGL:

- Compact four-door sedan, $15,470
- Midsized hybrid four-door sedan, $17,786
- Full-sized sedan, police pursuit, $23,417
- Small all-wheel-drive hybrid SUV, emergency services, $19,459
- Large four-wheel-drive SUV, emergency services, $33,948

Chief Johnson recommends that the FGL vehicle should be the same as other staff and command vehicles. In addition to the vehicle, equipment is needed to convert a small SUV into a fire department response vehicle, including emergency lights/siren, mobile fire radio and computer dispatch terminal, two portable fire radios, and a semi-automatic cardiac defibrillator. Depending on the local budget process, these items may need to be identified in a line-item format consistent with budget documentation protocol FIGURE 14-3. More than half of the FGL vehicle cost is devoted to the specialized equipment that goes on or into the vehicle.

FGL Administrative Office

Because the FGL program will be funded by a grant that requires incident reports and deployment research, the budget includes the cost of establishing two administrative workstations. The workstations will go into an available space at Fire Station 100. The capital equipment represents all of the durable items needed to establish this office: desks, chairs, computers, fax machine, and printer.

FGL Uniforms and Protective Clothing

The expense of fire fighter protective clothing and the initial uniform issue are included in the initial-year capital expenditures. Because the FGL program creates the new positions, the expense shows up here.

FGL SCBA Interface

As part of his or her personal protective equipment, every fire fighter is provided with a personally fitted SCBA face piece. The FGL interface is a small battery-powered device that is installed in this face piece. Fire fighters need to synchronize the SCBA interface with the PASS device during the morning equipment check. This allows the PASS device to identify the person using the SCBA to the FGL monitoring and mapping programs. Such matching facilitates quick identification of who is in trouble if a PASS alarm is activated or if the fire fighter's vital signs indicate a problem.

■ Ask for Everything You Need

Fire departments try to do the most they can with the fewest resources. When submitting a proposal for a new project or service, assume that there are no existing resources and ask for everything. This includes asking for enough tools to ensure continuing operation. For example, the laptops, as configured to work with the FGL communications, cost $6113 each. Only seven times in the past 10 years did this fire department respond to two simultaneous fire-ground operations. However, ordering three properly configured laptops provides the redundancy necessary to ensure continued operations in such a scenario.

■ Cost Recovery and Reduction

It will cost more than $525,000 per year to run the FGL program in a city with 55 to 70 on-duty fire fighters—an estimate obtained by applying the results of the Worcester Polytechnic Institute's Precision Indoor/Outdoor Personnel Location Project. The cost recovery is realized through faster completion of tasks in a high-risk environment as fire fighters use the mapping system to accomplish fire-ground tasks. In addition, safety is increased with identification of a fire fighter in medical or situational distress. A declining oxygen level will trigger an alert in the FGL much sooner than a fire fighter calling a mayday. This is important because carbon monoxide poisoning impairs perception and reaction times. Once a problem is detected, the FGL mapping will immediately locate the fire fighter, which in turn reduces the time needed to remove the

Fire Department - Operations
New program: **Fire Ground Location**

Year 1: Capital Expenditures		Unit	Total	Source or comments
FGL Vehicle				
1	Small AWD SUV hybrid, emergency services	$ 24,765	$ 24,765	State vehicle contract
1	Mobile fire radio and computer terminal	$ 15,525	$ 15,525	Communications
1	Fire emergency lighting package	$ 2,765	$ 2,765	Apparatus
1	Safety striping and graphics	$ 411	$ 411	Apparatus
1	FGL monitoring/mapping computer (vehicle)	$ 8,700	$ 8,700	University quote
3	FGL field antenna receiver system	$ 3,360	$ 10,080	University quote
1	Semi-automatic defibrillator	$ 875	$ 875	Operations
2	Portable 2-way fire radios	$ 6,346	$ 12,692	Communications
	Total FGL vehicle cost:		$ 75,813	
FGL Integrated PASS devices				
84	Integrated PASS device and FGL transmitter	$ 877	$ 73,668	Vendor/university quote
246	SCBA interface device	$ 375	$ 92,250	Vendor/university quote
60	Trade-in existing PASS device	$ (175)	$ (10,500)	Vendor quote
	FGL device deployment		$ 155,418	
FGL Field montoring stations				
3	Configured laptops with wireless broadband	$ 6,113	$ 18,339	Vendor/university quote
	FGL Field montoring stations		$ 18,339	
FGL administrative office at Fire Station 100				
2	cubicle work stations	$ 777	$ 1,554	Facilities catalog
2	Computer work stations - model B	$ 3,200	$ 6,400	IT catalog
2	Smartphones	$ 456	$ 912	IT catalog
1	Workstation scanner/printer - model B	$ 1,325	$ 1,325	IT catalog
1	File/storage cabinet	$ 200	$ 200	Facilities catalog
	FGL administrative office		$ 10,391	
New uniformed positions - maintain two person team 24 hours every day				
8	Protective clothing ensembles	$ 2,670	$ 21,360	Operations
4	Initial fire officer uniform issue	$ 865	$ 3,460	Operations
4	Initial firefighter uniform issue	$ 709	$ 2,836	Operations
	Initial uniform position PPE/uniforms:		$ 27,656	
	Year 1 capital expenses:		$ 287,617	

Annual Expenses
Personnel

		Unit	Total	Source or comments
1	Captain at pay step 28-7	$ 56,023	$ 56,023	Salary
3	Lieutenant at pay step 26-5	$ 46,785	$ 140,355	Salary
4	Technicians at pay step 22-3	$ 39,209	$ 156,836	Salary
	Fringe benefits		$ 101,372	28.7% of salary
	Total annual personnel expenditures:		$ 454,586	
Operational				
12	Replacement of damaged/destroyed PASS device	$ 877	$ 10,524	Vendor (12% of units)
81	Replacement of SCBA interface devices	$ 375	$ 30,375	Vendor (33% of units)
1	Replacement of FGL field antenna	$ 3,360	$ 3,360	University estimate
1	FGL vehicle maintenance/operating/replacement	$ 14,280	$ 14,280	$.84 per mile @17K/year
2	Telecommunications fee per line	$ 1,825	$ 3,650	intranet/phone service
2	Paper/office supplies per cubicle	$ 365	$ 730	Budget office
12	Mandated fire department training	$ 275	$ 3,300	Academy, per position
7	computer maintenance and replacement	$ 570	$ 3,990	Information technology
2	portable radio maintenance and replacement	$ 1,000	$ 2,000	Communications
1	Mobile fire radio/MDT maintenance/replace	$ 1,500	$ 1,500	Communications
	Total annual operating expenses:		$ 73,709	
	Annual operating expenses:		$ 528,295	
	Year One total expenditure:		$ 815,912	

FIGURE 14-3 Summary budget submission sheet for the FGL program.

distressed fire fighter from an IDLH environment. This capability may change a line-of-duty death to a near miss.

Overall, the cost recovery is achieved by reducing the severity level when a fire fighter is in distress, reducing hospital time, and avoiding a lasting disability. The safety of the rescue team is improved because knowing the location of the fire fighter reduces the amount of time spent searching within a hostile environment.

You Are the Fire Officer Conclusion

The Fire Ground Location (FGL) grant application is submitted for review by the city and university before it goes to the organization offering the grant. There is some concern that this system may be another toy that will not work as advertised. Chief officers are interested in the deployment but wonder what will happen when the grant runs out in three years.

Wrap-Up

Chief Concepts

- Every fire department has a budget that defines the funds that are available to operate the organization for one year.
- The budget document describes where the revenue comes from (input) and where it goes (output) in terms of personnel, operating, and capital expenditures.
- Most municipal governments use a base budget in the financial planning process. The base budget is the level of funding that would be required to maintain all services at the currently authorized levels, including adjustments for inflation and salary increases.
- The revenue stream depends on the type of organization that operates the fire department and the formal relationship between the organization and the local community.
- The mix of revenues available to local governments varies considerably because state governments set the rules for local governments. The fire officer needs to know the rules that apply to his or her jurisdiction.
- Many volunteer fire departments are organized as independent 501(c) nonprofit corporations. Some of these volunteer organizations raise their own operating funds through public donations or subscriptions and are entirely independent of local government.
- Planned expenditures must be balanced against anticipated revenues a year or more in advance.
- Sometimes the results of budget reductions drastically affect fire department operations. Fire departments have to make difficult choices when faced with declining revenues.
- Most agencies have established a standardized method of making purchases. Because most fire departments are either political entities or nonprofit organizations, they are accountable to the public for the wise use of funds.
- The petty cash system allows for a member of the fire department to be the custodian of an amount of cash that is provided by the organization.
- A purchase order typically requires an authorizing signature by an official who has control over the budget area from which the funds are paid.
- For purchases that exceed a predetermined amount, a requisition is required, rather than a purchase order.
- In an RFP, the fire department gives general information about what is desired and allows each vendor to determine how it will meet the need.
- Municipal and volunteer fire departments can also obtain funds by applying for grants. One of the largest sources of grants for the fire service is the Assistance to Firefighters Grant Program.
- Fire departments sometimes solicit donations from a targeted business or industry for a specific purpose.
- The format of the fire department's budget should comply with recommendations of the Governmental Accounting Standards Board, which is the standards-setting agency for governmental accounting.
- In most cases, more than 90 percent of the career fire department budget is allocated to salaries and benefits, including fringe benefits.
- The operating budget covers the basic expenditures that support the day-to-day delivery of municipal services.
- Capital expenditures comprise purchases of durable items that cost more than a predetermined amount and will last for more than one budget year.
- Large capital improvement projects, such as new fire station construction or major renovations, may be funded through bond programs.

- The budgetary process includes the following steps:
  - Developing a budget proposal
  - Considering expenditures
  - Creating a capital budget
  - Asking for everything you need
  - Determining cost recovery and reduction

Hot Terms

<u>Base budget</u> The level of funding required to maintain all services at the currently authorized levels, including adjustments for inflation, salary increases, and other predictable cost changes.

<u>Bond</u> A certificate of debt issued by a government or corporation; a bond guarantees payment of the original investment plus interest by a specified future date.

<u>Budget</u> An itemized summary of estimated or intended expenditures for a given period, along with proposals for financing them.

<u>Expenditure</u> The act of spending money for goods or services.

<u>Fire tax district</u> A special service district created to finance the fire protection of a designated district.

<u>Fire Ground Location (FGL) program</u> A two-person unit that responds to working incidents to track fire fighters operating in immediately dangerous to life and health (IDLH) environments and to monitor the medical status of members who rotate through the rehabilitation sector during an event.

<u>Fiscal year</u> A 12-month period during which an organization plans to use its funds. Local governments' fiscal years generally run from July 1 to June 30.

<u>Governmental Accounting Standards Board (GASB)</u> An organization whose mission is to establish and improve the standards of state and local governmental accounting and financial reporting, thereby resulting in useful information for users of financial reports, and to guide and educate the public, including issuers, auditors, and users of those financial reports.

<u>Line-item budget</u> A budget format in which expenditures are identified in a categorized line-by-line format.

<u>Revenues</u> The income of a government from all sources.

<u>Supplemental budget</u> Proposed increases in spending to provide additional services.

References and Additional Resources

NFPA reprinted material is not the complete and official position of the NFPA on the referenced subject, which is represented only by the standard in its entirety.

Antonellis, P. J. Jr. (2012). *Labor Relations for the Fire Service*. Tulsa, OK: PennWell/Fire Engineering.

Barnett, J. L., and P. M. Vidal. (2012, September). *State and Local Government Finances Summary: 2010*. Washington, DC: U.S. Census Bureau.

Bureau of Labor Statistics. (2012, February). *The Recession of 2007–2009*. Washington, DC: U.S. Bureau of Labor Statistics.

Compton, D. (2010). *Progressive Leadership Principles, Concepts and Tools*. Stillwater, OK: Fire Protection Publications.

Drucker, P. F. (1974). *Management: Tasks, Responsibilities, Practices*. New York, NY: Harper & Row.

Duckworth, R. J. (2012, September 18). *Precision Indoor/Outdoor Personnel Location Project*. Worcester, MA: Worcester Polytechnic Institute. Available at: http://www.wpi.edu/academics/ece/ppl/. Accessed July 23, 2013.

England, R. E., et al. (2012). *Managing Urban America*. Washington, DC: CQ Press.

Gormley, W. T., and S. J. Balla. (2012). *Bureaucracy and Democracy: Accountability and Performance*, 3rd ed. Washington, DC: CQ Press.

Hensler, B. (2011). *Crucible of Fire: Nineteenth-Century Urban Fires and the Making of the Modern Fire Service*. Washington, DC: Potomac Books.

Howitt, A. M., and H. B. Leonard, eds. (2009). *Managing Crises: Response to Large-Scale Emergencies*. Washington, DC: CQ Press.

Miller, K. (2010). *We Don't Make Widgets: Overcoming Myths That Keep Government from Radically Improving*. Washington, DC: Governing Books.

Mills, S. E. (2012). *Effective Emergency Operations Preparedness for Recurring Planned Public Events*. Emmitsburg, MD: Executive Fire Officer Program.

Mintzberg, H. (2011). *Managing*. San Francisco, CA: Berrett-Koehler.

National Fallen Firefighters Foundation. (2011). *Understanding and Implementing the 16 Firefighter Life Safety Initiatives*. Stillwater, OK: Fire Protection Publications.

O'Leary, R., and C. M. Gerard. (2013). "Collaborative Governance and Leadership: A 2012 Survey of Local Government Collaboration." In: *The Municipal Year Book 2013*. Washington, DC: International City/County Management Association, 57–70.

Page, J. O. (n.d.). "Chief Keith Klinger": Oral History Project. Bellflower, CA: Los Angeles County Fire Museum.

Schaenman, P. (2007). *Global Concepts in Residential Fire Safety, Part 1: Best Practices from England, Scotland, Sweden, and Norway*. Arlington, VA: Centers for Disease Control and Prevention, 101.

Smeby, L. C. Jr. (2014). *Fire and Emergency Services Administration: Management and Leadership Practices*, 2nd ed. Burlington, MA: Jones & Bartlett Learning.

Smith, K. B., and A. Greenblatt. (2013). *Governing States and Localities*. Washington, DC: CQ Press.

Snook, J. W., et al. (2011). *A Leadership Guide for Volunteer Fire Departments*. Sudbury, MA: Jones & Bartlett Learning.

Sprouse, C. B. (2012). *When Good People Make Bad Decisions: Assessing Decision Fatigue in Las Vegas Fire & Rescue*. Emmitsburg, MD: Executive Fire Officer Program.

Walker, W. E., et al., eds. (1979). *Fire Department Deployment Analysis: A Public Policy Analysis Case Study*. Publications in Operations Research Series. New York, NY: Elsevier North Holland.

Walters, J. (2007). *Measuring up 2.0: Governing's New, Improved Guide to Performance Measurement for Geniuses (and Other Public Managers)*. Washington, DC: Governing Books.

Walters, J. (2011, January). "Firefighters Feel the Squeeze of Shrinking Budgets: In Small and Large Cities Alike, Firefighters Have Gone from Heroes to Budget Bait." *Governing*. Available at: http://www.governing.com/topics/public-workforce/firefighters-feel-squeeze-shrinking-budgets.html. Accessed July 30, 2013.

Ward, M. J. (2011, April 22). "Hard Closures for LAFD Means 228 Fewer Firefighters On Duty Every Day." *Firegeezer.com*.

Yukl, G. A. (2013). *Leadership in Organizations*, 8th ed. New York, NY: Prentice Hall.

FIRE OFFICER in action

Municipal governments have yet to return to the 2006 level of funding, with many undergoing dramatic reduction of services and departmental reorganization. Many departments are in uncharted waters, as their worst-case scenarios were inadequate to handle today's economic reality. The fire officer who understands the budget process and clearly sees the revenue streams is in the best position when departments begin their economic recovery.

1. Which of the following represents a major local government revenue source?
   A. Fire lane parking tickets
   B. Property taxes
   C. State subsidy
   D. Federal grant

2. Which of the following is an early step in developing a grant proposal?
   A. Inventory the community apparatus and specialized equipment
   B. Determine which other jurisdictions are applying for the grant
   C. Get a copy of the last successful grantee's application
   D. Determine what the grant is designed to cover

3. Personal protective clothing falls under which category within the budget document?
   A. Personnel expenditures
   B. Capital expenditures
   C. Operating expenditures
   D. Facilities expenditures

4. When proposing a new service or technology, a good practice is to:
   A. request enough resources to assure continuing operations.
   B. ask for the minimum amount of resources to achieve the operational goal.
   C. determine the annual cost of operations and add 17 percent for inflation.
   D. always sole-source resources to assure vendor loyalty.

Fire Captain Activity

The initial Fire Ground Locator grant proposal is under review by the municipality, the university, and the organization offering the three-year grant.
- Year 1 expenditures total $815,912 to set the program up, including hiring 7 new uniformed positions.
- Years 2 and 3 will cost $528,295 each.

NFPA Fire Officer II Job Performance Requirement 5.3.1
Explain the benefits to the organization of cooperating with allied organizations, given a specific problem or issue in the community, so that the purpose for establishing external agency relationships is clearly explained.

Application of 5.3.1
1. Write a narrative describing how the fire department/university partnership will benefit the local community.

NFPA Fire Officer II Job Performance Requirement 5.4.1
Develop a policy or procedure, given an assignment, so that the recommended policy or procedure identifies the problem and proposes a solution.

Application of 5.4.1
1. The Assistant Chief of Operations wants to know how the Fire Ground Locator will interact within the command structure. Using the National Incident Management System (or a jurisdiction with which you are familiar), provide a procedure to follow if the FGL detects a low oxygen saturation in a fire fighter who has not declared a mayday.

NFPA Fire Officer II Job Performance Requirement 5.4.2

Develop a project or divisional budget, given schedules and guidelines concerning its preparation, so that capital, operating, and personnel costs are determined and justified.

Application of 5.4.2

1. The senior chiefs suggest that the FGL staffing consist of a technician and a university graduate student who is working on the FGL project. They suggest dropping the three lieutenant positions, saving about $150,000 per year. Provide a response.

NFPA Fire Officer II Job Performance Requirement 5.4.3

Describe the process of purchasing, including soliciting and awarding bids, given established specifications, in order to ensure competitive bidding.

Application of 5.4.3

1. The city purchasing office needs to know why the integrated PASS device and the SCBA interface device did not go through the request for proposal process. Justify why these devices were "sole sourced."

Managing Incidents

Fire Officer I

Knowledge Objectives

After studying this chapter, you will be able to:

- Explain how the Incident Command System was created. (pp 292–293)
- Describe the National Incident Management System **NFPA 4.6.1** **NFPA 4.6.2**. (pp 293–294)
- Describe the postincident review process **NFPA 4.6.3**. (pp 294–296)
- Describe the fire officer's role in incident management **NFPA 4.6** **NFPA 4.6.2**. (pp 296–297)
- Discuss strategic-level incident management **NFPA 4.6**. (pp 297–301)

Skills Objectives

After studying this chapter, you will be able to:

- Use the Incident Command System at an incident **NFPA 4.6** **NFPA 4.6.2**. (pp 296–297)
- Conduct a postincident review **NFPA 4.6.3**. (pp 294–296)

Fire Officer II

Knowledge Objectives

After studying this chapter, you will be able to:

- Describe the National Incident Management System **NFPA 5.6.1**. (pp 293–294)
- Develop and conduct a postincident analysis of a multiple-unit incident **NFPA 5.6.2**. (pp 294–296)
- Describe the National Response Framework. (pp 304, 306)
- Describe the tactical level of incident management **NFPA 5.6** **NFPA 5.6.1**. (pp 306–307)
- Describe the fire officer's greater alarm responsibilities **NFPA 5.6.1**. (pp 307–308)
- Describe the task level of incident management **NFPA 5.6** **NFPA 5.6.1**. (pp 308–309)

Skills Objectives

After studying this chapter, you will be able to:

- Manage an incident using the National Incident Management System **NFPA 5.6.1**. (pp 293–294, 306–309)

CHAPTER 15

Additional NFPA Standards

- NFPA 1026, *Standard for Incident Management Personnel Professional Qualifications*
- NFPA 1500, *Standard on Fire Department Occupational Safety and Health Program*
- NFPA 1521, *Standard for Fire Department Safety Officer*
- NFPA 1561, *Standard on Emergency Services Incident Management System*
- NFPA 1600, *Standard on Disaster/Emergency Management and Business Continuity Programs*

Principles of Fire and Emergency Service Administration (FESHE) Course Outcomes

3. Identify and explain the concepts of span and control, effective delegation, and division of labor. (pp 295–299, 302–304, 306–309)
6. Discuss the various levels of leadership, roles, and responsibilities within the organization. (pp 296–301, 306–309)
9. Identify the roles of the National Incident Management System (NIMS) and Incident Management System (ICS). (pp 292–294, 296–299, 302–304, 306–309)

You Are the Fire Officer

Engine and Medic 100 are dispatched for a vehicle crash with injury at an apartment complex. Captain Davis reports a single vehicle that has struck the side of a four-story multiple-family building. Davis calls for Quint 100 to respond to check on the stability of the structure and, if needed, provide heavy hydraulics for patient extrication. Walking up to the crash, Davis notes that the building's brick veneer has crumbled onto the car and briefly hears a hissing sound. An explosion occurs at the front of the car, followed by a pillar of fire that ignites the wood framing exposed by the fallen bricks.

1. As the first-arriving officer, what is your responsibility in the Incident Command System?
2. Which mode of command will you assume on this fire?
3. Which components of the Incident Command System would you use?

Introduction

A fire officer is expected to perform the duties of a first-arriving officer at any incident, including assuming initial command of the incident, establishing the basic management structure, and following standard operating procedures. A fire officer must also be fully competent at working within the Incident Command System (ICS) at every incident and function as a unit, group, or division leader. This chapter introduces the model procedures for incident management. The *Fire Attack* chapter focuses on the fire attack and application of the incident management procedures to this type of situation.

Fire Officer I and II

The Origin of Incident Management

In the past, incident management was a local activity, using unique terms and practices in each community. Eventually two different programs—the Southern California FIRESCOPE and the Phoenix, Arizona, Fire Ground Commander programs—provided the foundation for the National Incident Management System (NIMS).

■ FIRESCOPE and Fire Ground Commander

FIrefighting **RE**sources of **S**outhern **C**alifornia **O**rganized for **P**otential **E**mergencies (FIRESCOPE) was created in the wake of massive southern California wildfires. These fast-moving fires crossed boundary lines and burned for days within multiple jurisdictions. Because of the fires' scale and complexity, the community-based command and control systems proved ineffective at managing these operations. Resources were often deployed to combat a fire in one jurisdiction that could have been better used in another jurisdiction. Units from different agencies had problems communicating and coordinating their actions. Deficiencies in radio system compatibility, standardized terminology, and equipment compatibility led to an inefficient use of scarce and critical resources.

FIRESCOPE set out to resolve these jurisdictional, interoperability, and standardization issues. The program managers developed a standardized method of setting up an incident management structure, coordinating strategy and tactics, managing resources, and disseminating information. In 1982, this framework was adopted as a cornerstone of the National Interagency Incident Management System (NIIMS), adopted by the National Fire Academy as the model system for emergency management.

In Arizona, Phoenix chief officers Alan Brunacini, Bruce Varner, and Chuck Kime were developing a Fire Ground Commander (FGC) program to meet the needs of an all-hazards metropolitan fire department. The notes and handouts from these seminars eventually evolved into the NFPA Fire Command textbook.

The FGC program focused on small and medium-sized urban emergencies, such as structural fires, mass-casualty events, and hazardous materials events. FIRESCOPE, in contrast, handled the challenges at large-scale wildland fires. Multiple-jurisdiction events, campaign operations, and significant interagency coordination were significant issues in southern California.

According to Brunacini, "The California response-and-assistance network uses the Incident Command System as the basis for an incredible, statewide automatic-aid system. California's unique mutual response system serves as the large-scale, multi-agency organizational gold standard for our entire business" (Brunacini, 2002).

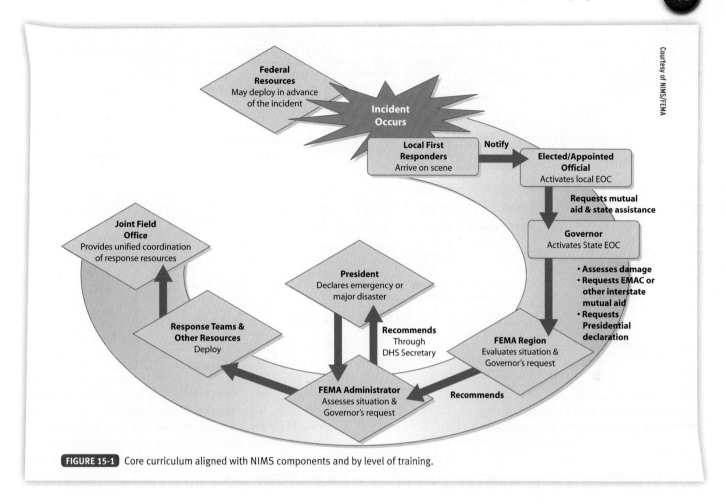

FIGURE 15-1 Core curriculum aligned with NIMS components and by level of training.

Developing One System

The first edition of NFPA 1561, *Standard on Emergency Services Incident Management System*, was issued in 1990. It was influenced by the model procedure guides developed by the National Fire Service Incident Management System Consortium. After the 2001 terrorist attacks, the federal government expanded the 1992 Federal Response Plan to engage state and local responders. Homeland Security Presidential Directive 5 (HSPD-5), *Management of Domestic Incidents*, further refined and nationalized NIMS for all national incidents.

Hurricanes Katrina and Rita identified gaps in national, state, and local emergency preparedness. In October 2006, the Post-Katrina Emergency Management Reform Act established structure, funding, and authority to improve national preparedness. This includes a National Response Framework and a National Incident Management System that supersedes earlier incident management practices (discussed later in this chapter).

Incident Command System

The Incident Command System (ICS) is part of the NIMS Command and Management component. Local emergency response agencies were required to adopt ICS to remain eligible for federal disaster assistance. Adopting this system required training in the core NIMS curriculum (listed in order of typical completion):

- IS 700: National Incident Management System (NIMS), an Introduction
- ICS 100: Introduction to the Incident Command System
- ICS 200: ICS for Single Resources and Initial Action Incidents
- IS 800: National Response Framework, an Introduction
- ICS 300: Intermediate ICS for Expanding Incidents
- ICS 400: Advanced ICS
- Appropriate position-specific training

Students attending courses at the National Fire Academy are required to have completed ICS 100 and 200 before arriving on campus **FIGURE 15-1**.

National Incident Management System

The 1974 Robert T. Stafford Disaster Relief and Emergency Assistance Act was amended in 2007 to provide federal government disaster and emergency assistance to state and local governments, tribal nations, eligible private nonprofit organizations, and individuals affected by a declared major disaster or emergency **FIGURE 15-2**. The Stafford Act covers all hazards, including natural disasters and terrorist events. To be eligible for Stafford Act funding, as laid out in HSPD-5, requires adoption and implementation of NIMS.

The National Incident Management System (NIMS) is a core set of doctrines, concepts, principles, terminology,

			Levels of Training		
			Awareness	Advanced	Practicum
Components of NIMS	Preparedness		IS 800 IS 705		
	Communications & Info Management		IS 704		
	Resource Management	IS-700	IS 703 IS 706 IS 707		
	Command & Management	ICS	ICS 100 ICS 200	ICS 300 ICS 400	Position-specific courses
		MACS	ICS 701		
		Public Info	ICS 702		
	Ongoing Management & Maintenance				

FIGURE 15-2 Overview of Stafford Act support to the states.

Courtesy of the National Response Plan (NRP)/United States Army Combined Arms Center.

and organizational processes. It allows for effective, efficient, and collaborative incident management across all emergency management and incident response organizations. Because it is used consistently nationwide, NIMS makes it possible for federal, state, tribal, and local governments; the private sector; and nongovernmental organizations to work together more easily. This cooperation facilitates preparing for, preventing, responding to, recovering from, and mitigating the effects of a variety of incidents. By tying disaster reimbursement with compliance, the federal government is compelling NIMS implementation by municipalities, hospitals, industries, and law enforcement.

NIMS has five components: preparedness, communications and information management, resource management, command and management, and ongoing management and maintenance.

Postincident Review

Some form of review should be conducted at the company level after every call in which the company performs emergency operations. In most cases, this can be an informal discussion conducted by the company officer to review the incident, discuss the situation, and evaluate team performance. Every situation should be viewed as a potential learning experience, and the company officer should provide feedback to the crew members to reinforce positive performance and identify areas where there is room for improvement. When a deficiency is noted, a plan for addressing the problem should be developed at the same time.

Situations that involve multiple-company operations should be reviewed from the perspective of how well the overall team performed, in addition to individual fire officers conducting company-level reviews. The format of the multicompany review necessarily depends on the nature and the magnitude of the incident, whether it was a one-room house fire that was addressed by three or four companies or a five-alarm blaze that involved dozens of companies and activated every component of the fire department. It is relatively easy to gather the companies that were involved in a one-alarm fire to conduct a basic review of the incident soon after the event. The critique of a large-scale incident, by comparison, can be a major event involving extensive planning, scheduling, and preparations.

Some organizations conduct critiques only for multiple-alarm or unusually large-scale incidents or for situations where something obviously went wrong. This creates the impression that nothing can be learned from the more ordinary-sized incidents, unless someone made a mistake. Whether it is conducted on a large or small scale, the primary purpose of a review process should always be to serve as an educational and training tool, not to place blame for improper or deficient actions. Approach every postincident review in a nonthreatening manner and provide an open format for discussion as an aid to future training.

The review process should be conducted routinely and in a systematic manner. The same basic factors should be examined and discussed. Many fire departments have a standard critique format and procedure.

■ Preparing Information for an Incident Review

A multiple-company incident review is conducted by the incident commander (IC). The preparatory work, however, is often delegated to one of the company officers. An administrative fire officer might assign the officer in charge of the first-due company to prepare the information and make the necessary arrangements for the review of a one-alarm incident. The battalion chief could be responsible for preparing the information for a critique of a multiple-alarm situation

> **Company Tips**
>
> **Addressing Poor Performance**
> Reviews should always focus on lessons and positive experiences; however, in some situations, errors must be addressed. The serious discussion between the fire officer and his or her immediate supervisor should occur in private, not in front of a room full of observers. At the company level, poor performance by one particular crew member should be discussed one-on-one in the office, not at the kitchen table. Areas where the whole crew needs to make improvements should be discussed with the entire crew.

that will be presented by a higher-ranking chief officer; however, the work of gathering and assembling the information might still be delegated to a company officer or staff officer.

Before the actual review occurs, information about the situation leading up to the incident and whatever occurred before the arrival of the first fire department unit should be obtained. This background sets the stage for a discussion of the actual event from an operational perspective and often reveals important factors that were unknown before the incident occurred. The presentation of this information allows individuals to understand what happened and what they observed at the time but did not have the opportunity to interpret.

In the case of a fire involving a building, this process should begin with a review of the building, including its size and arrangement, construction type, date of construction, known modifications or renovations, and any known history that could be significant. All built-in fire protection features, including firewalls, automatic sprinklers, standpipes, and alarm and detection systems, should be identified, with as much detail as possible being provided. Any structural factors that could affect fire spread, such as unsealed openings around pipes, open shafts, and concealed spaces, should be identified as well.

Other questions to be addressed deal with the nature of the occupancy. Was the building occupied, unoccupied, vacant, abandoned, or undergoing renovation? If it was a business, what did it do, make, or store on the premises? What was the fire load? Did the fire grow and spread more quickly because of the nature of the contents?

Some questions may not have answers at the time of the postincident review, but should still be considered in planning for this critique. Where, when, and how did the fire start? What was burning initially and how did the fire spread? Did the fire involve the contents or the structure, or both? Who discovered the fire, what did they observe, and when was the fire department called?

If the desired information about the building and occupancy is not readily available, the officer who is gathering this information can refer to several sources of information, such as the inspection file at the fire prevention bureau or the city department that issues building permits and keeps copies of the plans on file. Valuable information can often be obtained from preincident plans. Indeed, it is an excellent quality-control technique to compare the preincident plan information with the situation that was actually encountered. The postincident review is also a good opportunity to determine whether the information on file was useful to the incident commander.

The fire investigator can usually provide information about the fire's cause and origin, and the communications center can provide information about the source of the alarm and a complete listing of the resources that responded. These data should include all apparatus used as well as the staffing level on each vehicle, if this is a variable. A recording of the incident radio communications should be obtained, as should photographs or videos of the incident.

If water supply was a significant factor in the incident, additional information about the water system should be obtained and evaluated. The locations of hydrants and available flow data should be obtained. The main sizes and whether the area is within a distribution grid or at the end of a dead-end main could be important.

All of this information should set the stage for the analysis of the events that transpired from the time the first unit arrived at the scene. The analysis of the situation before arrival often reveals factors or circumstances that were unknown to the companies involved in the incident. The officer who is preparing for the incident review should be prepared to present all of this information in a manner that is factual and informative.

■ Conducting a Critique

The actual critique should be scheduled as soon as convenient after the event, while allowing adequate time to gather and prepare the necessary information. It is best to capture the information while it is still fresh in the minds of the participants. With a small-scale incident, it may be possible to have all of the involved companies in attendance at the review so that all of the crew members can be involved in the discussion and learning experience. The attendance at a critique of a larger-scale incident could be limited to company officers and command officers; however, it is a good idea to invite the entire crew of any company that played a significant role or was involved in some unusual occurrence.

Start the critique with an overview presentation of the background and basic information about the incident, including the timeline and the units that were dispatched. The first-arriving officer should then be asked to describe the situation as it was presented on arrival and the actions that were taken. This description should include the first officer's role as the initial incident commander as well as the tactical operations performed by that company.

Each successive company should then take a turn explaining what they saw and what they did. An effective visual method is to draw a plot layout of the area on a dry erase board and then have the various company-level officers locate where they positioned their apparatus and where they operated, and identify which actions they took. This input should gradually produce a complete diagram of the incident. Photographs, videos, and radio recordings can be inserted at pertinent points in the discussion. If a large number of companies were involved in the incident, the presentation could be made by division/group/unit officers instead of individual company officers.

During this process, the moderator should keep the analysis directed at key factors, including what the initial strategy was and how it changed (if such changes occurred), how the command structure was developed, how resources were allocated, and which special or unusual problems were encountered. The discussion should also focus on standard operating procedures, including whether they were followed and how well they worked in relation to the actual situation. The officer acting as moderator should help the participants differentiate between problems that occurred because procedures were not followed and areas where procedures should be changed or updated **TABLE 15-1**. The best way to evaluate the effectiveness of procedures is to determine whether following them actually produced the anticipated results. If a deviation from standard procedures occurred, the reasons and the impact should be discussed.

Once every company-level officer has had a chance to make a presentation, the officer directing the critique should provide his or her perspective on and assessment of the operation, including both positive and negative factors. This officer should make a point of evaluating the outcome in terms of where and when the fire was stopped and how much damage occurred. If the outcome was positive and everything met expectations, praise should be widely distributed. Where there is room for improvement, that fact should be noted in terms of valuable lessons learned.

Have the attendees identify the positive and negative points and list the actions that should be taken to improve future performance. These points should be listed on the board for everyone to see as the final product of the critique.

■ Documentation and Follow-up

The last step in conducting an incident review is to write a summary of the incident for departmental records. The results of this review should be documented in a standard format; many fire departments use a form to document the essential

Table 15-1	Incident Review Questions

- Did the preincident plan provide accurate and useful information?
- Are there factors that could have or should have been addressed by fire prevention before the incident?
- Were the appropriate units dispatched based on procedures and the information that was received?
- Were the units dispatched in a timely manner?
- Was the appropriate information obtained and transmitted to the responding units?
- What was the situation on arrival?
- What was the initial strategy as determined by the initial incident commander?
- How did the strategy change during the incident?
- How was the Incident Command structure developed?
- Were the resources provided adequate for the situation?
- How were the resources allocated and assigned?
- Were standard operating procedures followed?
- Do any standard operating procedures need to be changed?
- Which unusual circumstances were encountered, and how were they addressed?
- Is additional training needed?
- Did all support systems function effectively?

findings. Use of a form allows the information to be stored in a consistent manner and ensures that all relevant information is collected. The form should also list recommendations for changes in procedures, if necessary.

The completed package should then be forwarded to the appropriate section or sections within the department for review and follow-up. The training division should receive a copy of every incident review to determine training deficiencies and needs. Recommendations for policy changes should also be forwarded to the appropriate personnel for consideration.

The completed report should include all of the basic incident report information included in the National Fire Incident Reporting System (NFIRS) as well as the operational analysis information.

Fire Officer I

The Fire Officer's Role in Incident Management

Every fire officer is expected to function as an initial incident commander, as well as a company-level supervisor within the ICS. A supervising fire officer functioning as the incident commander must supervise the work of a group of fire fighters, report to a managing or administrative fire officer, and work within a structured plan at the scene of an incident. The incident commander position is the only position that is always filled. This officer is responsible for completing all tasks that are not delegated. The officer may also be required to initiate the procedure to assemble a multiple-agency response.

The first-arriving fire officer has the responsibility to establish command and manage the incident until relieved by a higher-ranking officer. What a first-arriving fire officer does in the first 5 minutes of the incident dictates how the scene will run for the next hour. This task must be accomplished in addition to supervising the members of the officer's own company. Managing an incident requires the fire officer to develop strategies and tactics, determine required resources, and decide how those resources will be used.

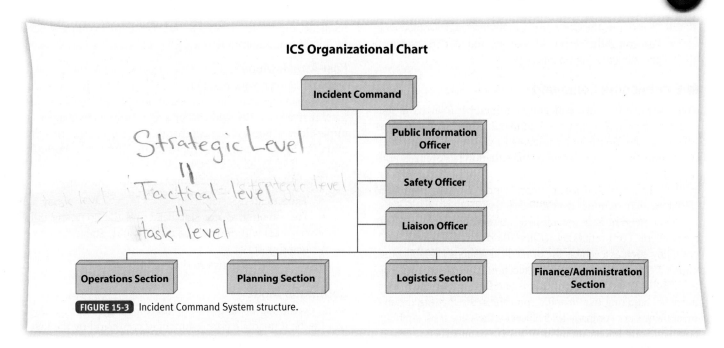

FIGURE 15-3 Incident Command System structure.

The ICS can be implemented in an incremental fashion. The command structure for an incident should be only as large as the incident requires. The goal is to use the model ICS structure to assign all of the functions that must be performed at that incident FIGURE 15-3. Most fire department tactical and task activities fall under the Operations section within the ICS.

The ICS allows the company officer to maintain a manageable span of control. One officer can provide effective supervision to a limited number of subordinates, whether those subordinates are individual workers or supervisors directing groups of workers. For emergency operations, the recommended span of control is three to five individuals reporting to one supervisor. The span of control is maintained by adding more levels of management as an officer's effective span of control is exceeded.

Levels of Command

The ICS includes three levels of command, with a set of responsibilities being assigned to each level. At small-scale incidents, or in the early stages of a larger incident, one company officer may cover all three levels simultaneously. When the incident expands, the management responsibilities are subdivided. A fire officer may have a role at any level.

The overall direction and goals are set at the strategic level. The incident commander always functions at the strategic level. A strategic goal, in a defensive situation, could be to stop the extension of a fire to any adjacent structure.

Tactical-level objectives define the actions that are necessary to achieve the strategic goals. A tactical-level supervisor would manage a group of resources to accomplish the tactical objective. In medium- to large-scale incidents, the tactical-level components would be called divisions, groups, or units, and each of these components could include several companies. A company-level officer would be in charge of each company, and all of the company-level officers would report to the tactical-level supervisor. Tactical assignments are usually defined by a geographical area (e.g., one part of a building) or a functional responsibility (e.g., ventilation), or sometimes by a combination of the two (e.g., ventilation in a particular part of the building).

In large incidents, another level of management may be added to maintain a reasonable span of control. Branches are established to group tactical components. An officer assigned to a branch would oversee some combination of divisions, groups, or units.

Task-level assignments are the actions required to achieve the tactical objectives. This is where the physical work is accomplished. Individual fire companies or teams of fire fighters perform task-level activities, such as searching for victims, operating hose lines, or opening ceilings.

Strategic-Level Incident Management

The first-arriving managing and supervising fire officers are required to focus on the strategic level as they arrive at an emergency. Depending on the size and complexity of the incident, the first-arriving officer may quickly move to the tactical or task level as the administrative fire officer and other departmental resources arrive.

Responsibilities of Command

The incident commander is the individual who is responsible for the management of all incident operations. This officer is responsible for the completion of three strategic priorities:

- Life safety
- Incident stabilization
- Property conservation

The incident commander is also responsible for the following aspects of operations:

- Building a command structure that matches the organizational needs of the incident

- Translating the strategic priorities into tactical objectives
- Assigning the resources that are required to perform the tactical assignments

Establishing Command

The first fire officer or fire department member to arrive at the scene is expected to assume command of the incident. The initial incident commander remains in charge of the incident until command is transferred or the situation is stabilized and terminated. If the incident expands, a higher-ranking officer is likely to arrive and assume command, at which point the initial incident commander's role changes.

Command must be established and the ICS must be used at every event. For events that require the services of a single fire company, such as a vehicle fire or medical response, establishing command may simply require notification to the dispatcher that the company is on the scene and is handling the situation. For incidents requiring two or more companies, the first fire department member or company-level officer on the scene must establish command and initiate an incident management structure that is appropriate for the incident.

Activating the command process includes providing an initial radio report and announcing that command has been established. This report should provide an accurate description of the situation for units that are still en route. It should also identify the actions that arriving units will take. In this report, the company officer should include the following information:

- Identification of the company or unit arriving at the scene
- A brief description of the incident situation, such as the building size, height, and occupancy, or the magnitude of a multiple-vehicle collision
- Obvious conditions, such as a working fire, multiple patients, or a hazardous materials spill, or a dangerous situation, such as a man with a gun
- A brief description of the action to be taken (e.g., "Engine 2 is advancing an attack line into the first floor")
- Declaration of the strategy to be used (offensive or defensive)
- Any obvious safety concerns
- Assumption, identification, and location of command
- Request for additional resources (or release of resources), if required

Command Options

The first-arriving company-level officer has three options when arriving at the incident and assuming command: investigation, fast attack, or command mode. The decision of which action to take is based on the situation that is presented to the fire officer.

Investigation Mode

There may be nothing showing when fire fighters arrive, or the incident may appear to be a very minor situation. In such a case, the first-arriving company will conduct an investigation. The other units assigned to the event will stage and remain uncommitted pending the results of that investigation. The first-arriving company-level officer performs the role of initial incident commander as well as supervising the company performing the investigation.

Fast-Attack Mode

Some situations require immediate action by the first-arriving fire company to save a life. An example is when a single company arrives at a fire with occupants in imminent danger at upper-floor windows. The company-level officer performs the initial incident command responsibilities through a portable radio while engaged in a fast attack.

The fast-attack mode ends when one of the following occurs:

- The situation is stabilized.
- The situation is not stabilized and the company officer must withdraw to the exterior and establish a command post.
- Command is transferred to another officer.

Command Mode

Some events are so large, complex, or dangerous that they require the immediate establishment of command by the first-

---

Safety Zone

"Nothing Showing" Calls Can Be Dangerous

Serious injuries and some fire fighter fatalities have occurred at incidents that started as "nothing showing" events. In many of these cases, after entering the building, the first-in fire company suddenly encountered a rapidly deteriorating situation. Vital firefighting equipment, such as self-contained breathing apparatus (SCBA) and tools, was left behind on the apparatus. Fire fighters were unprepared for the situation they encountered.

When incidents, particularly in commercial and industrial buildings, are investigated, consider the following factors:

- Assume that there is a fire in the building and bring and wear the appropriate personal protective equipment.
- Use thermal imagers.
- Go to the location of a monitored fire alarm activation. Fires in newer buildings are often detected by the fire alarm system while they are still in an incipient stage. Do not assume that the call is a false alarm if nothing is visible from the exterior. Check the fire alarm panel for the location of the activated device and be prepared for action when you go there to investigate.
- Use two or three companies to perform the initial investigation in larger buildings.
- Hold the entire first-alarm assignment until the investigation is completed. Many fire departments have the assisting companies reduce their response from emergency to nonemergency when the first-arriving company reports no indication of fire. This policy reduces the risk of accidents during emergency response.

FIRE AND RESCUE DEPARTMENTS OF NORTHERN VIRGINIA
INITIAL **INCIDENT COMMAND BOARD**

ADDRESS/COMMAND:					INITIAL RIT ENGINE:	
FIRST ALARM		Floor, Group, Division, Branch	Floor, Group, Division, Branch	Floor, Group, Division, Branch	OPERATIONS CHANNEL:	
ENGINE					TYPE OF OCCUPANCY:	
ENGINE					**TASKS**	**REQUESTS**
ENGINE					WATER SUPPLY	GAS COMPANY
ENGINE					PRIMARY SEARCH	POWER COMPANY
TRUCK					LADDERS	FIRE INVESTIGATOR
TRUCK					VENTILATION	POLICE
RESCUE					UTILITY CONTROL	LIGHT/AIR UNITS
MEDIC					**NOTES**	
AMB						
BFC						
EMS CAPT						
SECOND ALARM		Floor, Group, Division, Branch	Floor, Group, Division, Branch	Floor, Group, Division, Branch		
ENGINE						
ENGINE						
ENGINE						
ENGINE						
TRUCK						
TRUCK						
RESCUE						
MEDIC						
AMB					CHECK ALL OPERATING UNITS AIR SUPPLY. CREWS WITH *2000 PSI OR LESS* NEED TO BE **REPLACED** AND SENT TO THE *MEDICAL UNIT*.	
BFC						
EMS CAPT						

| RECEO-VS | CHECK EXPOSURES ⇒ | B3 | B2 | B1 | FIRE UNIT | D1 | D2 | D3 |

FIGURE 15-4 Tactical worksheet.

arriving company-level officer. In this case, the company-level officer's personal involvement in tactical operations is less important than the command responsibility. The company-level officer should establish a command position in a safe and effective location and initiate a <u>tactical worksheet</u> **FIGURE 15-4**. The role of the initial incident commander at this type of situation is to direct incoming units to take effective action. While the initial incident commander remains outside, the rest of the company members should do one of the following:

- Initiate fire suppression or emergency action with one of the members assigned as the acting company officer. The acting company officer must be equipped with a portable radio, and the crew must be capable of performing safely without the initial incident commander.
- After the initial incident commander assigns them, the remaining company members work under another company officer.
- Stay with the initial incident commander to perform staff functions that assist command.

■ Functions of Command

There are nine functions of command:

- Determining strategy
- Selecting incident tactics
- Setting the action plan

Getting It Done

Examples of Initial Radio Reports

- **For a single-company incident:** "Engine 2 is on the scene of a fully involved auto fire with no exposures. Engine 2 can handle."
- **For an EMS incident:** "Quint 618 is on the scene of a four-vehicle crash with multiple patients and one trapped. Transmit a special alarm for a heavy rescue company, a second suppression company, and an EMS task force. Quint 618 will be Palisade Command."
- **For an offensive structure fire:** "Engine 1034 has a two-story motel with a working fire in a second-floor room on side Charlie. Engine 1034 has a hydrant and is going in with a handline to start primary search. This will be an offensive fire attack. Engine 1034 will be Cedar Avenue Command."
- **For a defensive structural fire:** "Engine 22 is on the scene of a one-story strip shopping center with an end unit heavily involved. Engine 22 is laying an LDH supply line and will be using a master stream to cover the exposure on side Delta. This is a defensive fire. Strike a second alarm. Engine 22 will be Loisdale Command."

- Developing the ICS organization
- Managing resources
- Coordinating resource activities
- Providing for scene safety
- Releasing information about the incident
- Coordinating with outside agencies

Immediate command functions include determining strategy, selecting incident tactics, and setting an action plan. These three functions must be completed as part of the first-arriving officer's size-up and initial actions. The initial radio report, as described earlier, will cover these three functions.

The initial fire-ground assignments and company-level tasks may be predetermined by departmental SOP or practice. The company officer should ensure that the SOP is followed and look for conditions that may require a deviation from the standard initial actions.

Once the initial actions are under way, the incident commander works on the next four functions: developing the ICS organization, managing resources, coordinating resource activities, and providing for scene safety. Designate a safety officer as well as unit, group, or division commanders, as necessary. Request additional resources if required, and manage them through staging and assignments within the incident management system.

Once the incident action plan is fully operating, the incident commander works on the last two functions: releasing information about the incident and coordinating with outside agencies.

■ Transfer of Command

An administrative fire officer assumes command of significant incidents. At a large incident, command could be transferred more than once, depending on the situation and the chain of command, as successively higher-ranking or more qualified officers arrive at the incident scene.

Command should be transferred only to improve the quality of the command organization. It should not be transferred simply because a higher-ranking officer has arrived, and it should not be transferred more times than is necessary to manage the incident effectively.

Transfer of command should follow a standard procedure:

1. The officer assuming command communicates with the initial incident commander. This can occur over the radio, but a face-to-face meeting is preferred.
2. The initial incident commander briefs the new incident commander, including the following information in his or her report:
   - Incident conditions, such as the location and extent of fire, the number of patients, and the status of the hazardous materials spill or leak
   - Tactical worksheet and incident action plan
   - Progress toward completion of the tactical objectives
   - Safety considerations
   - Deployment and assignment of operating companies and personnel
   - Need for additional resources
3. Command is officially transferred only when the new incident commander has been briefed.
4. The fact that command has been transferred is communicated to the dispatch center and all units operating on the fire scene.

The initial incident commander should always review the tactical worksheet with the new incident commander. This worksheet outlines the status and location of personnel and resources in a standard form. It is especially important on those events where the first-arriving company initially started in the command mode.

After the transfer of command has occurred, the new incident commander determines the most appropriate assignment for the previous incident commander. A company-level officer may be assigned as a division or group supervisor or may remain with the new incident commander at the command post. The size and the complexity of the incident determine how much the management structure will need to be expanded.

■ Fire Fighter Accountability

Occurring concurrently with the evolution of incident management systems was development of systems for accounting for fire fighters and others working at an incident. Although the

Assessment Center Tips

Think Big as the Incident Commander
As part of fire officer candidate evaluations, anticipate that the incident management part of the assessment center will present a situation that requires you to call for additional help. A typical assessment center problem generally involves a challenging situation because the process is designed to determine whether the individual is ready for promotion to a level that involves higher-level responsibilities. A fire officer should expect that a fire problem will probably require more resources than initially assigned.

If the jurisdiction's first alarm assignment for a commercial fire is four engines, two trucks, and an ambulance, there will be an incident that uses all of the first alarm resources and requires the lieutenant candidate to call for more assistance. Captain candidates should expect a complex problem that would require a second or greater alarm assignment.

The items that are usually assessed in this evaluation include the candidate's ability to evaluate an emergency situation tactically, establish an incident command post, demonstrate knowledge of the local ICS, delegate assignments, and demonstrate follow-up. Many incident simulations include some type of challenge that requires the candidate to take immediate, corrective action. The challenges could include a company not performing its assignment, a crew not responding to command post messages, a missing or injured fire fighter, or a sudden change in the incident that requires a change in tactics.

concept of accountability had been used in the United Kingdom for several years, in the United States it became required with the first edition of NFPA 1500 in 1987. Legal sanctions accelerated adoption of fire fighter accountability practices.

The Seattle Fire Department, for example, lost six fire fighters between 1987 and 1995. The Washington State Department of Labor and Industry found the department was negligent in SCBA training and tracking of fire crews at large-scale operations. At one incident, a significant amount of time elapsed before the incident commander was aware that a fire fighter was missing. A judge allowed the spouses and estates of the deceased fire fighters to sue Seattle executive and administrative fire officers directly for $55 million worth of personal liability. As part of the court-mandated corrective action, the department purchased personal alert safety system (PASS) devices and implemented a passport-style accountability system.

Additional pressure to improve accountability came from a request by the International Association of Fire Fighters (IAFF) for a clarification of Occupational Safety and Health Administration (OSHA) respiratory protection regulation 29 CFR 1910.134. This section of the Code of Federal Regulations covered industrial employees operating in confined spaces, toxic environments, or oxygen-deficient atmospheres. These work areas are classified as immediately dangerous to life and health (IDLH).

OSHA ruled in 1996 that fire fighters working within a structure fire were operating in an IDLH atmosphere. Consequently, fire departments must comply with 29 CFR 1910.134 while SCBA is being used. A minimum of two fire fighters must enter the IDLH area together and remain in visual or voice contact with each other at all times. In addition, at least two properly equipped and trained fire fighters must be available to assist personnel working in the hazardous area:

- Be positioned outside the IDLH atmosphere
- Account for the interior teams
- Remain capable of rescue of the interior team or teams

This interpretation became known as the two-in/two-out rule and evolved into the rapid intervention team (RIT) or rapid intervention crew (RIC) concept that was incorporated into NFPA 1500. NFPA 1407, *Standard for Training Fire Service Rapid Entry Crews*, describes basic training procedures and information.

Near-Miss REPORT

Report Number: 12-0000010

Event Description: Our department responded to a residential structure fire reported from a neighborhood near our headquarters station. Upon arrival, we were faced with a heavily involved garage fire that was spreading into the attic of the house. We started a fire attack on the burning garage from the exterior with a 2½-inch hose and a team consisting of two fire fighters and myself. During the attack, our team positioned themselves on the empty driveway and moved to within several feet of the burning garage. Our assistant fire chief was functioning as an operations manager from the front yard, giving direction to crew members entering the front door. As he was doing so, he noticed from the side of the garage that the brick veneer wall above the garage door had begun to bow out and was leaning toward our location in the driveway.

From our vantage point, we could not see that the structure supporting the brick veneer had burned away. During our initial attack, the extent of damage to the structure was somewhat hidden by the flames and smoke. As we began to knock the fire down, the assistant fire chief recognized the signs of an unsupported brick veneer wall that was in imminent danger of collapse. He immediately came over and moved our attack team back roughly 15 feet away from the garage. Not more than 30 seconds after we were repositioned, the entire brick veneer wall pulled away from the destroyed framework and collapsed onto the driveway, with debris tumbling right up to the feet of the hose team. Both the chief and I ran up to the fire fighters, checking for injuries, and found them to be okay. From our vantage point in the driveway, the wall appeared flat and gave no signs of potential collapse.

Lesson Learned: The importance of recognizing the potential for structural collapse during a fire cannot be overstated. Adhering to the principle of knowing the collapse zone and staying clear becomes critical. Residential structures built using brick veneer should always be suspected of failure when a heavy fire condition exists in the area of the wood framing adjacent to the wall. Always play it safe and keep personnel away from such veneer walls should you discover those significant conditions, and consider the time the fire has acted upon them. Acting immediately and not waiting to see what happens is the final step. Step in, use a "lean forward" attitude, and correct the situation before something goes wrong.

Courtesy of the National Fire Fighter Near-Miss Reporting Systems (firefighternearmiss.com)

After the Transfer of Command: Building the Incident Management System

The incident management system can readily expand to handle larger and more complex incidents. Managing and supervising fire officers may be assigned to section, branch, division, and group assignments within the incident management system.

■ Command Staff

After command has been transferred, the company-level officer may be assigned to the command staff. Individuals on the command staff perform functions that are reported directly to the incident commander. The safety officer, liaison officer, and information officer are always part of the command staff; these duties cannot be delegated to other sections of the incident organization.

Aides, assistants, and advisors may be assigned to work directly for the incident commander. An aide is a fire fighter (sometimes a fire officer) who serves as a direct assistant to a command officer. In many fire departments, operational-level command officers have regularly assigned aides who drive the command vehicle and perform administrative support duties, as well as functioning as aides at incident scenes. In other departments, aides are assigned only as needed at incident scenes. The incident commander may assign more than one aide to perform support functions at a major incident. Aides can also be assigned to other officers in the command structure.

Safety Officer

The safety officer is responsible for ensuring that safety issues are managed effectively at the incident scene. The safety officer acts as the eyes and ears of the incident commander by identifying and evaluating hazardous conditions, watching out for unsafe practices, and ensuring that safety procedures are followed. Normally, the safety officer is appointed early during an incident. As the incident becomes more complex, additional qualified personnel can be assigned as assistant safety officers to subdivide the responsibilities.

The safety officer is an advisor to the incident commander but has the authority to stop or suspend operations when unsafe situations occur. This authority is clearly stated in national standards, including NFPA 1500, *Standard on Fire Department Occupational Safety and Health Program*; NFPA 1521, *Standard for Fire Department Safety Officer*; and NFPA 1561, *Standard on Emergency Services Incident Management System*. Several state and federal regulations require the assignment of a safety officer at hazardous materials incidents and certain technical rescue incidents. Making such an assignment is a good practice even if it is not explicitly required by a regulatory agency.

The safety officer should be a qualified individual who is knowledgeable in fire behavior, building construction and collapse potential, firefighting strategy and tactics, hazardous materials, technical rescue practices, and departmental safety rules and regulations. A safety officer should also have considerable experience in incident response and specialized training in occupational safety and health. Many fire departments have full-time safety officers who perform administrative functions relating to health and safety when they are not responding to emergency incidents.

Liaison Officer

The liaison officer is the incident commander's point of contact for representatives from outside agencies and is responsible for exchanging information with representatives from those agencies. During an active incident, the incident commander may not have time to meet directly with everyone who comes to the command post. The liaison officer position takes the incident commander's place by obtaining and providing information or directing people to the proper location or authority. The liaison area should be adjacent to, but not inside, the command post.

The incident commander can also assign a liaison officer to represent the fire department directly with another agency. At a complex incident involving extensive interaction between the police and fire departments, where unified command has not been established, the fire department could assign a liaison officer to work with the police, or the police commander could assign a liaison officer to the fire department command post.

Public Information Officer

The public information officer is responsible for gathering and releasing incident information to the news media and other appropriate agencies. At a major incident, the public wants to know what is being done. The public information officer serves as the contact person for media requests so that the incident commander can concentrate on managing the incident. A media briefing location should be established that is separate from the command post.

■ General Staff Functions

When an incident is too large or too complex for just one person to manage effectively, the incident commander may appoint officers to oversee major components of the operation. Four standard components are defined in the ICS model. Everything that occurs at an emergency incident can be divided among these four major functional components:

1. Operations
2. Planning
3. Logistics
4. Finance/administration

The incident commander decides which (if any) of these four components need to be activated, when to activate them, and who should be placed in each position. Recall that the blocks on the ICS organization chart refer to functional areas or job descriptions, not to positions that must always be staffed. The positions are assigned only when they are needed and, if a position is not assigned, the incident commander is responsible for managing that function.

The persons in charge of the four major sections are known as the ICS general staff. The four people on the ICS general staff, when they are assigned, may conduct their operations either from the main command post or from a different location. At a large incident, the four functional organizations may operate from different locations, but the general staff are always in direct contact with the incident commander **FIGURE 15-5**.

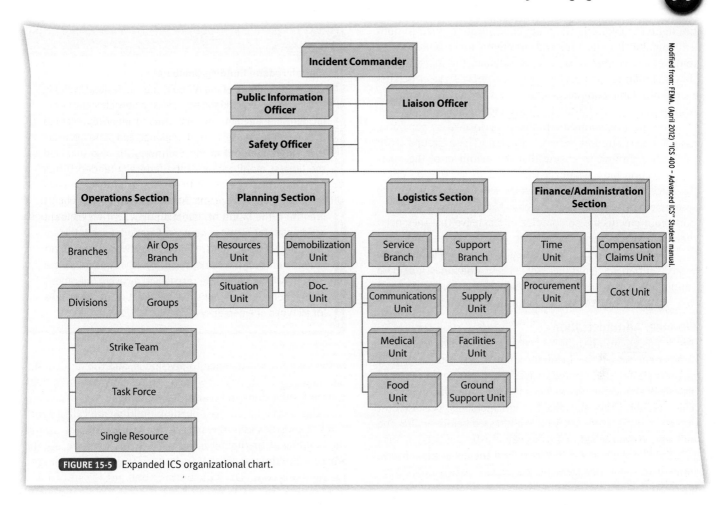

FIGURE 15-5 Expanded ICS organizational chart.

Operations

The operations section is responsible for the management of all actions that are directly related to controlling the incident. The operations section fights the fire, rescues any trapped individuals, treats the patients, and does whatever else is necessary to deal with the emergency situation. This part of the organization produces the most visible results.

For most structure fires, the incident commander directly supervises the functions of the operations section. A separate operations section chief is used at complex incidents so that the incident commander can focus on the overall situation while the operations section chief deals with the strategy and tactics that are required to get the job done.

Planning

The planning section is responsible for the collection, evaluation, dissemination, and use of information relevant to the incident. To do so, it works with status boards and preincident plans, as well as building construction drawings, maps, aerial photographs, diagrams, and reference materials.

The planning section is also responsible for developing and updating the incident action plan (IAP). In many ways, this component resembles the scheduling department at a company. It plans what needs to be done by whom and identifies which resources are needed. To perform this role, the planning section must be in close and regular contact with all of the other sections.

The incident commander activates the planning section when information needs to be obtained, managed, and analyzed. The planning section chief reports directly to the incident commander. Individuals assigned to planning functions examine the current situation, review available information, predict the probable course of events, and prepare recommendations for strategies and tactics. The planning section also keeps track of resources at large-scale incidents and provides the incident commander with regular situation and resource status reports.

The planning functions may be delegated to subunits. These include a resources unit, a situation unit, a documentation unit, a demobilization unit, and technical specialists.

Incident Action Plan

The incident action plan (IAP) is a basic component of ICS; all incidents require an action plan. The IAP outlines the strategic objectives and states how emergency operations will be conducted. At most incidents, the IAP is relatively simple and can be expressed by the incident commander in a few words or phrases. A written IAP is required for large or complex incidents that have an extended duration. The IAP for a large-scale incident can be a lengthy document that is regularly updated and used for daily briefings of the command staff.

Logistics

The logistics section is responsible for providing supplies, services, facilities, and materials during the incident. The logistics section chief reports directly to the incident commander. Among the responsibilities of this section are keeping apparatus fueled, providing food and refreshments for fire fighters, obtaining the foam concentrate needed to fight a flammable liquids fire, and arranging for a bulldozer to remove a large pile of debris. In essence, the logistics section is similar to the purchasing and support services departments of a large company—that is, it ensures that adequate resources are always available and functional.

Logistical functions are routinely performed by personnel assigned to a support services division. These groups work in the background to ensure that the members of the operations section have whatever they need to get the job done. Resource-intensive or long-duration situations may require assignment of a logistics section chief because service and support requirements are so complex or extensive that they need their own management component.

The logistics section may use subunits to provide the necessary support for large incidents. These units may include a supply unit, rehabilitation unit, facilities unit, ground support unit, communications unit, food unit, and medical unit.

Finance/Administration

The finance/administration section is the fourth major ICS component under the incident commander. This section is responsible for the administrative, accounting, and financial aspects of an incident, as well as any legal issues that may arise. This function is not activated at most incidents, because cost and accounting issues are usually addressed before or after the incident. Nevertheless, a finance/administration section may be needed at large-scale and long-term incidents that require immediate fiscal management, particularly when outside resources must be procured quickly.

A finance/administration section is usually established during a natural disaster, when state or federal expense reimbursements are expected, or during a hazardous materials incident in which reimbursement may come from the shipper, carrier, chemical manufacturer, or insurance company. The finance section is equivalent to the finance department in a company; it accounts for all activities of the company, pays the bills, ensures that there is enough money to keep the company running, and keeps track of the costs.

The finance/administration section may incorporate subunits to handle its duties efficiently. These include a time unit, procurement unit, compensation and claims unit, and cost unit.

■ Location Designators

For the purposes of firefighting operations, the exterior sides of a building are designated as sides A (Alpha), B (Baker), C (Charlie), and D (Delta). The front of the building is designated as side A, with sides B, C, and D following in a clockwise direction around the building. The companies working in front

> **Fire Marks**
>
> **Murrah Federal Building Bombing**
> The 1995 bombing of the Alfred P. Murrah Federal building in Oklahoma City, Oklahoma, added a new element to the incident management process. This incident represented the largest response of local, regional, and state agencies to a single incident in the Southwest. It also involved the largest mobilization of the federally funded Urban Search and Rescue Task Forces, with 11 of the 28 teams responding to the incident. The federal after-action report identified the value of establishing a written incident action plan for every 12-hour work period. Each IAP identified the strategic goals, tactical objectives, and support requirements for the work period. While written IAPs were used in wildland fires, the rescue and recovery operations at Oklahoma City established the practice of using IAPs for all types of emergency incidents.

of the building are assigned to division A, and the radio designation for their supervisor is "division A." Similar terminology is used for the sides and rear of the building.

The areas adjacent to a burning building are called exposures. Exposures take the same letter as the adjacent side of the building. A fire fighter facing side A can, therefore, see the adjacent building on the left (exposure B) and the building to the right (exposure D). If the burning building is part of a row of buildings, the buildings to the left are called exposures B, B1, B2, and so on. The buildings to the right are exposures D, D1, D2, and so on.

Within a building, divisions commonly take the number of the floor on which they are working. For example, fire fighters working on the fifth floor would be in division 5, and the radio designation for the chief assigned to that area would be "division 5." Crews performing different tasks on the fifth floor would all be part of this division. Common simple terminology should be used for designations whenever possible, such as "roof division" and "basement division."

> **Fire Marks**
>
> **Impact of Fires on Regulations**
> After the 1980 MGM Grand hotel fire, in which 85 people were killed, Nevada revised its fire sprinkler requirements, requiring them in buildings that were 55 feet or 5 stories high.

Fire Officer II

National Response Framework

The federal government established the National Response Framework (NRF) in March 2008. NRF is a comprehensive, national, all-hazards approach to domestic incident response that describes specific authorities and best practices for managing incidents. It builds upon the National Incident Management System, which provides a consistent template for managing incidents. The Homeland Security Act charged the Department of Homeland Security administrator with

VOICES
OF EXPERIENCE

Perhaps the most common ICS position for the company officer is that of incident commander. On every company response, the company officer is the IC. During larger, more complex responses, the company officer may find himself or herself functioning in a subordinate ICS position assigned by the IC. Sometimes it all happens during one event.

Engine 5 was finishing another routine shift when the company received an alarm for an unknown fire in an industrial district. The fire fighters responded as the first-in engine company and discovered that the fire was a grass/vegetation fire on some abandoned industrial property. While conducting the initial size-up, all indications identified the scene as a routine grass fire and suggested that no additional help would be needed. Additional units en route were told to disregard the call, and confinement and extinguishment actions were implemented by the crew of Engine 5.

The company officer was the IC and responsible for all decisions related to selecting strategy, implementing tactical options, supervising various work tasks, and ensuring the safety of everyone involved. Everything was routine. All command and control responsibilities were being performed by the company officer, and his crew was implementing his tactical plan of attack to extinguish the fire. Everything was going well.

> **After 48 hours of continuous emergency operations, Engine 5 could honestly claim they were "first in, last out."**

Then things changed. Smoke from the burning grass/vegetation became much darker and thicker than one would normally expect from this type of fire. As part of his continuing scene size-up process, the incident commander spoke with a security representative from an adjacent property and learned that the property involved was a registered Superfund site. The fire scene was really a hazardous materials scene.

Offensive fire suppression operations were immediately abandoned. A defensive strategy was adopted and additional units were dispatched to the scene to aid in safely confining and controlling the fire. When the battalion chief arrived on the scene, he was briefed and command was transferred. The chief officer became the new IC, and the company officer assumed duties as a group supervisor. The scene continued to evolve as more engine companies became involved in operations.

One operational period transitioned into another. Additional command staff and city emergency management officials arrived on scene, as did representatives from various state and federal agencies and a few mutual aid partners that were requested to respond. All of these additional parties were integrated into the incident management structure. The fire was extinguished, but the focus had shifted to assessing the hazardous materials threat. As the scene grew, many company officers found themselves reassigned to fulfill various ICS positions such as incident safety officer, operations section chief, hazardous materials branch director, and decontamination group supervisor.

Eventually, a comprehensive scene hazard and risk analysis was completed, and it was determined that the scene had been stabilized. The incident transitioned from an emergency mode to a recovery operation, where clean-up and mitigation efforts became the focus. Units that were no longer needed were demobilized, and the incident command structure began to collapse and get smaller and smaller. Eventually, all the chiefs had cleared the scene and the additional units that were needed for emergency operations went back into service. In the end, the same engine company that initially responded and established command was left as the last unit on scene and terminated the incident by returning to service. After 48 hours of continuous emergency operations, Engine 5 could honestly claim they were "first in, last out."

Robert L. Havens, CFPS
Captain/Paramedic
Port Arthur Fire Department
Port Arthur, Texas

Table 15-2 ESF Annexes Applicable to Fire Officers

Annex	Primary Agencies
ESF 4: Firefighting	Department of Agriculture/Forest Service
ESF 5: Emergency Management	Department of Homeland Security (DHS)/Federal Emergency Management Agency (FEMA)
ESF 6: Mass Care, Emergency Assistance, Housing and Human Services	DHS/FEMA
ESF 8: Public Health and Medical Services	Department of Health and Human Services
ESF 9: Search and Rescue	DHS/FEMA DHS/U.S. Coast Guard (USCG) Department of the Interior (DOI)/National Park Service (NPS) Department of Defense (DOD)
ESF 10: Oil and Hazardous Materials Response	Environmental Protection Agency, DHS/USCG
ESF 13: Public Safety and Security	Department of Justice (DOJ)

building a comprehensive National Incident Management System with federal, state, and local government personnel, agencies, and authorities to respond to attacks and disasters; consolidate existing federal emergency response plans into a single, coordinated national response plan; and administer and ensure the implementation of the NRF, including coordinating and ensuring the readiness of each Emergency Support Function under the NRF.

Federal, state, and some private-sector and nongovernmental entities organize their resources and capabilities under 15 Emergency Support Functions (ESFs). ESFs align categories of resources and provide strategic objectives for their use. The ESF annexes that are of interest to a fire officer are listed in **TABLE 15-2**.

Tactical-Level Incident Management

Divisions, groups, and units are where the strategy specified in the incident action plan is converted into action. Supervising and managing fire officers will be commanding these incident management assignments.

■ Divisions, Groups, and Units

Divisions, groups, and units are tactical-level management elements that are used to assemble companies and resources for a common purpose. The flexibility of the ICS enables organizational units to be created as needed, depending on the size and the scope of the incident. In the early stages of an incident, individual companies are often assigned to work in different areas or perform different tasks. As the incident grows and more companies are assigned to areas or functions, the incident commander can establish tactical-level units to place multiple resources under one supervisor. This organization keeps the incident commander's span of control within desired limits and provides direct coordination of common efforts.

A division represents a geographical operation, such as one floor or one side of a building. A group represents a functional operation, such as ventilation. Unit is a generic term that can be applied to either a geographical or functional component. An officer is assigned to supervise each division, group, or unit as it is created **FIGURE 15-6**.

These organizational elements are particularly useful when several resources are working near each other, such as on the same floor inside a building, or are doing similar tasks, such as providing ventilation. The assigned supervisor can directly observe and coordinate the actions of several crews.

Divisions

A division is composed of the resources responsible for operations within a defined geographical area. This area could be a floor inside a building, the rear of the fire building, or a geographical area at a brush fire. Divisions are most often used during routine fire department emergency operations. When all of the units in one area are assigned to one supervisor, the incident commander has to communicate with just that one individual. The assigned supervisor, in turn, coordinates the activities of all resources that are working within that area, in accordance with the incident commander's strategic plan.

The division structure provides effective coordination of the tactics being used by different companies working in the same area. For example, the division supervisor might coordinate the actions of a crew that is advancing a hose line into the

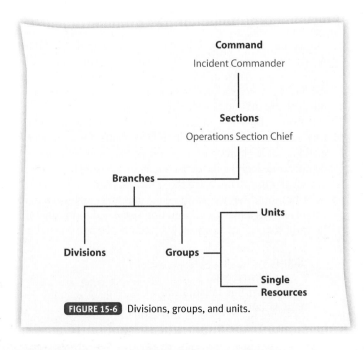

FIGURE 15-6 Divisions, groups, and units.

fire area, a crew that is conducting search-and-rescue operations in the same area, and a crew that is performing horizontal ventilation.

Groups

An alternative way of organizing resources is by function rather than by location. A group is composed of resources assigned to a specific function, such as ventilation, search and rescue, or water supply. The officer assigned to supervise the ventilation group, for example, would use the radio designation "ventilation group" to refer to this function and is required to coordinate activities with other division, group, or unit supervisors.

Groups are responsible for performing an assignment, wherever it may be required, and often work within more than one division. Sometimes groups are established with both functional and geographical designations, such as a west wing search-and-rescue group and an east wing search-and-rescue group.

Units

A unit is an organizational element with functional responsibility for a specific incident activity, such as planning or logistics, or a specific geographic assignment. The officer assigned to supervise the air resupply unit, for example, would use the radio designation "air resupply" to refer to it and is required to coordinate activities with other division, group, or unit supervisors.

The unit is the smallest organizational element within the incident management system. Units are more frequently used in large or complex incidents where many specialized groups must work together.

Division/Group/Unit Supervisor Responsibilities

In many cases, a company-level officer is assigned by the incident commander to function as a division/group/unit supervisor. When an incident is rapidly escalating, the company-level officers on the first alarm are often assigned to geographical or functional areas and designated as the division/group/unit supervisor. For example, the company officer in charge of the first-arriving ladder company could be assigned as the "ventilation group" supervisor. Additional companies assigned to ventilation would then become part of this group. When additional command officers arrive, the ventilation group supervisor might be relieved by a chief officer and revert to functioning as a company-level officer.

As a division/group/unit supervisor, the company-level officer directly supervises and monitors the operations of assigned resources and provides updates to command. Duties of a division/group/unit supervisor include the following:

- Using an appropriate radio designation, such as roof division, division A, or rescue group
- Completing the objectives assigned by command
- Accounting for all assigned companies and personnel
- Ensuring that operations are conducted safely
- Monitoring work progress
- Redirecting activities as necessary
- Coordinating actions with related activities and adjacent division/group/unit supervisors
- Monitoring the welfare of assigned personnel
- Requesting additional resources as needed
- Providing the incident commander with essential and frequent progress reports
- Reallocating or releasing assigned resources

Division, group, and unit supervisors all function at the same level within the ICS, regardless of the rank of the individual who is assigned. Divisions do not report to groups, and groups do not report to divisions; instead, the supervisors are required to coordinate their actions and activities with each other. A division supervisor must be aware of everything that is happening within the assigned geographical area, so a group supervisor must coordinate with the division supervisor when the group enters the division's geographical area, particularly if the group's assignment will affect the division's personnel, operations, or safety. Effective communication among divisions and groups is critical during emergency operations.

■ Branches

A branch is a supervisory level established in either the operations or logistics function to provide an appropriate span of control. At a major incident, several different activities may occur in separate geographical areas or several distinct but related functions may be performed. The span of control might still be a problem, even after the incident commander has established divisions/groups/units. In these situations, the incident commander can establish branches to place a branch director in charge of a number of divisions/groups/units. For example, after a tornado, if damage occurred in two separate areas across town, the incident commander could establish two branches, one for each area affected. If numerous injuries occurred, the incident commander could establish a medical branch to coordinate all aspects of caring for the victims, including triage, treatment, and transportation within each branch.

Fire Officer Greater Alarm Responsibilities

The incident commander calls for additional resources or greater alarms when more resources are needed at an incident. In many cases, the incident commander calls for a greater alarm to have stand-by resources immediately available, in case they are needed or in anticipation of a future need. A company-level officer may be called upon to perform many duties when responding as an additional resource:

- Reinforcing a fire attack strategy by adding resources.
- Relieving an exhausted crew from the first alarm and performing the fire-ground task that they were performing.
- Performing support activities, such as salvage, overhaul, moving equipment to a forward staging area, going door-to-door to evacuate a neighborhood, assisting occupants in recovering personal items, or providing runners to support a face-to-face communications system.
- Maintaining a ready reserve in a staging area.
- Performing additional related duties. This catchall phrase means that the officer and company may be performing tasks that are not traditional fire company tasks, but are time or mission critical. For instance, they could be assigned to an evacuation center to provide assurance and assessment of the occupants who fled from an apartment fire.

Additional resources may be dispatched as strike teams or task forces, as opposed to individual companies. In such a case, the companies are directed to combine forces to respond and operate as a group. The group members could be assigned to meet at a designated location and travel to the scene together or to assemble at the staging area.

■ Staging

Staging refers to a standard procedure to manage uncommitted resources at the scene of an incident. Instead of driving up to the incident scene and committing apparatus and crews without direction from the incident commander, staged resources stand by uncommitted and wait for instructions at a predetermined location, usually some distance away from the incident. These resources are immediately available for use where and when the incident commander needs them. This organization allows the incident commander to determine the most appropriate assignment for each unit.

Many fire departments have policies that automatically commit one or two first-arriving companies to the incident and direct all others to level I staging. In level I staging, only the predesignated units respond directly to the scene, and later-arriving units remain uncommitted and wait for instructions. For example, the second-arriving aerial ladder would stop about a block short of the fire building, where it would have the flexibility to go to whichever side it was needed. Standard operating procedures guide when units stage, where they stage, and how they communicate with the incident commander to indicate they are available for an assignment.

Level II staging, which is generally used for greater alarm incidents, directs responding companies to a designated stand-by location away from the immediate incident scene. A designated officer supervises the level II staging area with the "staging" radio identification and assigns units from the staging area as they are requested by the incident commander. The incident commander might direct the staging officer to maintain a minimum level of resources in the staging area.

Make sure that the terms you use and the assignments you make comply with the ICS used in your jurisdiction. As a fire officer candidate, you will be evaluated on how well you implement the local system.

Task-Level Incident Management

Individual companies, sometimes referred to as single resources, operate at the task level. A company is considered a single resource because it is the basic work unit assigned to perform most emergency scene tasks. Certain types of tasks are routinely assigned to engine and ladder companies. The normal role of a company-level officer is to supervise a group of fire fighters who are operating as a company at the task level.

At a simple incident, the company-level officers report directly to the incident commander. When divisions/groups/units are established, company-level officers report to their assigned division/group/unit supervisor and continue to supervise the operations of their respective companies in performing assigned tasks. Sometimes, the assigned supervisor is one of the company-level officers. Company-level officers advise the division/group/unit supervisor of work progress, preferably through face-to-face contact. All company-level officer requests for additional resources or assistance must go through the division/group/unit supervisor.

■ Task Forces and Strike Teams

Task forces and strike teams are groups of single resources that have been assigned to work together for a specific purpose or for a period of time. Grouping resources reduces the span of control by placing several units under a single supervisor.

A task force includes two to five single resources that are assembled to accomplish a specific task. For example, a task force could be composed of two engines and one truck company, or two engines and two brush units, or one rescue company and four ambulances. A task force operates under the supervision of a task force leader. In many cases, the designated task force leader is one of the company-level officers. All communications for the separate units within the task force are directed to the task force leader.

Task forces are often part of a fire department's standard dispatch philosophy. In the Los Angeles Fire Department, for example, a task force consists of two engines and one ladder truck, staffed by a total of 10 fire fighters. Some departments create task forces of one engine and one brush unit for response during wildland fire season. In this scenario, the brush unit responds with the engine company wherever it goes.

A strike team consists of five units of the same type with an assigned leader. A strike team could be five engines (engine strike team), five trucks (truck strike team), or five ambulances (EMS strike team) **FIGURE 15-7**. This team operates under the supervision of a strike team leader. In most cases, the strike team leader is a command officer.

Strike teams are commonly used for wildland fires, where dozens of companies may respond. During wildland fire season, many departments establish strike teams of five engine companies that are dispatched and work together on major fires. The assigned companies are assigned to rendezvous at

---

Assessment Center Tips

Rapidly Organize Your Greater Alarm
Anticipate the need to call for a special, additional, or greater alarm when handling a simulated emergency incident during a promotional examination. Practice identifying the staging location when calling for additional help. For example:

> Rosedale Command to Communications, transmit a second alarm. Staging for the second alarm units will be at the shopping center parking lot at 5th Street and Avenue D. Designate the officer on the first-arriving engine to be the staging officer.

FIGURE 15-7 A strike team is five units of the same type under one leader.

a designated location and then respond to the scene together. Each engine has an officer and a crew of fire fighters, and one officer—typically a battalion chief—is designated as the strike team leader. All communications for the strike team are directed to the strike team leader.

EMS strike teams, consisting of five ambulances and a supervisor, are often organized to respond to multiple-casualty incidents or disasters. Rather than requesting 15 ambulances and establishing an organizational structure to supervise 15 single resources, the incident commander can request three EMS strike teams and coordinate their operations through the three strike team leaders—a far more efficient arrangement.

■ Greater Alarm Infrastructure

Small fire departments have limited infrastructure support. When they are working at a major incident, a considerable delay might occur before a rehabilitation unit is established with rest and recovery supplies. Fire officers should prepare for this possibility by making sure that their apparatus carries enough water and food to support the fire company for a reasonable period of time. In wildland areas, many companies plan for a 2-day excursion, carrying 5 gallons of water and six energy bars per fire fighter. In the summer, 10 gallons of drinking water per fire fighter per day could be required.

You Are the Fire Officer Conclusion

Captain Davis lays out the situation to Lieutenant Williams when Quint 100 arrives. An outside natural gas fed fire is entering a four-story Type V multiple-family dwelling through exposed framing and failing outside windows. Fire is rapidly expanding due to burning vinyl siding. The driver and two passengers of the car that crashed into the gas meter require heavy hydraulic extrication. The collision has damaged the gas meter before the shut-off.

As the initial incident commander, Davis divides the incident into four divisions:

- Search: Apartment-by-apartment search and removal of apartment occupants.
- Hazardous materials: Stop the natural gas leak.
- Extrication: Removal of the three occupants trapped in the vehicle.
- Suppression: Extinguish the fire.

To accomplish the strategic and tactical objectives, this event will require a greater alarm assignment of resources and liaison with other agencies. Remember the rehabilitation unit.

Wrap-Up

Chief Concepts

- A fire officer is expected to perform the duties of a first-arriving officer at any incident, including assuming initial command of the incident, establishing the basic management structure, and following standard operating procedures.
- The incident management system evolved from the southern California FIRESCOPE and the Phoenix Fire Ground Commander programs.
- The first-arriving fire officer has the responsibility to establish command and manage the incident until relieved by a higher-ranking officer. This task is accomplished in addition to supervising the members of the officer's own company.
- There are three levels of command in the ICS: strategic, tactical, and task.
- The first-arriving managing and supervising fire officers are required to focus on the strategic level as they arrive at an emergency.
- The incident commander is responsible for addressing three strategic priorities: life safety, incident stabilization, and property conservation.
- Command must be established and the ICS must be used at every event.
- The first-arriving company-level officer has three options when arriving at the incident and assuming command: investigation, fast attack, or command mode.
- There are nine functions of command:
  - Determining strategy
  - Selecting incident tactics
  - Setting the action plan
  - Developing the ICS organization
  - Managing resources
  - Coordinating resource activities
  - Providing for scene safety
  - Releasing information about the incident
  - Coordinating with outside agencies
- At a large incident, command could be transferred more than once, depending on the situation and the chain of command, as successively higher-ranking or more qualified officers arrive at the incident scene.
- Negligence lawsuits in Seattle and an OSHA ruling on respiratory protection led to greater emphasis on fire fighter safety and accountability at an emergency incident.
- The incident management system can expand as necessary to handle larger and more complex incidents.
- Individuals on the command staff perform functions that are reported directly to the incident commander. The safety officer, liaison officer, and information officer are always part of the command staff.
- Everything that occurs at an emergency incident can be divided among four major functional components: operations, planning, logistics, and finance/administration.
- The exterior sides of a building are sides A (Alpha), B (Baker), C (Charlie), and D (Delta). The areas adjacent to a burning building are called exposures.
- Some form of review should be conducted at the company level after every call in which the company performs emergency operations.
- Multiple-company incident review is conducted by the incident commander. The preparatory work for this critique is often delegated to one of the company officers.
- The actual critique of an incident should be scheduled as soon as convenient after the event, after allowing sufficient time to gather and prepare the necessary information.
- The last step in conducting an incident review is to write a summary of the incident for departmental records. The results of this review should be documented in a standard format.
- The National Response Framework is a comprehensive, national, all-hazards approach to domestic incident response that describe specific authorities and best practices for managing incidents. It builds upon the National Incident Management System, which provides a consistent template for managing incidents.
- The Stafford Act was amended in 2007 to provide federal government disaster and emergency assistance to state and local governments, tribal nations, eligible private nonprofit organizations, and individuals affected by a declared major disaster or emergency.
- Divisions, groups, and units are where the strategy developed in the incident action plan is converted into action. Supervising and managing fire officers command these incident management assignments.
- Divisions, groups, and units are tactical-level management elements that are used to assemble companies and resources for a common purpose.
- A branch is a supervisory level established in either the operations or logistics function to provide for an appropriate span of control.
- The incident commander calls for additional resources or greater alarms when more resources are needed at an incident.
- Staging refers to a standard procedure to manage uncommitted resources at the scene of an incident.

Wrap-Up, continued

- Individual companies, sometimes referred to as single resources, operate at the task level. A company is considered a single resource because it is the basic work unit assigned to perform most emergency scene tasks.
- Task forces and strike teams are groups of single resources that have been assigned to work together for a specific purpose or for a period of time.
- Small fire departments have limited infrastructure support. Fire officers should prepare for the possibility that support will be delayed by making sure that their apparatus carries enough water and food to support the fire company for a reasonable period of time.

Hot Terms

Branch A supervisory level established in either the operations or logistics function to provide an appropriate span of control.

Branch director A supervisory position in charge of a number of divisions and/or groups. This position reports to a section chief or the incident commander.

Command staff Positions that are established to assume responsibility for key activities in the incident management system that are not a part of the line organization; these include safety officer, public information officer, and liaison officer.

Division A supervisory level established to divide an incident into geographical areas of operations.

Division supervisor A supervisory position in charge of a geographical operation at the tactical level.

Finance/administration section A section of the incident management system that is responsible for the accounting and financial aspects of an incident, as well as any legal issues that may arise.

Group A supervisory level established to divide the incident into functional areas of operation.

Group supervisor A supervisory position in charge of a functional operation at the tactical level. This position reports to a branch director, the operations section chief, or the incident commander.

ICS general staff The group of incident managers composed of the operations section chief, planning section chief, logistics section chief, and finance/administration section chief.

Incident action plan (IAP) The objectives reflecting the overall incident strategy, tactics, risk management, and member safety that are developed by the incident commander. Incident action plans are updated throughout the incident.

Incident Command System (ICS) A system that defines the roles and responsibilities to be assumed by personnel and the operating procedures to be used in the management and direction of emergency operations.

Incident commander The person who is responsible for all decisions relating to the management of the incident and is in charge of the incident site.

Liaison officer The incident commander's representative or a point of contact for representatives from outside agencies.

Logistics section Responsible for providing facilities, services, and materials for the incident. Includes the communications, medical, and food units within the service branch, as well as the supply, facilities, and ground support units within the support branch. The logistics section chief is part of the general staff.

Logistics section chief A supervisory position that is responsible for providing supplies, services, facilities, and materials during the incident. The person in this position reports directly to the incident commander.

National Incident Management System (NIMS) Provides a consistent nationwide template to enable federal, state, tribal, and local governments; the private sector; and nongovernmental organizations to work together to prepare for, prevent, respond to, recover from, and mitigate the effects of incidents, regardless of cause, size, location, or complexity, so as to reduce the loss of life, property, and harm to the environment.

National Response Framework (NRF) A comprehensive, national, all-hazards approach to domestic incident response that describes specific authorities and best practices for managing incidents; it builds upon the National Incident Management System, which provides a consistent template for managing incidents.

Operations section Responsible for all tactical operations at the incident. In the national model, the operations section can be as large as 5 branches, 25 divisions/groups or units, or 125 single resources, task forces, or strike teams.

Wrap-Up, continued

Operations section chief A supervisory position that is responsible for the management of all actions that are directly related to controlling the incident. This position reports directly to the incident commander.

Planning section Responsible for the collection, evaluation, dissemination, and use of information about the development of the incident and the status of resources. It includes the situation status, resource status, and documentation units as well as technical specialists. The planning section chief is part of the general staff.

Planning section chief A supervisory position that is responsible for the collection, evaluation, dissemination, and use of information relevant to the incident. This position reports directly to the incident commander.

Public information officer A command staff position that is responsible for gathering and releasing incident information to the news media and other appropriate agencies. This position reports directly to the incident commander.

Safety officer The person who is responsible for monitoring and assessing safety hazards and unsafe conditions; develops measures to ensure personnel safety.

Staging A specific function in which resources are assembled in an area at or near the incident scene to await instructions or assignments.

Strategic level Command level that entails the overall direction and goals of the incident.

Strike team A specific combination of the same kind and type of resources, with common communications and a leader.

Strike team leader A supervisory position that is in charge of a group of similar resources.

Tactical level Command level in which objectives must be achieved to meet the strategic goals. The tactical-level supervisor or officer is responsible for completing assigned objectives.

Tactical worksheet A form that allows the incident commander to ensure all tactical issues are addressed and to diagram an incident with the location of resources on the diagram.

Task force Any combination of single resources assembled for a particular tactical need, with common communications and a leader.

Task force leader A supervisory position that is in charge of a group of dissimilar resources.

Task level Command level in which specific tasks are assigned to companies; these tasks are geared toward meeting tactical-level requirements.

Two-in/two-out rule A guideline created in response to OSHA Respiratory Regulation (29 CFR 1910.134), which requires a two-person team to operate within an environment that is immediately dangerous to life and health (IDLH) and a minimum of a two-person team to be available outside the IDLH atmosphere to remain capable of rapid rescue of the interior team.

Unit Either a geographical or a functional assignment.

References and Additional Resources

NFPA reprinted material is not the complete and official position of the NFPA on the referenced subject, which is represented only by the standard in its entirety.

Brunacini, A. V. (2002). *Fire Command: The Essentials of Local IMS*. Phoenix, AZ: Heritage Publishers Inc.

Bugbee, P. (1971). *Man against Fire: The Story of the National Fire Protection Association, 1896–1971*. Boston, MA: National Fire Protection Association.

Cole, D. (2002). *The Incident Command System: A 25-Year Evaluation by California Practitioners*. Emmitsburg, MD: U.S. Fire Administration, Executive Fire Officer Program.

Coombs, W. T. (2012). *Ongoing Crisis Communication: Planning, Managing, and Responding* (3rd ed.). Thousand Oaks, CA: Sage.

Deal, T., et al. (2010). *Beyond Initial Response: Using the National Incident Management System's Incident Command System*. Bloomington, IN: AuthorHouse.

Federal Emergency Management Agency. (2011). *National Incident Management System: Training Program*. Washington, DC: U.S. Department of Homeland Security.

Federal Emergency Management Agency. (2012). *National Preparedness Report*. Washington, DC: U.S. Department of Homeland Security.

Federal Emergency Management Agency. (2013). *National Response Framework* (2nd ed.). Washington, DC: U.S. Department of Homeland Security.

FIRESCOPE. (2001). *Fire Service Field Operations Guide: ICS 420-1*. Rancho Cordova, CA: Governor's Office of Emergency Services.

Fowler, D. E. (1990). *Report on the Operations of the San Francisco Fire Department Following the Earthquake and Fire of October 17, 1989*. San Francisco, CA: San Francisco Fire Department.

Gasaway, R. B. (2013). *Situational Awareness for Emergency Response*. Tulsa, OK: Pennwell/Fire Engineering.

Griffins, J. S. (2012). *Fire Department of New York: An Operational Reference* (9th ed.). Los Alamos, NM: James S. Griffin.

Hensler, B. (2011). *Crucible of Fire: Nineteenth-Century Urban Fires and the Making of the Modern Fire Service*. Washington, DC: Potomac Books.

Kimball, W. Y. (1966). *Fire Attack 1: Command Decisions and Company Operations*. Boston, MA: National Fire Protection Association.

Kimball, W. Y. (1968). *Fire Attack 2: Planning, Assigning, Operating*. Boston, MA: National Fire Protection Association.

Layman, L. (1953). *Fire Fighting Tactics*. Boston, MA: National Fire Protection Association.

Maniscalco, P. M., and H. T. Christen. (2010). *Homeland Security: Principles and Practice of Terrorism* Response. Sudbury, MA: Jones and Bartlett.

Mills, S. E. (2012). *Effective Emergency Operations Preparedness for Recurring Planned Public Events*. Emmitsburg, MD: Executive Fire Officer Program.

Murtaugh, M. (1993). *Fire Department Promotional Tests. A New Direction: New Testing Components, New Testing Formats*. Pearl River, NY: Fire Tech Promotional Courses.

National Fallen Firefighters Foundation. (2011). *Understanding and Implementing the 16 Firefighter Life Safety Initiatives*. Stillwater, OK: Fire Protection Publications.

Page, J. O. (1973). *Effective Company Command*. Alhambra, CA: Borden.

Peaks, M. (2011). *Training Bulletin No. 76: Company Operations*. Los Angeles, CA: Los Angeles Fire Department.

Robinson, G. J. (2011). *Lasting Leadership: Preparing for the Transition to Company Officer*. Emmitsburg, MD: Executive Fire Officer Program.

Schilling, J-P. (2012). *Critical Decision Making: Improving Fire Officer Development*. Emmitsburg, MD: U.S. Fire Administration, Executive Fire Officer Program.

Smeby, L. C. Jr. (2014). *Fire and Emergency Services Administration: Management and Leadership Practices* (2nd ed.). Burlington, MA: Jones & Bartlett Learning.

Warman, J. A. (2011). *Executive Analysis of Fire Service Executive Leadership*. Emmitsburg, MD: U.S. Fire Administration, Executive Fire Officer Program.

FIRE OFFICER
in action

Although the federal government has established a National Incident Management System, it is the local fire department that uses its own standard operating procedures and actually organizes the resources to control the incident. For example, the vehicle collision in the apartment parking lot received a standard local response of an engine company and a paramedic ambulance.

After the natural gas explosion, Captain Davis reported the change in conditions and called for a first-alarm assignment for an apartment fire and gas leak. Local procedures dispatched four engines, two aerial apparatus, a heavy rescue company, and two command officers. The Type V constructed building had combustible vinyl siding, and flames entered the second- and third-floor apartments through windows that were broken by the explosion or fire.

Battalion Chief Johnson called for a second alarm, bringing another six fire engines to the incident. The regional hazardous materials response team was dispatched when command confirmed that the shut-off at the natural gas meter was destroyed. As part of the local protocol, a half-dozen support units were dispatched with the second alarm, including a rehabilitation unit, a safety officer, a fire investigator, a paramedic supervisor, a public information officer, and the on-duty executive fire officer.

Attack lines were deployed into six apartments to keep the fire from extending into the apartment building. The Incident Command System was expanded to include a branch for the relocation of displaced residents. The incident commander assigned the hazardous materials team to a sector to assist the gas company when difficulties were encountered in isolating and closing the gas service.

The incident management system provides a framework to handle all types of incidents at any level of complexity, ranging from an investigation of an odor in a building to a major incident.

1. Which of the following is *not* a responsibility of the incident commander?
   A. Stabilize the incident
   B. Stop property loss
   C. Maintain public order
   D. Protect life

2. Documentation of the incident strategy, tactics, and risk management during a specific time period at an incident is found in the:
   A. after-action report.
   B. incident action plan.
   C. public information officer's press release.
   D. Stafford Act reimbursement log.

3. A group of five pumpers operating together under an assigned leader is called a:
   A. wildland greater alarm assignment.
   B. task force.
   C. mutual aid disaster team.
   D. strike team.

4. The fast-attack mode ends when:
   A. the company officer must withdraw and establish a command post.
   B. level I staging is filled.
   C. the first administrative fire officer arrives at the incident scene.
   D. a greater alarm is requested by the company officer.

FIRE OFFICER *in action* (continued)

Fire Captain *Activity*

With the natural gas explosion and people trapped in the vehicle, Captain Davis was very engaged as the initial incident commander. When Chief Johnson arrived, both officers went to the back of the Battalion 2 vehicle to set up the command post.

Chief Johnson told Captain Davis, "Take your preincident plan and lay out an incident action plan." The chief's command vehicle had a whiteboard designed to be filled in with an incident action plan as well as track crew assignments. As Davis starts documenting the plan, the chief calls for a second alarm.

NFPA Fire Officer II Job Performance Requirement 5.6.1

Produce operational plans, given an emergency incident requiring multiunit operations, the current edition of NFPA 1600, and AHJ-approved safety procedures, so that required resources and their assignments are obtained and plans are carried out in compliance with NFPA 1600 and approved safety procedures resulting in the mitigation of the incident.

Application of 5.6.1

1. Using the incident description from this chapter's You Are the Fire Officer scenario, develop an incident action plan for operations. Details of the incident at the time of the IAP:
   - An active high-pressure natural gas leak is burning on side B of the apartment building. The gas meter is extensively damaged and the gas cannot be turned off at the meter.
   - There is a vigorous outside fire on side B, fueled by combustible vinyl siding. Fire is entering the four-story apartment through the exposed wood framing and broken windows on the second and third floors.
   - The building is an unsprinklered 15-year-old Type V multifamily building with central corridor, elevators, and standpipe. There are eight apartments per floor.
   - A sedan struck the building at a high rate of speed; the driver and two passengers are trapped in the vehicle.
   - The second alarm units have been requested but have not arrived. Need to show IAP assignments for the first and second alarm units.

Resources:
- Initial dispatch: Engine 100 and Medic 100.
- Special call (requested by Engine 100): Quint 100.
- Balance of first alarm (called by Engine 100 when gas explodes—10 minutes from initial dispatch): Engines 11, 57, and 41. Tower 03. Rescue 57. Medic 03. Battalion 2 and 7.
- Second alarm (called by Battalion 2—15 minutes after initial dispatch): Engine 84, 109, 76, and 5. Truck 57 and Quint 109. Rescue 76. Medic 41 and 109. Battalion 9 and 1. Rehab unit. Safety officer. Hazardous materials response team. Fire investigator. EMS supervisor. Building inspector. Public information officer. Operations chief.

NFPA Fire Officer II Job Performance Requirement 5.6.2

Develop and conduct a postincident analysis, given multiunit incident and postincident analysis policies, procedures, and forms, so that all required critical elements are identified and communicated and the approved forms are completed and processed.

Application of 5.6.2

1. Using an organization with which you are familiar, review an after-action report and describe the critical areas identified in the report and the way in which the organization addressed the issues.

Rules of Engagement

Fire Officer I

Knowledge Objectives

After studying this chapter, you will be able to:
- Discuss the origins of the Rules of Engagement (ROE). (pp 318–319)
- Describe the scope of the line-of-duty death problem. (p 319)
- Describe the elements of size-up (NFPA 4.6.1). (p 320)
- Discuss occupant survivability in terms of fire behavior and fire fighter safety (NFPA 4.6.1). (pp 320–321)
- Discuss the concept that lives should not be risked for lives or property that cannot be saved (NFPA 4.6.1). (pp 321–322)
- Discuss the level of risk extended to protect savable property (NFPA 4.6.1). (pp 322–323)
- Discuss the level of risk extended to protect savable lives (NFPA 4.6.1). (pp 323–324)
- Discuss the importance of fire fighters staying together (NFPA 4.6.1). (pp 324, 326)
- Discuss the importance of maintaining situational awareness (NFPA 4.6.1). (pp 326–327)
- Discuss the importance of fire-ground communications (NFPA 4.2.1). (pp 327–328)
- Discuss the importance of reporting unsafe practices or conditions. (pp 328–329)
- Discuss the value of retreat in unsafe conditions (NFPA 4.6.1). (p 330)
- Describe when a mayday should be declared. (pp 330–331, 333)

Skills Objectives

After studying this chapter, you will be able to:
- Conduct a size-up of an incident (NFPA 4.6.1). (p 320)
- Maintain fire-ground communications (NFPA 4.2.1). (pp 327–328)
- Report unsafe practices or conditions. (pp 328–329)

Fire Officer II

Knowledge Objectives

After studying this chapter, you will be able to:
- Discuss the origins of the Rules of Engagement (ROE). (pp 318–319)
- Describe the scope of the line-of-duty death problem. (p 319)
- Describe the elements of size-up (NFPA 5.6.1). (p 320)
- Discuss occupant survivability in terms of fire behavior and fire fighter safety (NFPA 5.6.1). (pp 320–321)
- Discuss the concept that lives should not be risked for lives or property that cannot be saved (NFPA 5.6.1). (pp 321–322)
- Discuss the level of risk extended to protect savable property (NFPA 5.6.1). (pp 322–323)
- Discuss the level of risk extended to protect savable lives (NFPA 5.6.1). (pp 323–324)
- Discuss the importance of fire fighters staying together (NFPA 5.6.1). (pp 324, 326)
- Discuss the importance of maintaining situational awareness (NFPA 5.6.1). (pp 326–327)
- Discuss the importance of fire-ground communications (NFPA 5.6.1). (pp 327–328)
- Discuss the importance of reporting unsafe practices or conditions (NFPA 5.6.1). (pp 328–329)
- Discuss the value of retreat in unsafe conditions (NFPA 5.6.1). (p 330)
- Describe when a mayday should be declared (NFPA 5.6.1). (pp 330–331, 333)

Skills Objectives

After studying this chapter, you will be able to:
- Conduct a size-up of an incident (NFPA 5.6.1). (p 320)
- Maintain fire-ground communications (NFPA 5.6.1). (pp 327–328)
- Report unsafe practices or conditions (NFPA 5.6.1). (pp 328–329)

CHAPTER 16

Additional NFPA Standards

- NFPA 1404, *Standard for Fire Service Respiratory Training*
- NFPA 1500, *Standard on Fire Department Occupational Safety and Health Program*
- NFPA 1561, *Standard on Emergency Services Incident Management System*
- NFPA 1852, *Standard on Selection, Care and Maintenance of Open-Circuit Self-Contained Breathing Apparatus (SCBA)*
- NFPA 1981, *Standard on Open-Circuit Self-Contained Breathing Apparatus for Emergency Services (SCBA) For Emergency Services*
- NFPA 1982, *Standard for Personal Alert Safety Systems (PASS)*

Principles of Fire and Emergency Service Administration (FESHE) Course Objectives

2. Recognize the need for effective communication skills both written and verbal. (pp 326, 328–331)
6. Describe the various levels of leadership, roles, and responsibilities within the organization. (pp 320–324, 326–330)
9. Identify the roles of the National Incident Management System (NIMS) and Incident Management System (ICS). (pp 320–322, 324, 326, 331–333)

You Are the Fire Officer

It was a somber dinner at Fire Station 100. The television news show was covering a greater alarm commercial fire 100 miles away where fire fighters were trapped and missing. Watching the live stream video from a website, it was clear that the department was experiencing a tragedy.

Chief Johnson was having dinner with the crew and shared his frustration that the department was experiencing far fewer major fires, but the ones they had seemed to be more dangerous. At an after-dinner meeting with Captain Davis and Lieutenant Williams, the chief shared the International Association of Fire Chiefs' "Rules of Engagement." Chief Johnson directed Station 100 to prepare an in-service "Rules of Engagement for Structural Firefighting" presentation for the battalion.

1. What are common factors found in fire-ground line-of-duty deaths?
2. Which situations should trigger a "no go" tactical response?
3. Which fire-ground behaviors need to be changed to reduce line-of-duty fire-ground deaths?

Introduction

The *Rules of Engagement for Structural Firefighting* (IAFC Safety, Health, and Survival Section 2012) focus on the fire fighter and fire officer working at the task level of fire suppression operations within a high risk environment **TABLE 16-1**.

Table 16-1 Rules of Engagement for Structural Firefighting

1. Size up your tactical area of operation.
2. Determine the occupant survival profile.
3. *Do not* risk your life for lives or property that cannot be saved.
4. Extend *limited* risk to protect savable property.
5. Extend *vigilant* and *measured* risk to protect and rescue *savable* lives.
6. Go in together, stay together, come out together.
7. Maintain continuous awareness of your air supply, situation, location, and fire conditions.
8. Constantly monitor fire-ground communications for critical radio reports.
9. You are required to report unsafe practices or conditions that can harm you. Stop, evaluate, and decide.
10. You are required to abandon your position and retreat before deteriorating conditions can harm you.
11. Declare a mayday as soon as you THINK you are in danger.

Reproduced from: IAFC Safety, Health, and Survival Section. (n.d.) "Rules of Engagement for Firefighter Survival." Rules you can LIVE By. http://www.iafcsafety.org/downloads/ROE_Poster_FINAL.pdf

Fire Officer I and II

How the Rules Came to Be

The Rules of Engagement (ROE) evolved from fire fighter survival "Rules" developed by the Seattle Fire Department and International Association of Fire Chiefs (IAFC) Safety Committee. In 2004, the IAFC established an expanded IAFC Safety, Health, and Survival Section. Many of the chief and company officer committee members working on the ROE had experienced a fire fighter fatality in their organization. The IAFC Board of Directors endorsed Rules of Engagement in August 2011 as "best safety practice model procedures." (IAFC 2012)

The *Rules of Engagement for Structural Firefighting* (Safety, Health, and Survival Section 2012) focus on the fire fighter and fire officer working at the task level of fire suppression operations. Covering high-risk operations in a high-hazard environment required that the rules meet the following criteria:

- Be a short, specific set of bullet points
- Be easily taught and remembered
- Define critical risk issues
- Define "go" and "no go" situations

Understanding the Scope of the Problem

A survey of 23 fire fighter fatalities compared the type of structures involved in the line-of-duty deaths (LODD). In *U.S. Firefighter Disorientation Study: 1979–2001* (Mora, 2003), Captain William R. Mora compared the results of opened and enclosed structures. An opened structure has windows or doors of sufficient number and size to provide for prompt ventilation and emergency evacuation. An enclosed structure is one lacking windows or doors of sufficient number and size to provide for prompt ventilation and emergency evacuation.

In the fires covered in Mora's study, fast and aggressive interior attacks were conducted within both types of structures. During those attacks in zero-visibility conditions, crew integrity was lost and fire fighters moved away from the hose line. Disorientation would then occur as fire fighters exceeded their air supply, were caught in flashovers or backdrafts, or were trapped by collapsing floor or roof. More than three-fourths of the disoriented fire fighter deaths occurred in enclosed structures.

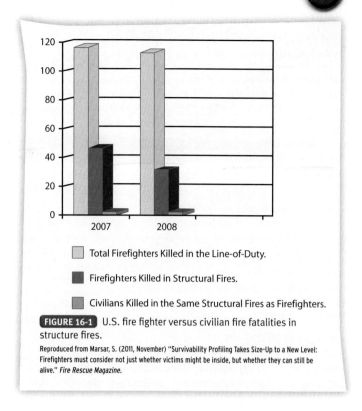

FIGURE 16-1 U.S. fire fighter versus civilian fire fatalities in structure fires.

Reproduced from Marsar, S. (2011, November) "Survivability Profiling Takes Size-Up to a New Level: Firefighters must consider not just whether victims might be inside, but whether they can still be alive." *Fire Rescue Magazine*.

■ Survivability Profiling

Captain Stephen Marsar explored the concept of survivability profiling while enrolled in the Executive Fire Officer program at the National Fire Academy. During his 19 years with the Fire Department of New York City, 32 members died while operating within a burning structure, yet no civilians perished in those fires (Marsar 2011). Using data from the National Fallen Firefighters Foundation, Marsar showed that this was a national experience **FIGURE 16-1**.

■ Fire Fighter Survivability inside Structure Fires

A 30-year review of NFPA annual fire fighter fatality reports by the Fire Analysis and Research Division identified a concerning trend (Fahy 2007, p 4):

> The one area that had shown marked increases during the period is the rate of deaths due to traumatic injuries while operating inside structures. In the late 1970s, traumatic deaths inside structures occurred at a rate of 1.8 deaths per 100,000 structure fires and by the late 1990s had risen to approximately 3.0 deaths per 100,000 structure fires. Since that time, the rate has fallen, and now stands at 1.9 deaths per 100,000 structure fires, a rate only slightly lower than that observed in the early 1980s. Almost all of these non-cardiac fatalities inside structure fires were the result of smoke inhalation (62.1 percent), burns (19.1 percent), and crushing or internal trauma (16.5 percent).

Reprinted with permission from NFPA's report, What's Changed Over the Past 30 Years? by Fahy, R. F., LeBlanc, P. R. and Molis, J. L., Copyright © 2007, National Fire Protection Association.

Clearly, the fire department as well as the company officer should minimize fire fighter exposure to unsafe conditions and stop unsafe practices. Doing so, however, will require a change in command perspective. The fire service is a paramilitary organization on the fire ground. As part of this organizational structure, the incident commander makes a decision and sends the order down through supervisors to the task group. Fire crews view these orders as top-down direction. Consequently, there may be little two-way discussion about options.

Fire crews have been trained to accept the order and to comply without discussion. Although valid when issued, these orders may sometimes involve inadequate size-up and risk assessment, placing fire fighters at extreme risk. What appeared to be safe at the time of the order may quickly become unsafe within minutes.

The ROE integrate the fire fighter into the risk assessment decision-making process. Fire fighters and the company officers are the members at greatest risk for injury or death and often will be the first to identify unsafe conditions and practices. In turn, these members should be the ultimate decision makers regarding whether it is safe to proceed with assigned objectives. In situations where it is not safe to proceed, the ROE allow a process for correct decision making while still maintaining command unity and discipline.

The development of the ROE integrates several nationally recognized programs and principles. They included risk assessment principles from NFPA 1500, *Standard on Fire Department Occupational Safety and Health Program*, and NFPA 1561, *Standard on Emergency Services Incident Management System*. They also integrate concepts and principles from crew resource management as well as data and lessons learned from the National Near-Miss Reporting System, the National Fallen Firefighters Foundation's 16 Life Safety Initiatives, and fire fighter fatality investigation reports from the National Institute of Occupational Safety and Health (NIOSH).

Rule 1. Size up Your Tactical Area of Operation

Objective: *To cause the company officer and fire fighters to pause for a moment and look over their area of operation, evaluate their individual risk exposure, and determine a safe approach to complete assigned tactical objectives* (IAFC, 2012).

NIOSH fire fighter fatality investigations often cite lack of a complete size-up as a contributing factor in fire fighter deaths. Members must conduct a rapid, deliberate size-up on their side of the incident, or operational area, to determine their risk exposure and select the safest approach to achieve the objectives assigned by command. If the risk that must be assumed to meet the objective is unacceptable, that decision must be communicated to the incident commander or supervising command officer, and the objective or action plan must be adjusted to make the situation safer.

Each side of the fire ground has unique fire conditions requiring risk assessment. The fire attack crew should take a few seconds to size up the total situation within their line of sight. They must evaluate what is burning, where it is, and where the fire is likely to go. Evaluating these factors can allow fire fighters to forecast future conditions and predict their individual risk.

As part of the size-up, the first company officer, or team leader, must cover each side of the fire ground (conduct a 360-degree walk-around) and provide an early progress report to the incident commander. If significant risks are identified or other important information is observed that will affect safety or the action plan in the operational area, that information must be communicated to the incident commander or other supervising command officer.

Fire Fighter Near-Miss Report #10-157 demonstrates how quickly the fire ground can change:

> When we arrived on scene, we found a restaurant on fire. I was the nozzle man on the initial attack. I noticed several HVAC units on the roof and their relation to our entry point. When we entered the kitchen, there was heavy fire underneath the HVAC units so I advised the crew to move back. As we left the building, there was a complete roof collapse in the area our crew had been working. Because of situational awareness . . . we were all able to go home. There was only a small flame visible over the back door when we made entry. What you see outside before entry can quickly change.

Courtesy of the National Fire Fighter Near-Miss Reporting Systems (firefighternearmiss.com) "Good 360 and Situational Awareness Saves Crew,10-0000157" Fairfax, VA: International Association of Fire Chiefs.

Safety Zone

No Go: Size-up
If the assigned objective cannot be achieved because existing conditions are unsafe and prevent success, stop and report the situation to the incident commander and revise the objective and action plan.

Fire Marks

Dramatic Fire-Ground Changes
Underwriters Laboratory (UL) compared fire development in legacy homes (dating from the 1950s to 1960s) and modern homes. The fires in the legacy homes took 29 minutes and 25 seconds to reach flashover. In the modern home, flashover occurred in just 3 minutes and 40 seconds (Kerber 2012). This rapid flashover time for the modern home quickly reduces the survival profile of any trapped victims and increases the risk to fire fighters.

If a fire fighter is in a room within a modern home that is approaching the flashover point, the time from untenable heat to flashover is less than 10 seconds. This does not allow much of a survival period for the fire fighter to exit a building. Some lightweight unprotected truss systems can collapse as quickly as 6.5 minutes after flame impingement, often without warning (Kerber 2012).

Rule 2. Determine the Occupant Survival Profile

Objective: *To cause the company officer and fire fighter to consider fire conditions in relation to possible occupant survival of a rescue event as part of their initial and ongoing individual risk assessment and action plan development* (IAFC, 2012).

An essential component in the size-up process is to determine whether any occupants are trapped and whether they can survive the current and projected fire conditions. Fire fighters save lives through aggressive search and rescue operations by first-arriving fire companies. Unfortunately, these activities also generate the greatest risk of line-of-duty injury and death. A safe and appropriate action plan cannot be accurately developed until company members first determine whether any occupants are trapped and can survive the fire conditions during the entire search and removal rescue event.

If survival is not possible for the entire search, locate, and removal period, a more cautious approach to fire operations must be taken. Fire control should be achieved before proceeding with the primary and secondary search efforts. NIOSH Fire Fighter Fatality Investigation F2010-10 provides an example of rapid fire development overwhelming the rescue effort (NIOSH 2010b):

Safety Zone

No Go: Unsavable Life
If the occupant(s) cannot survive the search *and* rescue event, do not commit to this operation. Achieve fire control before searching for victims.

[A 9-1-1] caller described an elderly man who was on an oxygen supply in a wheelchair that was on fire. Units arrived on scene to find heavy fire conditions at the rear of a house and moderate smoke conditions within the uninvolved areas of the house.

Thick, black rolling smoke banked down to knee level after the hose line was advanced 12 feet into the kitchen area. While ventilation activities were occurring, the search and rescue crew observed fire rolling across the ceiling within the smoke. They immediately yelled to the hose line crew to "get out." The [fire fighter] victim was found wrapped in the 2½-inch hose line that had ruptured and without his face piece on.

■ Today's Smoke Is More Toxic

The Providence Fire Department (R.I.) operated at three structure fires on March 23 and 24, 2006. At each incident, a fire fighter suffered a heart attack. Medical assessment included discovery of high levels of cyanide in the fire fighters' bloodstreams. Many other crew members operating at those fires complained of symptoms consistent with cyanide poisoning, including headaches, weakness and fatigue, nausea, and shortness of breath. Of the 27 members who were evaluated, 8 were found to have elevated levels of cyanide. *Report of the Investigation Committee into the Cyanide Poisonings of Providence Firefighters* determined that the cyanide was a frequent product of combustion and "concurs with the growing list of experts who have concluded that hydrogen cyanide poses a much more significant problem to firefighters than previously believed" (Varone, 2006).

The fire fighter and incident commander must factor growing fire conditions, resources on scene, and the time needed to complete a rescue into the decision to conduct and support primary search and rescue operations. Search and rescue and the related removal of any trapped victims from the fire building take time, and most often these operations are occurring while conditions continue to deteriorate quickly. This situation decreases the possibility of victim survivability while increasing risk to fire fighters. A search and rescue decision must be balanced against the time available and the current and evolving conditions. In some cases, search and rescue must be suspended because of deteriorating conditions until the fire is controlled FIGURE 16-2.

Rule 3. Do Not Risk Your Life for Lives or Property That Cannot Be Saved

Objective: *To prevent fire fighters from engaging in high-risk search and rescue and firefighting operations that may harm them when fire conditions prevent occupant survival and significant or total destruction of the building is inevitable (IAFC 2012).*

The company officer and the incident commander must recognize that they cannot always save a life. If conditions indicate no occupant can survive the current and projected fire conditions in the search compartment, then search and rescue operations should be suspended until the fire is controlled.

Numerous NIOSH fire fighter fatality reports cite cases where fire fighters were killed while operating in buildings where fire conditions would be clearly defined as defensive fires. Where such conditions exist, a defensive strategy must be considered by the incident commander at the outset of firefighting operations.

■ Lives That Could Not Be Saved

Three of four fire fighters perished during a rescue attempt at a residential house fire. A quint, staffed with the shift commander and apparatus operator, and an engine, staffed with a fire officer and engine operator, responded to a structure fire with a report of children trapped.

Arriving crews had smoke showing from the structure. A woman and child were trapped on a porch roof and said that three children were still in the house. The shift commander called for six additional fire fighters. As search and rescue started, the chief and a fire fighter arrived.

The chief established incident command and the search team brought out one of the trapped children to the incident commander in cardiac arrest. The chief started CPR on the child. A second child was brought out in cardiac arrest. The fire fighter who arrived with the chief approached the front door and noticed that the 1½-inch handline that was stretched by the first arriving crew had been burned through and water was free flowing. It is believed that the three victims were hit with a thermal blast of heat before the handline burned through.

One of the deceased fire fighters was found on the first floor, another on the stairway to the second floor, and the third at the top of the stairs with the third child (NIOSH 2001b).

The first child was removed in cardiac arrest—the first indication that the chance that additional victims might survive was small. As conditions continued to worsen, a second child

FIGURE 16-2 Example of a structure fire with no surviving occupants within the compartment.

> **Safety Zone**
>
> **No Go: Unsurvivable Conditions**
> - If fire conditions prevent occupant survivability of any rescue event
> - If the fire has, or will, destroy the building

Rule 4. Extend Limited Risk to Protect Savable Property

Objective: *To cause fire fighters to limit risk exposure to a reasonable, cautious, and conservative level when trying to save a building* (IAFC, 2012).

<u>Limited</u> is defined as "the point, edge, or line beyond which something cannot or may not proceed, confined or restricted within certain limits." There is a limit, or line, beyond which the company officer and incident commander may not allow fire fighters to be exposed to unsafe fire conditions. If a building can be saved, limited risk and carefully calculated operations should be employed and operations must be continuously monitored to ensure fire fighter safety.

The key concept in this rule is *savable*. No building is worth the life of a fire fighter. If conditions worsen and become unsafe during interior operations, other safe approaches must immediately be considered, or crews must be withdrawn from the building in a timely fashion and defensive exterior operations be employed. Most buildings lost to fire are rebuilt, so the loss is not always permanent.

Interior firefighting operations must be fully supported with adequate resources on scene. The risk of engaging in such operations must be closely and continually assessed during interior operations. A fire that cannot be controlled quickly will continue to eat away at the building's structural integrity, weakening it and thus increasing risk of collapse.

Where the building is deemed savable, deploy attack hose lines of proper size and number to achieve rapid fire control. Adequate staffing should be available to conduct effective operations. It may be appropriate to use large-caliber hose lines or apparatus-mounted monitor devices to knock down fire quickly before crews enter a building.

Safety Zone

No Go: Unsavable Building
If the building cannot be saved, it's a no go. Consider an exterior defensive strategy.

NIOSH Fire Fighter Fatality Investigation F2000-13 involved a double fire fighter fatality at a small, stand-alone fast-food restaurant (NIOSH 2001a):

On February 14, 2000, a 44-year-old male and a 30-year-old female, both career fire fighters, died in a restaurant fire. At 0430 hours, central dispatch received a call reporting fire in a building. The first fire companies on scene reported there was visible fire emitting through the roof of a closed fast-food restaurant. The engine officer radioed dispatch reporting that he and his crew were going to complete a "fast attack." After making forcible entry, the victims entered with a 1¾-inch hose line as their captain finished donning his gear.

Soon after a district chief arrived, he requested that all companies convert to a defensive attack and evacuate the structure due to heavy fire conditions. At this point the middle roof section (over the kitchen) of the building had collapsed. An interior evacuation took place, and neither of the victims exited. The IC sent several fire fighters inside to search for the victims. One fire fighter was located and removed and was transported to a local hospital, where he was pronounced dead. The fire fighters located the other victim sometime later and she was pronounced dead at the scene. **FIGURE 16-3**

Fire Fighter Near-Miss Report #09-578 illustrates the need for vigilance and coordination during long-duration operations (IAFC 2009):

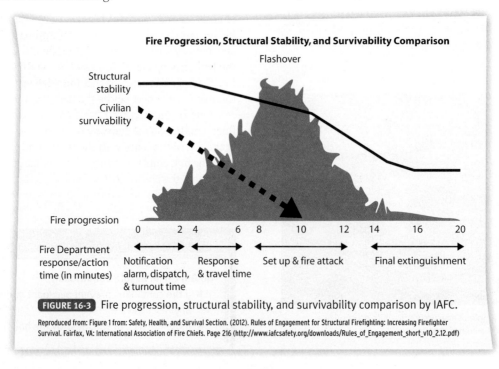

FIGURE 16-3 Fire progression, structural stability, and survivability comparison by IAFC.

Reproduced from: Figure 1 from: Safety, Health, and Survival Section. (2012). Rules of Engagement for Structural Firefighting: Increasing Firefighter Survival. Fairfax, VA: International Association of Fire Chiefs. Page 216 (http://www.iafcsafety.org/downloads/Rules_of_Engagement_short_v10_2.12.pdf)

Crews had been operating for 2 hours at a structure fire in a large manufacturing facility. We were operating in a defensive mode, and had been since our arrival. My crew was located about 50 feet away from the building on a hill above the building. I was summoned to meet face-to-face with the battalion chief. The battalion chief was accompanied by an acting battalion chief, and he directed me to move my crew and line closer to the building. He also directed me to open two doors on the B side of the building so we could get water directly on the fire.

While we were attempting to force one of the doors on the B side, the brick and cinder block wall collapsed, narrowly missing one crew member who was in the doorway. Crew members who were standing farther back were hit by the bricks. I reported to command that all personnel were okay. Command was completely unaware that any crew was operating in the collapse zone and had not given any order to open any doors. The battalion chief who gave the order was not in command and was not assigned as a sector officer. He had been conducting his own 360-degree walk-around of the building when he gave the order.

Courtesy of the National Fire Fighter Near-Miss Reporting Systems (firefighternearmiss.com). (2009 June 13) "Collapse of building injures FFs., 09-0000578" Fairfax, VA: International Association of Fire Chiefs.

Rule 5. Extend Vigilance and Measured Risk to Protect and Rescue Savable Lives

Objective: *To cause fire fighters to manage search and rescue and supporting firefighting operations in a calculated, controlled, and safe manner, while remaining alert to changing conditions, during high-risk primary search and rescue operations where lives can be saved IAFC, 2012).*

During search and rescue operations, crews must remain <u>vigilant</u> to changing fire conditions that may increase the risk to them or prevent survival of occupants. All members should be vigilant and <u>measured</u>—restrained, calculated, and deliberate—when applying strategy and tactics during a search and rescue event.

Being alert and watchful means continually assessing fire conditions throughout the rescue event. Conditions will either be deteriorating or improving. Vigilance also requires monitoring the radio for reports of conditions occurring elsewhere on the fire ground. Worsening conditions observed from the exterior or elsewhere on the fire ground can quickly increase the risk to fire fighters involved in search operations **FIGURE 16-4**. The term *savable lives* is used to describe occupants who may potentially survive both the fire conditions and the rescue event.

NIOSH Fire Fighter Fatality Investigation Report F2007-28 examines a double fire fighter fatality during a primary search and rescue (NIOSH 2009b):

On July 21, 2007, a 34-year-old career captain and a 37-year-old engineer (riding in the fire fighter position) died while conducting a primary search for two trapped civilians at a residential structure fire. The two victims were members of the first-arriving crew. They made a fast

FIGURE 16-4 Example of a deteriorating conditions.

attack and quickly knocked down the visible fire in the living room. After exiting the building, they requested vertical ventilation, grabbed a thermal imaging camera, and made reentry without a handline to search for the two residents known to be inside. Another crew entered without a handline and began a search for the two residents in the kitchen area. A positive-pressure ventilation fan was set up at the front door to increase visibility for the search teams. The second crew found and was removing one of the civilian victims from the kitchen area when rollover was observed extending from the hallway into the living room. Fire fighters became concerned for the air supply of both victims who were still in the structure. Other fire crews conducted a search for the two fire fighters and found them in a back bedroom, where they had been overcome by the rapid fire event.

Courtesy of the National Fire Fighter Near-Miss Reporting Systems (firefighternearmiss.com) (2007 March 10) "FF falls through burnt floor, 07-0000789" Fairfax, VA: International Association of Fire Chiefs.

Two crews without handlines were performing a primary search and occupant rescue. Introducing a positive pressure fan to clear the smoke resulted in accelerated fire development that overwhelmed one of the crews. A lack of vigilance and inadequate fire suppression resources were factors in this incident.

■ Deteriorating Conditions

Firefighter Near-Miss Report #07-789 provides a first-person perspective on searching in deteriorating conditions (IAFC 2007):

Responding to a residential structure fire, my crew was given the assignment of performing a primary search. Fire attack had already been assigned to another crew.

> **Safety Zone**
>
> **No Go: Inadequate Resources for Interior Operations**
> If you do not have the resources to conduct a safe search and rescue or firefighting operations, do not conduct an interior operation.

My officer assumed command while my crew member and I started with our task. The attack crew encountered heavy fire and heat shortly inside the front door off to the right. We started a left-hand search covering approximately two rooms.

At this point, the attack crew advised command that they had found a hole in the floor, and command relayed this information to all units on the scene. Stopping to check with my partner to make sure we were still together, I turned to resume our search. As soon as I started forward, I found myself falling about 10 feet into the basement. The area in which I fell turned out to be the main area of fire involvement.

My first actions were to collect myself and to get my face piece repositioned properly, as it had become partly dislodged. Surrounded by debris and fire, I had to get into a position to facilitate me communicating on the radio. My crew member stuck an axe handle in the hole for me to feel, which gave me comfort in knowing someone was with me. I tried to follow our department's standard operating procedures as best I could while maintaining calmness. I notified command of my unit identification, what had occurred, and my physical status.

Upon command acknowledging my situation, I then operated my PASS device in short bursts to allow crews to locate me as well as to be able to communicate verbally. The attack crew was at this same hole but in a different location. It should be noted, the "hole" was about 8 feet by 12 feet in size. The nozzle man had partly fallen into the hole but was pulled back by his crew. Because of their position, they were able to knock fire down around me once they realized I had fallen.

Command had advised the RIC (rapid intervention crew) to go in and assist the downed fire fighter; another crew entered with a thermal imaging camera. The crew with the camera was able to find my exact location. With that, an attic ladder was called for and I was able to climb out of the basement. I was fortunate that I was able to come away from this incident with only some strains, bumps, and bruises.

Rule 6. Go in Together, Stay Together, Come out Together

Objective: *To ensure that fire fighters always enter a burning building as a team of two or more members and no fire fighter is allowed to be alone at any time while entering, operating in, or exiting a building (IAFC, 2012).*

NIOSH Fire Fighter Fatality Investigation Reports frequently identify situations when a fire fighter is alone in a burning structure:

- Fire fighter separation from partners or crew members
- Single fire fighter freelancing
- Fire fighter entering a structure alone
- Fire fighter leaving his or her partner or crew to exit alone when low on SCBA air

A critical element for fire fighter survival is crew integrity. Crew integrity means fire fighters stay together as a team of two or more. They enter a structure together and remain together at all times while in the interior, and all members come out together. No fire fighter shall be allowed to be alone at any time while in a burning structure.

It is an individual responsibility of every fire fighter to stay connected with his or her partner or crew members at all times. Freelancing by any member must be strictly prohibited. Additionally, crews or buddy teams must never freelance. All fire fighters must operate under the direction of the incident commander or division/group supervisor or their company officer/lead buddy team member at all times.

The ultimate responsibility for enforcing the principle of crew integrity and ensuring that no members get separated or lost rests with the company officer or lead buddy team member. These individuals must maintain constant contact with their assigned members by voice, touch, or visual observation while in the hazard zone. They must ensure their team stays together. If any of these elements are not adhered to, crew integrity is lost and fire fighters are placed at great risk.

If a fire fighter becomes separated and cannot get reconnected with his or her partner, the fire fighter must immediately get on the radio and attempt to communicate with his or her company officer or partner. If reconnection is not accomplished after three radio attempts or reconnection does not take place within 1 minute, a mayday should be declared. As part of a mayday declaration, the fire fighter must next activate the radio's emergency alert button (where provided), then manually turn on the PASS alarm. It is critically important that no delay occur in declaring a mayday—even a 1-minute delay can be life threatening. If the fire fighter gets reconnected before a rapid intervention crew (RIC) arrives, the mayday can be cancelled.

NIOSH Fire Fighter Fatality Investigation Report F2008-07 illustrates the impact of crew separation (NIOSH, 2009c):

> On March 7, 2008, two male career fire fighters, aged 40 and 19 (Victims 1 and 2, respectively), were killed when they were trapped by rapidly deteriorating fire conditions inside a millwork facility in North Carolina. The captain of the hose line crew was also injured, receiving serious burn injuries. The victims were operating a hose line protecting a firewall in an attempt to contain the fire to the burning office area and keep it from spreading into the production and warehouse areas.
>
> The crew became separated when a fire fighter ran low on air and followed the hose line to the outside. Conditions continued to worsen, and the captain tried to request backup assistance but got no response from the incident commander. The captain sent a second fire

Safety Zone

No Go: Need Two for Interior Operations
If you don't have a partner, never enter a burning building.

VOICES
OF EXPERIENCE

In July 2001, I became the fire chief of the Seattle, Washington, Fire Department. A few weeks later, I had a visit from a lieutenant of a ladder company. He described nearly dying during an expanding multiple-alarm fire in a multiresident building. He and his proby [probationary fire fighter] were operating on the third floor with zero visibility and increasing heat when they ran out of air. Moments later, the lieutenant found a window and peered out to see a roof of another building one story below and he decided to bail out. Unfortunately, there was a 10-foot separation between the two buildings, and he fell three stories to the ground. He was severely injured but survived, and only recently returned to full duty. His proby found his way out safely. His departing request from the meeting was, "Chief, I don't want this to happen to another firefighter."

> "Chief, I don't want this to happen to another fire fighter."

Within two months of the lieutenant's visit, two additional near-fatal incidents occurred. One fire fighter became separated from his partner in the hold of a multiple-alarm ship fire and ran out of air. Neither he nor his partner declared a mayday. By chance, another crew found the fire fighter and removed him from the ship. He was transported to the hospital and treated for smoke inhalation.

A week or so later, a captain operating at a residential fire experienced a mechanical failure on his SCBA and collapsed from smoke inhalation. No mayday was declared. Another crew found the unconscious captain and rescued him. He was intubated and transported to the hospital, where he was hospitalized for three days for smoke inhalation.

Following these three near misses, I ordered a "safety stand-down." We next rapidly developed a "Fire Fighter Survival" training program. The previously scheduled training for the fourth quarter was dropped and replaced with the survival training. Over the next year, we developed a new SOP titled "Best Safety Practices in Decision Making." One of the bullets in the SOP stated: "Any Member Is Authorized to Say NO to Unsafe Practices or Conditions. Stop, Talk, and Decide."

[On the night of January 5, 1995, four Seattle fire fighters died when the first floor of a commercial warehouse collapsed into the basement.] During department-wide training on the new SOP, a survivor of the 1995 fire described arriving on a second-due engine company with heavy smoke issuing from the building. His crew stretched a hose line to the front door to back up the first crew, which had entered earlier. As he was putting on his face piece, he noted smoke pushing out of cracks in the sidewalk and thought that was awfully odd. The culture at the time (like many fire departments) made it awkward for any fire fighter to challenge a company officer on a decision. The crew entered the building, and moments later the floor fell away from them. The fire fighter survived by grabbing a window sill as the floor fell away and was pulled out by other fire fighters. He told the training officer that had the new policy had been in place that night in 1995, he would have been far more comfortable in raising an alert and perhaps all the fire fighters would have been evacuated from the building before the collapse.

Gary P. Morris
Fire Chief (ret)
Seattle, Washington

Assessment Center Tips

Maintain Crew Integrity
Scenarios used for promotional exams often include a missing or unresponsive fire fighter or subordinate team/task leader. The Rules of Engagement recommend declaring a mayday "if reconnection is not accomplished after three radio attempts or reconnection does not take place within 1 minute." (IAFC 2012)

fighter (Victim 2) outside to relay information about their condition. Victim 2 talked with the incident safety officer, and then returned to rejoin his crew. The captain, now alone and with conditions rapidly deteriorating, called for assistance, but fire fighters on the outside did not initially hear his mayday. Once it was realized that the crew was in trouble, multiple rescue attempts were made into the burning warehouse in an effort to reach the trapped crew as conditions deteriorated further. Three members of a rapid intervention team (RIT) were hurt rescuing the injured captain.

NIOSH Fire Fighter Fatality Investigation Report F2008-34 describes the outcome of a single fire fighter entry (NIOSH, 2008):

> On October 29, 2008, a 24-year old male volunteer fire fighter (the victim) was fatally injured while fighting a residential structure fire. The victim, one of three fire fighters on scene, entered the residential structure by himself through a carport door with a partially charged 1½-inch hose line; he became lost in thick black smoke. The victim radioed individuals on the fire ground to get him out. Fire fighters were unable to locate the victim after he entered the structure, which became engulfed in flames. The victim was caught in a flashover and was unable to escape the fire.

Rule 7. Maintain Continuous Awareness of Your Air Supply, Situation, Location, and Fire Conditions

Objective: *To cause all fire fighters and company officers to maintain constant situational awareness of their SCBA air supply and where they are in the building, as well as all that is happening in their area of operations and elsewhere on the fire ground that may affect their risk and safety (IAFC, 2012).*

NIOSH investigations often identify fire fighters running out of air, getting caught in rapidly deteriorating fire conditions, and becoming disoriented and lost in the building as major factors in their deaths. The SCBA air supply is the fire fighter's life support system. Without an air supply, a fire fighter has a very narrow window of survivability and can die in a few short minutes. For this reason, fire fighters must always confirm they have a full bottle of air before entering a burning structure and remain constantly aware of their air supply while in the building **FIGURE 16-5**.

FIGURE 16-5 The SCBA air supply is the fire fighter's life support system.

NFPA 1404, *Standard for Fire Service Respiratory Protection Training*, requires an air management program in which an individual exits an immediately dangerous to life and health (IDLH) environment before consumption of reserve air supply begins. NFPA 1500, *Standard on Fire Department Occupational Safety and Health Program*, requires standardized IDLH exiting be practiced when the SCBA cylinder reaches a level of 600 L or more. The air management plan should identify a turnaround point when operating within an IDLH environment and include a plan to exit when the low-air alarm activates.

Fire fighters must conduct a thorough and detailed checkout of their assigned SCBA at the start of each shift. This procedure must comply with NFPA 1852, *Standard on Selection, Care, and Maintenance of Open-Circuit Self-Contained Breathing Apparatus (SCBA)*, to confirm the SCBA is fully functional. Any malfunction detected before entry or while in a fire building requires an immediate exit from the structure.

Fire fighters must frequently check their air supply while in a fire-involved structure. Major benchmarks for checking air supply include the following points:

- Before entry
- After going up or down stairs
- Before entering and searching a room
- After exiting a room
- After going down a hallway or aisle
- Before and after doing a labor-demanding task

The faster rate of air consumption during laborious tasks can cut in half the minute rating for air consumption of SCBA

Getting It Done

Reserve SCBA Air Increased in 2013

The SCBA low-pressure alarm, called the end-of-service time indicator (EOSTI), moved from 25 percent to 33 percent of SCBA capacity in the 2013 edition of NFPA 1981, *Standard on Open-Circuit Self-Contained Breathing Apparatus for Emergency Services (SCBA) for Emergency Services*. SCBA sold after August 30, 2013, must comply with this requirement to be labeled as compliant with NFPA 1981.

The EOSTI increase was developed to increase the remaining air capacity once the low-air alarm activates. Using the tolerance of 0 to +5 percent, the low-air alarm will activate when the SCBA air supply reaches 33 to 38 percent of capacity.

---

bottles. Ideally, all fire fighters will recognize their individual air consumption rate. Additionally, the company officer must know the projected consumption rate of crew members to plan a scheduled exit. Fire fighters must provide frequent air supply status reports to their company officer. The company officer should include the lowest air supply status as part of progress reports to the incident commander or the division or group officer.

Fire Fighter Near-Miss Report #07-697 reinforces the need for an air management program (IAFC, 2007):

> While commanding a single-family residence house fire that started from a lightning strike, a fire fighter depleted his air and got lost in the residence. The residence was a single-story ranch home with brick and had two additions in the rear where the fire started. The fire fighter was assigned to the interior fire attack group with the assistant fire chief and three additional fire fighters. The fire fighter was removing sheetrock from the ceiling and extinguishing the fire in the attic. He stated that his low-air alarm was going off but he did not notify anyone. He continued to work until his air ran out. While trying to leave the residence, he became disoriented for a short period before finding the assistant chief, who showed him the way out. The fire fighter was equipped with his own radio and the SCBA was working properly,

including the heads-up display. Upon exiting the residence, the IC was notified that the fire fighter was going to the ambulance for rest and oxygen with a paramedic. After being evaluated and treated, he was transported to the local hospital for smoke inhalation.

Courtesy of the National Fire Fighter Near-Miss Reporting Systems (firefighternearmiss.com) (2007 January 24) "FF suffers smoke inhalation when he ignores low air warning, 07-0000697" Fairfax, VA: International Association of Fire Chiefs.

■ Air Management as a Situational Awareness Tool

Air supply status reports improve the logistical commitment of crews. With these reports, the incident commander will have early awareness of the approaching need for a fresh crew, as compared to getting surprised as the crew announces its exit from a position because they are running out of air. With ongoing air supply status reports, the incident commander can call fresh crews up from staging and conduct a transition at the operating position instead of at the door after exit.

The National Near-Miss Reporting System lists situational awareness as the most commonly reported cause of a life-threatening near-miss event. Situational awareness is the level of understanding and attentiveness one has regarding the reality of a set of conditions. When situational awareness is high, there are rarely surprises. When situational awareness is low or absent, unexpected events occur that can injure or kill fire fighters. Situational awareness is the relationship between what one perceives is happening and what is really happening.

Every fire fighter must observe or otherwise be aware of his or her surroundings, landmarks, windows, exits, and route taken when penetrating the building. These landmarks become important for self-survival if the fire fighter becomes separated or lost; they are critical when the fire fighter is facing a life-threatening emergency and needs to exit the building safely. The more detailed and accurate the location description when calling a mayday, the faster a RIC can reach a fire fighter.

Conditions encountered early on in the fire attack may become out of control as the fire evolves, placing the fire fighter at continued risk. Often structural integrity is compromised as a fire grows. For all these reasons, fire fighters must be aware of their work environment and in control of their actions.

Rule 8. Constantly Monitor Fire-Ground Communications for Critical Radio Reports

Objective: *To cause all fire fighters and company officers to maintain constant awareness of all fire-ground radio communications on their assigned channel for progress reports, critical messages, and other information that may affect their risk and safety* (IAFC 2012).

Portable radios are the lifeline for fire fighter survival. Both the IAFC and the IAFF support the position that every fire fighter operating within the hazard zone must be equipped with a portable radio or other approved voice communication device.

Portable radios are not just the fire crew's lifeline connection with the incident command and rescuers—they also allow

---

Safety Zone

No Go: Manage SCBA Air

There is a no-go situation:
- If you do not have a full SCBA bottle.
- If you do not know your air supply all the times.
- If you do not know where you are at all times. Don't continue; stop, reorient, or report.
- If you have reached your turnaround point on your SCBA air supply—exit the building.

FIGURE 16-6 Closely monitor radio traffic for critical progress reports.

the fire fighter to increase situational awareness by closely monitoring the assigned tactical channel. Radio communications from other points on the fire ground provide additional information about changing fire conditions elsewhere and help the fire fighter maintain situational awareness **FIGURE 16-6**.

Closely monitoring the radio traffic for critical progress reports also provides greater lead time for fire crews to evacuate the structure should fire conditions deteriorate rapidly. All crew members must monitor radio communications, even if the company officer has the only radio. Sometimes the company officer in charge of the crew may miss critical communications because of noise, and a crew member can alert him of the message.

The company officer or team leader should provide supervisors, or the incident commander, with frequent progress reports so that the command organization maintains real-time situational awareness and can adjust the action plan. These ongoing radio reports also keep other crews informed of progress or deteriorating conditions.

In situations where the crew may have only one radio and that radio fails while they are in a burning building, the crew must evacuate to a safe location and report their situation. A crew without a working radio is put at great risk because they lose their connection with the incident commander and rescuers. They will not hear the order to evacuate the building.

NIOSH Fire Fighter Fatality Investigation report F2009-11 provides an example (NIOSH, 2010a):

> Shortly after midnight on Sunday, April 12, 2009, a 30-year-old male career probationary fire fighter and a 50-year-old male career captain were killed when they were trapped by rapid fire progression in a wind-driven residential structure fire. The victims were members of the first-arriving company and initiated fast-attack offensive interior operations through the front entrance. Less than 6 minutes after arriving on scene, the victims became disoriented as high winds pushed the rapidly growing fire through the den and living room areas where interior crews were operating. Seven other fire fighters were driven from the structure, but the two victims were unable to escape.

Safety Zone

No Go: Every Task Group Gets a Portable Radio
- If your team is not equipped with a radio(s), do not enter a burning building.
- If your team is equipped with only one radio, the radio fails while in the building, and you have no other means to communicate, exit the building.

The investigation revealed the captain victim left his portable radio on his apparatus. Thus he was operating blind and without the ability to be aware of radio communications (in deteriorating conditions) and without the ability to provide the incident commander any progress reports, nor could he have declared a mayday.

His partner and victim number 2, a probationary fire fighter, was found to have his portable radio in the bunker coat pocket, but it was turned off and on the wrong channel. He also was operating blind to any radio communications or reports of fire conditions (or evacuation orders).

Rule 9. You Are Required to Report Unsafe Practices or Conditions That Can Harm You. Stop, Evaluate, and Decide.

Objective: *To prevent company officers and fire fighters from engaging in unsafe practices or exposure to unsafe conditions that can harm them and to allow any member to raise an alert about a safety concern without penalty and mandate the supervisor address the question to ensure safe operations (IAFC, 2012).*

The nature of firefighting nearly always places the company officer and crew members in the area of greatest risk. As such, they must remain alert to unsafe conditions or practices that can harm them and have a process for reporting and correcting them. Rule 9 accomplishes that goal.

Because of the significant risk faced by fire fighters, it is the responsibility of the company officer to minimize their exposure to unsafe conditions and to stop unsafe practices. NIOSH fire fighter fatality reports routinely describe incidents where unsafe conditions or practices existed that were observed and later became a contributing factor in the line-of-duty death. Chief Paul LeSage (2009) determined that 74 percent of errors on the fire ground happen when individuals fail to intervene in known or observed unsafe situations.

■ Learning from the Aviation Industry

The aviation industry experienced a similar problem of one-way decision making and communication in the late 1970s. The culture did not tolerate a challenge to the captain from crew members. As a result, post-crash investigations found that captains occasionally flew their planes into the ground, even when other crew members, including the co-pilot, knew something was wrong and tried to tell the pilot, only to have their input rejected.

The creation of a cockpit crew resource management system established two-way communication among all members of the flight crew, thereby involving all of them in risk assessment and decision making. The program resulted in a rather dramatic reduction in fatal accidents caused by pilot errors.

Rule 9 for the fire service applies the principles of crew resource management by encouraging all fire fighters to apply situational awareness and be responsible for their own safety and that of other fire fighters. The *Crew Resource Management and Leading Change* chapter covers the details of crew resource management. The intent of this rule is to allow any member to report a safety concern through a structured process without fear of penalty.

■ Raise the Red Flag

Rule 9 by no means suggests that a fire fighter is authorized to engage in insubordination. The fire ground is a site of fast-paced action and clearly must be managed by a well-disciplined and -structured command organization. This rule does, however, allow a "red flag" to be raised about a safety issue by any member. When such a red flag is raised, the supervisor is mandated to accept that concern, take a few seconds to assess it, talk with others, and make a safe decision (go or no-go). In some cases, the situation may affect other areas of the fire ground or the action plan, and must be communicated to the incident commander or other supervising officers.

NIOSH Fire Fighter Fatality Investigation Report F2008-03 provides an example (NIOSH, 2009a):

> On June 18, 2007, nine career fire fighters (all males, ages 27–56) died when they became disoriented and ran out of air in rapidly deteriorating conditions inside a burning commercial furniture showroom and warehouse facility. The first-arriving engine company found a rapidly growing fire at the enclosed loading dock connecting the showroom to the warehouse.
>
> The assistant chief entered the main showroom entrance at the front of the structure but did not find any signs of fire or smoke in the main showroom. Within minutes, the fire rapidly spread into and above the main showroom, the showroom addition, and the warehouse. The fire overwhelmed the interior attack, and the interior crews became disoriented when thick black smoke filled the showrooms from ceiling to floor. The interior fire fighters realized they were in trouble and began to radio for assistance as the heat intensified. One fire fighter activated the emergency button on his radio. The front showroom windows were knocked out, and fire fighters, including a crew from a mutual-aid department, were sent inside to search for the missing fire fighters. Fire raced through the main showroom. Interior fire fighters were caught in the rapid fire progression and nine fire fighters died. Nine other fire fighters barely escaped serious injury.
>
> Throughout this incident, leading up to flashover, there were numerous observable or described unsafe conditions and practices: lack of implementation of the incident command system, use of undersized fire attack lines (including a booster line) for the fire conditions/magnitude of fire, an attack on the fire on the interior with tank water and no hydrant supply, inadequate-sized supply (2.5-inch hose lines) and very long supply lines, no consistent buddy teams, and lack of ventilation, among others. Unsafe conditions included a continued advancing and expanding fire over which fire crews were never able to achieve any kind of fire control.

Courtesy of the National Fire Fighter Near-Miss Reporting Systems (firefighternearmiss.com) (2009 April 06) "Roof vent late in operation causes collapse., 09-0000364" Fairfax, VA: International Association of Fire Chiefs.

Even when unsafe conditions are identified, command intervention may be required, as described in Fire Fighter Near-Miss Report #09-364 (IAFC 2009):

> On the morning of [date and time deleted], we received an alarm of a structure fire about 0.5 mile from Station [1]. Engine [1], Engine [2], and Rescue [1] responded.
>
> Upon the arrival of Engine [1] and Rescue [1], we found heavy smoke showing from the structure. Engine [1] laid two 1¾-inch lines and caught the hydrant right across the street. The Engine [1] group went in through the front door, but due to extreme amount of heat, were not able to get to the seat of the fire. The Group 1 lieutenant called for vertical ventilation, but due to staffing issues, ventilation was not able to be performed and we moved to a defensive attack.
>
> Eventually, the incident commander decided to vertically ventilate the structure. This order was given 45 minutes into the fire and approximately 30 minutes after a defensive attack was ordered and the positive-pressure ventilation (PPV) fan was set up. The roof was already visibly sagging, and from the road you could see heavy fire in the attic. Three lieutenants and a captain on scene advised the incident commander that the task was unsafe.
>
> Group 4 ignored our plea and put a ladder on the building. Once they were on the roof and starting to ventilate, the roof gave way and collapsed. The two fire fighters in Group 4 were standing on the roof ladder and were able to roll off the house. The roof ladder sustained major fire damage and the fire fighter making the cuts was very lucky that his backup fire fighter was holding onto him and saw the roof start to give way. This allowed for the backup fire fighter to pull the [power saw operator] up and fall off the roof onto the ground instead of into the fire. After it was over, the fire fighters from Group 4 stated that the incident commander ordered us to ventilate and "we were not going to break an order."

Safety Zone

No Go: It Is Not Okay to Get Hurt Intentionally
If anything will harm you, it's a no-go. Report that hazard immediately to command or to your supervisor.

Rule 10. You Are Required to Abandon Your Position and Retreat Before Deteriorating Conditions Can Harm You.

Objective: *To cause fire fighters and company officers to be aware of fire conditions and cause an early exit to a safe area when they are exposed to deteriorating conditions, unacceptable risk, or a life-threatening situation (IAFC, 2012).*

Fire fighters may detect a rapidly deteriorating condition before the incident commander does. Flashovers can develop in seconds, and fire fighters may have only a few more seconds of survival time when they do occur. Where the situation creates a high potential for an injury or a life-threatening situation, no fire fighter needs approval from a supervisor or the incident commander to abandon a high-risk position, nor should fire fighters be required to report their intent to abandon their position if such reporting would impede or delay a rapid exit to a safe location.

Withdrawal from a life-threatening position must occur early enough to allow a safe exit from the building or to relocate to a safe location. Fire fighters should understand that an emergency exit from a building often takes longer than it took to get into the interior operating position, and that conditions will be worse. There should be no hesitation, because seconds can mean the difference between surviving and dying. If saving the hose line or any equipment will delay exit, the fire fighter should leave it behind and get out. It is far better for a crew to abandon their position early than to take a needless stand and be pushed out.

A radio report to the incident commander (or the division/group supervisor) on the decision to abandon the position should be made as soon as possible, but only when safe to do so and where it does not cause a delay in exiting.

■ Melted Helmets and Heat-Crazed Face Pieces

The culture of fire fighters "standing their ground" with a willingness to take on overwhelming flame or heat where the fire cannot be controlled with existing fire attack lines cannot be accepted. Fire fighters who engage in this risky behavior have frequently melted or heat-damaged personal protective equipment and suffered recurrent burn injuries. Where this behavior exists, the fire department management team must intervene and eliminate this unsafe behavior.

Fire Fighter Near-Miss Report #05-589 describes a survivor's experience after being overtaken by fire (IAFC, 2005):

> There was heavy fire coming from the B, C side; we pulled a 2½-inch line to the C side and knocked down the majority of the visible fire. We decided an interior attack would be performed. When we got to the top of the stairs, you could feel that it was getting hotter and we went to the right. There was still no visibility at this time. As we went down the hall trying to find a room to search, we heard the chief yelling that if we couldn't find it, then we need to start backing out. At this time, I felt a door handle and yelled that I had found one. I stood up on my knees to open the door and at that time, I could see fire come from behind me. The fire completely covered my body. I fell to the floor and started rolling around, trying to get the fire out. When I felt like I had it out, I felt my body to make sure I wasn't still on fire.
>
> I looked around and could not see anyone. I called a mayday. I started searching for another room or a window because the hallway I was in was on fire, the stairs were on fire, and I needed a way out. However, I didn't find anything at this time. The fire in the hall died down and I felt for the hose line. When I found it, I started to follow it to the stairs. I found the lieutenant and told him to get out, and he dove down the stairs, which were still on fire.
>
> I stooped and waited at the top of the stairs, trying to see if the fire would die down. When it didn't, I made the decision that I wasn't going to just sit there and wait to die. I dove head first through the fire down the stairs. I hit the wall where the stairs made a turn and started rolling down them. When I got to the bottom, I could see the lieutenant, who yelled at me, and we found the front door and ran to it. When we got there, it had shut. We were beating and pulling on it trying to get it open. After doing this for a while, I calmed down and realized that the door had locked after we went in. I unlocked it and opened it, and we went out and into the front yard. They stripped us of our gear. My helmet was damaged; my coat was damaged. Both parts of the gear were damaged due to the high heat conditions.

Courtesy of the National Fire Fighter Near-Miss Reporting Systems (firefighternearmiss.com) 2005 October 26) "Several problems occur during residential house fire., 05-0000589" Fairfax, VA: International Association of Fire Chiefs.

Rule 11. Declare a Mayday As Soon As You Think You Are in Danger.

Objective: *To ensure the fire fighter is comfortable with, and there is no delay in, declaring a mayday when a fire fighter is faced with a life-threatening situation, and to ensure the mayday is declared as soon as a fire fighter thinks he or she is in trouble (IAFC, 2012).*

Firefighting inevitably places firefighters in constantly changing hazardous environments. Even where the best safety practices are applied, the unexpected can happen. Fire fighters can be exposed to rapidly deteriorating conditions that can quickly become life threatening, or they may become separated from fellow crew members and lost in the building. Any delay in declaring a mayday can lead to a lethal event.

The need for this rule is illustrated by a survey conducted among 339 attendees at the National Fire Academy. Only 82 percent would declare a mayday if faced with "zero visibility,

Safety Zone

No Go: Do Not Be Overtaken by Fire
- If the fire is about to overtake you, retreat to a safe location or exit the building before you are harmed.
- If you lose radio communications, exit the building.

> **Safety Zone**
>
> **No Go: Do Not Do It by Yourself**
> If you are lost, separated, or in trouble, do not try to find your way out by yourself. Declare a mayday.

no contact with hose line or lifeline, do not know direction to an exit"; only 58 percent would declare a mayday if they could not find an exit in 60 seconds (Clark 2002).

There is a very narrow window of survivability when a fire fighter gets in trouble with a life-threatening situation. In such a scenario, an early mayday must be declared and fire fighters should provide the incident commander with their name, company, location, air supply, and situation, along with any other critical information that will aid rescuers in locating them quickly.

When declaring a mayday, the fire fighter should activate the portable radio's emergency alert button, if one is provided. New technology also allows the radio to be programmed to go to a designated emergency channel, clear of the tactical radio traffic, to declare the mayday. This capability allows the fire fighter to talk directly to the incident commander free of interference. Once the mayday is declared, the fire fighter should manually activate the PASS device.

NIOSH Fire Fighter Fatality Investigation Report #F2004-04 describes the results of not declaring a mayday (NIOSH 2005):

> On December 16, 2003, a 30-year-old male fire fighter (the victim) died after he became separated from his crew members while searching for the seat of a fire on the second floor at a furniture warehouse. His crew exited due to worsening conditions, and a missing member announcement was made. After an evacuation order was given, and as engine crew members were exiting, the victim's officer mistakenly identified one of them as his missing member and cancelled the emergency message. Once fire fighters had exited, a personnel accountability report (PAR) was taken on the street, which revealed that the victim was still missing. By this time, the second floor had become fully involved. The victim's officer initiated a second emergency message for a missing member and a search was begun. The victim, who had a working radio, was found lying face down, with his face piece dislodged, under 2 feet of debris, and 900 psi remaining in his self-contained breathing apparatus (SCBA). The victim did not declare a mayday and did not activate his radio's emergency alert button.

> **Getting It Done**
>
> **New Standardized PASS Alarm**
> The 2013 edition of NFPA 1982, *Standard on Personal Alert Safety Systems (PASS)*, established a universal PASS alarm with a single, distinctive sound for pre-alarm and full alarm.

Courtesy of the National Fire Fighter Near-Miss Reporting Systems (firefighternearmiss.com) (2008 November 07) "Lieutenant falls through floor, 08- 0000577" Fairfax, VA: International Association of Fire Chiefs.

■ Mayday versus Emergency Traffic

Mayday is a term reserved for a situation where a fire fighter is experiencing a life-threatening emergency. The term *emergency traffic* should be utilized for other emergencies on the fire ground, such as evacuating a building, live wires down, or discovery of a hazardous situation.

Training must emphasize the immediate declaration of a mayday when the fire fighter is facing a life-threatening situation. If the emergency is resolved before the RIC reaches the fire fighter in trouble, the RIC operation can be cancelled.

There is a common link in fire fighters delaying to declare a mayday and military pilots delaying to eject from a crippled aircraft. According to Richard Leland, President of the National Aerospace Training and Research Center, there are 10 reasons for failure or delayed ejection that must be addressed in ejection training (Leland, 1999):

1. Temporal distortion (time seems to speed up or slow down)
2. Reluctance to relinquish control of one's situation
3. Channeled attention (continuing with a previous selected course of action because other, more significant information is not perceived)
4. Loss of situational awareness (controlled flight into terrain)
5. Fear of the unknown (reluctance to leave the security of the cockpit)
6. Fear of retribution (loss of the aircraft)
7. Lack of procedural knowledge
8. Attempting to fix the problem
9. Pride (ego)
10. Denial ("this isn't happening to me")

Fire Fighter Near-Miss Report #08-577 describes a mayday response (IAFC, 2008):

> Crews were fighting a fire in a four-unit rowhouse/townhouse, wood-frame dwelling. Fire was visible from the C side (exterior). Crews reported fire in the walls and ceilings on the first floor, and fire was moving up to the second floor. Initial crews were containing the fire, while an additional crew moved to the second floor for reconnaissance and search. When the reconnaissance crew reached the second floor, two members (a lieutenant and a fire fighter) entered a bathroom that was located directly above the fire. Almost immediately upon their entry into the bathroom, a 6- to 7-foot section of the bathroom floor collapsed, with the lieutenant falling through the floor. The lieutenant was able to catch himself on a floor joist and nearby debris. As the floor collapsed, the fire fighter jumped into a tub and did not fall through the floor.
>
> Immediately upon falling through the floor, the lieutenant called a mayday and provided a clear and

Near-Miss REPORT

Report Number: 13-0000321

Event Description: At 2353 hours, our city dispatch center received the report of a fire in the apartments above a tavern, located in the downtown area. This particular block consists of buildings erected in the late 1800s and early 1900s. Typical of buildings constructed in the Midwest during this period, they are of ordinary construction and share common walls. The building of origin was occupied by a tavern on the ground floor and three apartments on the second floor, and it included a basement. The initial 911 call was placed from inside the tavern.

The first fire units arrived on scene within a few minutes of being dispatched. The officer of Engine [1], the first-arriving fire unit, reported seeing smoke as soon as he left the station (Station [1] is only a few blocks away from this fire scene). Upon arrival, the lieutenant reported smoke visible from the "A" side.

City police department units arriving from the west side of the building reported that flames were visible at the rear of the building, or "C" side. Flames were not visible on the "A" side at this time. Rescue [1], the second fire unit to arrive on scene, led by another lieutenant (the report submitter), observed a large volume of smoke coming from the roof, and subsequently requested a second alarm.

The Engine [1] crew, which was staffed with two personnel, had entered the tavern by this time. They attempted to clear the 15 or so occupants out of the tavern and relocate them across the street. The police personnel assisted in this effort. Engine [1] informed Rescue [1] that occupants within the tavern stated that there might still be people in the apartments above. The Rescue [1] crew, staffed with a driver and myself, prepared to make entry, and conducted a primary search without a handline, due to the urgency of the situation.

Access to the second-floor apartments was via a ground-level doorway on the "A" side, near the "A-D" corner of the building. The stairway to the apartments rose about 10–12 feet, from street level to a second-floor landing. After masking up, I stood at the base of the stairs preparing to enter. I observed a moderate volume of black smoke about halfway up the stairway. The smoke was actively moving, but not remarkably so at this time. I climbed the stairway alone and ascended to the landing, searching for signs of life. At the landing, the smoke was much thicker and very black in color. It had descended down to floor level, and continued to grow in density and velocity. I took a moment to assess visibility. I extended my arm in front of me and could not see my elbow. I heard the sound of embers hitting my helmet. I now knew there was fire above me. I waited at that position until my partner arrived by my side.

Since the stairwell rose along the "D" side of the building, I knew we had to go toward the "B" side to find the apartments. Conditions were deteriorating quickly.

I crawled through the smoke with my partner now directly behind me. We maintained physical contact from this point on. While conducting a right-hand search, I located the first apartment door and found it to be locked. Based upon this apartments' proximity to the stairwell, as well as the direction the smoke was moving, I felt that this was not the apartment of origin. We moved on.

My partner pulled up next to me and located a second apartment a few feet away that had its door cracked open an inch or so. Black smoke was pouring out of this open door with great velocity. He pushed the door open farther so that we could gain entry to the apartment. Instantly, the smoke, heat, and energy overwhelmed us like a tidal wave. I could actually feel the pressure wave of the smoke with my body as it rushed past us. We were completely enveloped. I got the sensation that we were drowning in smoke. This was the thickest, darkest, blackest, most energetic smoke I had ever encountered. It was moving so quickly around us that I began to get dizzy watching it.

Just as my partner broke the plane of the doorway into the apartment, I put my hand on his shoulder to stop him and said that we were getting out of here. All my senses told me that this building was about to flash over. We quickly made our way back to the stairwell and descended to street level. I estimate the time we spent conducting the search was a total of 30 seconds.

Immediately outside, I met with the IC. He wanted a status update regarding conditions on the second floor. As I began to relay the information we had, I turned to look back at the stairwell. The entire second floor had erupted into flames. The flashover was violent enough to fully involve not only the area we were just in, but flames were forced halfway down the stairwell and through the roof as well. I estimate that from the time I gave the order to evacuate until the time we met with the IC in the street, 10–15 seconds had elapsed. Looking back now, I realize that we had barely escaped with our lives.

All units on scene were ordered to switch to a defensive mode, and we kept the fire contained to the building of origin. All occupants were later accounted for and found to be safe.

(continued)

Lessons Learned:

1. There is a certain level of comfort you experience when you are familiar with the layout of the building in which you are battling a fire. Know the buildings in your area.
2. Be aware of your surroundings. Use all your senses to gather information. Constantly reevaluate conditions. Don't get tunnel vision.
3. When your senses tell you that you're in trouble, believe it.
4. Know your limitations and your time frame. We had two people, no attack line, and no backup crew; encountered severe smoke and heat; and knew that there was fire in front of us and above us. Conditions rendered survival for a trapped civilian unlikely. Know when to get out.
5. As an officer, you are responsible for protecting not only civilian life, but also the lives of your crew members. Risk a lot to save a lot, but risk nothing to save nothing.
6. Remember that smoke is fuel. The thicker the smoke and the blacker the smoke, the more fuel you are dealing with.

I have witnessed a few flashovers in my career. All of them had these three characteristics:

1. There was black smoke—not brown, not gray, but black: the blackest black you can imagine.
2. The smoke was extremely dense. You can see nothing through it—not a wall, not your arm, not even light.
3. The smoke is packed with energy. It moves with incredible speed, and in this case, sounded like a freight train moving past us.

If you pull up to a scene and see smoke like this, I would think twice about sending crews inside to fight the fire. If you are in a fire and witness smoke with these qualities, flashover is imminent. Get out immediately. I hope that this information will help you make sound decisions at the scene of your next working fire.

Courtesy of the National Fire Fighter Near-Miss Reporting Systems (firefighternearmiss.com)

concise report detailing his unit, his location, the situation, and his immediate needs. Operations acknowledged the mayday, quickly confirmed the situation, and deployed the RIT to the location of the trapped lieutenant. Minutes later, the lieutenant had been extricated and self-evacuated from the structure. Immediately upon hearing that the trapped lieutenant had been located, extricated, and removed from the building, command removed all personnel from the structure and ordered a PAR. The PAR revealed that all members were accounted for and firefighting operations recommenced.

You Are the Fire Officer Conclusion

Lieutenant Williams develops eight drills to cover the Rules of Engagement:

- Conduct size-up of assigned operation area or objective: risk assessment and situation reports. (Scenario classroom)
- Determine the occupant survival before committing to a high-risk rescue effort.
- Determine your SCBA air consumption for crew members and turnaround point. (Scenario practice)
- Situational awareness and crew integrity during search operations. (Scenario practice)
- Throwing the red flag for a no-go situation. (Scenario classroom)
- Evacuating a structure. (Scenario practice)
- Calling a mayday. (Scenario practice)
- Responding to a mayday. (Scenario practice)

Chief Johnson approves the training plan and will work with the operations chief and fire academy to schedule the scenario practice sessions at an acquired structure or the drill tower.

Wrap-Up

Chief Concepts

- The *Rules of Engagement for Structural Firefighting* were developed by the IAFC's Safety, Health, and Survival Section as best safety practice model procedures.
- A survey of 23 fire fighter fatalities compared the type of structures involved in the line-of-duty deaths, and found that more than three-fourths of the disoriented fire fighter deaths occurred in enclosed structures.
- Fire fighters are much more likely to die in fires than civilians are.
- The fire department as well as the company officer should minimize fire fighter exposure to unsafe conditions and stop unsafe practices. This requires a change in command perspective.
- NIOSH fire fighter fatality investigations often cite lack of a complete size-up as a contributing factor in fire fighter deaths.
- An essential component in the size-up process is to determine whether any occupants are trapped and whether they can survive the current and projected fire conditions.
- The fire fighter and the incident commander must factor growing fire conditions, resources on scene, and the time needed to complete a rescue into the decision to conduct and support primary search and rescue operations.
- The company officer and the incident commander must recognize that fire fighters cannot always save a life.
- There is a limit beyond which the company officer and the incident commander may not allow fire fighters to be exposed to unsafe fire conditions.
- During search and rescue operations, crews must remain alert to changing fire conditions that may increase risk to them or prevent survival of building occupants.
- A critical element for fire fighter survival is crew integrity—fire fighters staying together as a team of two or more.
- The SCBA air supply is the fire fighter's life support system. Without an air supply, a fire fighter has a very narrow window of survivability and can die in a few short minutes.
- Air supply status reports improve the logistical commitment of crews. They give the incident commander early awareness of the approaching need for a fresh crew.
- Portable radios are the lifeline for fire fighter survival. Both the IAFC and the IAFF support the position that every fire fighter operating within the hazard zone must be equipped with a portable radio or other approved voice communication device.
- The nature of firefighting nearly always places the company officer and crew members in the area of greatest risk. As such, they must remain alert to unsafe conditions or practices that can harm them and have a process for reporting and correcting those hazards.
- The program known as cockpit crew resource management established a two-way communication between all members of the flight crew, making them all involved in

risk assessment and decision making; such a system can be adapted to the fire service.

- When a red flag is raised by any fire crew member, the supervisor is mandated to accept that concern, take a few seconds to assess it, talk with others, and then make a safe decision (go or no-go).
- Fire fighters may detect rapidly deteriorating conditions before the incident commander does. Flashovers can develop in seconds, and fire fighters may have only a few more seconds of survival time when they do occur.
- The culture of fire fighters "standing their ground" and taking on overwhelming flame or heat where a fire cannot be controlled with existing fire attack lines cannot be accepted.
- Any delay in declaring a mayday can lead to a lethal event.
- Training must emphasize the immediate declaration of a mayday when the fire fighter faces a life-threatening situation. If the emergency is resolved before the RIC reaches the fire fighter in trouble, the RIC can be cancelled.

Hot Terms

<u>Crew integrity</u> A system in which fire fighters stay together as a team of two or more members.

<u>Enclosed structure</u> A building that lacks windows or doors of sufficient number and size to provide for prompt ventilation and emergency evacuation.

<u>Limited</u> At the point, edge, or line beyond which something cannot or may not proceed; confined or restricted within certain limits.

<u>Mayday</u> A situation where a fire fighter is experiencing a life-threatening emergency.

<u>Measured</u> Careful, restrained, calculated, and deliberate.

<u>Opened structure</u> A building that has windows or doors of sufficient number and size to provide for prompt ventilation and emergency evacuation.

<u>Rules of Engagement</u> Rules developed by the International Association of Fire Chiefs to promote safety for fire fighters and fire officers working at the task level of fire suppression operations.

<u>Situational awareness</u> The level of understanding and attentiveness one has regarding a set of conditions.

<u>Vigilant</u> On the alert and watchful.

References and Additional Resources

NFPA reprinted material is not the complete and official position of the NFPA on the referenced subject, which is represented only by the standard in its entirety.

Baldwin, T., et al. (2008, November). *NIOSH Firefighting Fatality Investigation and Prevention Program: Leading Recommendations for Preventing Firefighter Fatalities, 1998–2005.* Cincinnati, OH: National Institute for Occupational Safety and Health.

Clark, B. (2002, October 28). "When Would You Call Mayday–Mayday–Mayday?" *Firehouse Magazine.* Atkinson, WI: Cygnus Business Media. http://www.firehouse.com/article/10574008/when-would-you-call-mayday-mayday-mayday.

Fahy, R. F., P. R. LeBlanc, and J. L. Molis. (2007). *What's Changed over the Past 30 Years?* Quincy, MA: National Fire Protection Association.

Fahy, R. F., P. R. LeBlanc, and J. L. Molis. (2013). *Firefighter Fatalities in the United States—2012.* Quincy, MA: National Fire Protection Association.

International Association of Fire Chiefs, Safety, Health, and Survival Section. (2012). *Rules of Engagement for Structural Firefighting: Increasing Firefighter Survival.* Fairfax, VA: International Association of Fire Chiefs.

Kerber, S. (2012). "Analysis of Changing Residential Fire Dynamics and Its Implications on Firefighter Operational Timeframes." *Fire Technology* 48(4): 865–891. Quincy, MA: National Fire Protection Association.

Leland, R. (1999). *Ejection Seat Training Operations and Maintenance Manual.* Southampton, PA: Environmental Tectonics Corporation.

LeSage, P., J. Dyar, and B. Evans. (2009). *Crew Resource Management: Principles and Practice.* Sudbury, MA: Jones and Bartlett.

Morris, Gary P. (2005, September). "Too Little, Too Late." *Fire Chief Magazine.* Chicago, IL: Penton.

Morris, Gary P. (2005, October). "By Any Other Name." *Fire Chief Magazine.* Chicago, IL: Penton.

Morris, Gary P. (2007, December). "History Revealing." *Fire Chief Magazine.* Chicago, IL: Penton.

Morris, Gary P. (2009, June). "Empower All Firefighters to Stop Unsafe Practices." *Firehouse Magazine.* Atkinson, WI: Cygnus Business Media.

Morris, Gary P. (2009, July). "Come Up for Air." *Fire-Rescue Magazine.* Tulsa, OK: PennWell Corporation.

Marsar, S. (2009). *Can They Be Saved? Utilizing Civilian Survivability Profiling to Enhance Size-up and Reduce Firefighter Fatalities in the Fire Department, City of New York.* Emmitsburg, MD: U.S. Fire Academy, Executive Fire Officer Program.

Marsar, S. (2011, November). "Survivability Profiling Takes Size-up to a New Level: Firefighters Must Consider Not Just Whether Victims Might Be Inside, But Whether They Can Still Be Alive." *Fire Rescue Magazine.* http://www.firefighternation.com/article/strategy-and-tactics/survivability-profiling-takes-size-new-level

Mora, W. R. (2003). *U.S. Firefighter Disorientation Study, 1979–2001.* San Antonio, TX: San Antonio Fire Department.

National Fallen Firefighters Foundation. (2011). *Understanding and Implementing the 16 Firefighter Life Safety Initiatives.* Stillwater, OK: Fire Protection Publications.

National Fire Fighter Near Miss Reporting System. (2005 October 26). *Several problems occur during residential house fire, 05-0000589.* Fairfax, VA: International Association of Fire Chiefs.

National Fire Fighter Near Miss Reporting System. (2007 January 24). *FF suffers smoke inhalation when he ignores low air warning, 07-0000697.* Fairfax, VA: International Association of Fire Chiefs.

National Fire Fighter Near Miss Reporting System. (2007 March 10). *FF falls through burnt floor, 07-0000789.* Fairfax, VA: International Association of Fire Chiefs.

National Fire Fighter Near Miss Reporting System. (2008 November 07). *Lieutenant falls through floor, 08- 0000577*. Fairfax, VA: International Association of Fire Chiefs.

National Fire Fighter Near Miss Reporting System. (2009 April 06). *Roof vent late in operation causes collapse. 09-0000364* Fairfax, VA: International Association of Fire Chiefs.

National Fire Fighter Near Miss Reporting System. (2009 June 13). *Collapse of building injures FFs., 09-0000578*. Fairfax, VA: International Association of Fire Chiefs.

National Fire Fighter Near Miss Reporting System. (2010 January 20). *Good 360 and Situational Awareness Saves Crew, 10-0000157*. Fairfax, VA: International Association of Fire Chiefs. http://www.firefighternearmiss.com/Resources/ROTW_PDF/ROTW_PDF_062112.pdf

National Institute for Occupational Safety and Health (NIOSH). (2001a, February 7). *Restaurant Fire Claims the Life of Two Career Fire Fighters—Texas*. Report #F2000-13." Atlanta, GA: Centers for Disease Control and Prevention.

National Institute for Occupational Safety and Health (NIOSH). (2001b, April 11). *Structure Fire Claims the Lives of Three Career Fire Fighters and Three Children in Iowa*. Report #F2000-04. Atlanta, GA: Centers for Disease Control and Prevention.

National Institute for Occupational Safety and Health (NIOSH). (2005, March 31). *Career Fire Fighter Dies of Carbon Monoxide Poisoning after Becoming Lost While Searching for the Seat of a Fire in Warehouse—New York*. Report #F2004-04. Atlanta, GA: Centers for Disease Control and Prevention.

National Institute for Occupational Safety and Health (NIOSH). (2008, June 11). *Volunteer Fire Fighter Dies while Lost in Residential Structure Fire—Alabama*. Report #F2008-34. Atlanta, GA: Centers for Disease Control and Prevention.

National Institute for Occupational Safety and Health (NIOSH). (2009a, February 11). *Nine Career Fire Fighters Die in Rapid Fire Progression at Commercial Furniture Showroom—South Carolina*. Report #F2008-03. Atlanta, GA: Centers for Disease Control and Prevention.

National Institute for Occupational Safety and Health (NIOSH). (2009b, March 30). *A Career Captain and an Engineer Die while Conducting a Primary Search at a Residential Structure Fire—California*. Report #F2007-28. Atlanta, GA: Centers for Disease Control and Prevention.

National Institute for Occupational Safety and Health (NIOSH). (2009c, August 17). *Two Career Fire Fighters Die and Captain Is Burned when Trapped during Fire Suppression Operations at a Millwork Facility—North Carolina*. Report #F2008-07. Atlanta, GA: Centers for Disease Control and Prevention.

National Institute for Occupational Safety and Health (NIOSH). (2010a, April 8). *Career Probationary Fire Fighter and Captain Die as a Result of Rapid Fire Progression in a Wind-Driven Residential Structure Fire—Texas*. Report #F2010-10."Atlanta, GA: Centers for Disease Control and Prevention.

National Institute for Occupational Safety and Health (NIOSH). (2010b, November 17). *One Career Fire Fighter/Paramedic Dies and a Part-Time Fire Fighter/Paramedic Is Injured when Caught in a Residential Structure Flashover—Illinois*. Report #F2010-10. Atlanta, GA: Centers for Disease Control and Prevention.

National Institute of Standards and Technology. (2001). *International Study of the Sublethal Effects of Fire Smoke on Survivability and Health (SEFS): Phase I Final Report*. NIST Technical Note 1439. Washington, DC: U.S. Department of Commerce.

Rochford, R. (2008). *Hydrogen Cyanide: New Generation Concerns Resulting in Firefighting Tactics and Medicine*. Jacksonville, FL: Jacksonville Fire Rescue Department. Accessed June 24, 2013, at http://www.everyonegoeshome.com/resources/HydrogenCyanide.pdf.

Varone, J. C., T. N. Warren, K. Jutras, J. Molis, and J. Dorsey. (2006, May 30). *Report of the Investigation Committee into the Cyanide Poisonings of Providence Firefighters*. Providence, RI: Providence Fire Department.

FIRE OFFICER in action

The Rules of Engagement scenario practical drills are combined into a 4-hour multiple-company drill at an abandoned middle school. Lieutenant Williams breaks the crew into two groups to perform the activities. One group will determine their SCBA air consumption and identify their turnaround point while practicing situational awareness and crew integrity during a simulated search. The other group will perform evolutions to evacuate a structure, to call a mayday, and to respond to a mayday.

1. A search team consisting of four fire fighters has different SCBA turnaround times. When should the team start to leave the structure?
   A. When the shortest turnaround time is reached
   B. When the second low-air alarm activates
   C. When the average turnaround time is reached
   D. When all four low-air alarms are activated

2. During a search, Quint 100's crew deploys two teams containing two fire fighters each. Team Bravo has a portable radio failure. What should they do?
   A. Complete their search assignment and conduct a face-to-face report with the sector commander
   B. Complete their search assignment and find Team Alpha to continue the task as a four-person team
   C. Immediately evacuate the structure and conduct a face-to-face update with the group commander
   D. Immediately evacuate the structure and report to the rehabilitation group

3. Who can raise a red flag during search and rescue operations?
   A. Individuals assigned to command or IMS group assignments
   B. First-line supervisors and chiefs
   C. Task team leaders
   D. Anyone involved in the operation

4. Engine 100 is the first-arriving company at a four-story multiple-family dwelling. Heavy black smoke with increasingly large flickers of orange flame is pulsing out of a third-floor balcony door. Occupants who are still in that compartment:
   A. Are likely to survive if removed from the compartment in less than 5 minutes.
   B. Are likely to survive if rapidly removed and admitted to a burn center within 20 minutes.
   C. Have a small chance of surviving if rapidly removed and admitted to a burn center within 20 minutes.
   D. Are beyond successful resuscitation.

Fire Captain Activity

Battalion Chief Johnson meets with Captain Davis. "The new NFPA 1981–compliant 2216 psi '30-minute' SCBA bottles we are getting issue a low-air warning at 730 psi, instead of 550 psi. We need to update our fire-ground procedures."

NFPA Fire Officer II Job Performance Requirement 5.6.1
Produce operational plans, given an emergency incident requiring multiunit operations, the current edition of NFPA 1600, and AHJ-approved safety procedures, so that required resources and their assignments are obtained and plans are carried out in compliance with NFPA 1600 and approved safety procedures resulting in the mitigation of the incident.

Application of 5.6.1
1. Provide the chief with a plan to identify a turnaround point when using NFPA 1981 (2013)–compliant SCBA in IDLH environments.
2. Using a department with which you are familiar, develop a mayday procedure that reflects the Rules of Engagement and existing AHJ-approved safety procedures.
3. Using a department with which you are familiar, describe the difference in duties between the incident safety officer and the rapid intervention crew leader when responding to a mayday situation.

Fire Attack

Fire Officer I

Knowledge Objectives

After studying this chapter, you will be able to:

- Discuss the results of the full-scale structure fire experiments conducted by the National Institute of Standards and Technology (NIST) and Underwriters Laboratories (UL). (pp 340–343)
- Describe a fire officer's role in supervising a single company (NFPA 4.2.1). (pp 342–344)
- Describe how to size up an incident. (pp 344–348)
- Discuss how to develop an incident action plan (IAP) (NFPA 4.6.1) (NFPA 4.6.2). (pp 348–349)
- Discuss tactical safety considerations when fighting fires. (pp 349–350, 352)

Skills Objectives

After studying this chapter, you will be able to:

- Supervise a single fire company (NFPA 4.2.1). (pp 342–344)
- Size up an incident. (pp 344–348)
- Develop an incident action plan (IAP) (NFPA 4.6.1) (NFPA 4.6.2). (pp 348–349)

Fire Officer II

Knowledge Objectives

After studying this chapter, you will be able to:

- Discuss the results of the full-scale structure fire experiments conducted by the National Institute of Standards and Technology (NIST) and Underwriters Laboratories (UL). (pp 340–343)
- Discuss the fire officer's role in supervising multiple companies (NFPA 5.6.1). (pp 352–354)
- Identify and describe general structure fire considerations (NFPA 5.6.1). (pp 354, 356)

Skills Objectives

After studying this chapter, you will be able to:

- Supervise multiple fire companies (NFPA 5.6.1). (pp 352–354)

CHAPTER 17

Additional NFPA Standards

- NFPA 1407: *Standard for Fire Service Rapid Intervention Crews*
- NFPA 1710: *Standard for the Organization and Deployment of Fire Suppression Operations, Emergency Medical Operations and Special Operations to the Public by Career Fire Departments*
- NFPA 1720: *Standard for the Organization and Deployment of the Fire Suppression Operations, Emergency Medical Operations and Special Operations to the Public by Volunteer Fire Departments*

Principles of Fire and Emergency Service Administration (FESHE) Course Objectives

6. Discuss the various levels of leadership, roles, and responsibilities within the organization. (pp 342–344, 347–350, 352–354)
9. Identify the roles of the National Incident Management System (NIMS) and Incident Management System (ICS). (pp 344–345, 350, 352, 356)

You Are the Fire Officer

Engine and Quint 100 are responding to a fire in an assisted-living facility on Dover Road. The first-arriving pumper, Engine 7, reports smoke showing from a second-floor balcony in a five-story ordinary construction building. Engine 100 is the second-arriving pumper, and Captain Davis takes "Dover Command." Following the department's standard operating procedures (SOPs), Davis calls for a second alarm.

Lieutenant Williams leads the members of Quint 100 and Engine 100 to the second floor to perform a primary search. The building is divided by fire doors into three sections, with the fire-involved apartment in the middle. The fire fighters encounter residents and staff coming down the hallway in the center section on the second floor. Williams has the residents and staff shelter in place in the Side D section. The fire fighters encounter moderate smoke conditions as they advance to the fire apartment.

1. Which information is needed for an accurate size-up?
2. What are the strategic priorities?
3. Which tactics are appropriate for those strategic priorities?

Introduction

Structural firefighting is a practice built upon experience and experiments. The findings from recent experiments have dramatically changed the understanding of fire dynamics and identified the importance of controlling the flow path of high-momentum fire gases. In turn, traditional ventilation and interior firefighting practices have changed.

Fire Officer I and II

Structure Fire Research

Full-scale structure fire experiments conducted by the National Institute of Standards and Technology (NIST) and Underwriters Laboratories (UL) in partnership with fire departments provide the fire officer with an opportunity to provide a more effective and safer incident action plan by understanding how the build environment has changed and by applying the lessons learned about fire dynamics within a compartment.

■ New Fire Behavior Graph

In 1908, the American Society for Testing and Materials (ASTM) conducted full-scale fire experiments that led to the development of a standard time–temperature curve to guide the testing of building partitions and floors for fire resistance. By 1917, ASTM E 119, *Standard Test Methods for Fire Tests of Building Construction and Materials*, provided a time–temperature curve that was used as a source when the fire service described fire behavior as applied to structural firefighting practices **FIGURE 17-1**.

One hundred years ago, when these standards were developed, fires within a structure were fuel limited, meaning that, without intervention, fire would consume all of the fuel. Recent full-size fire experiments, however, have shown that modern fires are ventilation limited, resulting in a different time–temperature curve **FIGURE 17-2**. NIST (2010) provided this explanation for the changes:

> The Fire Behavior in a Structure curve demonstrates the time history of a ventilation limited fire. In this case the fire starts in a structure which has the doors and windows closed. Early in the fire growth stage there is adequate oxygen to mix with the heated gases, which results in flaming combustion. As the oxygen level within the structure is depleted, the fire decays, the heat release from the fire decreases and as a result the temperature decreases. When a vent is opened, such as when the fire department enters a door, oxygen is introduced. The oxygen mixes with the heated gases in the structure and the energy level begins to increase. This change in ventilation can result in a rapid increase in fire growth potentially leading to a flash-over (fully developed compartment fire) condition.

■ Modern versus Legacy Single-Family Dwellings

UL has documented changes within the residential fire environment that impact occupant survival and fire fighter safety. In *Analysis of Changing Residential Fire Dynamics and Its Implications on Firefighter Operational Timeframes*, Stephen Kerber

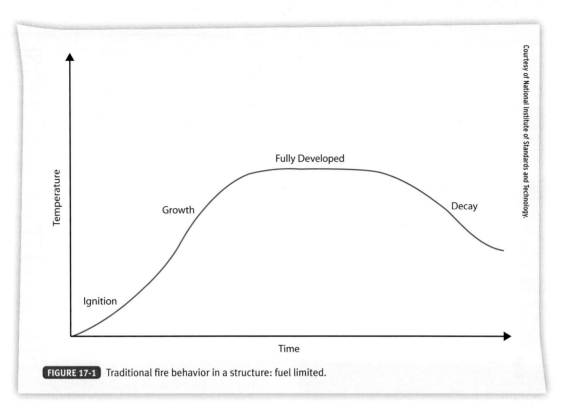

FIGURE 17-1 Traditional fire behavior in a structure: fuel limited.

identifies four factors that distinguish <u>modern dwellings</u> from <u>legacy dwellings</u>, meaning structures built before 1980:

- Larger homes
- Open house geometries
- Increased fuel loads
- New construction materials

Modern dwellings are almost twice as large as legacy single-family dwellings. Open geometry in modern homes reduces compartmentalization and allows more air to support rapid fire propagation. The rate of heat released by burning contents is exponentially higher, resulting in the fire within a room rapidly running out of air. Once a new source of air is introduced, such as from the opening of a door or failure of a window, the fire will quickly develop.

Taking the fire test data and applying a fire department response timeline, UL identified a dramatic reduction in available time when suppressing a modern single-family dwelling fire, as depicted in **FIGURE 17-3**.

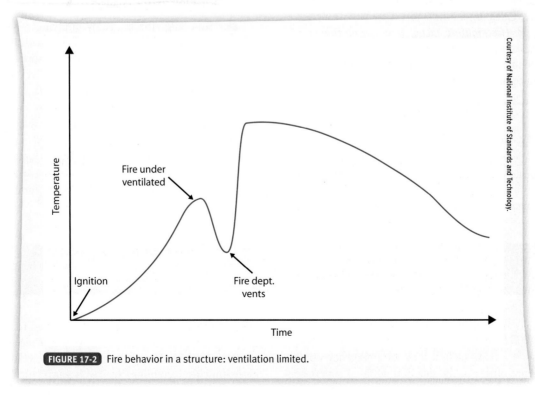

FIGURE 17-2 Fire behavior in a structure: ventilation limited.

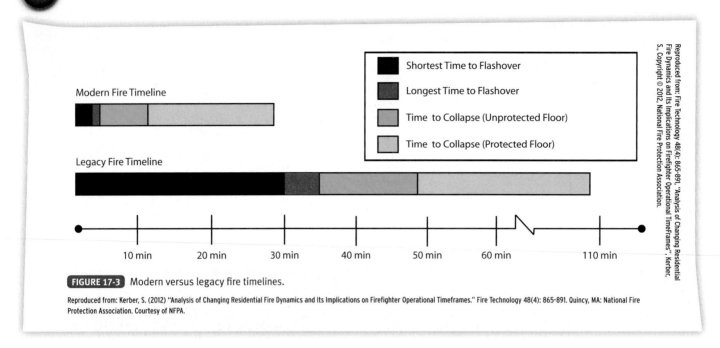

FIGURE 17-3 Modern versus legacy fire timelines.

Reproduced from: Kerber, S. (2012) "Analysis of Changing Residential Fire Dynamics and Its Implications on Firefighter Operational Timeframes." Fire Technology 48(4): 865-891. Quincy, MA: National Fire Protection Association. Courtesy of NFPA.

Fire Marks

Researching Line-of-Duty Deaths

Fire researchers at the National Institute of Standards and Technology began their work to identify fire fighter survival factors when assisting the District of Columbia's investigation of a 1999 Cherry Road townhouse fire. NIST created a computer simulation of the smoke and heat flow conditions that resulted in the deaths of Fire Fighters Anthony Phillips and Louis Matthews.

Fire protection engineers Dan Madrzykowski and Stephen Kerber continued to research the impact of wind-driven fires, including conducting field experiments in Toledo, Ohio; Chicago; and New York City. NIST Technical Note 1629, *Fire Fighting Tactics under Wind-Driven Fire Conditions*, provided the basis for a revolutionary change in firefighting operations aimed at suppressing fires within a compartment or structure.

■ Flow Path

Flow path is the volume between an inlet and an exhaust that allows the movement of heat and smoke from a higher-pressure area within the fire area toward lower-pressure areas accessible via doors, windows, and other openings. The New York Fire Department conducted a series of full-scale exercises on Governor's Island in 2012 to explore alternative ventilation techniques based on the findings in earlier wind-driven research. Both UL and NIST participated in these experiments. Observations at the Governor's Island experiments included the following points when combatting ventilation-limited compartment fires, as listed in UL's *Study of the Effectiveness of Fire Service Vertical Ventilation and Suppression Tactics in Single Family Homes*:

- It is essential to control the access door to restrict introduction of air into the fire room and thereby delay flashover.
- The only way to go from a ventilation-limited to a fuel-limited fire is through application of water before vertical ventilation.
- "Softening" the target by applying 30–90 seconds of water into the compartment dramatically reduces fire development and improves conditions.
- You cannot make a big enough ventilation hole to localize fire growth or reduce temperatures in ventilation-limited structure fires.

Operations conducted in the flow path place fire fighters at significant risk due to the increased flow of fire, heat, and smoke toward their position **FIGURE 17-4**. Limiting flow paths until fire suppression water is ready to be applied is an important factor in limiting heat release and temperatures in the house.

Fire Officer I

Supervising a Single Company

A fire company is a basic tactical unit for emergency operations, with the fire officer in the role of a working supervisor **FIGURE 17-5**. When the company is assigned to advance an attack line into a structure, the fire officer is alongside the crew members, directing and leading them as well as continually evaluating the environment for hazards such as backdraft,

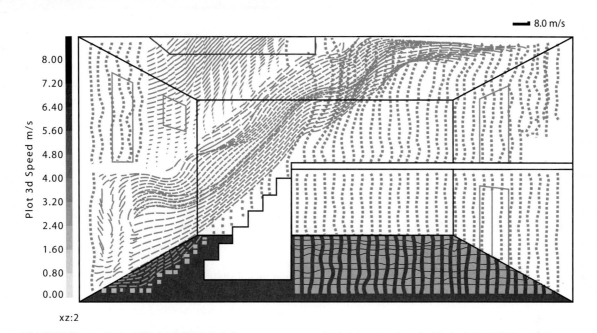

FIGURE 17-4 NIST Fire Dynamics Simulator model of the Cherry Road fire. The fire fighter victims were operating at the top of the stairs, in the middle of the flow path of high-momentum fire gases.

Reproduced from: Madrzykowski, Daniel (2013) Fire Dynamics: The Science of Fire Fighting. International Fire Service Journal of Leadership and Management. Vol 7. Number 2, page 7–15. Tulsa, OK: Oklahoma State University. Courtesy of National Institute of Standards and Technology.

flashover, or structural collapse. The officer's personal and physical involvement in fire suppression activities should never override his or her supervisory duties.

As well as leading and participating in company-level operations, the fire officer constantly evaluates their effectiveness. Is the fire stream making progress or losing ground against the fire? What is the impact of horizontal and vertical ventilation? It is up to the fire officer to identify and mitigate fire-ground problems. For example, if the fire attack appears to have stalled, the officer determines if an additional or larger line is needed or if the attack should be made from a different direction.

The fire officer relays relevant information to the branch, group, or incident commander. At a structure fire, this responsibility would include informing command when the company's assignment is completed or when the task is delayed or cannot be accomplished. It also includes informing the supervisor immediately when problems or hazards are encountered, such as signs of structural collapse. Each company officer serves as the eyes and ears for the incident commander.

■ Closeness of Supervision

The inherent risks associated with emergency operations demand close supervision at all times. Fire fighters who are assigned specific tasks, such as forcing open a door, often concentrate on the task and ignore everything else. The fire officer's role is to review the entire area of responsibility, monitor progress, coordinate with other companies, and look out for hazards.

The level of supervision should be balanced with the experience of the company members and the nature of the assignment. An inexperienced crew performing a high-risk task requires more direct supervision than an experienced crew performing a more routine task. The location of the task also affects the level of supervision. A crew that is advancing a 2½-inch attack line inside a large warehouse will require much closer supervision than a crew operating a master stream 15 feet away from the same burning warehouse. The company officer must know where the crew members are located and what they are doing. When performing a task in a high-risk situation, such as an immediately dangerous to life or health (IDLH) environment or below-grade rescue, the fire officer needs to see and directly communicate with all of the members.

FIGURE 17-5 Supervising emergency operations is a core fire officer task.

Situational Leadership

The fire officer needs to adopt the appropriate leadership style for the specific situation. During nonemergency situations, a participative leadership approach can develop group cohesiveness and productivity. In such a case, there is time for discussion and feedback before a decision is made. The consequences seldom involve imminent risk of injury or death.

In contrast, time is often of the essence at an emergency incident. Decisions are needed quickly and actions must be implemented in a timely manner. These situations do not allow a long discourse on what needs to be done or development of a group consensus. The fire officer must frequently make decisions with little or no input from subordinates, and every member of the company must be prepared to perform a specific set of tasks with a minimum of instruction.

When the fire officer gives an order to lay a supply line, the fire fighter who is assigned to catch the hydrant must know exactly what to do. While the fire fighter is dragging the hose around the hydrant, the fire officer should be gathering information to develop the next step in the incident action plan. The other fire fighter, who will be advancing a fire attack line, can anticipate which task comes next without needing a lengthy explanation from the fire officer.

Success using an authoritative style comes when the officer has developed the trust and confidence of his or her subordinates before the incident. Officers who demonstrate sound knowledge of the technical aspects of firefighting and demonstrate trustworthiness are well prepared to lead fire fighters into a high hazard task. Fire fighters do not trust the judgment of a fire officer who does not fully understand the job and has not made fire fighter safety a priority.

Standardized Actions

Emergency incident operations must be conducted in a structured and consistent manner. This goal is accomplished by placing a strong emphasis on standard operating procedures (SOPs). SOPs provide a framework to allow activities to be efficiently completed through the efforts of everyone involved in the event. Just like a football team playbook, SOPs explain the standard approach that should be followed in a particular situation.

Standardized approaches are especially helpful in volunteer organizations. NFPA 1720, *Standard for the Organization and Deployment of the Fire Suppression Operations, Emergency Medical Operations and Special Operations to the Public by Volunteer Fire Departments*, describes demand zones (urban, suburban, rural, remote, and special) and staffing response times. For example, 10 fire fighters are expected to respond within 10 minutes to 80 percent of the incidents in the suburban demand zone.

Such a standardized approach facilitates the development of an incident action plan and promotes a consistent approach to risk and safety. Utilizing an Incident Command System (ICS) facilitates a consistent development of efficient incident management with effective control.

Command Staff Assignments

Command staff assignments include the roles of safety officer, liaison officer, and public information officer. While working

> **Assessment Center Tips**
>
> **Do Not Forget Your Crew!**
> During a simulated incident exercise, single-company supervisors are assigned to command staff positions as the Incident Command System ramps up. Remember to identify where the rest of your fire company will be assigned.
>
> Often the company will be assigned to work with another fire company. Make sure that your crew is working under another supervisor. In the "You Are the Fire Officer" scenarios in this text, the lieutenant of Quint 100 is the safety officer, and the crew of Quint 100 is teamed with the crew of Truck 7 in the Search Division, working under Truck 7's commander.

in one of these positions, the fire officer reports directly to the incident commander. The incident commander could assign any available and qualified officer to perform one of these roles.

The safety officer is responsible for overseeing the incident from a safety perspective, keeping the incident commander informed of safety concerns, and taking preventive action when an immediate hazard is identified. When assigned as liaison officer, the fire officer functions as the link between the incident commander and representatives from various agencies. At a hazardous materials incident, such representatives could include the property owner or a chemical manufacturer. When operating as the public information officer (PIO), the fire officer is responsible for gathering information to be released to the general public, developing news releases, and giving interviews or press conferences. The PIO acts as the spokesperson for the incident commander.

Sizing up the Incident

Size-up is a systematic process of gathering and processing information to evaluate the situation and then translating that information into a plan to deal with the situation. The art of sizing up an incident requires a diverse knowledge about emergency incidents. The fire officer could be faced with a house fire, a hazardous materials release, and a major automobile collision in one shift. Each situation requires the ability to recognize and analyze the important factors, develop a plan of action, and then implement the plan. Each type of incident also involves specialized knowledge.

Inexperienced fire officers may think that size-up begins when they get their first glimpse of the incident scene and ends when they have decided on a plan of action. In reality, size-up begins long before arrival and continues until the incident is stabilized. Deciding on an initial plan of action after arriving on the scene is only one step in a continuous process. As the plan is executed, new information must be gathered and processed to determine whether the plan is working as anticipated and to make any adjustments that are required.

The initial size-up at the scene of an incident must often be conducted under intense pressure to "do something." This pressure could be coming from the public, from fire companies that are arriving and anxious to get into action, and from within the fire officer who feels the anxiety of being responsible for deciding what to do quickly. The end result of a good size-up is an incident action plan (IAP) that considers all the pertinent information, defines strategies and tactics, and assigns resources to complete those tactics.

A comprehensive evaluation of a situation requires information that often is not available to the fire officer. Consequently, the officer must look carefully and assess what can be seen, make reasonable assumptions about what cannot be seen, and anticipate what is likely to happen. An experienced officer will develop an initial plan and then adjust that plan as more information becomes available. Such additional information could, for example, require a major change in the plan. The incident commander must consciously differentiate among what is known, what is assumed, and what is anticipated in processing size-up information.

As noted earlier, the first phase of size-up begins long before the incident occurs. Everything the fire officer learns, observes, and experiences goes into his or her memory bank, and those memories are then used to make size-up decisions. Knowledge of the types of building construction used in the area, the available water sources, and the available resources are all factors that go into the size-up. Whenever a preincident plan is updated, including a familiarization visit, the fire officer is performing part of a future incident size-up.

■ Prearrival Information

The specific size-up for an incident begins with the dispatch. The name, location, and reported nature of the incident all help the fire officer anticipate what might be happening at the scene. Knowledge gained from a preincident plan and awareness of what would be normally found at that day and time are factored into the prearrival data.

■ On-Scene Observations

The ability to size up a fire situation quickly requires both a systematic approach and a solid foundation of reference information. SOPs list the essential size-up factors, which include building size and arrangement, type of construction, occupancy, fire and smoke conditions, and other factors, such as weather and time of day. The SOPs should guide the officer's systematic thinking to ensure that all of the important factors are considered.

A fire officer must understand and recognize basic fire dynamics. It is easy enough to identify fire coming out of a window and smoke pushing out from the eaves of a house. An understanding of conduction, convection, and radiation, however, is needed to predict where the fire is burning and where it will spread. An officer who has studied and observed fire dynamics is able to process this information very quickly, without having to spend an excessive amount of time looking at the building and thinking about what is happening inside.

Visualization is one of the most significant factors in size-up. Every previous situation that the fire officer has experienced or observed, including some that might have been observed only in training sessions, photographs, or videos, is stored in the individual's memory. When a new situation is observed, the mind subconsciously looks for a matching image to create a template for this new observation. Instead of methodically processing the information, the brain can instinctively jump to a similar observation and apply the stored experience to the new set of circumstances. The same process works for other human senses, particularly smell, taste, and touch, fostering rapid recognition and interpretation of many situations.

Experienced fire officers know vigorous dark smoke churning means a high heat release rate, indicating flashover conditions are present. Darker smoke is generally closer to the seat of the fire than lighter-colored smoke. By evaluating where the smoke is coming from in the building, the fire officer may be able to predict where the fire will be traveling, unless mitigating circumstances occur.

The volume and color of the smoke aid the officer in determining ventilation timing. Dark smoke is also an indication that there are more carbon particles suspended within the smoke and less oxygen. Before introducing additional air into a compartment that is belching dark, billowing smoke, the fire officer needs to "soften the target" with a quick application of water into the compartment and coordinate ventilation efforts with the fire hose crew.

Be cautious when approaching a modern structure with no smoke showing. UL's full-scale experiments showed that once a fire became ventilation limited, the smoke being forced out of the gaps of the structure greatly diminished or completely stopped. No smoke showing during size-up should increase the officer's caution and awareness of conditions inside the structure.

An understanding of fire dynamics is also needed to develop action plans. Fires in the incipient phase may be extinguished with a portable fire extinguisher. Fires in the free burning phase may require a small handline or even multiple large handlines or master streams.

Fuel load is another important factor during size-up. A fire in Class A combustibles normally dictates a direct attack with water. Fires involving Class B combustibles require the use of foams. When using foams, the officer must determine which type and how much foam will be needed to complete the job before beginning the application.

■ Lloyd Layman's Five-Step Size-up Process

In 1940, Chief Lloyd Layman authored *Fundamentals of Fire Tactics*, which used the military training model with an emphasis on general principles. Layman presented a five-step process for analyzing emergency situations:

1. Facts
2. Probabilities
3. Situation
4. Decision
5. Plan of operation

Facts

Facts are the things that are known about the situation. Chief Layman was an early advocate of preincident planning. This process allows the first-arriving officer to determine quickly the type of construction, the presence of any fire protection systems, fire flow requirements, water supply sources, special hazards, and other significant factors. Additional facts become known upon dispatch, such as the time of day, location of incident, weather, and specific resources that the fire department is sending. Visible smoke and fire conditions are additional facts. The more facts the fire officer can secure, the more accurately the situation can be sized up.

Probabilities

Probabilities are things that are likely to happen or can be anticipated based on the known facts. For example, the fire officer might know that the structure has a fire sprinkler system but cannot be sure that it will work. The sprinkler system may have been turned off for maintenance or damaged by an explosion. During size-up, the fire officer must consider the most likely possibilities based on the known facts.

The fire officer has to anticipate where the fire is likely to spread. If the fire is burning freely in the attic, how is it likely to affect the stability of the building? This skill requires an in-depth knowledge of factors such as building construction, fire dynamics, hazardous materials, sprinkler system performance, and human behavior in fires.

Another aspect of considering probabilities is the amount of time it takes to accomplish fire-ground tasks. How far is the fire likely to advance before the hose lines are in place to attack? How long will it take to get ground ladders to Charlie side? How fast can a crew with a backup hose line get to your location? How large will the fire be in 10 minutes? Which type of damage is being done to the structural components of the building?

Situation

The situation assessment involves three considerations. The first consideration is whether the resources on scene and en route will be sufficient to handle the incident. The incident commander will be able to determine whether additional companies will be needed to control the problem. The incident commander also has to anticipate the time lag between requesting resources and having them on the scene available to use.

Many fire departments have policies that require an incident commander to request additional resources whenever all of the units will be committed to firefighting operations, so as to provide an on-scene reserve. Some departments require the first-arriving company to call for a second alarm immediately if any smoke or fire is showing from a high-rise, hospital, or "big box" mercantile building.

The second consideration is the specific capabilities and limitations of the responding resources in relation to the problem. Consider the amount of water carried, pump capacity, amount and diameter of hose, and length and type of aerial devices. Staffing is an important factor. A company staffed with two fire fighters cannot perform as much work as a company of six fire fighters. The incident commander might have to call for additional resources or adjust performance expectations downward based on the resources at hand.

The third consideration consists of the capabilities and limitations of the personnel, based on training and experience. Consider your personal experience with this type of incident, as well as the experience and capabilities of your crew members. If two fire fighters out of a four-person crew are rookies, the company requires more direct supervision.

Decision

The fourth step in Layman's size-up procedure is making fire attack decisions. This step requires the fire officer to make specific judgment decisions based on the known facts and probabilities, as well as the situation evaluation. The officer needs to answer four questions:

1. Are there enough resources responding to and on the scene to extinguish the fire or mitigate the situation?
2. Are sufficient resources available, and do conditions allow for an interior attack?
3. What is the most effective assignment of on-scene resources?
4. What is the most effective assignment of responding resources?

The answers to these questions allow the fire officer to develop the overall strategy and the underlying tactics to mitigate the incident.

Plan of Operation

The final step in Layman's decision process is to develop the actual plan that will be used to mitigate the incident. Most other methods of size-up do not include this step as part of the size-up; they refer to the information that is obtained during size-up being used to develop a plan of action.

■ National Fire Academy Size-up Process

The National Fire Academy (NFA) has developed a size-up system that includes three phases:

1. Preincident information
2. Initial size-up
3. Ongoing size-up

Phase One: Preincident Information

Phase one considers what you know before the incident occurs. This consideration closely mirrors the facts step identified by Layman. Preincident plans often provide valuable information for this phase. Information about the building and the occupancy, such as the building layout and construction type, built-in protection systems, type of business, and nature of the contents, is clearly needed to perform an accurate size-up. Preincident information should also identify water supply sources, including their location, accessibility, and capacity.

If preincident plans are not available, this information must be determined through on-scene observation and research. The fire officer's intuition and experience may have to be relied on until this information can be obtained.

Fire officers' memory stores a large amount of useful information that figures into the size-up process. Environmental information includes factors such as heat and cold extremes, humidity, wind, and snow or ice accumulations. The time of day can also be a critical piece of information; responding to fire in a multifamily residential building at 2:00 A.M. involves different considerations from responding to the same fire at 2:00 P.M. A fire officer also needs to be familiar with departmental and mutual aid resources.

Phase Two: Initial Size-up
The second phase of the size-up begins on receipt of an alarm. Three questions need to be answered:

1. What do I have?
2. Where is it going?
3. How do I control it?

Start with the information that was established in phase one and build upon it. Phase two considers the specific conditions that are present at the incident. The incident commander's initial actions should include a 360 walk-around of a fire building to observe conditions, such as the location and the size of the fire, as well as the volume, color, movement, and location of the smoke. The size-up should also assess the size and construction of the building, identify any exposures that are present, and look for indications of special concerns, such as the possibility that lightweight construction could be involved.

To determine where the fire is going, the fire officer must consider the current location and stage of development of the fire, the direction in which it is likely to spread, and the likely impact of fire suppression efforts. Is this a ventilation-limited fire? Is there an imminent risk of structural collapse? The accuracy of these predictions is heavily dependent on the fire officer's knowledge and experience. This stage is closely associated with the probabilities step noted by Layman.

Answering the third question, "How do I control it?", requires the fire officer to consider different alternatives in relation to the available resources, including apparatus, equipment, and personnel. The development of strategy and tactics must be based on the resources that are available and, if additional resources are needed, the time it will take for them to arrive. This stage is closely associated with the situation step noted by Layman.

Phase Three: Ongoing Size-up
The third phase addresses the need to continually size up the situation as it evolves. This phase includes an ongoing analysis of the situation and an ongoing evaluation of the effectiveness of the plan being executed. The incident commander should always be prepared to modify the plan if the situation changes, including switching between offensive and defensive strategies. Additional resources could be required to address unanticipated needs or to replace tired crews. The ongoing evaluation could also indicate that resources can be redeployed from one task to another where they would be more effective.

The ongoing size-up requires a constant flow of feedback information to the incident commander. The incident commander needs to know when:

- An assignment is completed
- An assignment cannot be completed
- Additional resources are needed
- Resources can be released
- Conditions have changed
- Additional problems have been identified
- Emergency conditions exist

■ Risk–Benefit Analysis
Risk–benefit analysis is a key factor of size-up when selecting the appropriate strategic mode. The strategic mode dictates the degree of risk to fire fighters that is acceptable in a given situation. The degree of risk that is acceptable is determined by the realistic benefits that can be anticipated by taking a particular course of action.

Potential benefits may include the possibility of saving lives or preventing injury to persons who are in danger, preventing property damage, and protecting the environment from harm. The major risk factor is the possibility of death, disability, or injury to fire fighters. Potential occurrences, such as flashover, backdraft, or structural collapse, also have to be evaluated. Each alternative action plan involves a different combination of risks and benefits.

The fire officer can manage these risks. Training, experience, protective clothing and equipment, communications equipment, SOPs, accountability systems, rules of engagement, and the rapid intervention team are all factors that allow crews to work with a reasonable level of safety in a hazardous environment.

Fire fighters accept a higher personal risk when it is necessary to save a life; however, there is no justification for risking the lives of fire fighters when no one requires rescue or when interior fire conditions are unsurvivable, such as in fires in vacant or abandoned buildings or situations where significant damage is already done. Building construction features, such as lightweight trusses, can also cause the risks of offensive operations to outweigh the potential benefits. When defensive operations are conducted, the risks to fire fighters are significantly reduced.

The risk analysis determines the appropriate strategy for an incident: offensive, defensive, or transitional. An <u>offensive operation</u> consists of an advance into the fire building by fire fighters with hose lines or other extinguishing agents to overpower the fire. Offensive activity drives most fire department training, operations, and organizational structures.

When this mode is selected, the incident commander believes that the benefits associated with controlling and extinguishing the fire outweigh the risks to the fire fighters. To have any chance of success, such an operation requires sufficient resources to mount an attack that is powerful enough to overwhelm the fire within the time that is available to operate safely inside the burning building. The risks of an offensive attack can be justified only when it may generate realistic benefits. Specifically, there must be lives or valuable property that can be saved to justify taking this risk.

A <u>defensive operation</u> is used when the risks outweigh the expected benefits. In this mode, fire fighters are not allowed to enter the structure or to operate from positions that involve avoidable risks. Defensive operations are conducted from the exterior, using large streams to contain and, if possible, overwhelm a fire **FIGURE 17-6**. A defensive approach is used when

FIGURE 17-6 Defensive operations.

there is a risk of a structural collapse or there are inadequate resources to conduct an adequate interior attack to control the fire. A defensive strategy is also the appropriate choice in a situation where the building and contents would be a total loss even if an aggressive interior attack could control the fire.

In a transitional operation, the firefighting operation is changing or preparing to change. A transitional strategy applies where a defensive operation is initiated, but with the recognition that the situation could change to an offensive approach when interior conditions improve. The offensive operation is conducted in a manner that allows interior crews to be withdrawn quickly if structural instability is noted. At the same time, backup resources are positioned to reinforce the offensive operations. This strategy could also be used to conduct a search and rescue operation quickly, ahead of a fire, knowing there is only a limited time for crews to get in and get out. A transitional attack could also be applied in a situation where an initial attack is made with an exterior master stream to knock down a large body of fire, while crews prepare to conduct an offensive interior operation.

Developing an Incident Action Plan

After size-up, the incident commander develops an IAP based on the incident priorities. There are two major components to the IAP:

Company Tips

How Aggressive Should Suppression Be?
The sweet spot for effective interior fire operations depends on the built environment as well as fire department resources, experience, and training. What is appropriate for a big city department that can deliver 36 fire fighters, 12 company officers, and 3 command officers in 15 minutes is not appropriate for a rural fire department that can assemble 6 fire fighters and 1 company officer within the same time frame.

1. The determination of the appropriate strategy to mitigate an incident
2. The development of tactics to execute the strategy

Strategies are general, whereas tactics are specific and measurable. In nonfire terms, strategies are equivalent to goals, whereas tactics are the objectives used to meet the goals. The incident commander identifies the strategic goal and then identifies the tactical objectives that are required to reach this goal. Resources (companies) are assigned to perform individual tasks that allow the tactical objectives to be achieved.

SOPs are used to provide a consistent structure to the process of establishing strategies, tactics, and tasks. Without SOPs, every situation might be managed differently, depending on the individual who was in command and his or her interpretation of the situation, as well as the decisions made by individual company officers. SOPs guide the decision-making process and ensure consistency between officers and events. On arrival at a house fire with smoke showing, the SOP might require the first-arriving officer to assume command, the first-arriving engine company to begin a fire attack, and the second-arriving engine to provide a supply line for the first engine. Thus SOPs assist everyone involved in the process by outlining which actions to take when presented with a given situation. In effect, SOPs are preestablished components of an incident action plan.

■ Incident Priorities

There are three basic priorities for an incident action plan:

1. Life safety
2. Incident stabilization
3. Property conservation

Incident stabilization and property conservation are often addressed simultaneously, although property conservation is ranked lower on the priority list. If the incident commander does not have sufficient resources to perform both of these operations, property conservation measures might have to be delayed until the incident has been stabilized. In reality, many of the actions that are taken to stabilize an incident also work toward assuring life safety.

The life safety priority refers to all people who are at risk due to the incident, including the general public as well as fire department personnel. The incident stabilization priority is directed toward keeping the incident from getting any worse. If one structure is fully involved, protecting the exposures is part of incident stabilization. In such a case, the burning structure cannot be saved, but fire fighters can ensure that the fire will not spread beyond that building.

Property conservation is directed toward preventing any additional damage from occurring. Such measures could include minimizing water damage by covering building contents with salvage covers or removing valuable property from harm's way. Property can also be protected by venting smoke and heat from a building and then covering the openings to keep wind and rain from causing additional damage.

Tactical Priorities

At the scene, the incident commander faces a number of issues that require attention. If an abundance of resources arrived simultaneously, it would be a simple matter to assign them in a manner that would get everything done immediately. Because this rarely happens, fire officers need a method to address the most critical concerns first. Tactical priorities provide a list and an order of priority for dealing with them. The tactical priorities and the information obtained in the size-up are used to develop the IAP.

Lloyd Layman developed a list of seven factors to assist in developing an action plan. Although other techniques exist, Layman's method remains one of the most popular, 50 years after it was first published.

RECEO VS is an acronym that covers the critical factors in developing a strategy. The first five factors are listed in a priority order, whereas the last two may be used at any point to support the first five:

- **R**escue
- **E**xposures
- **C**onfinement
- **E**xtinguishment
- **O**verhaul
- **V**entilation
- **S**alvage

Rescue

Life safety is fire fighters' highest incident priority, so rescue is at the top of Layman's list. The rescue category includes all functions related to searching for potential victims and removing them from danger to a location of safety, as well as any other activity that is necessary to reduce the risk of death or injury. Medical care after the occupants have been removed is a rescue priority.

The best method of protecting the occupants from harm is to extinguish the fire quickly. Whenever interior search and rescue operations are under way, the fire must be attacked to give the rescuers time to enter the structure and search for occupants.

Exposures

The next priority is to keep the incident from getting bigger. Exposure protection is most important in a situation in which the incident commander does not immediately have sufficient resources to control or extinguish the fire fully. The best method of protecting an exposed property is to make an aggressive attack and to extinguish the fire before it can spread. In the same manner that quickly extinguishing the fire reduces the rescue problem, so rapid extinguishment alleviates the exposure problem.

If the fire is beyond the capabilities of an initial attack, the best decision a fire officer can make is to protect the adjacent properties rather than waste resources attempting an inadequate attack on the original fire building. If there are insufficient resources to save the fire building, use the available resources to protect adjacent buildings.

Confinement

The third priority is to prevent the fire from spreading to uninvolved areas of the same property. If the fire is in one room, the objective should be to confine the fire to that room of origin. If confining the fire to one room is not possible, then the objective could be to confine it to the floor or area of origin.

Confinement can also be directed toward supporting rescue efforts. Confining the fire to the room of origin can provide time for other crews to perform rescue. Confinement efforts could also be directed toward protecting exit stairways and corridors so as to remove occupants safely.

Extinguishment

The next priority is to extinguish the fire or mitigate the incident. Rapid extinguishment eliminates or significantly reduces the need for rescue, exposure, and confinement priorities. The priority order becomes important when rapid extinguishment is not feasible—the higher priorities must always be considered ahead of extinguishment. Many tactical options may be employed to accomplish extinguishment, using either offensive or defensive strategies.

Recent NIST and UL research shows the dramatic impact of initial application of 30–60 seconds of water into a compartment: dramatic reduction of heat within the flow path, improved occupant survival, and safer interior suppression conditions.

Overhaul

The fifth and last of the tactical priorities is overhaul, the activity that makes sure the fire is completely out. After all of the visible fire has been extinguished, the area must be checked for residual fire, including any fire that could have extended into areas not originally involved, such as walls, ceilings, and attic spaces. Incomplete overhaul may allow the fire to rekindle.

Ventilation

Ventilation is designed to remove the products of combustion from a fire area and allow cool, fresh air to enter. Effective ventilation can improve the survival chances for occupants as well as the conditions that would allow fire fighters to enter and operate inside a building. When operating at ventilation-limited fires, controlling the door and coordinating ventilation with water application are vital to controlling the flow path early in suppression operations.

Salvage

Salvage is the other tactical activity that does not have a specific place on the priority list. Salvage includes protecting or removing property that could be damaged by the fire, smoke, water, or firefighting operations. It also includes securing the building and protecting it from weather. The salvage operation can be placed wherever it is appropriate in a particular situation and is often performed in parallel with the other priorities, if resources are available. Salvage often involves placing salvage covers over property to protect it during overhaul. In addition, it can include actions such as using only enough water to extinguish the fire to limit water damage.

Tactical Safety Considerations

Fighting fires is an inherently dangerous activity that exposes fire fighters to a wide variety of risks and hazards. Many advances have been made, including improved protective clothing and equipment, more effective methods of fighting fires, and better procedures for managing fire incidents. The hazards still exist, however, and new threats continue to appear.

Protective clothing and self-contained breathing apparatus (SCBA) provide better protection than was available to earlier generations of fire fighters. To benefit from this protection, the full ensemble of personal protective clothing and equipment must be worn whenever fire fighters are exposed to hazardous conditions.

Personal protective equipment (PPE) allows fire fighters to enter deeper into burning buildings and stay longer, often without sensing the level of heat in that environment. PPE has saved many lives and prevented disabling injuries. Unfortunately, it also exposes fire fighters to an increased risk of being trapped by a sudden flashover, becoming disoriented inside a burning building, or running out of air in a highly toxic atmosphere.

The PPE's weight, bulk, and thermal properties must be considered during extreme weather conditions. Fire fighters who are wearing PPE need sufficient rehabilitation between work periods.

■ Scene Safety

Many fires occur at night, leading to a dark fire scene full of tripping hazards and obstacles. The use of lights to illuminate the incident scene adequately is an important safety consideration. During cold weather, the fire officer may also need to have abrasive materials spread about the scene to improve traction on ice.

Many emergency operations are conducted in the street, exposing fire fighters to traffic hazards. Apparatus should be positioned to protect the scene, fire fighters need to wear high-visibility safety vests or jackets that comply with ANSI 207, and the fire officer should request traffic control to reduce the risk of fire fighters being struck by vehicles.

Hazardous areas at the incident scene, such as holes and potential collapse zones, should be clearly identified. Some fire departments use barrier tape to establish exclusion zones around hazards. Many departments utilize hot/warm/cold zones to identify the level of hazard:

- **Hot zone:** Highest hazard; minimum number of crew allowed in the area. The hot zone could be an area of structural weakness, a hazardous environment, or an area requiring use of full PPE including SCBA. A hot zone is under direct supervision of a unit, branch, or group leader with an identified entry/exit point.
- **Warm zone:** Moderate hazard; restricted number of emergency service members allowed in the area. This area remains part of an active incident and requires the use of some protective equipment (helmet, boots, or respirator) or special vigilance.
- **Cold zone:** Minimum hazard; no restrictions on number of crew or nonemergency service individuals allowed in the area.

The fire officer should encourage fire fighters to stash a spare pair of socks, gloves, a towel, and a knit cap on the fire apparatus. Getting soaked in a winter fire is a common occupational hazard. If they add a T-shirt, shorts, and coveralls to this stash, fire fighters can have dry clothes after a soaking or a technical decontamination.

> ### Getting It Done
>
> **On-Duty Wallet, Phone, and Keys**
> The fire officer should also encourage the crew to carry an on-duty wallet. The on-duty wallet can be lost during decontamination without much disruption to the fire fighter's personal life.
>
> The on-duty wallet includes a duplicate driver's license, copies of any state- or department-mandated certification cards (e.g., paramedic), and, if desired, a credit card. The fire fighter should place enough money in this wallet to get through the day and leave his or her regular wallet locked in the car or the fire department locker. The same goes for a wireless or smart phone. Consider using a disposable or spare phone while on duty.
>
> Create a set of duty keys. Make duplicates of the keys you need to get in the fire station, your locker, and your car. As with the on-duty wallet, you can lose your duty keys and not have your life disrupted to a great extent. Do not forget to put a durable identification tag on your duty keys so that you can get them back after they are sanitized! Lock up your regular keys, smart/cell phone, wedding ring, and other valuable jewelry before you report for duty. It may be a good idea to use a combination lock to secure your storage space.

Exposure to asbestos building components, drug laboratories, and chemical/biological weapons are three situations in which a fire crew may have to disrobe at the incident scene. A technical decontamination may result in the fire fighters losing their uniforms in the warm zone. Without the stash, the only option may be wearing a disposable Tyvek gown.

Fire fighters should change out of their duty uniforms and clean up before they leave the fire station, even after a slow duty day. Infectious diseases, asbestos, industrial chemicals, and products of combustion are hazards that should stay at the fire station and not be brought home to expose family members.

■ Rapid Intervention Crews

The rapid intervention crew (RIC) is "a dedicated crew of firefighters who are assigned for rapid deployment to rescue lost or trapped members." NFPA 1407, *Standard for Training Fire Service Rapid Intervention Crews*, describes required capabilities of the RIC. An initial rapid intervention crew (IRIC) is two members from the initial attack crew, as defined in NFPA 1710, *Standard for the Organization and Deployment of Fire Suppression Operations, Emergency Medical Operations and Special Operations to the Public by Career Fire Departments*. A RIC, by comparison, consists of four members, and is sometimes identified as a rapid intervention team (RIT). This dedicated crew is not to be confused with the IRIC. Most fire departments add an additional fire company to the first alarm assignment for a structure fire, either as part of the initial dispatch or as soon as a working fire is reported, to function as the RIC.

Near-Miss REPORT

Report Number: 10-0000277

Event Description: Fire department units responded to a reported residential fire [just after midnight]. Size-up indicated heavy smoke showing, with confirmation of the residents being out of the structure. The residents also reported that the fire appeared to be in the basement laundry area, around the dryer. Entry was made into the structure simultaneously with some ventilation under way. Further ventilation operations were ordered following reports from inside.

The basement stairway was located next to the main floor kitchen, with the seat of the fire located directly below the kitchen in the basement. The incident progressed as expected through the first 20-minute PAR. Water supply was established, utilities were ordered for disconnect, the RIT team was established, and ventilation was well under way.

About 30 minutes into the incident, a request was made from interior crews to have someone bring another 1¾-inch line in through the garage to access the basement. The RIT team was assigned to perform this task and then stopped, as the garage was too packed full of stuff to even make their way inside. That crew was then outside the structure, but had not reassembled for RIT assignment (IC's call).

The IC was making another 360-degree walk-around to check on a utility worker when dispatch notified IC of the 40-minute operational mark. During this PAR, command heard a PASS device activate and yelled in the direction of the activation, thinking that someone might have been standing still and it activated. The PPV fan was still operating so the noise level was elevated. Soon after hearing the PASS device, dispatch also reported the radio emergency alarm activation of a radio. The IC was on Side A looking in the open front door of the structure and could see the faint blinking of the PASS strobe in the direction of the sounding PASS device.

Immediately, one of the personnel originally assigned to RIT was told face-to-face to get that person out of the building. At the time, the IC was extremely unhappy, thinking that someone had just let their PASS device activate and didn't bother to stop it. The RIT member followed the hose line in a short distance, approximately 30 feet, toward the strobe and dragged the downed firefighter out. Upon exiting the structure, he was helped to his feet, immediately assessed for injury, and then relocated to the ambulance for further evaluation. Subsequently, he was transported to the hospital, as a precaution, for further testing.

In interviewing the [downed fire fighter], he stated that he was with his crew member in the basement on fire attack, along with another two-person crew. His low-air alarm had activated and he continued to work, thinking he had plenty of time. After a time, he told his partner that he was going to run outside and get another bottle. He then left his partner and headed out of the laundry area in the basement, following the hose line around the corner and up the stairs. Partway up the stairs, he completely ran out of air. In a condition of "high motivation," he started to hurry. Staying low and following the hose line, he became disoriented and ended up reversing his direction. He then fell back down the stairs, knocking his face piece off.

The conditions were still untenable. He repositioned his face piece so his Nomex hood would give him some filtering action. He then activated his PASS device and activated the emergency button on his radio. He stated that he was unable to speak due to the heavy smoke conditions.

As a side note, his partner thought he heard a PASS device activate, but he stopped hearing it (the captain went back up the stairs to attempt exit), so he assumed that it was an accidental activation. As the captain cleared the top of the stairs, he had to keep his face close to the floor. It was at this point that the RIT person located him and pulled him to the exit.

Lesson Learned: One important lesson learned that must be addressed is that the "SCBA lost and disoriented fire fighter" training conducted by this department worked in its most basic form. When the situation became less than ideal, the captain controlled his emotions, remained calm, activated his PASS device and his radio emergency button, took steps to get the best-quality air he could find, and was actively involved in rescuing himself. Remembering those important training points is very commendable. However, although the outcome of this near-miss incident was positive, this particular incident itself was completely preventable.

(continued)

Several opportunities for improvement have been identified in response to this incident:

1. Strict adherence to the two-in/two-out rule should be enforced.
2. Strict adherence to the department's air-management protocol (when the low-air alarm activates, you call for relief and then immediately exit with your assigned partner or crew) should be required. Additionally, this should be reinforced with a department SOP/SOG regarding SCBA operational procedures and air management.
3. A *zero tolerance* departmental policy regarding PASS device activations should be implemented and enforced. When a device activates, it gets immediate attention. Anything less than this response creates a potential environment of dangerous complacency when hearing the devices activate.
4. Incident command should diligently track at any moment where personnel are and maintain a good communication link with anyone on the fire ground.
5. Once a RIT team is assigned, they should not be reassigned unless activated or relieved by a replacement, until the incident de-escalates.
6. Regular training should be conducted on the "lost and disoriented fire fighter" procedures, along with SCBA air management training.
7. Training should be regularly given to reinforce the importance of the SOPs/SOGs that are in place to keep personnel safe on the fire ground.

Courtesy of the National Fire Fighter Near-Miss Reporting Systems (firefighternearmiss.com)

Many departments have developed RIC training and deployment procedures. The standard radio term *mayday* is used to indicate that a fire fighter is lost, missing, or in life-threatening danger. The incident commander's standard response to a mayday is to activate the RIC. When this occurs, the RIC operation has priority over all other tactical functions.

The RIC is generally positioned outside the building, near an entrance, ready for immediate action. SOPs governing RICs generally specify a list of equipment that will be prepared for use, including forcible entry tools, a thermal imager, a search rope, medical equipment, and a spare SCBA or full replacement cylinder. Every RIC team member should have a portable radio and should be wearing full PPE. While standing by, the RIC members should closely monitor fire-ground radio communications to keep track of activities and listen for a mayday or any indication of a problem.

Personnel Accountability Report

A personnel accountability report (PAR) is a systematic method of accounting for all personnel at an emergency incident. When the incident commander requests a PAR, each fire officer physically verifies that all assigned members are present and confirms this information to the incident commander. A PAR should also be requested at tactical benchmarks, such as a change from an offensive strategy to a defensive strategy.

The fire officer must be in visual or physical contact with all company members to verify their status. The fire officer then communicates with the incident commander by radio to report a PAR. Anytime a fire fighter cannot be accounted for, he or she is considered missing until proven otherwise. A report of a missing fire fighter always becomes the highest priority at the incident scene.

If unusual or unplanned events occur at an incident, a PAR should always be performed. For example, if there is a report of an explosion, a structural collapse, or a fire fighter missing or in need of assistance, a PAR should immediately take place so that it can be determined how many personnel might be missing, the missing individuals can be identified, and their last known location and assignment can be established. This last location would be the starting point for any search and rescue crews.

Fire Officer II

Supervising Multiple Companies

The first-arriving officer at a fire incident assumes the role of incident commander. When operating as the incident commander, the fire officer has an even greater level of responsibility, as the incident commander is responsible for every company on the scene and for management of the overall operation. As covered in the *Managing Incidents* chapter, the initial incident commander's responsibilities include conducting a size-up, developing an action plan to mitigate the situation, assigning the resources to execute the plan, developing a command structure to manage the plan, and ensuring that the plan is completed safely. The fire officer who is functioning as the incident commander is responsible for all of these functions.

The fire officer might also be assigned as a division/group/unit leader, supervising multiple units, or even as a branch director within the Incident Command System (ICS). In each of these situations, the officer serves as a relay point within the command structure. Direction comes down through the system, from the incident commander, through the intermediate levels, to the individual companies or units. At the same time, information is transmitted upward from subordinates to the officer, who either acts on the information directly or relays the information up to the next level. For example, a division supervisor could be assigned to oversee operations directed toward keeping the fire out of an exposure. The division supervisor would be expected to manage that operation and supervise the assigned companies, giving regular progress reports to the incident commander.

■ Determining Task Assignments

Tactical priorities are subdivided into tasks and assigned to companies. Tasks are specific assignments that are typically performed by one company or a small number of companies working together.

During the early stages of an incident, there are likely to be more tasks to be performed than there are companies available to do the work. The incident commander makes assignments based on tactical priorities and available resources. He or she must prioritize assignments and distribute them to companies as they arrive or become available. Companies that have completed one task assignment may have to be reassigned to another.

The incident commander should use the tactical priorities to determine the relative importance of each task that needs to be performed, in the context of the specific situation **FIGURE 17-7**. SOPs often guide these decisions, as do the staffing and standard equipment on each piece of apparatus.

The normal function of the company should also be considered, when possible. In most cases, an engine company would be assigned to attack the fire, whereas a paramedic ambulance provides medical care and a truck company is assigned to ventilation duties. Sometimes, however, the incident commander has to assign companies to perform tasks that might not fit their normal role, simply because of the importance of the task and the limited resources available. For example, if no medic unit is on scene, the incident commander might have to assign an engine company to provide medical care. If the truck company will be delayed, an engine company could be assigned to perform ventilation.

The rescue priority could be assigned to different companies. Such tasks might include performing primary and secondary searches of the structure, raising ladders, removing trapped occupants, providing medical care and transport, and establishing a rapid intervention crew.

Task assignments for the exposure priority typically include establishing a water supply and setting up master streams for use on the fire building or on the exterior of the exposure. Placing a handline on the unburned side of a firewall or in the cockloft of an exposure building could also be a task assignment. Removing combustible material from the windows of the exposed buildings is another example.

Task assignments for fire confinement typically include advancing handlines to the room of origin, into stairways, and into the attic or the floor above the fire. Ventilation tasks could be performed to support confinement efforts.

Task assignments for extinguishing a fire typically include establishing a water supply, advancing a handline to the seat of the fire, and applying water or other extinguishing agents. When making assignments for hose lines, the fire officer should indicate the line size as well as placement, unless these decisions are predetermined by departmental SOPs.

Task assignments for overhauling a fire typically include pulling ceilings and walls in the burned areas, removing door and floor trim where charred, checking the attic and basement for extension, checking the floors above and below the fire, and removing or wetting all burned material.

Task assignments for ventilating a fire typically include vertical ventilation, horizontal ventilation, positive-pressure ventilation, negative-pressure ventilation, and natural ventilation. When making assignments for ventilation, the officer should indicate the location of entry points, the potential victims, and the location of the fire to ensure that proper techniques are used.

Task assignments for salvage typically include throwing salvage covers over large, valuable items; removing lingering smoke; soaking up water from floors; deactivating sprinklers; and removing important documents or memorabilia.

■ Assigning Resources

The resources assigned to an incident vary greatly and are influenced by history, tradition, and budgets. One department may respond to a structure fire with a single engine and tanker, whereas another might send four engines, two aerial apparatus, a rescue or squad company, and two command officers.

Some situations exceed the capabilities of any organization and require assistance from other agencies or jurisdictions. In most cases, the first level of assistance for fires is mutual aid from surrounding fire departments. Most fire departments also have working relationships with other local agencies, such as law enforcement and public works, to obtain the resources that are likely to be needed in most situations.

When the need for resources exceeds the normal capabilities of the fire department and involves numerous other agencies, a fire officer may have to activate the local emergency plan. This plan defines the responsibilities of each responding agency and outlines the basic steps that must be taken for a

FIGURE 17-7 Incident commanders should use tactical priorities when assigning tasks.

particular situation. For example, firefighting is normally the responsibility of the fire department, along with response to technical rescue and hazardous materials incidents. Traffic control, crowd control, perimeter security, and terrorist incidents are usually the responsibility of the police or sheriff's department. Every fire officer should be familiar with the local plan and the role of the fire department within it.

The method most commonly used by a fire officer to activate the local emergency plan is to notify the dispatch center, which should have the information and procedures needed to either activate the local emergency plan or notify the emergency management office. The emergency management office normally is responsible for maintaining and coordinating the plan, as well as facilitating the response to emergency situations.

The local emergency plan usually includes an evacuation component that can be used for any situation where the residents or occupants of an area must be protected from a dangerous situation. A hazardous materials release or major fire event, for example, could require the evacuation of a local area. In many cases, the evacuation plan can be activated on a limited scale, without activating the overall local emergency plan.

The police department is the agency that has the primary responsibility for notifying people within an evacuation area of what to do. The use of the public address feature on vehicle sirens, the emergency broadcast system, or a reverse 911 telephone system can aid in this process. For large-scale or long-term evacuations, the Red Cross may need to be contacted to establish emergency shelters for the residents. Depending on the situation, fire crews could be available to assist evacuating residents, or they might be occupied with fighting a fire or other mitigation actions. Mutual aid companies or other agencies could be needed to assist with evacuation.

Consider the nature of the event when establishing an evacuation area. A natural gas leak in a street generally requires short-term evacuation of residents in the immediate area. An apartment building fire could result in several residents being displaced for a long period of time, but it affects only those who live in the building or complex. A major hazardous materials release could result in the evacuation of a large area and thousands of people. The fire officer should consult the *Emergency Response Guidebook*, the local hazardous materials team, or CHEMTREC for guidance in determining evacuation distances, based on the product, quantity, and environmental conditions.

General Structure Fire Considerations

The *Preincident Planning and Code Enforcement* chapter divided buildings into use groups and briefly discussed common firefighting hazards that are associated with some of those groups. This section provides an overview of typical firefighting hazards, listed by traditional fire-ground classifications.

■ Single-Family Dwellings

More civilian fire deaths occur in one- and two-family dwellings than in any other type of occupancy in the United States. These dwellings are not required to undergo regular fire safety inspections. Such structures may range from a single-story frame house of less than 1000 ft^2 (305 m^2) to a mansion with more than 15,000 ft^2 (4572 m^2) on a private estate in a non-hydrant area. Balloon frame construction, which was common until the 1930s, incorporated void spaces in the exterior walls, interconnected with the void spaces in the floors and ceilings, where a fire could spread rapidly from the basement to the attic. Single-family dwellings constructed since 1980 often have an open geometry design with lightweight structural components supporting floors and roofs. Rapid fire spread will occur in both modern and balloon frame buildings.

An extensive variety of hazards can be found in single-family dwellings, including improper storage of gasoline, faulty electrical wiring, and poor housekeeping habits. Each of these conditions increases the probability that a fire will occur at that home. Many homes contain storage areas for a retail business in the basement, attic, garage, or spare bedroom. Other home-based businesses, such as vehicle repair businesses or home-cleaning services, store and use chemical products.

A trend in some older communities is to convert buildings that were originally designed as single-family homes into apartments. Another trend is multiple families living together in one crowded house. In such a case, a fire company could respond to a fire in a single-family dwelling and encounter a dozen trapped residents.

■ Low-Rise Multiple-Family Dwellings

The low-rise multiple-family dwellings that have been built since the 1980s are often Type V (wood frame) construction, with lightweight wood truss components in the floors and roof. Creative building designs often create voids as well as architectural features like cathedral ceilings that limit the ability of fire companies to open up ceilings to check for fire extension. Like newer single-family dwellings, newer or renovated multiple-family dwellings usually have extensive insulation.

Often, the first-arriving fire companies responding to an activated fire alarm in a garden-style apartment report nothing showing on arrival. When they force the door of the apartment that has the activated smoke alarm, they encounter thick smoke boiling out of the apartment.

High peak heat release furniture and tighter insulation means that this type of building retains much more heat than older buildings. Internal exposures will suffer more heat damage and there is a higher chance of fire spread. Lightweight wood trusses require minimal fire damage to collapse—an event that often occurs suddenly when the first-arriving company starts its operation.

Multiple-family dwellings with NFPA-compliant 13R automatic sprinkler systems provide greater protection to the occupants but may increase the degree of difficulty for the fire fighters. Buildings with a 13R system will be taller with longer hallways. Apparatus access may be restricted due to creative landscaping. Even with these features, a fire within the 13R-protected area will be small with less damage. The preincident plan should identify the best apparatus access positions and the hose lays required to get to every space in the building. Make sure that the fire apparatus can actually get into gated communities, and make provisions to allow prompt access to the property every hour of every day.

VOICES OF EXPERIENCE

Fire attack . . . seems so simple: put the wet stuff on the red stuff. If it was just that simple, as a fire officer arriving on scene, it may be up to you to make that happen. But as you have learned in your training, it isn't quite that simple. Development of an incident action plan (IAP) is required to take a systematic approach to a fire, so the type of fire attack must also be systematic and dynamic to reach our objective: putting out the fire.

Whether you're making a positive-pressure attack, offensive or defensive, remember that the fire attack has to be carefully considered and then maintained. Stay flexible in the face of the possible changes that can occur during fire operations. I teach that the single most important component of fire attack is to maintain situational awareness. Paying attention to your surroundings at all times is imperative to being safe—that is, noticing changing situations ranging from structure stability to fire behavior. Changes in these and other situations can mean the difference between safely mitigating the situation and having a disaster to deal with.

> "What I saw changed my thought process for fire attack forever."

I'm reminded of one of my first calls. I was on interior attack. I can still see the outlines of the 2 × 4s burning bright red (the building was under some remodeling), and, of course, the heat and smoke were rather intimidating as well. But my partner and I made our attack, and with the help of another attack team we brought the fire under control. It wasn't until overhaul that I realized I had not been paying attention to everything. We went in, and moving my flashlight around and retracing our steps inside, I looked up—and my eyes must have grown to the size of baseballs. What I saw changed my thought process for fire attack forever: right above our heads was a balcony-type section that was leaning severely enough that even under postfire conditions it made me back up. Right above exactly where we were standing on that leaning part of the structure was a small workout area, with a weight machine and a couple of hundred pounds of weights. If that section had gone . . . well, I stopped thinking about what could have happened and made it a point to make sure that I was never in that situation again. All aspects of the fire attack plan are important: ICS, IAP, safety, ventilation, watching for abnormal fire behavior—and the list goes on and on. But knowing where you are during fire attack—maintaining situational awareness—is so important.

As an officer, you might conduct training or run an academy. Remember that the basic fundamentals of fire attack must include every aspect of that attack and how better to control our scenes and personnel to make the best outcome happen every time: everyone goes home.

Jim Lovell
Captain
Fire Training Division
Santa Fe County Fire Department
Santa Fe, New Mexico

Company Tips

Zero-Clearance Chimneys
A type of construction that was popular decades ago, a zero-clearance chimney has an inside steel flue and an outside flue, separated by 1 inch (2.5 cm) of air. The owner may have added glass doors to the fireplace, increasing the temperature inside the flue beyond its design parameters. To further complicate this scenario, an artificial log may be used in place of a traditional log.

The artificial log plus fireplace doors plus an aging flue may cause a failure of the flue that manifests itself by heating up the structural wood components around the flue. In some cases, homeowners discover the problem when someone knocks on their front door on a bitter cold winter night to tell them that their attic is on fire.

The 13R system does not protect the attic area or the void spaces. Consequently, fires that start in these areas, or start on the outside combustible vinyl siding, are likely to spread rapidly.

High-Rise Considerations

A working fire within a high-rise structure requires more fire fighters and an expanded Incident Command System. The incident commander is responsible for the overall management of the incident. Because of the complexity of a high-rise fire, the incident commander often expands the ICS to include planning, operations, and logistics sections. The incident commander also usually appoints staff to fill the positions of safety officer, liaison officer, and public information officer on the command staff.

High-rise fires require a tremendous number of personnel deployed in a carefully concerted effort to achieve the strategic objectives. NIST's *Report on High-Rise Fireground Field Experiments: Technical Note 1797* documents the dramatic effect crew size has on the ability to accomplish tasks. In responding to a medium-growth-rate fire on the 10th floor, the three-person crews confronted an environment where the fire had released 60 percent more heat energy than the fire encountered by the six-person crews doing the same work. Put simply, the smaller crew encountered a larger fire. It took the three-person crews

Fire Marks

Sprinklers in High Rises
NFPA estimates that there is no automatic sprinkler system in:
- 41 percent of high-rise office buildings
- 45 percent high-rise hotels
- 54 percent of high-rise apartment buildings

In addition, built-in sprinkler systems fail in about one in 14 fires (NIST 2013).

Reproduced from: NIST (2013) Report on High-Rise Fireground Field Experiments: Technical Note 1797. Courtesy of National Institute of Standards and Technology.

1 hour to complete their fire response, while six-person crews required 40 minutes to accomplish the same response tasks.

Managing a high-rise fire requires dividing the incident into manageable units (i.e., branches), each with a supervisor. The last operations section position that a fire officer might be called on to fill at a high-rise incident is the staging area supervisor. The staging area supervisor is responsible for managing all activities within the staging area, including layout, check-in functions, and traffic control within the area. All fire officers must understand their relationship within the ICS. The large-scale operation required to respond to a high-rise fire necessitates a large ICS structure to maintain a manageable span of control and a margin of safety and accountability.

The fire officer may also be called on to fill a position within the logistics section. Like the operations section, the logistics section is divided into branches. Typically, it includes a support branch and a service branch. The service branch is headed by the service branch director and is responsible for communications and fire fighter rehabilitation. The support branch is headed by the support branch director and is responsible for ensuring adequate supplies, personnel, and equipment are available. A high-rise incident requires large amounts of personnel and support equipment, so a support branch director must be established as an early part of the command staff.

The support branch director is often located at a base. A base is an area where the primary logistics functions are coordinated and administered. The incident command post may be co-located with the base, and there is only one base per incident. Spare SCBA cylinders are the first priority for the support branch.

In a high-rise fire, it is important to gain control of the lobby quickly. The lobby control officer controls the entry and exit of both civilians and fire fighters in the lobby. In addition, the lobby control officer oversees the use of the elevators; operates the local building communication system; and assists in the control of the heating, ventilating, and air-conditioning systems. The lobby control officer reports to the logistics section chief or, if that position is not established, the incident commander.

A particularly labor-intensive task is undertaken by the stairwell support group: these fire fighters move equipment and water supply hose lines up and down the stairwells of the high-rise building. The stairwell support unit leader reports to the support branch director or the logistics section chief. One practice is to position fire fighters on every third floor. Equipment moves up the high rise in an assembly-line fashion, as the third-floor fire fighter moves equipment up to the sixth-floor fire fighter.

Residential high-rise buildings present extra challenges to fire officers; three wind-driven fires have killed members of the first-arriving fire company since 1983. Recommended practices include:

- Comply with your organization's SOPs.
- Consider bringing the big attack line first.
- Beware of weather conditions.
- Assemble an adequate crew.

You Are the Fire Officer Conclusion

Captain Davis fills out positions within the Incident Command System as the second-alarm resources arrive at the fire ground. The additional resources are split to assist the occupants and mitigate the incident, including an EMS task force assigned to the second-floor shelter-in-place area. Additional municipal and community resources arrive, along with the facility's emergency response team.

The fire is quickly extinguished within the apartment of origin. The fire department demobilizes the incident and works with the facility to maintain business continuity.

Wrap-Up

Chief Concepts

- The findings from recent experiments have dramatically changed the understanding of fire dynamics and identified the importance of controlling the flow path of high-momentum fire gases.
- New fire research challenges many strategy and tactic traditions with the fire service.
- Recent full-size fire experiments have shown that modern fires are ventilation limited, resulting in a different time–temperature curve than had been used for 100 years.
- UL has documented changes within the residential fire environment that impact occupant survival and fire fighter safety.
- Operations conducted in the flow path place fire fighters at significant risk due to increased flow of fire, heat, and smoke toward their position.
- A fire company is a basic tactical unit for emergency operations, with the fire officer in the role of a working supervisor.
- The inherent risks associated with emergency operations demand close supervision at all times.
- The fire officer needs to adopt the appropriate leadership style based on the specific situation at hand.
- Emergency incident operations must be conducted in a structured and consistent manner.
- Command staff assignments include the safety officer, liaison officer, and public information officer positions.
- The art of sizing up an incident requires a diverse knowledge base about emergency incidents.
- The specific size-up for an incident begins with the dispatch. The name, location, and reported nature of the incident all help the fire officer anticipate what might be happening at the scene.
- The ability to size up a fire situation quickly requires both a systematic approach and a solid foundation of reference information.
- Chief Lloyd Layman presented a five-step process for analyzing emergency situations:
  - Facts
  - Probabilities
  - Situation
  - Decision
  - Plan of operation
- The National Fire Academy (NFA) has developed a size-up system that includes three phases:
  - Preincident information
  - Initial size-up
  - Ongoing size-up
- Risk–benefit analysis is a key factor of size-up when selecting the appropriate strategic mode.
- To extinguish a fire, the volume of applied water must be sufficient to absorb the heat that is being released. The estimated fire flow is an approximation of the rate of water application that would be required to control a fire in a particular building or section of a building.
- After size-up, the incident commander develops an incident action plan based on the incident priorities.
- The three basic priorities for an incident action plan are life safety, incident stabilization, and property conservation.
- Tactical priorities provide a list and an order of priority for dealing with these priorities. The tactical priorities and the information obtained in the size-up are used to develop the incident action plan.

Wrap-Up, continued

- Fighting fires is an inherently dangerous activity that exposes fire fighters to a wide variety of risks and hazards.
- Many fires occur at night, creating a dark fire scene full of tripping hazards and obstacles. The use of lights to illuminate the incident scene adequately is an important safety consideration.
- Many departments have developed RIC training and deployment procedures. The standard radio term *mayday* is used to indicate that a fire fighter is lost, missing, or in life-threatening danger.
- When the incident commander requests a PAR, each fire officer physically verifies that all assigned members are present and confirms this information to the incident commander.
- When operating as the incident commander, the fire officer has an even greater level of responsibility, as the incident commander is responsible for every company on the scene and for management of the overall operation.
- Tactical priorities are subdivided into tasks and assigned to companies. Tasks are specific assignments that are typically performed by one company or a small number of companies working together.
- The resources assigned to an incident vary greatly and are influenced by history, tradition, and budgets.
- More civilian fire deaths occur in one- and two-family dwellings than in any other type of occupancy in the United States.
- Low-rise multiple-family dwellings that have been built since the 1980s are often Type V (wood frame) construction, with lightweight wood truss components in the floors and roof.
- A working fire within a high-rise structure requires more fire fighters and an expanded Incident Command System.

Hot Terms

Base The location at which the primary logistics functions are coordinated and administered. The incident command post may be co-located with the base. There is only one base per incident.

Defensive operation Conduct of suppression operations outside the fire structure; these operations feature the use of large-capacity fire streams placed between the fire and the exposures to prevent fire extension.

Flow path The volume between an inlet and an exhaust that allows the movement of heat and smoke from a higher-pressure area within the fire area toward lower-pressure areas accessible via doors, windows, and other openings.

Fuel limited Fire in a compartment (or building) that has adequate air supply. Without intervention, all of the fuel will be consumed by the fire.

Incident action plan (IAP) The objectives reflecting the overall incident strategy, tactics, risk management, and member safety that are developed by the incident commander. Incident action plans are updated throughout the incident.

Initial rapid intervention crew (IRIC) Two members from the initial attack crew who are assigned for rapid deployment to rescue lost or trapped members.

Legacy dwellings Single-family dwellings constructed before 1980.

Lobby control officer The fire officer who controls the entry and exit of both civilians and fire fighters in the lobby at a high-rise fire incident; this officer also oversees the use of the elevators, operates the local building communication system, and assists in the control of the heating, ventilating, and air-conditioning systems.

Modern dwellings Single-family dwellings constructed since 1980; they are typically larger structures with an open house geometry, lightweight construction materials, and exponentially increased fuel load.

Offensive operation An advance into the fire building by fire fighters with hose lines or other extinguishing agents to overpower the fire.

Personnel accountability report (PAR) A systematic method of accounting for all personnel at an emergency incident.

Rapid intervention crew (RIC) A dedicated crew of four fire fighters who are assigned for rapid deployment to rescue lost or trapped members.

Service branch A major division within the logistics section of the ICS; it oversees the communications, medical, and food units.

Size-up A systematic process of gathering and processing information to evaluate the situation and then translating that information into a plan to deal with the situation.

Stairwell support group A group of fire fighters who move equipment and water supply hose lines up and down the stairwells at a high-rise fire incident. The stairwell support unit leader reports to the support branch director or the logistics section chief.

Standard time–temperature curve A recording of fire temperature increase over time.

Support branch A major division within the logistics section of the ICS; it oversees the supply, facilities, and ground support units.

Transitional operation A situation in which an operation is changing or preparing to change.

Ventilation limited Fire in a compartment (or building) that has inadequate air supply. It will flare up when air is introduced into the compartment.

References and Additional Resources

NFPA reprinted material is not the complete and official position of the NFPA on the referenced subject, which is represented only by the standard in its entirety.

American Society for Testing and Materials. (2000). *ASTM E119: Standard Test Methods for Fire Tests of Building Construction and Materials*. West Conshohocken, PA: ASTM International.

Babrauskasm V., and R. B. Williamson. (1978). "The Historical Basis of Fire Resistance Testing: Part I and Part II." *Fire Technology* 14(3): 184–194 and 4: 184–251. Quincy, MA: National Fire Protection Association.

Brunacini, A. V. (2002). *Fire Command: The Essentials of Local IMS*. Phoenix, AZ: Heritage Publishers Inc.

Fahy, R. F., P. R. LeBlanc, and J. L. Molis. (2007). *What's Changed over the Past 30 Years?* Quincy, MA: National Fire Protection Association.

Fahy, R. F., P. R. LeBlanc, and J. L. Molis. (2013). *Firefighter Fatalities in the United States—2012*. Quincy, MA: National Fire Protection Association.

Federal Emergency Management Agency. (2011). *National Incident Management System: Training Program*. Washington, DC: U.S. Department of Homeland Security.

Gann, R. G., and N. P. Bryner. (2008). Combustion Products and Their Effects on Life Safety. In: *The Fire Protection Handbook*. Quincy, MA: National Fire Protection Association: 6-11 to 6-34.

Goldstein, J. (2012, July 1). "As Furniture Burns Quicker, Firefighters Reconsider Tactics." *The New York Times*. Accessed October 3, 2013: http://www.nytimes.com/2012/07/02/nyregion/nyc-fire-dept-rethinking-tactics-in-house-fires.html?pagewanted=all&_r=0.

Healy, G., D. Madrzykowski, S. Kerber, and K. Ceriello. (2013, April 30). *FDNY, UL and NIST: Scientific Research for the Development of More Effective Tactics*. Indianapolis, IN: Fire Department Instructor's Conference.

International Organization for Standardization. (2012). *Life-Threatening Components of Fire: Guidelines for the Estimation of Time to Compromised Tenability in Fires* (ISO #13571). Geneva, Switzerland: ISO Central Secretariat.

Kerber, S. (2010). *Impact of Ventilation on Fire Behavior in Legacy and Contemporary Residential Construction*. Northbrook, IL: Underwriters Laboratories.

Kerber, S. (2012). "Analysis of Changing Residential Fire Dynamics and Its Implications on Firefighter Operational Timeframes." *Fire Technology* 48(4): 865–891. Quincy, MA: National Fire Protection Association.

Kerber, S. (2013). *Study of the Effectiveness of Fire Service Vertical Ventilation and Suppression Tactics in Single Family Homes*. Northbrook, IL: Underwriters Laboratories Firefighter Safety Research Institute.

Kerber, S., and D. Madrzykowski. (2009). *Fire Fighting Tactics under Wind Driven Fire Conditions*. NIST Technical Note 1629. Gaithersburg, MD: National Institute of Standards and Technology.

Kimball, W. Y. (1966). *Fire Attack 1: Command Decisions and Company Operations*. Boston, MA: National Fire Protection Association.

Kimball, W. Y. (1968). *Fire Attack 2: Planning, Assigning, Operating*. Boston, MA: National Fire Protection Association.

Lawson, J. R. (2009). *A History of Fire Testing*. NIST Technical Note 1628. Gaithersburg, MD: National Institute of Standards and Technology.

Layman, L. (1953). *Fire Fighting Tactics*. Boston, MA: National Fire Protection Association.

Layman, L. (1955). *Attacking and Extinguishing Interior Fires*. Boston, MA: National Fire Protection Association.

Madrzykowski, D. (2013). "Fire Dynamics: The Science of Fire Fighting." *International Fire Service Journal of Leadership and Management* 7(2): 7–15. Tulsa, OK: Oklahoma State University.

Myers, B. (2004). *Mentoring: Preparing Company Officers*. Emmitsburg, MD: Executive Fire Officer Program.

National Fallen Firefighters Foundation. (2011). *Understanding and Implementing the 16 Firefighter Life Safety Initiatives*. Stillwater, OK: Fire Protection Publications.

National Fire Protection Association. Reproduced with permission from NFPA 1407-2010, Standard for Training Fire Service Rapid Intervention Crews, Copyright © 2009, National Fire Protection Association, Quincy, MA.

National Institute of Standards and Technology. (2010). *Report on Residential Fireground Field Experiments: Technical Note 1661*. Washington, DC: U.S. Department of Commerce.

National Institute of Standards and Technology. (2010). *Fire Dynamics*. Washington, DC: U.S. Department of Commerce. Accessed August 24, 2013, at http://www.nist.gov/fire/fire_behavior.cfm.

National Institute of Standards and Technology. (2013). *Report on High-Rise Fireground Field Experiments: Technical Note 1797*. Washington, DC: U.S. Department of Commerce.

Urban Fire Forum. (2013). *Fire Service Deployment: Assessing Community Vulnerability. High Rise Implementation Guide* (2nd ed.). Quincy, MA: Metropolitan Fire Chiefs.

FIRE OFFICER in action

Chief Johnson conducts an after-action review with the companies that responded to the assisted-living facility apartment fire.

1. A fire officer uses _____ supervision during incident size-up.
   A. authoritarian
   B. participative
   C. bureaucratic
   D. inclusive

2. _____ ensure(s) consistent and repeatable procedures at a local emergency incident.
   A. NFPA 1710
   B. The National Response Framework
   C. Standard operating procedures
   D. The Model Incident Command System

3. Arriving at a fire-in-progress at the assisted-living facility, Captain Davis knows that this type of 1970s-era structure frequently experiences vertical fire extension through pipe-chases, resulting in well-developed attic fires. This is an example of:
   A. a layman probability.
   B. defective building codes.
   C. a preincident walkthrough discovery.
   D. a NIMS unit assignment.

4. Which of the following components are part of a rapid intervention crew's equipment?
   A. Thermal imager
   B. Spare SCBA or full replacement cylinder
   C. Portable radio for each member
   D. All of the above

FIRE OFFICER *in action* (continued)

Fire Captain *Activity*

Captain Davis and Lieutenant Williams are preparing to conduct a preincident survey at a nursing home. When they contact the nursing home representative to schedule the survey, they learn that a new management group bought the home and the facility's emergency manager is completing a revised emergency plan.

NFPA Fire Officer II Job Performance Requirement 5.6.1

Produce operational plans, given an emergency incident requiring multiunit operations, the current edition of NFPA 1600, and AHJ-approved safety procedures, so that required resources and their assignments are obtained and plans are carried out in compliance with NFPA 1600 and approved safety procedures, resulting in the mitigation of the incident.

Application of 5.6.1

1. Using a target hazard (hospital, residential high-rise, nursing home) in a community with which you are familiar, develop an IAP to cover a well-developed fire that requires a third-alarm assignment and shelter for 50–100 occupants. Include the facility's existing emergency management plan.
2. Describe how to conduct a PAR check when all fire department units are engaged in operations at a three-alarm fire in a residential or health care target hazard.
3. Using a department with which you are familiar and third-alarm operations at a residential or health care target hazard, describe the tasks of each sector chief within the unified Incident Command System.

Fire Cause Determination

Fire Officer I

Knowledge Objectives

After studying this chapter, you will be able to:

- Identify the common causes of fire. (pp 364–365)
- Explain when to request a fire investigator. (p 365)
- Describe how to find the point of origin of a fire (NFPA 4.5). (pp 365–366)
- Discuss the legal considerations of fire cause determination (NFPA 4.5) (NFPA 4.5.3). (pp 366–368)

Skills Objectives

After studying this chapter, you will be able to:

- Determine the point of origin of a fire (NFPA 4.5). (pp 365–366)
- Demonstrate how to secure the scene to prevent unauthorized persons from entering the incident scene (NFPA 4.5) (NFPA 4.5.3). (p 367)

Fire Officer II

Knowledge Objectives

After studying this chapter, you will be able to:

- Discuss the nature of fire investigation. (p 368)
- Describe how to find the point of origin of a fire (NFPA 5.5) (NFPA 5.5.1). (pp 368–369)
- Describe how to determine the cause of the fire (NFPA 5.5) (NFPA 5.5.1). (pp 369–374)
- Describe the fire cause classifications (NFPA 5.5.1). (pp 374–375)
- Describe the indicators of incendiary fires (NFPA 5.5.1). (pp 375, 377)
- Discuss arson (NFPA 5.5.1). (pp 377–378)
- Describe the documents and reports a fire officer must complete (NFPA 5.5.1). (pp 378–380)
- Describe how a fire investigation continues after fire official involvement has ended. (p 380)

Skills Objectives

After studying this chapter, you will be able to:

- Determine the point of origin of a fire (NFPA 5.5). (pp 368–369)
- Determine the cause of the fire (NFPA 5.5). (pp 369–374)
- Complete a fire investigation report (NFPA 5.5.1). (pp 378–380)

CHAPTER 18

Additional NFPA Standards

- NFPA 921: *Guide for Fire and Explosion Investigations*
- NFPA 1033: *Standard for Professional Qualifications for Fire Investigator*

Principles of Fire and Emergency Service Administration (FESHE) Course Outcomes

2. Recognize the need for effective communication skills, both written and verbal. (pp 370, 374–375, 378–380)
6. Discuss the various levels of leadership, roles, and responsibilities within the organization. (pp 368–374, 379–380)
8. Identify the importance of ethics as it relates to fire and emergency services. (pp 379–380)
9. Identify the roles of the National Incident Management System (NIMS) and Incident Management System. (pp 366–368)

You Are the Fire Officer

Engine 100 is the first-due company to an automobile fire that has extended to become a commercial structure fire. The burning automobile is located inside a service bay that is part of an electronics store. The front part of the store is a retail operation; the rear includes a six-bay service area used to install automobile electronics and sound systems. It appears that part of the shop is also used to modify vehicles, because acetylene torches, flammable liquid cleaning tanks, and large tool chests are present in one of the six vehicle installation bays. The point of origin of the fire seems to be in the bay used to modify vehicles, below the charred remains of a sports car that is on a portable hydraulic lift.

There is extensive damage to three service bays, two vehicles are destroyed, and two vehicles are damaged. Three employees have suffered from minor burns and smoke inhalation.

The fire investigator is tied up on another call and asks Captain Jean Davis to determine the need to call in another investigator on overtime. Captain Davis and Lieutenant Taylor Williams review the preincident plan. Neither officer remembers seeing automotive repair and modification equipment during a facility walk-through 7 months ago, nor is the presence of such equipment documented on the preincident plan. Lieutenant Williams notes three 55-gallon drums of racing fuel stored in the corner of the service bay and wonders if the business has applied for a fire prevention permit for flammable liquid storage and for the renovations for the current use of the space.

1. How will you determine the origin of this fire?
2. How will you determine the cause of this fire?
3. How will you secure the scene?

Introduction

An investigation must be conducted to determine how a fire started. This activity is undertaken once the fire is extinguished and before the property is turned back over to the owner. One of the best starting points for efforts to prevent future fires is understanding the causes of fires that have occurred in the past. In addition, investigation is important to establish the responsibility for fires that could have been caused by criminal acts, negligence, or fire code violation. The results of an investigation may lead to legal actions, ranging from prosecution for arson or related crimes to civil litigation over deaths, injuries, or property damage.

The incident commander is responsible for conducting the investigation as well as completing the National Fire Incident Reporting System (NFIRS) documents or a local equivalent report. Sometimes, the incident commander delegates this responsibility to another fire officer, typically the officer in charge of the first-arriving company.

The first goal of the fire officer assigned this responsibility is to determine whether a formal fire investigation is needed. The authority having jurisdiction (AHJ) provides criteria for requesting an investigator. The fire officer must consider all of the circumstances as well as the probable cause of the fire. Any fire that results in a serious injury or fatality meets the criteria for a formal investigation. Any fire that appears to be arson or related to a criminal act also meets the criteria. Some agencies automatically dispatch a fire investigator to all working structure fires.

The legal responsibility for conducting fire investigations is defined by state or local legislation or regulations. The fire officer should determine which laws apply to fire investigations and which agency is responsible for conducting fire investigations in different circumstances. This information is available from the fire marshal's office or from an equivalent agency.

In situations where no formal investigation occurs and a fire investigator does not respond to the scene to determine the cause and origin, the incident commander is responsible for determining and reporting the fire cause.

Fire Officer I

Common Causes of Fires

Listing all possible fire causes would be challenging, but years of firefighting experience have shown that a relatively few causes are responsible for a large number of fires **TABLE 18-1**. Although these data do not help in determining the cause of a particular fire, it is important to understand why fires occur.

Table 18-1	Major Causes of Home Structure Fires (2006–2010)
Cooking	42%
Heating	17%
Intentional	8%
Electrical malfunction	6%
Smoking	5%
Clothes dryer or washer	4%
Exposure to other fire	3%
Candle	3%
Playing with heat source	2%

Reprinted with permission from NFPA: Fire Analysis and Research Division. "An Overview of the U.S. Fire Problem". National Fire Protection Association, September 2012.

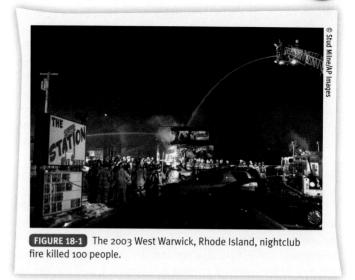

FIGURE 18-1 The 2003 West Warwick, Rhode Island, nightclub fire killed 100 people.

Requesting an Investigator

The fire officer should be able to determine a point of origin and a cause, or probable cause, for most fires. On small or routine incidents, this is the only investigation that is conducted, and in those cases, the fire officer must carefully document the findings.

A qualified fire investigator has specialized training in determining the cause and the origin of fires and, in most cases, is certified in accordance with NFPA 1033, *Standard for Professional Qualifications for Fire Investigator*. The investigator could be a fire department member, an employee of a county or state fire marshal's office, or a law enforcement officer. In some cases, a fire department investigator is responsible for determining the cause and the origin of fires, and a law enforcement agency is responsible for any subsequent criminal investigation. State and federal law enforcement agencies become involved in some investigations, particularly where arson is suspected.

The agency or organization that is responsible for conducting fire investigations will have established a set of guidelines on when to request an investigator. Such an investigator has extensive expertise in determining and documenting origin, gathering evidence, and interviewing. His or her primary responsibility is to develop a properly documented case and, if needed, to forward it to the prosecutor. The investigator must have credibility and experience in courtroom procedures to ensure that the facts are presented accurately.

A fire investigator should also be called when a death or serious burn injury occurs. If the cause of the fire was deliberate, a crime has occurred and must be fully investigated. If the cause was accidental, the information gathered in the investigation is important as a means to identify and evaluate fire prevention strategies. When any fatality occurs as a result of fire, the coroner or medical examiner's office must be contacted as well **FIGURE 18-1**.

Fire investigators often respond to large-loss fires, even if the cause is known. A fire investigator might also be called in situations that could have caused great harm, such as a fire in a hospital or college dormitory, even if the fire was quickly controlled.

The fire officer must evaluate the circumstances of the situation in relation to local guidelines and policies when considering how best to conduct the investigation. It is a good practice to request an investigator whenever the facts do not seem to make sense or there is a compelling reason to know the exact cause of the fire.

In many cases, the fire officer is able to determine the exact cause of a fire. Sometimes the cause cannot be established with certainty, but a probable cause can be identified. Nevertheless, the causes of many fires are never determined (or at least are not reported).

Fire Growth and Development

To determine the point of origin, the fire officer must understand fire behavior, growth, and development. Fire fighters learn the basic concepts of fire behavior in their Fire Fighter I and II courses. The fire officer should also understand the three methods of heat transfer: conduction, convection, and radiation. He or she must then be able to apply these basic concepts to understand fire growth and interpret the spread of a fire.

Consider a fire in a sofa. The heat and products of combustion rise until they reach the ceiling. They then spread out laterally until they reach the walls. Next, the heat and smoke bank down lower and lower into the room until the room is filled or until an open doorway or window is encountered. At that point, the heat and smoke tend to flow outward through the opening into the adjoining room or space, filling that area from the ceiling down. This process continues until the entire structure is filled with heat and smoke or until an opening to the outside is found, allowing the smoke and heat to escape from the structure.

While this process is occurring, the fire is continuing to develop and grow in the area of origin. The fire follows the same growth pattern as the smoke and heat, rising toward the ceiling, spreading out, and banking down. When a room fire reaches the point that flames are spreading across the ceiling, tremendous heat energy is radiated down onto the combustible contents within the room, such as sofas, chairs, and tables. As these items are heated, they begin to release combustible fuels into the atmosphere and eventually reach their ignition temperatures. As each item ignites, the rate of heat release increases dramatically and the temperature increases even faster. Within a few seconds, the entire room flashes over, or becomes fully involved in fire. The fire then quickly spreads to adjoining rooms. This process often causes windows to break or holes to burn through to the outside of the structure, releasing heat and smoke and making the fire visible on the exterior of the structure.

If the fire does not have a fresh supply of oxygen, the fire slowly dies down to a smoldering phase. This may occur either before or after flashover has occurred. At this point, the fire sometimes completely dies out and self-extinguishes. In other cases, a fresh supply of oxygen is introduced, and the fire resumes the fully developed phase.

■ Disabled Built-in Fire Protection

A disabled sprinkler system may be encountered in fires involving large industrial or commercial occupancies. Look for damaged or vandalized sprinkler hook-ups, hose cabinets, hard-wired smoke detectors, and high-rise communication systems. At one high rise under construction, a tractor-trailer–sized dumpster was placed in front of the standpipe/sprinkler connection that prevented the fire department from being able to use the built-in fire protection system.

If a building is protected by a central station monitoring company, that service should be one of the first entities to call 911 if a device becomes activated. If the call came from another source, particularly witnesses outside the property, the system might have been disabled or impaired. Part of the investigation is identifying the status of fire suppression and alarm systems at the time the fire started.

■ Delayed Notification or Difficulty in Getting to the Fire

One of the reasons businesses use a central station monitoring service is to ensure prompt notification of the fire department when a smoke or heat detector, water-flow, or manual pull station is activated. In an urban environment, an interior fire could have been burning for 20–25 minutes before a passerby observed flames coming from a window. If the business is in a rural setting or an isolated industrial park, the first notification may occur only after the roof collapses and flames begin shooting up into the sky.

Be alert for conditions or situations that might potentially delay the fire department's ability to get to the fire. Malfunctioning keys and key cards, vandalized doors, and materials blocking access are conditions to note. In addition, points of origin that are in the attic, basement, or closet should receive special consideration. Arsonists prefer these places to start a fire because such fires are prone to going unnoticed for longer periods of time. Sometimes an arsonist starts another fire to divert resources to a different location, thereby delaying the response to the real target property.

> **Safety Zone**
>
> **Gasoline and the Amateur Arsonist**
> Gasoline is the flammable liquid of choice of arsonists. Many of these individuals have been burned while trying to ignite poured gasoline.

■ Tampered or Altered Equipment

Document unusual conditions noticed at the fire scene. Look for indications of a forcible entry before the fire department arrived. If the fire origin was electrical, look for electrical devices that could have been altered or that appear unusual. An arsonist wants to try to make a fire appear accidental.

Use of excessive fuel creates a dangerous situation for the first-arriving fire fighters. The fire sometimes flashes but then dies down due to insufficient oxygen, leaving only parts of the fire area burning. Because of the fuel-rich conditions, however, the fire area may still contain unburned flammable liquid puddles that can reignite or flash over when additional air flows into the fire area. When the fire attack team makes entry and begins to advance, this mixture could be ripe for reignition.

Fire fighters in full protective clothing and SCBA cannot smell a flammable liquid. They must be alert for explosive fire development, stay with the hose line, and be prepared to make a hasty exit if the fire reignites.

Legal Considerations

Although the law states that there is a public interest in determining the cause and origin of a fire, investigators must also respect the competing interests of a citizen's rights to privacy and due process. The fire officer who investigated a fire is often called to testify in court and may be challenged on issues of proper procedure.

■ Searches

When a fire has occurred and the fire department has been called, the fire department has the right to determine the cause and origin of the fire. This process must be accomplished in accordance with the law. In *Michigan v. Tyler* (1978), the U.S. Supreme Court held: "Fire officials are charged not only with extinguishing fires, but with finding their cause. Prompt determination of the fire's origin may be necessary to prevent its recurrence, as through the detection of continuing dangers such as faulty wiring or a defective furnace. Immediate investigation may also be necessary to preserve evidence from intentional or accidental destruction."

The fire officer must take care to avoid an unlawful search and seizure, which is prohibited by the Fourth Amendment of the U.S. Constitution. Typically, no search warrant is needed to enter a fire scene and collect evidence when the fire department remains on scene for a reasonable length of time to determine the cause of the fire and as long as the evidence is in plain view of the investigator. This principle was reaffirmed by the U.S. Supreme Court in *Michigan v. Clifford* (1984).

The aftermath of a fire often presents exigencies that will not tolerate the delay necessary to obtain a warrant or to secure the owner's consent to inspect fire-damaged premises. Because determining the cause and origin of a fire serves a compelling public interest, the warrant requirement does not apply to such cases.

The plain view doctrine allows for potential evidence to be seized during the processing of a fire scene, if the fire investigator had a legal right to be there and the evidence is in plain view. Under *Michigan v. Clifford*, the court held that as fire fighters remove rubble or search other areas where the cause of a fire is likely to be found, an object that comes into view during such process may be preserved. In the same decision, the court found that once the investigator has determined where the fire started, the scope of the search authority is limited to that area. After the cause and the origin have both been determined, a search warrant or consent is required for any further search.

If reentry is needed after the fire department leaves the scene or to conduct a search for evidence of a crime after the fire's cause and origin have been determined, the investigator must obtain a search warrant or receive permission from the occupant to conduct such a search. The general provisions related to this process were clarified in *Michigan v. Tyler*:

- No search warrant is needed when fighting a fire or remaining on scene for a reasonable period of time to determine the cause of a fire, and any evidence is admissible under the plain view doctrine.
- Administrative search warrants are needed for reentry that is not a continuation of a valid search when the purpose is to determine the cause of the fire.
- A criminal search warrant is needed when reentry is not a continuation of a valid search and the purpose is to gain evidence for prosecution.

■ Securing the Scene

A fire officer who conducts a preliminary fire cause investigation and suspects that a crime has occurred should immediately request the response of a fire investigator. In these circumstances, the scene must be secured to protect any evidence that exists. If the fire department leaves the scene unsecured, the validity of any evidence that is collected after that point could be called into question. The fire officer must ensure that fire department personnel maintain custody of the scene until the investigator arrives.

Protecting the scene includes preventing unauthorized personnel from entering the scene. To create a security perimeter, the fire officer can use fire line or police crime scene tape secured to objects, such as trees and fence posts. Natural barriers can also be used to aid in securing the area. Objects such as fences or hedges can be used along with barrier tape to completely surround an area.

All access to and from the area must be controlled. To be certain that no unauthorized personnel enter the area, a fire fighter or law enforcement officer may need to be posted to limit access. Posting a guard preserves the chain of custody over the scene and any evidence that is present until the fire investigator arrives. Failure to take these steps may require the fire department to get a warrant to return to the fire scene.

Barrier tape may also be placed across a doorway to prevent unauthorized entry into a room or building. To secure and protect smaller areas of evidence, the fire officer may decide to cover them with a plastic sheet or tarp.

In any case, the number of fire personnel who are allowed into the secured area should be limited. The area should be treated as a crime scene, and only activities that are essential to control the emergency or protect the scene should be conducted. Fire fighters should not collect artifacts as souvenirs of their fire-fighting adventure, particularly when a scene is being secured.

■ Evidence

Evidence includes material objects as well as documentary or oral statements that are admissible as testimony in a court of law. Evidence proves or disproves a fact or issue. The fire officer must consider three types of evidence:

- Demonstrative evidence: Tangible items that can be identified by witnesses, such as incendiary devices and fire scene debris
- Documentary evidence: Evidence in written form, such as reports, records, photographs, sketches, and witness statements
- Testimonial evidence: Witnesses speaking under oath

If the fire officer has determined that the fire requires a formal investigation, then every effort should be made to protect and preserve the fire scene evidence. The structure, contents, fixtures, and furnishings should remain in their prefire locations, as intact and undisturbed as possible.

Evidence plays a vital role in the successful prosecution of arson cases. To prove that the crime of arson occurred, the fire investigator must rule out all potential accidental and natural causes of the fire. The investigator must consider all possible circumstances, conditions, or agencies that could have brought together a fuel, an ignition source, and an oxidizer, resulting in a fire or combustion explosion.

Artifacts, in the context of fire evidence, could include the remains of the material first ignited, the ignition source, or other items or components that are in some way related to the fire ignition, development, or spread. An artifact could also be an item on which fire patterns are present, in which case preservation of the artifact is not focused on the item itself, but rather the fire pattern that appears on the item.

■ Protecting Evidence

One part of the fire investigator's job is to dig out the fire scene. Like an archaeologist, the fire investigator removes each layer of fire debris. The investigator's ultimate goal is to identify the point of origin and the cause of the fire. Because fire follows the rules of science, an analysis of how the fire spread assists in determining where it originated and whether the cause was accidental or intentional.

Fire scene reconstruction is the process of re-creating the physical scene before the fire occurred, either physically or theoretically. As debris is removed, the contents and structural elements are replaced in their prefire positions, as much as possible.

If the damage and destruction are too extensive to physically restore the scene, whatever information is available is used to fill in the blanks. The investigator interprets the fire scene and documents the fire development by examining the damage to objects, devices, and surfaces. Throughout this process, the investigator must concentrate on locating, examining, and preserving evidence.

The fire officer must determine when to stop fire suppression or overhaul operations as part of the effort to preserve evidence for the investigator. From the fire investigator's viewpoint, the less the fire fighters disturb, the more intact the scene remains.

The worst case occurs when the fire investigator arrives at an apartment fire and discovers that the fire company has removed all of the fire debris, including the fire-damaged ceilings, walls, and doors. These actions prevent the investigator from being able to evaluate the evidence in the context of the fire. The situation would be akin to the police trying to reconstruct a motor vehicle collision from which both cars have been removed to the junkyard and no witnesses remain on the scene: It is possible, but very difficult.

The fire officer is responsible for protecting the fire scene evidence both from the public and from excessive overhaul and salvage FIGURE 18-2. In addition, the fire officer is the first step in the chain of evidence that is vital to successful prosecution of arson cases. Once the fire department arrives, it is responsible for preventing evidence contamination—a duty that requires fire fighters to stay until the fire investigator arrives. The chain of evidence requires that evidence remain secured and documented, from the fire scene to the courtroom. Most investigators document all physical evidence before collecting it by taking high-resolution photographs.

FIGURE 18-2 The fire officer must protect the fire scene evidence from the public and from excessive overhaul and salvage.

Fire Officer II

The Nature of Fire Investigation

NFPA 921, *Guide for Fire and Explosion Investigations*, describes the nature of fire investigation:

> A fire or explosion investigation is a complex endeavor involving skill, technology, knowledge, and science. The compilation of factual data, as well as an analysis of those facts, should be accomplished objectively, truthfully, and without expectation bias, preconception, or prejudice. The basic methodology of the fire investigation should rely on the use of a systematic approach and attention to all relevant details. The use of a systematic approach often will uncover new factual data for analysis, which may require previous conclusions to be reevaluated. With few exceptions, the proper methodology for a fire or explosion investigation is to first determine and establish the origin(s), then investigate the cause: circumstances, conditions, or agencies that brought the ignition source, fuel, and oxidant together.

Reproduced with permission from NFPA 921, Guide for Fire and Explosion Investigations, Copyright © 2008, National Fire Protection Association.

Finding the Point of Origin

The first step in fire cause determination is to identify the point of origin. The point of origin is the exact physical location where a heat source and a fuel come in contact with each other. It is usually determined by examining fire damage and fire pattern evidence at the fire scene. Flames, heat, and smoke leave distinct patterns that can often be traced back to identify the area or the specific location where the heat ignited the fuel and the burning first occurred. A fire investigator usually starts in the area where the least amount of damage occurred and follows the patterns back toward the area of greatest fire damage.

Eyewitnesses can often provide valuable information that can assist in determining the point of origin. Sometimes, a witness actually saw what happened and can describe where it occurred. In other cases, witness accounts can identify the area where the fire was first observed or where indications such as visible smoke or a burning odor were noted. Witnesses may also be able to describe what was in the suspected area of origin before the fire and what normally occurred in that area.

FIGURE 18-3 Often, the point of a V-shaped pattern is near or at the point of a fire's origin.

Determining the point of origin requires the analysis of information from four sources:

- The physical marks, or <u>fire patterns</u>, left by the fire.
- The observations reported by persons who witnessed the fire or were aware of conditions present at the time of the fire.
- Analysis of the physics and chemistry of fire initiation, development, and growth as an instrument to related known or hypothesized fire conditions capable of producing those conditions.
- Noting the location where electrical arcing has caused damage, as well as the electrical circuit involved. An <u>electrical arc</u> is a luminous discharge of electricity from one object to another, typically leaving a blackening of objects in the immediate area.

■ Fire Patterns

The point of origin can be identified by interpreting fire patterns. The fire pattern provides the fire officer with a history of the fire. Any flaming fire produces a plume of smoke, heat, and flame. As a fire burns up against a wall, it spreads up and out, creating a V- or U-shaped pattern. The origin of the fire is typically at the base of this V- or U-shaped pattern **FIGURE 18-3**. This type of pattern is also known as a movement pattern because it allows the fire officer to trace the fire and smoke patterns back to their origin.

The second type of pattern indicates the intensity of the fire. The intensity pattern indicates how much heat (energy) was transferred to the surrounding area and objects. The intensity is indicated by the response of various materials to the fire's rate of heat release and heat flux. It may produce a line of demarcation, which can indicate the area closest to the point where the greatest amount of heat was produced.

The analysis of <u>char</u> is closely related to the fire intensity pattern. Char is the blackened remains of a carbon-based material after it has been burned. For the fire officer, the depth of char can assist in determining the direction of fire spread; generally, the deeper the char, the longer the fire burned and therefore the closer the area of origin **FIGURE 18-4**.

FIGURE 18-4 Depth of char.

Depth of char is only one indicator of the apparent duration and intensity of a fire. It may be influenced by a variety of factors, including different types of wood and the fire's intensity, such that deep charring can occur in locations that are remote from the area of origin. For example, deep charring may occur at a location where a fire vents through a window or doorway because of the intensity of the fire as it combines with the fresh air in that area.

Determining the Cause of the Fire

Cause refers to the particular set of circumstances and factors that were necessary for the fire to have occurred. Cause determination can be approached as a three-step process:

1. Determine the source of ignition. This generally includes some device or piece of equipment that was involved in the ignition. A competent ignition source must be present to ignite the fuel.
2. Determine the fuel that was first ignited. Both the type of material and the form of the material should be identified.
3. Determine the circumstances or human actions that allowed the ignition source and the fuel to come together, resulting in a fire.

This three-step process describes only those factors that must be determined to conclusively establish the cause of a fire. NFPA 921 provides a systematic and scientific process for fire cause determination. It is not sufficient to identify and focus on a possible cause that fits the circumstances that were noted; the fire officer or fire investigator must also eliminate any alternative theories or explanations.

Before declaring a fire cause as incendiary or intentional, all accidental causes must be considered and eliminated. Potential causes should be ruled out only if definite evidence shows that they could not have caused the fire. For example, an electric heater can be ruled out if it was unplugged.

The fire officer does not have to consider every possible fire cause in relation to every fire that occurs. The facts of the situation can be considered in relation to previous knowledge to identify a list of potential causes for the fire. The fire officer must then apply deductive reasoning, considering each possible cause one by one and sequentially eliminating each cause that is not supported by the evidence.

For example, when investigating a house fire where an outside wall was ignited from the exterior in proximity to the electrical service, the fire officer must consider any potential source of ignition that could start a fire in that location. For example, the fire officer would identify lightning as a potential cause of the fire. The officer would then determine whether lightning was occurring in the area at the time of the fire. If it can be established that there was no lightning, then lightning can be eliminated from the list of possible causes. The officer might also consider that the fire could have been caused by a short circuit. If careful examination of the electrical equipment indicates that no short circuit occurred, that possibility can be eliminated.

The cause of a fire cannot be established until all potential causes have been identified and considered and only one cannot be eliminated. After all investigative possibilities have been exhausted, if two or more potential causes remain, the cause of the fire is considered undetermined.

Source and Form of Heat Ignition

The source of ignition is the energy source that caused the material to ignite. It must have been located at or near the point of origin. The fire officer may be able only to infer a probable ignition source; some sources of ignition remain at the point of origin in recognizable form, whereas others may be altered, destroyed, or removed.

If the equipment that provided the source of ignition was a cigarette lighter, for example, the form of the heat of ignition would be an open flame. The person who started the fire could have removed the source of ignition by placing it back in his or her pocket and leaving the scene. If the pilot light flame from a gas-fueled water heater ignited the fire, however, the source of ignition is likely to be found unaltered at the point of origin.

The source of ignition must have enough energy and must remain in contact with the fuel long enough to cause it to ignite. A competent ignition source has three components:

- *Generation*: The ignition source must produce sufficient heat energy to raise the fuel to its ignition temperature.
- *Transmission*: Sufficient heat energy must be transmitted from the source to the fuel to raise the fuel to its ignition temperature. Heat can be transferred through conduction, convection, or radiation.
- *Heating*: The heat transfer from the source to the fuel must continue long enough for the fuel to be heated to its ignition temperature.

Material First Ignited

The type of material first ignited refers to the nature of the material itself. For example, the type of material might be cotton. The form of material tells how that material is used. For example, cotton could be in the form of cotton plants in a field, baled cotton fibers, rolls of cotton thread, woven cloth, or clothing.

The physical configuration of the fuel is an important characteristic in ignition. A 12-inch by 12-inch (30-cm by 30-cm) beam and a pile of shavings are both wood; however, the beam is much more difficult to ignite. This information is significant when the investigator is trying to determine which type of heat source could have ignited the material. A pile of wood shavings could be ignited by a dropped match, whereas the wooden beam would have to be exposed to a more powerful source of heat.

Ignition Factor or Cause

The third important factor in determining the cause of a fire is the sequence of events that brought together the source of ignition and the fuel. The cause of a fire could be a human act that was either accidental or deliberate. In the case of an accidental cause, negligence—that is, failure to exercise appropriate care to avoid an accident—could be a factor. The fire cause could also be related to a mechanical failure, a poor design or improper assembly of a device, a worn-out piece of equipment, or a natural force, such as lightning.

Failure analysis is a logical, systematic examination of an item, component, assembly, or structure and its place and function within a system, conducted to identify and analyze the probability, causes, and consequences of potential and real failures. If the cause of a fire is determined to be related to some malfunction that occurred within a device or system, failure analysis is performed to identify what happened and why. This analysis could identify a design error, malfunctioning component, inadequate maintenance, operator error, or some other factor.

Fire Analysis

Fire analysis is the scientific process of examining a fire occurrence to determine all of the relevant facts, including the origin, cause, and subsequent development of the fire, as well as to identify the responsibility for whatever occurred. Fire analysis brings together all of the available information that can be obtained to examine what happened. In many cases, the fire officer may need to construct a timeline of events to establish the sequence of events that led up to the fire **FIGURE 18-5**.

Conducting Interviews

The fire officer may have to interview fire victims, witnesses, fire fighters, and suspected perpetrators. In doing so, the officer should make every effort to conduct separate interviews to provide the most accurate representation of what each person observed or experienced. When interviews are done in a

Near-Miss REPORT

Report Number: 11-0000155

Event Description: Just after midnight, a call was received from an occupant of a large building asking for assistance in securing a fire alarm, which was going off. I got my driver and medic crew and proceeded to the building without notifying the shift commander, as I should have. Upon arrival at the facility, we were met by a worker who stated that there had been an unusual smell inside for a few days and now the alarm had activated. He reported seeing no signs of smoke or fire as he left the structure.

My driver and I proceeded into the building with the civilian as an escort, due to the need for keyed entry throughout the building. Upon our entry, there was an audible alarm as well as a visual notification. We proceeded up to the second floor and into the computer area where the smell had been reported. At this time, a haze was noticed at the ceiling extending down approximately 2 feet. The civilian said that when he left the room, it was clear. My partner and I were looking around the room for possible causes of the haze when we began to feel heat coming from behind us. When I turned to my partner to tell him we needed to vacate the building, I noticed my voice had deepened and slowed considerably. I knew at this time we had been exposed to something.

A full commercial response was struck upon evacuation of the structure. The building was searched and found to have a circuit breaker taped opened as well as having had a release of FM 200 (an agent much like Halon) in the computer room. All crew members were medically cleared and felt no ill effects. The structure was released back to the occupants upon completion of operations.

Lesson Learned: I learned many lessons from this near-miss incident. The first involved the lack of communications to the shift commander. His lack of notification of where crews were and in what manner they were operating was cause for surprise and delay.

Further learning occurred from not following the given departmental policies on alarm responses and staffing for entry into a structure with an activated alarm. While there were four personnel on scene, our department's policy states that a rapid intervention team is to be established where there are no imminent lives in peril. I showed a complacent attitude in not being properly prepared to enter the structure, establish command, and communicate to those outside.

Another lesson learned was not to be disarmed by a civilian understating the possible problem and allowing myself to be lulled into a false sense of security.

Lessons were also learned in regard to the building itself. It was a secured government building, and we had no real plans to identify possible hazards within the structure as well as the ability to access areas throughout it. Furthermore, the light above the computer room door, which should have identified that there had been a release, was not activated.

Courtesy of the National Fire Fighter Near-Miss Reporting Systems (firefighternearmiss.com)

---

group, it is human nature for individuals to change their stories to match what others have said.

Each individual may have relevant information that will help to identify the cause and origin of a fire. For example, a witness might indicate that she saw lightning strike the building a few seconds before seeing smoke and flames. In another case, a witness might report that a can of gasoline was spilled accidentally and a few minutes later the basement erupted in flames.

The first-arriving fire team might report that the front door was broken open when they arrived. They also have valuable information about the color of the smoke or flames or the area where the fire was burning when they arrived.

The fire officer should review the facts that are already known or believed to be known before conducting an interview. During the interview, determine whether the witness is corroborating that information or providing conflicting information. The fact that the witness tells a different version of the story does not necessarily mean that the person is lying; the person could have made a different observation or have an inaccurate memory. Alternatively, the original information might have been inaccurate and the witness is providing good information. The collection of conflicting information means that further investigation is required for the actual facts to be determined. A witness statement should be disregarded only if it can be established with certainty that the information is incorrect—for example, if a witness indicates that he saw a man running from the back of a house before the fire started, yet there are no footprints in the snow.

Open-ended questions allow witnesses to tell what they saw or know, whereas questions that seek a "yes" or "no" answer limit the exchange of information. For example, the fire officer should ask the first-arriving crew to describe what they saw when they arrived, not ask whether there were flames showing through the front windows. After a witness provides an overall description, more direct questions can be asked to clarify the facts.

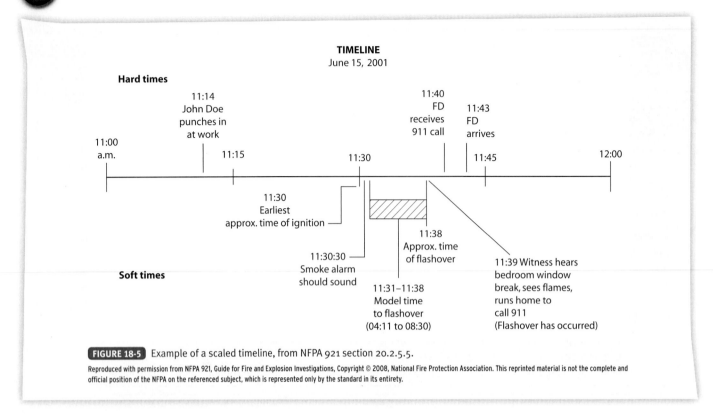

FIGURE 18-5 Example of a scaled timeline, from NFPA 921 section 20.2.5.5.

Reproduced with permission from NFPA 921, Guide for Fire and Explosion Investigations, Copyright © 2008, National Fire Protection Association. This reprinted material is not the complete and official position of the NFPA on the referenced subject, which is represented only by the standard in its entirety.

Cause and origin interviews are typically conducted at the fire scene. The officer should introduce himself or herself to the witness and note the person's name and contact information in case a follow-up interview is needed. As noted earlier, interviews are normally conducted with only one individual at a time so that the comments made by one person do not influence the responses of another. The fire officer should not interrupt the witness unless it is essential for clarification. After the interview is complete, the officer should thank the individual for providing assistance and provide his or her contact information in case the witness thinks of anything else.

Some witnesses may be reluctant to make a full disclosure of information. A special type of interview, called an interrogation, is used when an individual who is attempting to conceal information is being questioned. Interrogations require special knowledge and skills that are beyond the scope and duties of this text. Only a trained fire or police investigator should conduct interrogations.

Interview information must be documented. The documentation could be in the form of the interviewer's notes or a handwritten statement from the witness. An alternative is to record the witness statement with a tape recorder or video camera.

Vehicle Fire Cause Determination

Fire departments respond to more vehicle fires than structure fires. Chapter 25 of NFPA 921 provides a standardized procedure to conduct a vehicle fire investigation.

According to the U.S. Fire Administration, 47 percent of vehicle fires were caused by mechanical failure or malfunction. Some cars are burned to defraud insurance companies, however, whereas others are burned for revenge or to conceal other crimes.

The fire officer should trace the fire's development back to the point of origin, looking for the area where the most fire damage occurred and the areas of low burn. As in structural fires, there is often a distinctive burn pattern, such as a "V," that helps identify the area of origin; however, this pattern should be viewed with caution because of the characteristics of the materials that are present within most vehicles. Fires involving the fuel or tires tend to be particularly intense.

Once the point of origin is established, the specific cause must still be determined. If the cause is related to a mechanical malfunction, the exact cause may be difficult to establish without the assistance of an automotive expert. Additional factors should be considered to determine whether more extensive investigative techniques are required. Most fire departments do not conduct extensive investigations on vehicle fires unless arson is suspected because of the time and expense involved.

The first objective in most vehicle fire investigations is to look for indications of arson. Because some expensive vehicles are burned when the owners are unable to keep up with the high payments, insurance companies often conduct a more thorough investigation if they suspect that the fire was not accidental. The vehicle should be examined for its general condition. Is it a new, expensive vehicle burning at the end of a dead-end street in an unpopulated area? Have items of value, such as an expensive stereo, been removed? Are there indications of accelerant use inside the vehicle? Fuel lines should be checked to see whether they have been loosened. The condition of the tires might indicate that new tires were removed and replaced with old tires before the fire.

Several potential sources of accidental ignition must be considered for a vehicle fire. Mechanical sources may include the electrical system, exhaust system, catalytic converter, and turbocharger **FIGURE 18-6**. In older-model cars, a carburetor backfire can cause a fire within the engine compartment. Smoking materials can cause passenger compartment fires.

FIGURE 18-6 Mechanical failures associated with internal combustion engines are frequent causes of vehicle fires.

Note the make, model, and year of the vehicle, as well as the vehicle identification number (VIN). This information allows for a review of previous fires that have occurred in vehicles of the same make, model, and year. Some makes and models have a history of fires, and the reporting of this information can be useful in identifying product defects.

A diagram of the scene should be drawn. Photographs should be taken before the vehicle is removed.

Interview the driver/owner to determine when the vehicle was last driven and how far, what the total vehicle mileage is, whether there have been operating abnormalities, when the last service took place and when the vehicle was last fueled, how the vehicle was equipped (e.g., sound system, custom wheels), and which personal items were in the vehicle.

If the vehicle was being driven, determine the speed and any loads that were being pulled, any abnormal indications before the fire, the time and location at which the smoke and/or fire was observed, actions taken by the driver, and the amount of time that elapsed before the fire department arrived.

■ Wildland Fire Cause Determination

According to the National Interagency Fire Center, there has been a significant increase in the amount and cost of wildland fire suppression efforts as a result of dramatic changes in weather conditions, fuel buildup, and growth in the wildland–urban interface (WUI). The 2006 and 2007 fire seasons set modern historical records for both suppression costs and number of acres affected by wildfires. There were 1270 large or significant wildfires reported in 2012, part of the 7774 wildfires that burned 9,326,238 acres. While the number of wildland fires has since declined from the 2006–2007 spike, the number of acres burned per fire has increased, with 2012 representing the second largest number of acres lost since 1960 (Wildland Fire Lessons Learned Center, 2013).

The characteristics of wildland fires are quite different from those of structural fires. Wildland fires are influenced by environmental conditions, including topography, fuel load, wind, and weather. Such fires tend to spread vertically through convection, from lower vegetation to taller vegetation, and horizontally

FIGURE 18-7 Fire burns faster up a hill than it does down a hill, but is also dramatically affected by wind.

through radiation. The rate of spread varies, depending on the type of material burning and its density, the wind speed and direction, the humidity and fuel moisture content, the slope of the terrain, and natural features, such as valleys.

When a fire burn pattern on the side of a hill is investigated, the point of origin is most likely on the lower part of the slope, but not necessarily at the lowest point. The fire burns up the hill very rapidly, but it also burns down the hill, albeit at a slower rate. The wind also dramatically affects fire progression, either on flat ground or on a slope **FIGURE 18-7**.

Evaluating the degree of burn on the fuels helps establish the direction of travel. Ash residue also can be evaluated to establish the direction in which it was blown, which indicates the wind direction. When a tree burns and falls, the remaining trunk is usually burned at an angle, creating a point. This point generally appears on the side of the stump opposite the direction of fire approach.

Once the area of origin has been determined, the investigator can look for clues to help determine the cause of the fire. The following are examples of evidence that might be found at the point of origin:

- Campfire remains
- Time-delay devices
- Cigarette remains

- Lighters
- Multiple ignition points
- Splintered trees (indicating lightning strikes)
- Fulgurites (glassy, rootlike residue resulting from lightning strikes)
- Piles of fuel that are prone to spontaneous combustion
- Barrels used to burn trash
- Fallen electrical wires
- Trees on power lines
- Railroad tracks

Although none of these observations definitively confirms the exact cause of a fire, they are helpful in identifying potential causes. The National Wildfire Coordinating Group has developed professional qualification standards for wildland fire investigators.

Fire Cause Classifications

Once the facts are known, the circumstances of the fire are generally divided into four classifications according to NFPA 921:

1. *Accidental fire cause*: All of those fires for which the proven cause does not involve a deliberate human act to ignite a fire or spread fire into an area where it should not go.
2. *Natural fire cause*: Fires caused by lightning, earthquakes, wind, and other natural forces without human intervention.
3. *Incendiary fire cause*: Any fire that is deliberately ignited under circumstances in which the person knows that the fire should not be ignited.
4. *Undetermined fire cause*: Any fire in which the cause cannot be proved is classified as undetermined. The fire may still be under investigation.

Accidental Fire Causes

Many accidental fires result from human activities that are not intended to start or spread a fire. Finding evidence of candle use near the point of origin, for example, may suggest that an unattended open flame could be an ignition factor.

The most frequent ignition cause in residential fires is unattended cooking. When a fire originates on the stove, a pan is found on one of the burners, and the burner knob is on, the likely cause of the fire is accidental **FIGURE 18-8**. The second most common cause is smoking materials. Heating is the third leading cause of residential fires, with the two primary ignition causes being improper maintenance and combustibles located too close to the heating device. Other accidental fire causes include refueling of gasoline-powered equipment near an ignition source, placement of fireplace ashes in a combustible container, and sparks from welding.

When wood is continuously subjected to a moderate level of heat below its normal ignition temperature, but over a long period, it starts to break down into carbon. This chemical decomposition, called <u>pyrolysis</u>, results in a gradual lowering of the ignition temperature of the wood until autoignition occurs. Pyrolysis should be considered if the area of origin includes steam pipes, fluorescent light ballasts, flue pipes for a fireplace, or a wood-burning stove. The use of zero-clearance fireplace flues has caused an increase in this type of ignition.

The most common electrical fire scenario is misuse by the occupant, such as overloading electrical circuits, using

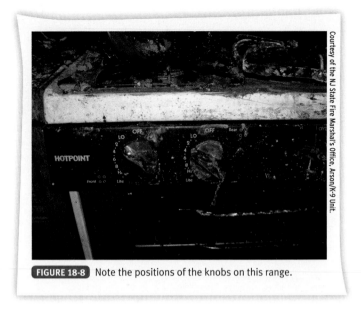

FIGURE 18-8 Note the positions of the knobs on this range.

lightweight extension cords for major appliances, or operating too many devices for the electrical service. Any of these errors can overload the wiring or the device, causing heat and eventually igniting a fire.

Large appliances, such as dishwashers and clothes dryers, may fail if their high-temperature controls or timing mechanisms fail. Excessive lint buildup that impedes air exchange and overheats the dryers is a frequent accidental scenario at multiple-family buildings with community laundry rooms. Dishwashers are susceptible to fire caused by plastic dishes and utensils falling onto the heating element—a consideration that should be taken into account when the dishwasher is the point of origin of a fire.

Electrical devices and appliances that start fires usually produce evidence of electrical damage on their power supply cord relatively close to the device or appliance. The wire insulation near the appliance often melts because of the heat produced by the burning appliance or other materials ignited by it. The melted insulation usually results in energized conductors contacting each other, generating a short circuit in the cord. An exception would be an appliance or device malfunction that results in the circuit breaker or fuse tripping so that the power cord is de-energized before the insulation melts.

Natural Fire Causes

A lightning strike in a forest can ignite a tree, and a strike on a building can have a tremendous destructive effect, sending several thousand volts through the electrical wiring system. This surge can destroy televisions, computers, telephone systems, and anything else that is plugged into the electrical system. Lightning can also follow antenna wires, telephone wires, metal plumbing, or the steel structure of a building and ignite combustibles that come in contact with those conduits.

When lightning is suspected, the fire officer should look for a contact point, usually near the top of the structure, concentrating on roof peaks with metal edging, antennas, or large metal objects, such as air-handling units. If the building's electrical system is involved, the surge usually fuses or melts the main fuses or circuit breakers. The surge can travel through television cables, phone lines, power lines, or even natural gas

> **Safety Zone**
>
> **Energized Aluminum Siding**
> Be cautious with aluminum-sided houses where the electrical service is mounted on the side of the house. The entire house exterior could be energized if a lightning strike has fused the circuit breaker box.

> **Assessment Center Tips**
>
> **When to Call for a Fire Investigator**
> One of the goals of an assessment center is to see whether the candidate understands the roles and responsibilities of other agencies that interact with the fire department. On either the incident scenario or a problem-solving scenario, the candidate may face a situation in which he or she is asked to identify a condition that requires action by the fire investigation unit. The fire officer should be able to recite, from memory, the situations and conditions for which a fire investigator must be notified. Be prepared to explain the actions that the candidate will take while waiting for the investigator, such as scene security and evidence preservation.

or propane lines, so the point of entry for each utility should be examined for scorching. The joints of aluminum rain gutters may show the same signs.

If there are indications that lightning could have caused a fire, the fire officer should check with weather services to determine whether any lighting strikes were recorded in the area. Some fire investigators commonly include a notation, such as "It was a clear and calm night with no lightning observed," in the introduction to their reports to establish that this potential source of ignition was eliminated.

Earthquakes, tornadoes, floods, and hurricanes can all cause fires. As a result of these natural forces, power lines may contact trees or fall onto structures, or sheets of metal may be blown across power lines. Earthquakes may also cause gas mains to break and release flammable gases into structures, where they find an ignition source. Lava from a volcano can easily start a forest fire or incinerate a community.

Incendiary Fire Causes

Because an incendiary fire is an intentional occurrence, the direct cause is a person; however, the methods and the reasons for starting fires vary tremendously. Given the complexity in identifying a fire as having an incendiary origin, a later section of this chapter is devoted to explaining why people intentionally set fires and which evidence might indicate that a fire was set intentionally.

An incendiary fire is one that is intentionally started when the person knows it should not be started. An incendiary fire is not necessarily arson. Arson is the crime of maliciously and intentionally or recklessly starting a fire or causing an explosion. The legal definition of arson as a criminal act is uniquely defined within different jurisdictions. The fire officer may be involved in making the determination of a fire's cause and origin and classifying it as incendiary, but ultimately the prosecuting attorney or grand jury will decide whether it should result in arson charges. The fire officer should be aware of the local or state judicial system that governs his or her jurisdiction and recognize how that system affects the use of the fire investigation results.

Undetermined Fire Causes

No matter how much training and experience a fire investigator has acquired, how many resources are available, or how much effort is expended, sometimes the cause of a fire cannot be determined. Perhaps the extensive damage caused by the fire or the firefighting activities makes it impossible to say with certainty what happened. This may occur when there are multiple reasons that the fire could have started that cannot be ruled out. In such a case, the evidence may not exist to make a definitive determination of the cause.

For example, if a gas can is found next to the water heater, it may be impossible to determine whether the vapors were accidentally ignited by the pilot light or intentionally ignited by an arsonist. Without additional evidence, either explanation may be feasible, so the cause cannot be determined with certainty.

The absence of any logical cause is another reason for classifying the cause of a fire as undetermined. Something caused the fire, to be sure, but the lack of evidence or knowledge precludes the fire investigator from determining the cause.

Indicators of Incendiary Fires

The fire officer needs to eliminate both accidental and natural causes before making a determination of an incendiary cause. Failure to eliminate accidental and natural causes makes it impossible to prove an incendiary cause.

Once the other causes have been eliminated, many conditions or factors may indicate that incident involved an intentional fire. These typically fall under five general categories:

- Disabled built-in fire protection
- Delayed notification or difficulty in getting to the fire
- Accelerants and trailers
- Multiple points of origin
- Tampered or altered equipment

Accelerants and Trailers

Accelerants are agents used to initiate a fire or increase the rate of fire growth. Trained canines or survey instruments can detect ignitable liquid accelerants. In addition, ignitable-liquid–fueled fires often leave distinct burn and char patterns **FIGURE 18-9**.

Trailers are materials used to spread a fire from one area of a structure to another, causing a fire to grow more quickly. The following materials are most often used as trailers:

- Paper towels
- Black gunpowder
- Film wrapped in paper rags
- Kerosene or other combustible liquids
- Gasoline or other flammable liquids

VOICES OF EXPERIENCE

As a fire officer, it is important to remember that the initial fire scene assessment conducted by you will set the stage for the rest of the investigation, and that it is also probably the most important part of any investigation.

On many occasions as a fire investigator, when I arrived at the scene to begin my investigations, what the fire officer had done prior to my arrival made my job either much easier or much harder. The fire officer in the performance of his or her duties will make or break the investigation. One of the primary responsibilities of the fire officer is the protection of evidence and security of the scene. Once the scene has been compromised, the value of evidence is lost.

> On many investigations, what greeted me was a pile of smoldering furniture on the front lawn.

On many investigations, what greeted me was a pile of smoldering furniture on the front lawn. When I asked the fire officer what had happened, I was told that they were doing overhaul. Excessive overhaul operations can cause the investigator to be unable to properly investigate the fire for origin and cause. As an investigator, we used to refer to the suppression crews as "Evidence Eradication Teams". When suppression and overhaul may destroy evidence, the fire officer must decide whether it's possible to preserve the evidence and, if so, how it can be done. Once the evidence has been removed or altered, fire investigators lose continuity in the chain of custody. Once the chain of custody is broken, we may have to say the cause of the fire is undetermined.

The favorite question I always asked the fire officer once I arrived on the scene was, "So what happened?" I got many varied answers, some of which I cannot mentioned in this essay. The fire officer should have a sound knowledge of how to determine the origin and cause of the fire. Establishing the room of origin and the possible points of origin are essential. As an investigator I really didn't care whether the fire officer was able to distinguish between a V or U pattern. As long has he or she could say that there is a pattern in this room and that the room was preserved was all I needed.

As a fire officer you are setting the investigation in motion. How you preserve evidence and reduce the destruction of evidence during overhaul operations will go a long way into determining the origin and cause of a fire. To be able to identify patterns and areas of origin are critical first steps in any investigation. Often a fire officer, prior to conducting overhaul operations, can stop for two minutes and take photos inside the room of origin. That officer is preserving evidence, and when those photos reach the fire investigator, he or she might not have to say the cause is undetermined.

What you do during suppression and overhaul operations will in many cases make or break the investigation.

Doug Goodings
Manager of Academic Standards and Evaluations
Office of the Fire Marshal/Ontario Fire College
Gravenhurst, Ontario, Canada

CHAPTER 18 Fire Cause Determination

FIGURE 18-9 Ignitable-liquid–fueled fires leave distinct burn and char patterns.

- Decorative streamers
- Cotton batting
- Paper
- Sheets or rolls of fabric softener
- Newspapers
- Combinations of these items

Trailers usually leave a distinct fire pattern that resembles the material's shape and often runs from one room to the next. Liquids typically have irregular edges and areas that are burned where the liquids pooled because of low spots. Sometimes, a trailer of ignitable liquid is poured from one electrical outlet to another to simulate an electrical fire in an attempt to mislead the investigator.

■ Multiple Points of Origin

Arsonists frequently utilize multiple ignition points inside a building. This is done in case one fire burns out prematurely as well as to maximize the amount of fire growth before the fire department can respond. In other cases, the arsonist is attempting to cause confusion or to trap the occupants by blocking access to all of the exits.

Igniting fires in multiple buildings at the same time presents the fire department with a complicated problem. The responding units may be sufficient to manage one fire, but two or three buildings on fire create a challenge.

Company Tips

NFPA Statistical Snapshot
1. An estimated 306,300 intentional fires of all types were set in 2005–2009.
2. These fires resulted in 440 deaths and $1.3 billion in property damage.
3. Three-fourths of intentional fires were set outside.
4. An estimated 53,700 structure fires were deliberately set or suspected of having been deliberately set.
5. The number of intentional fires and associated losses has remained stable since 2000.

Multiple points of origin do not mean that a fire was intentional. In some cases, material falling from the ceiling and burning at the floor level can create a secondary "U" or "V" pattern, resembling an additional point of origin. There may be nothing at that location that could have caused the fire to ignite.

Multiple points of origin may also occur when an electrical surge causes ignitions at different locations in a building's electrical system. In one instance, an overpressurized natural gas system started fires simultaneously in several kitchens and water heater closets, where each pilot light became a torch.

■ Arson

Arson is the crime of maliciously and intentionally, or recklessly, starting a fire or causing an explosion. National statistics indicate that one in every four fires is of incendiary origin (National Fire Data Center, 1997). Arson consistently has the highest rate of youth involvement when compared with all other crimes logged in the Federal Bureau of Investigation's index of the most serious felonies.

■ Arson Motives

There are six basic motives for arson:

1. Profit
2. Crime concealment
3. Excitement
4. Spite/revenge
5. Extremism
6. Vandalism

Profit

Most often, the plan behind arson committed for profit is to collect insurance money. Indicators of insurance fraud include inability to meet payments, failure to complete business contracts, poor sales volume, or lack of supplies and inventory. Investigators should verify the insurance coverage amounts and look for recent changes in the policy. In some cases, the insured makes claims for burned inventory that was actually removed before the fire or that never existed.

Elaborate schemes have been conducted to manipulate supply and demand for various products by destroying inventory or manufacturing facilities to cause a shortage and drive up prices. Arson has also been used for extortion or to eliminate competition. Sometimes, a contractor or individual starts a fire and then offers to repair the damaged property. Many vacant buildings have been burned to save the cost of their demolition.

Crime Concealment

A business owner or employee may elect to burn the business to destroy records that show embezzlement of cash, supplies, or inventory. Burglars might set fire to a building to destroy evidence of their entry, eliminate fingerprints, or even conceal the fact that a theft occurred before the fire. Arson is also used to destroy evidence of other crimes, including murder, or to create a distraction while a crime is taking place in another location.

Excitement

Sometimes, a fire is started for the excitement of the arsonist, who could be seeking thrills, attention, or recognition. The arsonist could be planning to make a dramatic rescue to gain praise or to obtain recognition for discovering or extinguishing the fire. The arsonist may also simply enjoy the excitement and spectacle that are generated by a fire.

Spite/Revenge

Spite and revenge fires involve intense emotions. The arsonist may intentionally set a fire in the exit stairway of an occupied building in the middle of the night. Frequently, this action is triggered by hatred, jealousy, or other uncontrolled emotions following a lover's quarrel, divorce, or bar fight. In many cases, the individual has consumed alcohol or drugs before starting the fire.

Extremism

Extremism on behalf of a variety of causes has been the rationale for setting fires for many centuries. Abortion clinics, religious institutions, businesses that were ecologically damaging, and businesses undergoing labor disputes have all been arson targets. An extremist may want to cause a monetary loss to the person or business or to bring attention to a cause. Radical activists know that a large loss of life focuses attention on their cause. Incendiary devices are frequently used in these crimes.

Vandalism

The motive for vandalism is simply to cause damage for its own sake. Vandalism is most often directed toward schools, abandoned structures, vegetation, and trash containers. In most cases, the fire setter commits this crime within walking distance of his or her home.

Documentation and Reports

All fires must be properly documented and reported according to the fire department's standard procedures. Most fire departments use the NFIRS reporting system or some variation of it. The basic report includes the incident number, alarm time and date, location of the incident, property ownership, building construction and occupancy type, weather conditions, responding units and personnel, and numerous other factors that are required for full documentation of an incident.

The data collected through NFIRS provide important information that the fire department can use to identify risk factors and trends and to plan the most efficient utilization of resources to prevent fires and respond to emergencies. This information also flows into the state and national database systems to provide a better understanding of the overall fire problem. The fire service and elected officials may also use these data to determine where resources should be directed and when laws should be changed.

■ Preliminary Investigation Documentation

In addition to the basic incident report, the fire officer often writes up a special narrative report if the cause of the fire is incendiary or if unusual circumstances are involved. Such a narrative report is particularly valuable if the fire officer is later called to testify in court about the incident. The *Fire Officer Communications* chapter discusses communications in more detail, but fire officers could use the following format to document activities and observations for the fire investigator:

A. Receipt of the alarm
   1. Who reported the fire?
   2. What time was the alarm received?
   3. Who discovered the fire?
B. Response to the incident
   1. Did the fire company encounter any suspicious activity while responding to the fire?
   2. How much time elapsed from dispatch to arrival?
   3. What were the observed weather conditions?
C. Accessibility at the scene
   1. What was the general condition of the fire?
      a. What were the extent and the intensity of the fire?
      b. What was the location of the fire or fires?
      c. Did anyone meet the fire fighters on arrival?
      d. Were any familiar spectators at the scene?
   2. Circumstances on arrival
      a. Was anything unusual, considering the fire load?
      b. Where were smoke and fire coming from?
      c. What were the rapidity and the spread of the fire?
   3. Gaining access to the building
      a. How did the fire fighters get into the structure?
      b. If entry was forced, how did this occur and who forced entry?
D. Fire suppression
   1. What were your immediate observations on entry?
      a. Where was the fire centered?
      b. Were there any unusual flame, smoke, or odors?
   2. What were the conditions of fire extinguishment?
      a. Did the fire flash when it was hit by water?
      b. Was the fire difficult to extinguish?
   3. Were there obstructions to fire suppression?
      a. Were fire protection systems tampered with?
   4. Was the alarm system functioning properly?
E. Civilian contacts
   1. Did witnesses make statements to fire fighters?
   2. Was the owner at the scene?
   3. What were the names of persons who were allowed into the fire scene?

F. Scene integrity
  1. Were any physical evidence or artifacts removed from the scene?
  2. Were any photos or videos taken during fire suppression?
  3. Did any fire department members make holes in walls or ceilings?

The report must be clear, complete, and factual. The fire officer must make no assumptions or speculations. The best report is a narrative that accurately and completely describes what the fire officer observed and what the fire company did.

■ Investigation Report

The fire officer or investigator who conducts the investigation will also submit a narrative report. In this report, the information is provided in chronological order, beginning with a description of the structure before the event occurred. This description would include the building height and dimensions, construction type, structural condition, occupancy, and utility services. It would then describe the alarm notification information, including the time of the call, the name of the caller, and what the caller said.

The report should fully describe the results of the fire scene examination, beginning with a description of the exterior damage. This is followed by a description of the interior damage, including the determination of fire origin and cause, as well as the examination and elimination of any other possible fire causes. This section closes with the officer's opinion and conclusion as to the cause and origin of the fire.

Attached to this report would be the information obtained from interviews and witnesses as well as statements from responders. Attachments should also include statements of evidence that were collected, warrants, and sketches.

■ Legal Proceedings

A company officer may occasionally be called on to testify in court proceedings as a witness. Whereas a witness can provide the court with testimony based only on his or her personal knowledge and observations, an expert witness has scientific, technical, or other specialized knowledge that can be relied on to interpret the facts. The role of an expert witness is to assist the judge and jurors to understand the evidence or to determine the true facts in an issue. An expert witness is allowed to give an opinion based on the facts or the data of the case **FIGURE 18-10**.

The first step in serving as a witness is to fully prepare for giving testimony. This includes reviewing all reports, photographs, and diagrams of the incident. Also, review any previous depositions or testimony that you have given on the incident. The prosecutor will review your testimony and qualifications.

When testifying:
- Dress appropriately.
- Follow the prosecutor's directions.
- Sit up with both feet on the floor.
- Avoid gesturing.
- Keep answers short and to the point.
- Use language a jury can understand.
- Be courteous and patient.
- Be honest.
- Do not hesitate or avoid answering questions.

FIGURE 18-10 An expert witness can assist the judge and jurors in understanding the evidence or to determine the facts.

- Speak clearly and loudly.
- If you do not remember, do not guess.

Remember that the defense counsel's job is to try to discredit you and your statements. It is essential to remain calm and cool, even when the lawyer asks questions or makes

Getting It Done

Coordinated Overhaul and Salvage
Although it is important not to destroy evidence during fire suppression operations, the fire officer's first responsibility is to maintain a safe fire ground. That means promptly controlling and extinguishing the fire. After the searches are completed and the fire is declared under control, the fire officer must consider evidence preservation. Overhaul and salvage operations should be coordinated with the fire investigator, if the investigator is on the scene. For example, a fire investigator may wish to photograph or document the condition of a ceiling before it is removed during overhaul operations. If the investigator's response is delayed, the fire officer should attempt to identify and protect the area of origin, limiting overhaul in that area to the absolute minimum required to ensure that the fire does not rekindle.

NFPA 921 makes the following recommendations:
- Use caution with straight- or solid-stream water patterns; they can move, damage, or destroy physical evidence.
- Restrict the use of water for washing down, and try to avoid possible areas of fire origin.
- Refrain from moving any knobs or switches.
- Use caution with power tools in the fire scene. Refuel this equipment away from the fire scene.
- Limit the number of fire fighters performing overhaul and salvage until the fire investigator is finished documenting the scene.

statements that are designed to make you look bad. Your role is to present factual information that you can support. Every fire officer should be familiar with NFPA 921.

After the Fire Officials Are Gone

Many fire investigations continue long after the fire department has cleared the event. For example, the insurance company may engage in a continuing investigation. This investigation is not intended to undermine or question the fire investigator's conclusions, but rather to look at the fire from the insurance company's point of view.

The roles of the fire investigator and the insurance company investigator differ in many cases. The fire investigator could be interested in determining the fire cause and origin to help prevent future fires or to help prosecute criminal actions. The insurance investigator might also be looking at the factors that contributed to the loss, such as the absence or inadequacy of fixed fire protection systems and adherence to or disregard of the applicable codes.

The insurance company has a financial interest in any fire loss. Consequently, it is often interested in determining which other parties could have liability for the loss, even if there is no criminal responsibility. Although a criminal prosecution requires proof beyond a reasonable doubt, a civil action involving the insurance company is decided by a preponderance of the evidence. Many insurance cases are not settled for several years after the fire is extinguished.

You Are the Fire Officer Conclusion

Working from the unburned area toward the burned area, it becomes clear that the point of origin of the fire was below the charred sports car that is still a few feet above the garage floor on a hydraulic lift. The technician's story matches what you see: A quarter-full gasoline tank was dropped or fell from the sports car and the fumes found an ignition source, perhaps the electric-powered space heater that is near where the tank fell.

Davis and Taylor agree to classify the fire as accidental and contact the on-duty fire investigator, who is still tied up at another fire. The investigator tells Davis to call for the overtime investigator because of the injured civilians, the dollar loss, and what appears to be an unpermitted operation of a vehicle repair shop within an electronics installation store. The investigator also suggests that Davis have the hazardous materials team assist with handling the heat-damaged racing fuel drums, the acetylene torch, and the other hazardous materials found in the store.

Wrap-Up

Chief Concepts

- One of the best starting points for efforts to prevent future fires is understanding of the causes of fires that have occurred in the past.
- The leading cause of home structure fires is cooking.
- The fire officer should be able to determine a point of origin and a cause (or probable cause) for most fires. The officer should also know his or her jurisdiction's criteria on when to call for a fire investigator.
- To determine the point of origin, the fire officer must understand fire growth and development and the three methods of heat transfer: conduction, convection, and radiation.
- A disabled sprinkler system may be encountered in fires involving large industrial or commercial occupancies.
- Be alert for conditions or situations that might delay the fire department's ability to get to the fire. Malfunctioning keys and key cards, vandalized doors, and materials blocking access are conditions to note.
- Document unusual conditions noticed at the fire scene. A professional arsonist wants to try to make a fire appear accidental.
- Although the law states that there is a public interest in determining the cause and origin of a fire, the fire investigator must take into account the competing interests of a citizen's rights to privacy and due process.

- When a fire has occurred and the fire department has been called, the fire department has the right to determine the cause and origin of the fire. This process must be accomplished in accordance with the law.
- A fire officer who conducts a preliminary fire cause investigation and suspects that a crime has occurred should immediately request the response of a fire investigator and secure the scene to protect any evidence.
- The fire officer must consider three types of evidence: demonstrative, documentary, and testimonial.
- Fire scene reconstruction is the process of re-creating the physical scene before the fire occurred, either physically or theoretically.
- According to NFPA 921, "The compilation of factual data, as well as an analysis of those facts, should be accomplished objectively, truthfully, and without expectation bias, preconception, or prejudice."
- The first step in fire cause determination is to identify the point of origin—the exact physical location where a heat source and a fuel came in contact with each other.
- U- and V-shaped fire patterns can lead the fire officer to the point of origin for a fire.
- Cause refers to the particular set of circumstances and factors that were necessary for the fire to have occurred.
- The source of ignition must have been located at or near the point of origin. The fire officer may be able only to infer a probable ignition source.
- The physical configuration of the fuel is an important characteristic in ignition.
- Failure analysis is a logical, systematic examination of an item, component, assembly, or structure and its place and function within a system, conducted to identify and analyze the probability, causes, and consequences of potential and real failures.
- Fire analysis is the scientific process of examining a fire occurrence to determine all of the relevant facts, including the origin, cause, and subsequent development of the fire, as well as to identify the responsibility for whatever occurred.
- As part of the fire investigation, the fire officer may have to interview fire victims, witnesses, fire fighters, and suspected perpetrators.
- Fire departments respond to more vehicle fires than structure fires.
- Wildland fires are influenced by environmental conditions, including topography, fuel load, wind, and weather.
- The four fire cause classifications are accidental, natural, incendiary, and undetermined.
- Many accidental fires result from human activities that are not intended to start or spread a fire.
- If lightning might have caused a fire, the fire officer should check with weather services to determine whether any lighting strikes were recorded in the area.
- An incendiary fire is a fire that is intentionally started when the person knows it should not be started, but it is not necessarily arson.
- No matter how much training and experience a fire investigator has acquired, how many resources are available, or how much effort is expended, sometimes the cause of a fire cannot be determined.
- When considering whether a fire might have been set intentionally, look for disabled built-in fire protection, delayed notification/difficulty in getting to the fire, accelerants and trailers, multiple points of origin, and tampered/altered equipment.
- Arson is the crime of maliciously and intentionally, or recklessly, starting a fire or causing an explosion.
- There are six basic motives for arson: profit, crime concealment, excitement, spite/revenge, extremism, and vandalism.
- All fires must be properly documented and reported according to the fire department's standard procedures. Most fire departments use the NFIRS reporting system or a variation of it for this purpose.
- In addition to the basic incident report, the fire officer often writes up a special narrative report if the cause of the fire is incendiary or if unusual circumstances are involved.
- In the investigation report, the information is provided in chronological order, beginning with a description of the structure before the event occurred.
- Whereas a witness can provide the court with testimony based only on his or her personal knowledge and observations, an expert witness has scientific, technical, or other specialized knowledge that can be relied on to interpret the facts.
- Many fire investigations continue long after the fire department has cleared the event.

Hot Terms

<u>Accelerant</u> An agent—often an ignitable liquid—used to initiate a fire or increase the rate of growth or spread of fire.

<u>Arson</u> The crime of maliciously and intentionally, or recklessly, starting a fire or causing an explosion.

<u>Artifacts</u> The remains of the material first ignited, the ignition source, or other items or components in some way related to fire ignition, development, or spread. An artifact may also be an item on which fire patterns are present, in which case the preservation of the artifact is not focused on the item itself, but rather on the fire pattern that appears on the item.

<u>Char</u> Carbonaceous material that has been burned and has a blackened appearance.

<u>Demonstrative evidence</u> Tangible items that can be identified by witnesses, such as incendiary devices and fire scene debris.

<u>Documentary evidence</u> Evidence in written form, such as reports, records, photographs, sketches, and witness statements.

Wrap-Up, continued

Electrical arc Luminous discharge of electricity from one object to another, typically leaving a blackening of objects in the immediate area.

Evidence The documentary or oral statements and the material objects admissible as testimony in a court of law.

Failure analysis A logical, systematic examination of an item, component, assembly, or structure and its place and function within a system, conducted to identify and analyze the probability, causes, and consequences of potential and real failures.

Fire analysis The process of determining the origin, cause, development, and responsibility for a fire or explosion, as well as the failure analysis of a fire or explosion.

Fire patterns Physical marks left on an object by the fire.

Fire scene reconstruction The process of re-creating the physical scene during fire scene analysis through the removal of debris and the replacement of contents or structural elements in their prefire positions.

Form of material What the ignited material is being used for; for example, the form of cotton material might be clothing or bales of cotton.

Point of origin The exact physical location where a heat source and a fuel come in contact with each other and a fire begins.

Pyrolysis The destructive distillation of organic compounds in an oxygen-free environment that converts the organic matter into gases, liquids, and char.

Source of ignition Devices or equipment that, because of their intended modes of use or operation, are capable of providing sufficient thermal energy to ignite flammable gas–air mixtures.

Testimonial evidence Witnesses speaking under oath.

Trailers Materials used to spread fire from one area of a structure to another.

Type of material What the ignited material is made of; for example, the type of material might be cotton.

References and Additional Resources

NFPA reprinted material is not the complete and official position of the NFPA on the referenced subject, which is represented only by the standard in its entirety.

Cantor, D., and L. Almond. (2002). *The Burning Issue: Research and Strategies for Reducing Arson*. London, UK: Office of the Deputy Prime Minister: 38.

Corey, M., ed. (1996). *Motive, Means and Opportunity: A Guide to Fire Investigation*. Princeton, NJ: American Re-insurance Company.

Craford, J. (2013). Comprehensive prevention program. In A. Thiel and C. R. Jennings, *Managing Fire and Emergency Services*. Washington, DC: International City/County Management Association. Chapter 6, pages 155–184.

Deal, T., M. de Betterencourt, and V. Deal. (2012). *Beyond Initial Response: Using the National Incident Management System's Incident Command System* (2nd ed.). Bloomington, IN: AuthorHouse.

DeHann, J. D., and D. J. Icove. (2011). *Kirk's Fire Investigation* (7th ed.). New York, NY: Prentice Hall.

England, R. E., et al. (2012). *Managing Urban America*. Washington, DC: CQ Press.

Evarts, B. (2012). *Intentional Fires*. Quincy, MA: National Fire Protection Association.

Fahy, R. F., et al. (2012). *Firefighter Fatalities in the United States—2011*. Quincy, MA: National Fire Protection Association.

Geiman, J. A., and J. M. Lord. (2012). "Systematic Analysis of Witness Statements for Fire Investigation." *Fire Technology* 48: 219–231.

Hill, H. J. (2012). *Failure Point: How to Determine Burning Building Stability*. Tulsa, OK: Pennwell.

IFSTA. (2005). *Introduction to Fire Origin and Cause* (3rd ed.). Stillwater, OK: Fire Protection Publications.

IFSTA. (2010). *Fire Investigator* (2nd ed.). Stillwater, OK: Fire Protection Publications.

Karter, M. J. Jr. (2012). *Patterns of Firefighter Fireground Injuries*. Quincy, MA: National Fire Protection Association.

Karter, M. J. Jr., and J. L. Molis. (2012). *U.S. Firefighter Injuries—2011*. Quincy, MA: National Fire Protection Association.

Karter, M. J. Jr., and G. P. Stein. (2012). *U.S. Fire Department Profile through 2011*. Quincy, MA: National Fire Protection Association.

Kerber, S. (2012). "Analysis of Changing Residential Fire Dynamics and Its Implications on Firefighter Operational Timeframes." *Fire Technology* 48(4): 865–891.

Maniscalco, P. M., and H. T. Christen. (2010). *Homeland Security: Principles and Practice of Terrorism Response*. Sudbury, MA: Jones and Bartlett.

Michigan v. Tyler, 436 U.S. 499 - Supreme Court (1978).

National Fire Data Center. (1997). *Arson in the United States*. Washington, DC: U.S. Fire Administration.

National Fire Data Center. (2012). *National Estimates Methodology for Building Fires and Losses*. Washington, DC: U.S. Fire Administration.

National Fire Protection Association. (2011). *NFPA 921: Guide for Fire & Explosion Investigations*. Quincy, MA: National Fire Protection Association.

National Fire Protection Association. (2012). *Fire Investigator: Principles and Practice to NFPA 921 and 1033* (3rd ed.). Burlington, MA: Jones & Bartlett Learning.

National Fire Protection Association. (2014). *NFPA 1033: Standard for Professional Qualifications for Fire Investigator*. Quincy, MA: National Fire Protection Association.

National Fire Protection Association. Reproduced with permission from NFPA 921, Guide for Fire and Explosion Investigations, Copyright © 2008, National Fire Protection Association.

National Wildfire Coordinating Group Executive Board. (2009). *Quadrennial Fire Review 2009*. Washington, DC: U.S. Department of Interior.

Poynter, D. (2007). *The Expert Witness Handbook: Tips and Techniques for the Litigation Consultant* (3rd ed.). Santa Barbara, CA: Para Publishing.

Redsicker, D. R., and J. J. O'Connor. (1996). *Practical Fire and Arson Investigation* (2nd ed.). Boca Raton, FL: CRC Press.

Roberts, C. D. (1985). "Fire Investigation and the Private Investigator." *National Fire and Arson Report* 3(3).

Thomas, D. S., et al. (2011). "Enticing Arsonists with Broken Windows and Social Disorder." *Fire Technology* 47: 255–273.

TriData. (1998). *An NFIRS Analysis: Investigating City Characteristics and Residential Fire Rates*. Washington, DC: U.S. Fire Administration.

Varone, J. C. (2011). *Legal Considerations for Fire and Emergency Services*. Clifton Park, NY: Delmar/Cengage Learning.

Wildland Fire Lessons Learned Center. (2013). *2012 Incident Review Summary*. Washington, DC: U.S. Department of Interior.

FIRE OFFICER in action

Captain Davis was the first-arriving commander at an apartment where food was left unattended on a stove. The fire spread to materials next to the burning pot and extended up into the cabinets and exhaust ductwork. The responders found a 10-year-old child, an 8-year-old child, and a 6-year-old child in the apartment with no adult supervision. The apartment was a mess, and there was no battery in the smoke alarm. Davis called for police assistance because of the unsupervised children.

The managing or supervising fire officer who handled the fire is in the best position to ensure that evidence is preserved and an initial investigation into the cause of the fire is started. An accurate fire cause investigation benefits all.

1. Which of the following is *not* a fire cause as defined by NFPA 921?
   A. Accidental
   B. Arson
   C. Undetermined
   D. Natural

2. You have started a cause and origin investigation and have determined that you need a fire investigator. It will take 4 hours for the investigator to get to the fire scene. What should you do next?
   A. Make sure that the incident scene perimeter is completely wrapped in fire-line tape before leaving. Post a "Stay out" sign and notify the owner or representative when the fire investigator will arrive.
   B. Secure the perimeter of the incident scene with fire-line tape and arrange for a continuous and uninterrupted fire department on-scene presence until the fire investigator arrives.
   C. Cancel the fire investigator and make sure that you completely document the incident scene with a digital camera.
   D. Obtain a contact number from the owner or representative before leaving the incident scene, and schedule a follow-up time when the fire investigator arrives.

3. What is the first objective of a cause and origin investigation?
   A. Determine the source of ignition
   B. Find the point of origin
   C. Determine the fuel that was first ignited
   D. Deduce how the ignition source and fuel came together

4. What is the most common cause of fires in residential structures?
   A. Children playing with matches
   B. Carelessly discarded smoking materials
   C. Cooking
   D. Congested dryer duct

Fire Captain Activity

Battalion Chief Johnson reviews the report of the automobile fire in the unpermitted race-car repair shop. The chief is pleased with the report and suggests that Taylor and Davis sharpen their fire cause determination skills, as the recovering economy and declining municipal revenues mean that there will not be funding available for an overtime fire investigator in the near future.

NFPA Fire Officer II Job Performance Requirement 5.5.1
Determine the point of origin and preliminary cause of a fire, given a fire scene, photographs, diagrams, pertinent data, and/or sketches, to determine if arson is suspected.

Application of 5.5.1
Pick one of the following assignments:
1. When provided with a description of a fire scene, photographs, diagrams, and pertinent data, determine if arson is suspected.
2. Using data from an agency or jurisdiction where you have access to fire incident data, describe the local arson experience.
3. Research and describe the national wildland arson problem.
4. Research and describe the national multiple-family dwelling arson problem.
5. Research and describe the national urban arson problem.
6. Research and describe the national vehicle arson problem.

Crew Resource Management and Leading Change

Fire Officer I

Knowledge Objectives

After studying this chapter, you will be able to:

- Discuss the origins of crew resource management (CRM). (p 386)
- Discuss the concepts involved in researching and validating CRM. (pp 386–387)
- List Dupont's "dirty dozen" human factors that contribute to tragedy. (p 387)
- Describe the six-point CRM model that can be used in the fire service. (pp 388–389, 391–394)
- Discuss the fire officer's role in recommending change within a department (NFPA 4.4.1) (NFPA 4.4.4). (p 395)

Skills Objectives

There are no Fire Officer I skills objectives for this chapter.

Fire Officer II

Knowledge Objectives

After studying this chapter, you will be able to:

- Discuss the origins of crew resource management (CRM). (p 386)
- Discuss the concepts involved in researching and validating CRM. (pp 386–387)
- List Dupont's "dirty dozen" human factors that contribute to tragedy. (p 387)
- Describe the six-point CRM model that can be used in the fire service. (pp 388–389, 391–394)
- Discuss the fire officer's role in recommending change within a department (NFPA 5.4.6). (p 395)

Skills Objectives

There are no Fire Officer II skills objectives for this chapter.

CHAPTER 19

Principles of Fire and Emergency Service Administration (FESHE) Course Objectives

4. Select and implement the appropriate disciplinary action based upon an employee's conduct. (pp 387–389)
5. Explain the history of management and supervision methods and procedures. (pp 386–388, 392–393)
6. Discuss the various levels of leadership, roles, and responsibilities within the organization. (pp 388–389, 391–395)
9. Identify the roles of the National Incident Management System (NIMS) and Incident Management System (ICS). (pp 391–392)

You Are the Fire Officer

Quint 100 is on the road when its crew members see a rapidly expanding mushroom of black smoke a few blocks away. Arriving at the incident, Lieutenant Taylor Williams reports a gasoline tank truck has overturned, ruptured its contents, and ignited. The burning petroleum has splashed into the lobby of an eight-story community hospital. The black boiling smoke is entering the open windows of an assisted-living high rise on Exposure D.

Establishing command, Lieutenant Williams learns that most of the first-alarm companies are held up at a railroad crossing. Apparatus Operator Karen Rollo reports a catastrophic failure of the rig's powertrain. Neither the fire pump nor the aerial is operational. Behind Quint 100, the lieutenant sees smoke and flames coming from manhole covers. The Quint 100 crew will be working alone for 10–15 minutes.

1. When faced with multiple tasks, how can you best allocate resources to ensure that the emergency situation is mitigated and everyone returns home safely afterward?
2. How can the error management model improve fire-ground operations?
3. How do assertive statements affect fire-ground operations?

Introduction

This chapter looks at two fire officer practices that affect success. Crew resource management (CRM) is a behavioral approach to reducing human error in high-risk or high-consequence activities. Leading change describes the process of recommending changes (as a lieutenant) and developing a process to establish a change (as a captain).

Fire Officer I and II

Origins of Crew Resource Management

On December 28, 1978, United Airlines Flight 173 was making a routine flight from Denver, Colorado, to Portland, Oregon, with 189 passengers and crew onboard. During the final approach, an unfamiliar "thump" was felt as the landing gear deployed, and the "gear down and locked" light on the cockpit instrument panel did not illuminate. The captain decided to circle the airport while he, the first officer, and the flight engineer attempted to figure out the problem. The flight engineer told the captain that 15 minutes of circling would really run the plane low on fuel.

One hour passed as Flight 173 circled the Portland area. Aviation tradition held that the captain was the infallible head of the ship and was never questioned. While the crew continued to fuss with the lights, the plane's engines coughed, sputtered, and finally went quiet. In the ensuing crash, 10 people were killed, including the flight engineer, and 23 were critically injured.

The McDonnell Douglass DC-8 aircraft used by Flight 173 was a fully functional, mechanically sound airframe that crashed because the humans flying the machine became over-engrossed in a burned-out light bulb. In response to this incident, a behavioral modification training system known as crew resource management (CRM) was developed in a 1979 National Aeronautics and Space Administration (NASA) workshop examining the role of human error in aviation accidents.

Resistance to mandatory CRM training continued until United Airlines Flight 232 experienced a catastrophic failure of its center engine in 1989. All three hydraulic lines necessary for controlling flaps, rudders, and other flight controls were severed. This damage robbed the crew of both the primary and redundant safety features that are built into every airframe. When these conditions were presented within a flight simulation exercise, it always resulted in an unrecoverable spin with all lives lost.

The flight crew and a check ride pilot, using engine controls alone, managed to bring the crippled plane into the Sioux City, Iowa, airport. One hundred eighty-four of the 295 people on board survived the fiery crash onto the runway. The crew attributed their success to CRM training as they initiated behaviors to overcome the five factors that contribute to human error. United Airlines Flight 232 was the landmark event that validated CRM's worth.

Researching and Validating CRM Concepts

Part of the 80 percent reduction in the aviation industry's accident rate is attributed to the development, refinement, and system-wide adoption of CRM. Lieutenant Colonel Tony Kern

Getting It Done

Creating a Safety Culture
Every fire department should take the following five steps to create a safety culture:
1. Provide honest sharing of safety information without the fear of reprisal from superiors.
2. Adopt a nonpunitive policy toward errors.
3. Take action to reduce errors in the system. Walk the talk.
4. Train fire fighters in error avoidance and detection.
5. Train fire officers in evaluating situations, reinforcing error avoidance, and managing the safety process.

(retired, U.S. Air Force), writing in *Controlling Pilot Error: Culture, Environment and CRM*, states: "CRM is designed to train team members how to achieve maximum mission effectiveness in a time-constrained environment under stress. That is a concept with nearly universal utility and timeless applicability." Dr. Kern received the Distinguished Leadership Award from *Aviation Week and Space Technology* in 2003 after making controversial decisions to ground nine air tankers during the 2002 U.S. wildland fire season after fatal accidents involving a Lockheed C-130A and a Consolidated PB4Y2, to bar future aerial firefighting contracts for these two airplane types, and to restrict aircraft operations in cooperation with other firefighting agencies. At the time, Kern was the national aviation director of the U.S. Forest Service.

Prof. Robert Helmreich and his staff at the University of Texas Human Factors Research Project (HFRP) developed CRM. They showed the dramatic value of applying CRM to aerospace, aviation, the military, maritime operations, and the medical profession. Their project started in the early 1980s as the Aerospace Crew Research Project, exploring the relationship among personality, group culture, and performance.

Human Error

Gordon Dupont spent 7 years reviewing aviation accidents as a technical investigator for the Canadian Aviation Safety Board. As the Special Programs Coordinator for Transport Canada, he was responsible for developing programs that would reduce maintenance error. During his study, Dupont considered the similarities between errors that occur in the cockpit and those that occur in the maintenance hangar. Dupont's "dirty dozen" are a comprehensive list of reasons and ways that humans make mistakes (Dupont, 1997):

1. Lack of communication
2. Complacency
3. Lack of knowledge
4. Distraction
5. Lack of teamwork
6. Fatigue
7. Lack of resources
8. Pressure
9. Lack of assertiveness
10. Stress
11. Lack of awareness
12. Norms

Whereas Dupont considered the human factor in the error-making process, Dr. James Reason looked at the systems approach to human error management. In "Human Errors: Models and Management," an article that appeared in *British Medical Journal* (2000), Reason stated:

> High technology systems have many defensive layers: some are engineered (alarms, physical barriers, automatic shutdowns, etc.), others rely on people (surgeons, anaesthetists, pilots, control room operators, etc.), and yet others depend on procedures and administrative controls. Their function is to protect potential victims and assets from local hazards. Mostly they do this very effectively, but there are always weaknesses.

He points out that each layer of defense is more like a slice of Swiss cheese than a solid barrier. The presence of a hole in one defensive layer does not create a bad outcome event, but when the holes in all of the levels of defense align, a bad or catastrophic outcome becomes more likely **FIGURE 19-1**.

Active Failures and Latent Conditions

Reason cites two reasons explaining why holes appear in the layers of defense: active failures and latent conditions.

<u>Active failures</u> are the unsafe acts committed by people who are in direct contact with the situation or system. They have direct and short-lived effects on the integrity of the defenses. Not wearing a seat belt while in a moving vehicle is an example of an active failure.

<u>Latent conditions</u> are the inevitable "resident pathogens" within the system. These conditions have two kinds of

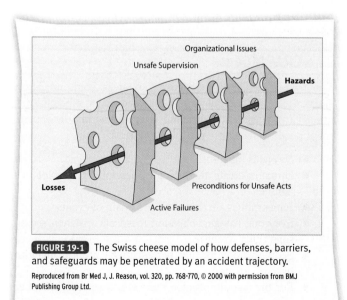

FIGURE 19-1 The Swiss cheese model of how defenses, barriers, and safeguards may be penetrated by an accident trajectory.

Reproduced from Br Med J, J. Reason, vol. 320, pp. 768-770, © 2000 with permission from BMJ Publishing Group Ltd.

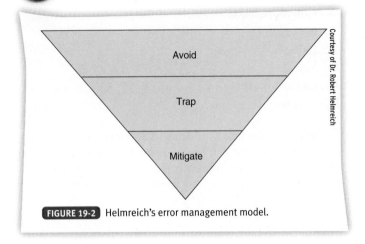

FIGURE 19-2 Helmreich's error management model.

adverse effects. First, they can translate into error-provoking conditions within the local workplace. Examples include time pressure, understaffing, inadequate equipment, fatigue, and inexperience. Second, they can create long-lasting holes or weaknesses in the defenses. Examples include untrustworthy alarms, unworkable procedures, and design and construction deficiencies.

Latent conditions may lie dormant within the system for many years before they combine with active failures and local triggers to create an accident opportunity. The ancient planes used for wildland firefighting that were grounded by Dr. Kern in 2002 provide an example of latent conditions.

■ Error Management Model

CRM is an error management model that incorporates three activities: avoidance, entrapment, and mitigating consequences. Error avoidance provides the greatest opportunity for trapping and preventing errors from becoming a catastrophe. Errors that are not avoided are trapped at the second level. Errors that slip through the first two levels require mitigation. Mitigation is the action taken by emergency responders to minimize the effect of an emergency on a community **FIGURE 19-2**.

The CRM Model

The fire service CRM model covers six areas: communication skills, teamwork, task allocation, critical decision making, situational awareness, and debriefing. Everyone within the department needs to recognize the following facts:

- *No one* is infallible.
- Humans create technology; therefore, technology is fallible.
- Catastrophes are the result of a chain of events.
- Everyone has an obligation to speak up when seeing something wrong.
- People who work together effectively are less likely to have accidents.
- For the team to become more effective, every member of the team must participate.

■ Communication Skills

Communication is the successful transfer and understanding of a thought from one person to another. The *Fire Officer Communications* chapter provides detailed information about how to communicate successfully.

Review of airline disaster cockpit recordings shows that flight crews typically engaged in several forms of miscommunication that contributed to catastrophic events. These problems have included misinterpretations of take-off, altitude, and landing instructions; airlines that fostered and condoned "fighter pilot" mentalities in their captains; a lack of assertiveness on the part of crew members who recognized problems before the captain; and distractions in the cockpit that inhibited focused communication, such as idle chitchat during pre-flight checks.

Developing a standard language and teaching appropriate assertive behavior are the keys to reducing errors resulting from miscommunication. In the aviation application, CRM advocates maintaining a "sterile cockpit"—that is, a cockpit environment where all communication and focus are devoted to flight operations.

When a fire apparatus is en route to an alarm, the crew members are sitting in an enclosed box, surrounded by switches, gauges, wheels, pedals, and noise and moving down the road. For these machines to move efficiently and safely, the people inside them need to focus on the mission and communicate effectively. The entire crew should be exchanging only information that is pertinent to responding to and arriving safely at the scene of the alarm.

The CRM-enriched environment generates a climate where the freedom to question is encouraged. Members are encouraged to speak up respectfully when they see something that causes concern, by using clear, concise questions and observations. A discrepancy between what is going on and what should be occurring is often the first indication of an error.

The freedom for all members to question something that appears unusual or is not understood is a delicate subject for many organizations. Members need to speak directly in a manner that does not challenge the authority of a superior.

Inquiry and advocacy are discrete, learnable skills that promote synergy between the mechanical element and the human players in a scenario. Inquiry is the process of questioning a situation that causes concern. Advocacy is the statement of opinion that recommends what the person believes is the proper course of action under a specific set of circumstances. Using the two skills effectively requires practice and patience on the part of all crew members. The key factor to keep in mind is that communication should not focus on *who* is right, but rather on *what* is right.

One effective tactic is to use specific buzzwords, such as "red light" and "red flag," to signal discomfort with a situation. These terms are cues that open the door to inquiry and advocacy. They should be reserved for situations that involve an immediate risk of injury.

Assertive Statement Process

Todd Bishop, from the Error Prevention Institute, teaches a five-step assertive statement process that encompasses the inquiry and advocacy communications steps. The assertive statement runs like this:

1. *Use an opening/attention getter.* Address the other individual: "Hey, Chief," or "Bill," or whatever appropriate moniker is used to get the individual's attention.
2. *State your concern.* Use an owned emotion: "That smoke is really pushing from those windows. I have a bad feeling about it."
3. *State the problem as you see it.* "It looks like it is going to flash. I think any crews entering that place are going to take a beating."
4. *State a solution.* "Why don't we vent the roof and give the place a minute or two to vent before we send in the attack team?"
5. *Obtain agreement or buy-in.* "Does that sound good to you?"

The inquiry and advocacy process and the assertive statement are essential components of the communication segment of CRM. They are the toughest lessons to impart and adopt because they often require a wholesale change in interpersonal dynamics. Once mastered, however, inquiry and advocacy can enhance performance, prevent mishaps, and save lives.

Effective listening, covered in the *Fire Officer Communications* chapter, is an important CRM communication skill. One effective listening technique is to purposely refrain from making any response or counterargument until the other individual has drained his or her emotional bubble.

■ Teamwork

CRM promotes members working together for the common good. Achieving this level of cohesiveness requires developing effective teams through buy-in by all members, leaders, and followers, in the effort to be efficient and safe. Leaders will always be in charge, but they must also be open to suggestion and constructive criticism. The Incident Command System, covered in the *Managing Incidents* chapter, provides the formal structure for CRM at the task, tactical, and strategic levels.

Leadership

Fire officers formally exercise leadership of the team through a combination of rank and authority. For officers to become truly effective leaders, they must earn the trust and respect of their subordinates and demonstrate the skills of effective leadership. These three components comprise the triangle of leadership **FIGURE 19-3**.

The informal authority to lead is derived through respect. True respect is based on three competencies: personal, technical, and social. Personal competence refers to an individual's own internal strengths, capabilities, and character. Technical competence refers to an individual's ability to perform tasks that require specific knowledge or skills. Social competence refers to the person's ability to interact effectively with other people. All three competencies are essential for leaders to function effectively in a CRM environment.

The fact that a leader must demonstrate effective social skills does not mean that a leader must always be deferential.

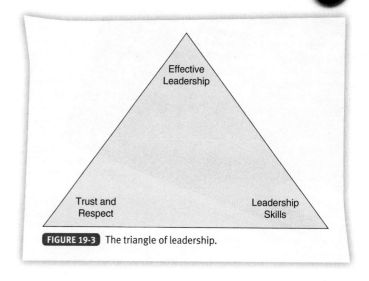

FIGURE 19-3 The triangle of leadership.

Social competence requires knowing how to speak to people respectfully, whether praising or chastising them.

Mentoring

Mentors pay special attention to helping others develop their skills. The *Training and Coaching* chapter provides detailed information on training, coaching, and mentoring. CRM provides an effective vehicle for leaders to impart knowledge and skills to their subordinates.

In particular, leading by example is a highly effective technique. Crews notice a supervisor's personal habits, pick up on his or her thought processes, and constantly evaluate the leader's performance. Traits that are admired and respected by fire fighters have a major influence on their future behavior. The crew usually sees more than the supervisor may think. Leading by example requires constant effort on the part of the leader, with no opportunities to turn the positive role model on or off at different times.

Human error is a normal, inevitable occurrence. A leader must be willing to admit to making a mistake. Superior leaders readily admit when they have made mistakes, accept responsibility, and focus their attention on moving forward. This trait creates an environment that fosters open communication, promotes safety, and encourages subordinates' belief in the team.

Mentoring also requires sharing knowledge. Knowledge is often closely associated with power. Leaders who are insecure about their positions tend to withhold knowledge and fail to share lessons as they are learned. Mistakes are often repeated in such an environment, and errors that could have been avoided are permitted to occur. A fire officer must maintain technical competence, stay on top of technological advances, and ensure that others have the information that will keep them from being killed or injured.

Technical competency requirements vary from position to position. The street-level fire fighter does not expect the fire chief to pull hose and throw ladders, but the chief must be able to command a fire, mass-casualty incident, or weapons of mass destruction event with the same degree of skill that the fire fighter is expected to use to pull hose and throw ladders.

VOICES OF EXPERIENCE

As we worked our way through summer afternoon traffic in Engine 66, sirens blaring, my apparatus operator, Doug, interrupted my concentrated study of the building pre-plan. "Wow, that's quite a column of smoke," he said. Quickly looking up, I saw a large, black roiling cloud pushing into the clear sky a half mile distant. We were responding to a popular fast food restaurant, and it was apparent the fire had gained significant ground in just a few short minutes.

Pulling into the lot, with Ladder 67 close behind, we were approached by the store manager, who yelled into my open window that he wasn't sure if everyone was out of the building. Heavy fire poured from an open rear kitchen doorway and from a gambrel-style roof with an attached tile-roofed facade. After providing a size-up for incoming units, I passed command to the arriving battalion chief and my crew of four assumed the role of fire attack.

> **Seconds later, the entire overhang fell to the ground in a large cloud of dust and debris.**

Dragging a line from the engine, we hit the fire hard through the open doorway, then directed the stream into the roof overhang, darkening the fire. As we prepared for entry, I could hear loud "pops" as clay tiles slid off the roof onto the pavement. Suddenly, I felt a hand on my shoulder and looked over to see Fire Fighter Mark from Truck 67, a seasoned veteran. "What's the plan?" he asked. "We're going to make entry, Engine 51's pulling a back-up line, and you can search," I said. Mark took a quick look overhead, leaned into my ear and said, "I think we should back out." I briefly struggled to reconcile my feelings. I was the officer, I had given an order, and we seemed to be making progress. I turned and asked Mark "Why's that?"

"I'm concerned about that roof; it looks like its peeling back," he said. Bowing to his experience, and wanting to take a second look at the roof, I reached forward and grabbed the coat of my lead fire fighter, and we pulled back. Seconds later, the entire overhang fell to the ground in a large cloud of dust and debris, completely obscuring the place where two crews had just been preparing for entry.

Astonished, my crew and I immediately credited Mark with recognizing something none of us had seen. The roof was made of real ceramic tile, not a lightweight composite, and Mark had seen from underneath that it appeared to be simply nailed to the façade. I'd learned a powerful lesson as a new lieutenant. Conflicting views present you with a chance to be *curious* instead of *defensive*.

From that moment forward, I realized that curiosity is a key component of good leadership. If you remain open to input, and respond to queries with curiosity, you will demonstrate to your team that you are interested in their input. This will allow your team to sustain a shared understanding of the risks, resources, and plans during dynamic incidents, and build your crew resource management skills.

Paul LeSage
Assistant Chief
Tualatin Valley Fire & Rescue
Aloha, Oregon

Handling Conflict

The focal point of CRM in conflict resolution is to focus on *what* is right, not *who* is right. This approach allows the leader to focus on the best, safest outcome. The *Handling Problems, Conflicts, and Mistakes* chapter provides detailed information and recommended skills for handling problems and conflicts.

The fire officer needs to establish an open climate for error prevention. The leader who perceives a subordinate's comment or question as a threat to his or her own authority is part of the problem instead of an essential component of the solution. Leaders are required to keep their own egos in check. A leader who rules by intimidation can often win an argument, but the forces of nature always prevail over words. That is, an officer can order a crew to advance into a dangerous position, but the building will not follow an order to remain standing. An officer who fails to listen or refuses to listen is simply dangerous.

Responsibility

Skeptics suggest that CRM advocates "management by committee," in which leadership gives way to consensus. In reality, that is not an accurate statement. Leaders need to keep their eyes and ears open, especially for input from subordinates. For decision making to be efficient, someone must be in charge—that is, someone must have ultimate responsibility for decisions and outcome.

In the cockpit, the captain retains ultimate authority for decision making and ultimate responsibility for getting the plane from airport to airport. Subordinates are encouraged to provide input, but the final decision rests with the recognized authority. Lines of authority are maintained and that ultimate authority rests with the legitimate ranking officer.

Mission analysis requires fire service leaders to look at all situations with a risk-versus-gain mentality. Brunacini's mantra—risk a lot to save a savable life, take a calculated risk to save savable property, and risk nothing to save what is already lost—establishes a concrete foundation on which leaders can manage emergency operations.

Followership

<u>Followership</u> encompasses the appropriate actions of those who are led. All followers should perform a self-assessment of their ability to function as part of a team. The self-assessment should consider four critical areas:

1. *Physical condition.* People in good physical condition are more aware, alert, and oriented to their surroundings. Maintaining good physical health is critical to the success of any endeavor; however, it is absolutely critical in the fire service.
2. *Mental condition.* We are constantly pulled in a variety of directions; however, the challenging and dangerous environments where fire fighters must perform demand their full attention. Fire fighters must ask themselves, "Am I free of distractions that could divert my attention from the task at hand?"
3. *Attitude.* To be an effective team member, a fire fighter must be willing to follow orders and be part of a cohesive team.
4. *Understanding human behavior.* The effectiveness of CRM is based on understanding human behavior and interpersonal dynamics in a team environment. To be effective using CRM, the team members must understand how their individual behavior relates to one another's actions and to the team as a whole.

To be effective team players and maximize CRM benefits, each individual must have the following characteristics:

- A healthy appreciation for personal safety
- A healthy concern for the safety of the crew
- A respect for authority
- A willingness to accept orders
- A knowledge of the limits of authority
- A desire to help their leader be successful
- Good communication skills
- The ability to provide constructive, pertinent feedback
- The ability to admit errors
- The ability to keep one's ego in check
- The ability to balance assertiveness and authority
- A learning attitude
- The ability to perform demanding tasks
- Adaptability

These qualities of a good follower are remarkably similar to those required of effective leaders. Effective followership enhances leadership, which in turn promotes effective teams. Teams that practice CRM make fewer mistakes and are able to recognize and correct errors before they cause tragic outcomes.

■ Task Allocation

Task allocation refers to dividing responsibilities among individuals and teams in a manner that allows for their effective accomplishment. Task overload occurs when the fire officer exceeds his or her capacity to manage all of the simultaneous functions and responsibilities. Safety is compromised with task overload.

Think about the individual who is driving down the interstate at rush hour, texting with a colleague. He looks up and realizes that traffic has stopped. His smart phone falls to the floor as he slams on the brakes, too late to avoid a collision. Task overload has occurred—with disastrous results.

Knowing one's own limits and the capacity of the team is the first step in the CRM task allocation phase. Everyone has a point at which outside stimuli override the ability to process information and perform effectively. If resources are not available to carry out all tasks at once, then tasks must be prioritized to identify those that can be done safely and effectively.

Fire officers fall into three categories when it comes to multitasking ability. Some officers are reluctant to admit that they are ever overwhelmed and believe that they become more effective as the situation becomes more hectic. A second group of officers becomes overwhelmed before the full complexity of the event is even recognized. The third group of leaders effectively assesses the incident, calls for additional resources early, and manages to stay ahead of the incident and balance the span of control.

NASA noted that pilots believed they were capably handling multiple tasks, even when mistakes began to appear and performance deteriorated. *The Multi-tasking Myth: Handling Complexity in Real World Operations* (2009) reported that the pilots were so busy that they failed to recognize when the situation was getting out of control.

Each fire officer has to evaluate his or her personal capacity to manage complicated situations and identify weak spots. Training and practice can improve performance, but everyone still has limitations. When these are known, fire officers can take action early to compensate for them.

The fire officer must also know the crew's limits. In some fire departments, the same crew works together as a team for long periods, allowing the fire officer to evaluate strengths and weaknesses and focus on team development. In contrast, this understanding can be difficult to develop when the make-up of a crew changes from day to day or the crew is assembled at the time of alarm.

The strengths of individuals and crews should be evaluated and enhanced in multiple nonemergency settings. Performance is enhanced through training classes, live training exercises, table-top modeling, didactic presentations by experts, mentoring, and exchanging FIGURE 19-4.

FIGURE 19-4 Training exercises help the officer determine the strengths of individuals.

Critical Decision Making

Although CRM promotes the concept of team involvement in all aspects of an operation, emergency scenes often demand rapid decision making by the crew leader. Input is welcome from all team members; however, the final responsibility for decision making resides with the crew leader, and experience and training often play pivotal roles in successful outcomes.

Gary Klein, a cognitive psychologist who studies decision making under pressure, found that fire officers and military combat officers rely on past experiences when developing plans of action. Urban fire commanders told Klein that they made fire-ground decisions based on previous experiences, rather than referring to traditional decision-making models. The commanders reported that they often did not even consider the options available; instead, they simply looked at the incident, recalled previous experiences of a similar nature, and applied the processes that had been used to mitigate the previous incidents successfully. Many of the fire officers interviewed by Klein had tremendous urban firefighting experience during the 1970s and 1980s.

Klein observed that the fire-ground and combat commanders must often make decisions in environments that are time compressed and dynamic, involve missing or ambiguous data and multiple players, and lack real-time feedback. His research identified two decision-making models: recognition-primed decision making (RPD), which describes how commanders can recognize a plausible plan of action, and naturalistic decision making, which describes how commanders make decisions in their natural environment.

Decision making is improved through gaining experience, training constantly, improving communication skills, and engaging in preincident planning. Practicing enhanced communication skills and taking advantage of improved crew interaction at all levels provide the following benefits:

- Problem identification is enhanced.
- Supervisors at all command levels maintain better incident control.
- Situational awareness is improved.
- Hazards are more rapidly identified.
- Resource capability is rapidly assessed.
- Potential solutions are more rapidly developed.
- Decision making is improved.
- Surprises and unanticipated problems are reduced.

Situational Awareness

Situational awareness is the accurate perception of what is going on around you. NASA has expanded this definition to "the awareness of acknowledging and assessing; [situational awareness] is the basis for choosing courses of action of the situation both now and in the future."

Situational awareness affects performance and decision making. When such awareness is maintained, operations are completed without a hitch. When it is not maintained, errors occur, performance suffers, and catastrophic events often result. Situational awareness is a human factors element that is somewhat difficult to explain but is all too easy to recognize after an error occurs. It also carries over into nonincident operations. Failing to note at shift change that the SCBA cylinders are down to 1500 psi, for example, is a situational awareness failure that can have unfortunate consequences.

One of the common human behavior factors that leads to a loss of situational awareness is the tendency to ignore or disregard information that is out of context. If an incident commander is busy thinking about an attack strategy, a brief observation or comment about unusual smoke conditions can be easily brushed aside and forgotten. A critical radio message might be acknowledged without really being understood or placed in context in the situation at hand.

The fire officer needs to constantly check and cross-check the situation and operational performance. This constant review helps ensure that the individual and the entire team are focused on concluding the incident in an efficient and injury-free fashion.

Maintaining Emergency Scene Situational Awareness

Six steps should be followed to maintain emergency scene situational awareness:

1. *Fight the fire.* All members of the crew must focus their attention on the details of the incident, while keeping the larger picture in mind. Crash investigations show that air crews have sometimes become distracted by events that took their attention away from flying the plane. In some cases, these distractions resulted in CFIT (controlled flight into terrain) events.
2. *Assess problems in the time available.* Emergency scenes do not always permit leisurely, time-unconstrained periods for decision making. Nevertheless, there has to be a balance between rushing headlong into a burning building and waiting until every possible hazard is evaluated before attacking an incipient fire. Seasoned incident commanders know that taking an extra 30 to 60 seconds to absorb and process as much information as possible often results in better and more confident decisions.
3. *Gather information from all sources.* Information gathering at the emergency scene, by necessity, has to be a rapid process. Of course, one person cannot see everything, know everything, hear everything, or smell everything. Using the rest of the crew as information resources, as well as additional arriving command officers and people familiar with the building or terrain, should enhance decision making and help keep situational awareness current.
4. *Choose the best option.* Once all of the factors have been weighed, choose the option that maximizes results and minimizes risk.
5. *Monitor results and alter the plan as necessary.* Maintaining situational awareness requires continual evaluation of the effectiveness of decisions. Having a "plan B" or "plan C" ready for implementation may be necessary. An extra set of eyes at a strategic location on the incident scene can often provide valuable information.
6. *Beware of situational awareness loss factors.* The following loss factors must be taken into account:
   - *Ambiguity:* An event, order, or message that has more than one potential meaning.
   - *Distraction:* Anything that takes attention away from the larger mission. Distraction can come in the form of outside influences (the slamming of a door down the hall while you are watching television) or inside influences (cultural and personal biases).
   - *Fixation:* Tunnel vision.
   - *Overload:* More things are happening than one person can process. For the fire officer, the first few minutes after arrival at an exceptionally chaotic scene can lead to overload.
   - *Complacency:* When humans perform the same process repeatedly, they tend to become bored due to the hypnotic effect of repetitive action. Inattentiveness breeds carelessness. The perception or description of any structure fire as routine is the pinnacle of complacency.
   - *Improper procedure:* Advancing a handline into a structure fire before ventilating and positioning apparatus downwind at a hazardous materials release are examples of improper procedures. Immediate action is required to correct these mistakes.
   - *Unresolved discrepancy:* Suppose you are heading westbound on an interstate and you see a vehicle approaching eastbound in the same lane you are traveling in. There is a brief period where your eyes send the signal to the brain, but you mutter, "I don't believe this."

Fire Marks

CRM Contributes to the "Miracle on the Hudson"
Hitting a flock of geese on takeoff from New York's LaGuardia Airport completely shut down both engines on U.S. Airways Flight 1549's Airbus A320 in 2009. Jerry Mulenberg, writing in NASA's *ASK Magazine*, describes the sequence of events:

> With more than 19,000 hours of flight time, including flying gliders, Captain Chesley (Sully) Sullenberger's experience and training helped prepare him for the once-in-a-lifetime decision he faced. In less than 3 minutes, Captain Sullenberger:
> - Requested permission to return to LaGuardia—approved (normal procedure)
> - Performed engine restart procedure—engines didn't restart (multiple attempts would have resulted in the same outcome and time was short)
> - Asked for alternate airport-landing location in New Jersey—granted (wanted to avoid densely populated area near LaGuardia)
> - Made final decision to land in the Hudson River—successful in saving all onboard (co-pilot provided airspeed and altitude readings for pilot to glide aircraft in)

During a talk about his experiences, Captain Sullenberger mentioned that he owed a lot to the CRM training he went through (Mulenberg, 2011).

Safety Zone

Mental Joggers

A solid strategy for improving situational awareness requires the use of mental joggers for the individual and crew. The crew mental joggers ask:

- "What do we have here?"
- "What's going on here?"
- "How are we doing?"
- "Does this look right?"

The personal mental joggers ask:

- "What do I know that they need to know?"
- "What do they know that I need to know?"
- "What do we all need to know?"

Indications and observations that are in conflict with expectations must not be ignored. Fire fighters must be trained to recognize unresolved discrepancies as needing attention immediately.

- *Nobody fighting the fire:* An incident commander may find his or her attention drawn to the actions of one specific company, causing the leader to ignore the rest of the incident. Emergency scene commanders cannot allow themselves to be drawn away from the big picture.

An aviation practice that can help to maintain situational awareness is to say the checklist out loud when preparing for a low-frequency/high-risk task. Memorizing a checklist and going through it by yourself is not as safe or as effective as two people looking at a checklist and speaking the items out loud. This practice has also been used to reduce paramedic medication errors.

Near-Miss REPORT

Report Number: 11-0000330

Event Description: We were on our second air bottle and had just left rehabilitation and relieved another company operating inside the structure. We proceeded to a bedroom where the fire damage was significant, in search of hotspots. One crew member ascended a scuttle-hole ladder in an adjacent bedroom to better assess the attic space, while the rest of us began accessing the ceiling with pike poles. I had momentarily walked to the hall doorway to acquire a PAR when I heard a tremendous noise behind me.

Turning around, I saw a large pile of debris consisting of heavy roof material in the center of the room where we were operating. One fire fighter was against an interior wall, with no sign of the second crew member. The fire fighter against the wall cried out that the ceiling had fallen on a team member. A mayday call was immediately transmitted, and command initiated RIT procedures. We instantly began to dig through the rubble in search of the downed fire fighter. The fire fighter on the ladder informed me he could not acquire a visual on the downed fire fighter, and raised the possibility that the crew member might have fallen into the basement. I radioed this information to command and continued our search.

Surrounding fire-ground operations continued as we located the trapped fire fighter underneath heavy debris. He was "on air" and conscious, but was unable to free himself throughout the ordeal. Extrication ensued, and he was taken to an awaiting ALS unit where, after assessment, he was found to have only minor injuries. Department procedure then dictated that he be taken to an appropriate medical facility for observation. The fire fighter was treated and released later that day.

Lessons Learned: Anytime a company is operating on the interior of a building, regardless of the phase of firefighting operations, structural integrity is always a consideration. Advanced burn depths, the presence of gusset plates, and other factors should sound internal alarms, not only for the company officer, but also for all team members operating on the interior. In my opinion, training as a team and keeping our team together on the fire ground was paramount in the quick and successful extrication of the trapped fire fighter. Having procedures in place and training extensively for mayday-type situations and RIT scenarios gave us confidence in our actions when the need arose to save one of our own.

Courtesy of the National Fire Fighter Near-Miss Reporting Systems (firefighternearmiss.com)

Fire Officer I

Recommending Change

The fire officer is in the best position to lead change within the fire department. Working at the point where the services are delivered, the fire officer has a vivid view of opportunities, challenges, and barriers. The CRM assertive statement process provides a model for discussing sensitive or high-consequence issues.

In discussing the challenge of implementing the National Fallen Firefighters Foundation's Life Safety Initiatives, Bill Manning (n.d.) describes five "disconnects":

1. "Culture change" is viewed by some as a threat.
2. Bad (unsafe) behaviors and attitudes are allowed to leach into what the membership see as part of "tradition."
3. Safety and mission within organizational cultures are imbalanced.
4. The voices (and actions) of safety leadership are either subconsciously muffled or consciously subdued.
5. The lessons from behavioral safety science have not been embraced by fire service leaders, much less blended into everyday operations.

The fire officer who wants to recommend changes to existing policies or create a new policy should start with the results of a post-incident debriefing. Debriefings should capitalize on the positives, identify shortcomings, discuss strategies for improvement, and allow participants an opportunity to see and hear the incident through different sets of eyes and ears. Okray and Lubnau recommend a five-step model:

Step 1: Just the facts.
Step 2: What did you do?
Step 3: What went wrong?
Step 4: What went right?
Step 5: What are you going to do about it?

Actions taken by the department to correct performance and mechanical issues identified during such a debriefing are a measure of organizational effectiveness. The *Fire Officer Communications* and *Handling Problems, Conflicts, and Mistakes* chapters provide many concepts and tools for the fire officer who wants to change an existing policy or implement a new one.

Fire Officer II

Implementing Change

The fire officer has the best opportunity to dramatically improve the safety and effectiveness of fire operations by implementing changes in policies and procedures. The CRM approach concentrates on the conditions under which individuals work and tries to build defenses to avert errors or mitigate their effects. High-reliability organizations, which have less than their fair share of accidents, recognize that human variability is a force to harness in averting errors, but they work hard to focus that variability and are constantly preoccupied with the possibility of failure.

Firefighting shares some features in common with other high-reliability organizations, such as nuclear-powered aircraft carriers, air traffic control towers, nuclear power plants, and invasive medicine. Specifically, these organizations share these characteristics:

- Are complex, internally dynamic, and intermittently intensely interactive
- Perform exacting tasks under considerable time pressure
- Carry out these demanding activities with low incident rates and an almost complete absence of catastrophic failures over several years

High-reliability organizations that have embraced CRM have seen reductions of 70 to 89 percent in their rates of incidents, accidents, and injuries (LeSage, 2009). Implementing the error management model provides specific practices in six areas: communication skills, teamwork, task allocation, critical decision making, situational awareness, and debriefing. The formal structure, tradition, and practices of the fire service provide a foundation for any fire officer to make significant improvements in fire fighter safety and incident effectiveness.

The chapters *Leading the Fire Company*, *Working in the Community*, and *Budgeting* provide concepts, practices, and tips when the captain wants to develop an effective change in a department's policy or procedure.

You Are the Fire Officer Conclusion

Lieutenant Williams is faced with a significant life-threatening situation with very few resources. Clearly describing the situation, calling for more resources, and starting task allocations will begin to impose order on the chaos. Quint 100's primary task is to obtain the information needed for situational awareness and to avoid task overload.

The crew can immediately address some life-saving issues by assuring that the fire doors are closed and the sprinkler system is operating at the hospital. Fire Fighter James Grynski suggests using the standpipe system and high-rise hose packs to control the fire spread in the hospital.

With the delay of the first alarm, Lieutenant Williams needs to think what the situation will be 20 minutes from now and make tactical decisions based on those assumptions.

Wrap-Up

Chief Concepts

- Crew resource management (CRM) is a behavioral approach to reducing human error in high-risk or high-consequence activities.
- Leading change describes the process of recommending changes (as a lieutenant) and developing a process to establish a change (as a captain).
- CRM was developed as a behavioral modification training system in a 1979 National Aeronautics and Space Administration (NASA) workshop examining the role of human error in aviation accidents.
- The aviation industry has reduced its accident rate by 80 percent through the development, refinement, and system-wide delivery of CRM.
- Gordon Dupont determined that a "dirty dozen" of human factors contribute to tragedy:
  - Lack of communication
  - Complacency
  - Lack of knowledge
  - Distraction
  - Lack of teamwork
  - Fatigue
  - Lack of resources
  - Pressure
  - Lack of assertiveness
  - Stress
  - Lack of awareness
  - Norms
- James Reason identified two precursors to holes appearing in the layers of defense: active failures and latent conditions.
- CRM is an error management model that includes three activities: avoidance, entrapment, and mitigating consequences. Error avoidance provides the greatest opportunity for trapping errors and preventing them from evolving into full-blown catastrophes.
- A six-point CRM model serves the fire service well:
  - Communication skills
  - Teamwork
  - Task allocation
  - Critical decision making
  - Situational awareness
  - Debriefing
- CRM suggests that developing a standard language, maintaining a "sterile cockpit," and teaching appropriate assertive behavior are the keys to reducing errors resulting from miscommunication.
- CRM promotes the idea that members must work together for the common good. This requires developing

- effective teams through buy-in of all members, leaders, and followers, in the effort to be efficient and safe.
- Task overload occurs when the fire officer exceeds his or her capacity to manage the various simultaneous functions and responsibilities.
- CRM promotes the concept of team involvement in all aspects of operation. In the area of emergency scene decision making, time, experience, and training play pivotal roles in successful outcomes.
- The loss of situational awareness is frequently the first link in a chain of errors that leads to calamity.
- The fire officer is in the best position to lead change within the fire department.
- The CRM approach concentrates on the conditions under which individuals work and tries to build defenses to avert errors or mitigate their effects.

Hot Terms

Active failures Unsafe acts committed by people who are in direct contact with the situation or system.

Crew resource management (CRM) A behavioral modification training system developed by the aviation industry to reduce its accident rate.

Followership The act or condition of following a leader; adherence.

Latent conditions Inevitable "resident pathogens" within the system.

Naturalistic decision making The process by which commanders make decisions in their natural environment.

Recognition-primed decision making (RPD) The process by which commanders can recognize a plausible plan of action.

Situational awareness The process of evaluating the severity and consequences of an incident and communicating the results.

References and Additional Resources

NFPA reprinted material is not the complete and official position of the NFPA on the referenced subject, which is represented only by the standard in its entirety.

Brunacini, A. V. (2004). *Command Safety: The IC's Role in Protecting Firefighters*. Stillwater, OK: Fire Protection Publications.

Caulfield, H. J., and D. Benzaia. (1985). *Winning the Fire Service Leadership Game*. New York, NY: Fire Engineering.

Dupont, Gordon, "The Dirty Dozen Errors in Maintenance," 11th Symposium on Human Factors in Aviation Maintenance. (1997) San Diego, CA. Reprinted by permission of the author.

Gasaway, R. B. (2013). *Situational Awareness for Emergency Response*. Tulsa, OK: Pennwell/Fire Engineering.

Helmreich, R.L., A. C. Merritt, and J. A. Wilhelm. (1999). "The Evolution of Crew Resource Management Training in Commercial Aviation." *International Journal of Aviation Psychology*, 9(1), 19–32.

International Association of Fire Chiefs. (2002). *Crew Resource Management: A Positive Change for the Fire Service*, 3rd edition. Fairfax, VA: International Association of Fire Chiefs.

Kanki, B. G., R. L. Helmreich, and J. Anca (eds.). (2010). *Crew Resource Management*, 2nd edition. London, UK: Academic Press.

Kern, A. T. (2001). *Controlling Pilot Error: Culture, Environment and CRM*. New York, NY: McGraw-Hill Professional.

Klein, G. (2009). *Streetlights and Shadows: Searching for the Keys to Adaptive Decision Making*. Cambridge, MA: MIT Press.

LeSage, P., J. T. Dyar, and B. Evans. (2009). *Crew Resource Management: Principles and Practice*. Sudbury, MA: Jones and Bartlett.

Loukopoulos, L. D., R. K. Dismukes, and I. Barshi. (2009). *The Multi-tasking Myth: Handling Complexity in Real World Operations*. Ashgate Studies in Human Factors for Flight Operations. Burlington, VT: Ashgate.

Manning, B. (n.d.). "Creating the 'New' Fire Service Safety Culture: A Perspective, Part 1." *Everybody Goes Home*. Emmitsburg, MD: National Fallen Firefighters Foundation. Accessed September 16, 2013, at http://www.everyonegoeshome.com/partners/fsculture_p1.html.

Marx, D. (2009). *Wack-a-Mole: The Price We Pay for Expecting Perfection*. Plano, TX: By Your Side Studios.

McAniff, E. P., and J. J. Cunningham, (1974). *Leadership in the Fire Service*. Bayside, NY, McAniff Associates, Inc.

Miller, K. (2010). *We Don't Make Widgets: Overcoming Myths That Keep Government from Radically Improving*. Washington, DC: Governing Books.

Mulenberg, J. (2011, March 11). "Crew Resource Management Improves Decision Making." *ASK Magazine*, 42. Washington, DC: U. S. National Aeronautics and Space Administration. http://appel.nasa.gov/2011/05/11/crew-resource-management-improves-decision-making/.

Myers, B. (2004). *Mentoring: Preparing Company Officers*. Emmitsburg, MD, Executive Fire Officer Program.

National Fallen Firefighters Foundation. (2011). *Understanding and Implementing the 16 Fire-Fighter Life Safety Initiatives*. Stillwater, OK: Fire Protection Publications.

National Institute of Standards and Technology. (2013). *Report on High-Rise Fireground Field Experiments: Technical Note 1797*. Washington, DC: U.S. Department of Commerce.

National Institute of Standards and Technology. (2010). *Report on Residential Fireground Field Experiments: Technical Note 1661*. Washington, DC: U.S. Department of Commerce.

Nurnberg, E. J. (2013) *Gaining Initiative: Rapid Tactical Decision Making in the Iowa City Fire Department*. Emmitsburg, MD: U.S. Fire Academy, Executive Fire Officer Program.

Okray, R. and Lubnau T. (2004). *Crew Resource Management for the Fire Service*. Tulsa, OK: Pennwell/Fire Engineering.

Page, J. O. (1973). *Effective Company Command*. Alhambra, CA: Borden Publishing.

Pfeifer, J. W. (2012). Adapting to Novelty: Recognizing the Need for Innovation and Leadership. In: *WNYF: With New York Firefighters*. New York, NY: Fire Department City of New York. Volume 1/2012. Page 20.

Reason, J. (2000, March 18). "Human error: Models and Management." *British Medical Journal*, 320(7237): 768–770.

Reinhart, R. (2007). *Basic Flight Physiology* (3rd ed). New York, NY: McGraw Hill Professional.

Wrap-Up, continued

Ross, K. G., et al. (2004, July–August). "The Recognition-Primed Decision Model." *Military Review*, pp. 6–10.

Safety Regulation Group. (2002). *An Introduction to Aircraft Maintenance Engineering Human Factors for JAR 66*. Norwick, UK: Civil Aviation Authority.

Smith, G. (2012). *Identifying the Knowledge and Skills Needed for Successful Critical Thinking and Decision Making on the Fire Ground*. Emmitsburg, MD: U.S. Fire Academy, Executive Fire Officer Program.

Snook, J. W., et al. (2011). *A Leadership Guide for Volunteer Fire Departments*. Sudbury, MA: Jones and Bartlett/IAFC.

Sproule, C. B. (2012). *When Good People Make Bad Decisions: Assessing Decision Fatigue in Las Vegas Fire & Rescue*. Emmitsburg, MD: U.S. Fire Academy, Executive Fire Officer Program.

Yukl, G. A. (2013). *Leadership in Organizations*, 8th edition. New York, NY: Prentice-Hall.

FIRE OFFICER in action

Battalion Chief Johnson arrived and organized the incident response into four divisions: Hospital, High-rise, Tanker, and Underground. Quint 100's crew has been reassigned to the Hospital Division and is completing the initial search and rescue.

Crew resource management provides practices and procedures that allow a crew to operate effectively in a hostile and dangerous environment.

1. There are pockets of fire within the first floor of the hospital. Fire Fighter Ed Schultz says, "I am concerned that there is a lot of gasoline still in the basement. We should take a gas monitoring meter." This is an example of:
   A. taskmanship.
   B. an assertive statement.
   C. insubordination.
   D. a quality circle.

2. Error management involves all of the following activities except:
   A. entrapment.
   B. avoidance.
   C. establishing expectations.
   D. mitigating consequences.

3. Not wearing a seat belt in a moving vehicle is an example of:
   A. active failure.
   B. incomplete task allocation.
   C. failed followership.
   D. a latent condition.

4. What is the difference between CRM and traditional fire supervision?
   A. CRM's requirement for committees and consensus
   B. Effective listening
   C. Increased authority of the informal leader
   D. Application of error management to human behavior

FIRE OFFICER *in action* (continued)

Fire Captain *Activity*

Once the tanker-into-hospital event was mitigated, Chief Johnson asked Captain Davis to walk around the incident scene to assess the damage and review the incident. Johnson praised the initial actions taken by Lieutenant Williams.

The chief asked the captain what went wrong with the operation. Davis mentioned the difficulties in moving patients away from the fire area, especially on the second floor. "Our preincident plan was for handling a fire in a single patient room, not fire impinging an entire wing!"

Johnson told Davis, "Write down a description of your experience, and propose a policy for a mass evacuation of a hospital or nursing home."

NFPA Fire Officer II Job Performance Requirement 5.4.6
Develop a plan to accomplish change in the organization, given an agency's change of policy or procedures, so that effective change is implemented in a positive manner.

Application of 5.4.6
1. Using a department with which you are familiar, develop a policy for a mass evacuation of a hospital, jail, or nursing home.
2. Using a department with which you are familiar, describe the process of implementing a change in fireground procedures.

An Extract from: NFPA 1021, Standard for Fire Officer Professional Qualifications, 2014 Edition

Chapter 4 Fire Officer I

4.1* General. For qualification at Fire Officer Level I, the candidate shall meet the requirements of Fire Fighter II as defined in NFPA 1001, Fire Instructor I as defined in NFPA 1041, and the job performance requirements defined in Sections 4.2 through 4.7 of this standard.

4.1.1* General Prerequisite Knowledge. The organizational structure of the department; geographical configuration and characteristics of response districts; departmental operating procedures for administration, emergency operations, incident management system and safety; fundamentals of leadership; departmental budget process; information management and recordkeeping; the fire prevention and building safety codes and ordinances applicable to the jurisdiction; current trends, technologies, and socioeconomic and political factors that affect the fire service; cultural diversity; methods used by supervisors to obtain cooperation within a group of subordinates; the rights of management and members; agreements in force between the organization and members; generally accepted ethical practices, including a professional code of ethics; and policies and procedures regarding the operation of the department as they involve supervisors and members.

4.1.2 General Prerequisite Skills. The ability to effectively communicate in writing utilizing technology provided by the AHJ; write reports, letters, and memos utilizing word processing and spreadsheet programs; operate in an information management system; and effectively operate at all levels in the incident management system utilized by the AHJ.

4.2 Human Resource Management. This duty involves utilizing human resources to accomplish assignments in accordance with safety plans and in an efficient manner. This duty also involves evaluating member performance and supervising personnel during emergency and nonemergency work periods, according to the following job performance requirements.

4.2.1 Assign tasks or responsibilities to unit members, given an assignment at an emergency incident, so that the instructions are complete, clear, and concise; safety considerations are addressed; and the desired outcomes are conveyed.

(A) Requisite Knowledge. Verbal communications during emergency incidents, techniques used to make assignments under stressful situations, and methods of confirming understanding.

(B) Requisite Skills. The ability to condense instructions for frequently assigned unit tasks based on training and standard operating procedures.

4.2.2 Assign tasks or responsibilities to unit members, given an assignment under nonemergency conditions at a station or other work location, so that the instructions are complete, clear, and concise; safety considerations are addressed; and the desired outcomes are conveyed.

(A) Requisite Knowledge. Verbal communications under nonemergency situations, techniques used to make assignments under routine situations, and methods of confirming understanding.

(B) Requisite Skills. The ability to issue instructions for frequently assigned unit tasks based on department policy.

4.2.3 Direct unit members during a training evolution, given a company training evolution and training policies and procedures, so that the evolution is performed in accordance with safety plans, efficiently, and as directed.

(A) Requisite Knowledge. Verbal communication techniques to facilitate learning.

(B) Requisite Skills. The ability to distribute issue-guided directions to unit members during training evolutions.

4.2.4 Recommend action for member-related problems, given a member with a situation requiring assistance and the member assistance policies and procedures, so that the situation is identified and the actions taken are within the established policies and procedures.

(A)* Requisite Knowledge. The signs and symptoms of member-related problems, causes of stress in emergency services personnel, adverse effects of stress on the performance of emergency service personnel, and awareness of AHJ member assistance policies and procedures.

(B) Requisite Skills. The ability to recommend a course of action for a member in need of assistance.

4.2.5* Apply human resource policies and procedures, given an administrative situation requiring action, so that policies and procedures are followed.

(A) Requisite Knowledge. Human resource policies and procedures.

(B) Requisite Skills. The ability to communicate orally and in writing and to relate interpersonally.

4.2.6 Coordinate the completion of assigned tasks and projects by members, given a list of projects and tasks and the job requirements of subordinates, so that the assignments are prioritized, a plan for the completion of each assignment is developed, and members are assigned to specific tasks and both supervised during and held accountable for the completion of the assignments.

(A) Requisite Knowledge. Principles of supervision and basic human resource management.

(B) Requisite Skills. The ability to plan and to set priorities.

4.3 Community and Government Relations. This duty involves dealing with inquiries of the community and communicating the role, image, and mission of the department to the public and delivering safety, injury, and fire prevention education programs, according to the following job performance requirements.

4.3.1 Initiate action on a community need, given policies and procedures, so that the need is addressed.

(A) **Requisite Knowledge.** Community demographics and service organizations, as well as verbal and nonverbal communication, and an understanding of the role and mission of the department.

(B) **Requisite Skills.** Familiarity with public relations and the ability to communicate verbally.

4.3.2 Initiate action to a citizen's concern, given policies and procedures, so that the concern is answered or referred to the correct individual for action and all policies and procedures are complied with.

(A) **Requisite Knowledge.** Interpersonal relationships and verbal and nonverbal communication.

(B) **Requisite Skills.** Familiarity with public relations and the ability to communicate verbally.

4.3.3 Respond to a public inquiry, given policies and procedures, so that the inquiry is answered accurately, courteously, and in accordance with applicable policies and procedures.

(A) **Requisite Knowledge.** Written and oral communication techniques.

(B) **Requisite Skills.** The ability to relate interpersonally and to respond to public inquiries.

4.4 Administration. This duty involves general administrative functions and the implementation of departmental policies and procedures at the unit level, according to the following job performance requirements.

4.4.1 Recommend changes to existing departmental policies and/or implement a new departmental policy at the unit level, given a new departmental policy, so that the policy is communicated to and understood by unit members.

(A) **Requisite Knowledge.** Written and oral communication.

(B) **Requisite Skills.** The ability to relate interpersonally and to communicate change in a positive manner.

4.4.2 Execute routine unit-level administrative functions, given forms and record-management systems, so that the reports and logs are complete and files are maintained in accordance with policies and procedures.

(A) **Requisite Knowledge.** Administrative policies and procedures and records management.

(B) **Requisite Skills.** The ability to communicate orally and in writing.

4.4.3 Prepare a budget request, given a need and budget forms, so that the request is in the proper format and is supported with data.

(A) **Requisite Knowledge.** Policies and procedures and the revenue sources and budget process.

(B) **Requisite Skill.** The ability to communicate in writing.

4.4.4 Explain the purpose of each management component of the organization, given an organization chart, so that the explanation is current and accurate and clearly identifies the purpose and mission of the organization.

(A) **Requisite Knowledge.** Organizational structure of the department and functions of management.

(B) **Requisite Skills.** The ability to communicate verbally in a clear and concise manner.

4.4.5 Explain the needs and benefits of collecting incident response data, given the goals and mission of the organization, so that incident response reports are timely and accurate.

(A) **Requisite Knowledge.** The agency's records management system.

(B) **Requisite Skills.** The ability to communicate both orally and in writing.

4.5* Inspection and Investigation. This duty involves conducting inspections to identify hazards and address violations, performing a fire investigation to determine preliminary cause, securing the incident scene, and preserving evidence, according to the following job performance requirements.

4.5.1 Describe the procedures of the AHJ for conducting fire inspections, given any of the following occupancies, so that all hazards, including hazardous materials, are identified, approved forms are completed, and approved action is initiated:

(1) Assembly
(2) Educational
(3) Health care
(4) Detention and correctional
(5) Residential
(6) Mercantile
(7) Business
(8) Industrial
(9) Storage
(10) Unusual structures
(11) Mixed occupancies

(A) **Requisite Knowledge.** Inspection procedures; fire detection, alarm, and protection systems; identification of fire and life safety hazards; and marking and identification systems for hazardous materials.

(B) **Requisite Skills.** The ability to communicate in writing and to apply the appropriate codes and standards.

4.5.2 Identify construction, alarm, detection, and suppression features that contribute to or prevent the spread of fire, heat, and smoke throughout the building or from one building to another, given an occupancy, and the policies and forms of the AHJ so that a pre-incident plan for any of the following occupancies is developed:

(1) Public assembly
(2) Educational
(3) Institutional
(4) Residential
(5) Business
(6) Industrial
(7) Manufacturing
(8) Storage
(9) Mercantile
(10) Special properties

(A) **Requisite Knowledge.** Fire behavior; building construction; inspection and incident reports; detection, alarm, and suppression systems; and applicable codes, ordinances, and standards.

(B) **Requisite Skills.** The ability to use evaluative methods and to communicate orally and in writing.

4.5.3 Secure an incident scene, given rope or barrier tape, so that unauthorized persons can recognize the perimeters of the scene and are kept from restricted areas, and all evidence or potential evidence is protected from damage or destruction.

(A) **Requisite Knowledge.** Types of evidence, the importance of fire scene security, and evidence preservation.

(B) **Requisite Skills.** The ability to establish perimeters at an incident scene.

4.6* Emergency Service Delivery. This duty involves supervising emergency operations, conducting pre-incident planning, and deploying assigned resources in accordance with the local emergency plan and according to the following job performance requirements.

4.6.1 Develop an initial action plan, given size-up information for an incident and assigned emergency response resources, so that resources are deployed to control the emergency.

(A)* Requisite Knowledge. Elements of a size-up, standard operating procedures for emergency operations, and fire behavior.

(B)* Requisite Skills. The ability to analyze emergency scene conditions; to activate the local emergency plan, including localized evacuation procedures; to allocate resources; and to communicate orally.

4.6.2* Implement an action plan at an emergency operation, given assigned resources, type of incident, and a preliminary plan, so that resources are deployed to mitigate the situation.

(A) Requisite Knowledge. Standard operating procedures, resources available for the mitigation of fire and other emergency incidents, an incident management system, scene safety, and a personnel accountability system.

(B) Requisite Skills. The ability to implement an incident management system, to communicate orally, to manage scene safety, and to supervise and account for assigned personnel under emergency conditions.

4.6.3 Develop and conduct a post-incident analysis, given a single unit incident and post-incident analysis policies, procedures, and forms, so that all required critical elements are identified and communicated, and the approved forms are completed and processed in accordance with policies and procedures.

(A) Requisite Knowledge. Elements of a post-incident analysis, basic building construction, basic fire protection systems and features, basic water supply, basic fuel loading, fire growth and development, and departmental procedures relating to dispatch response tactics and operations and customer service.

(B) Requisite Skills. The ability to write reports, to communicate orally, and to evaluate skills.

4.7* Health and Safety. This duty involves integrating health and safety plans, policies, and procedures into daily activities as well as the emergency scene, including the donning of appropriate levels of personal protective equipment to ensure a work environment that is in accordance with health and safety plans for all assigned members, according to the following job performance requirements.

4.7.1 Apply safety regulations at the unit level, given safety policies and procedures, so that required reports are completed, in-service training is conducted, and member responsibilities are conveyed.

(A) Requisite Knowledge. The most common causes of personal injury and accident to members, safety policies and procedures, basic workplace safety, and the components of an infectious disease control program.

(B) Requisite Skills. The ability to identify safety hazards and to communicate orally and in writing.

4.7.2 Conduct an initial accident investigation, given an incident and investigation forms, so that the incident is documented and reports are processed in accordance with policies and procedures of the AHJ.

(A) Requisite Knowledge. Procedures for conducting an accident investigation and safety policies and procedures.

(B) Requisite Skills. The ability to communicate orally and in writing and to conduct interviews.

4.7.3 Explain the benefits of being physically and medically capable of performing assigned duties and effectively functioning during peak physical demand activities, given current fire service trends and agency policies, so that the need to participate in wellness and fitness programs is explained to members.

(A) Requisite Knowledge. National death and injury statistics, fire service safety and wellness initiatives, and agency policies.

(B) Requisite Skills. The ability to communicate orally.

■ Chapter 5 Fire Officer II

5.1 General. For qualification at Level II, the Fire Officer I shall meet the job performance requirements defined in Sections 5.2 through 5.7 of this standard.

5.1.1* General Prerequisite Knowledge. The organization of local government; enabling and regulatory legislation and the law-making process at the local, state/provincial, and federal levels; and the functions of other bureaus, divisions, agencies, and organizations and their roles and responsibilities that relate to the fire service.

5.1.2 General Prerequisite Skills. Intergovernmental and interagency cooperation.

5.2 Human Resource Management. This duty involves evaluating member performance, according to the following job performance requirements.

5.2.1 Initiate actions to maximize member performance and/or to correct unacceptable performance, given human resource policies and procedures, so that member and/or unit performance improves or the issue is referred to the next level of supervision.

(A) Requisite Knowledge. Human resource policies and procedures, problem identification, organizational behavior, group dynamics, leadership styles, types of power, and interpersonal dynamics.

(B) Requisite Skills. The ability to communicate orally and in writing, to solve problems, to increase teamwork, and to counsel members.

5.2.2 Evaluate the job performance of assigned members, given personnel records and evaluation forms, so that each member's performance is evaluated accurately and reported according to human resource policies and procedures.

(A) Requisite Knowledge. Human resource policies and procedures, job descriptions, objectives of a member evaluation program, and common errors in evaluating.

(B) Requisite Skills. The ability to communicate orally and in writing and to plan and conduct evaluations.

5.2.3 Create a professional development plan for a member of the organization, given the requirements for promotion, so that the individual acquires the necessary knowledge, skills, and abilities to be eligible for the examination for the position.

(A) Required Knowledge. Development of a professional development guide and job shadowing.

(B) Required Skills. The ability to communicate orally and in writing.

5.3 Community and Government Relations. This duty involves dealing with inquiries of allied organizations in the community and projecting the role, mission, and image of the department to other organizations with similar goals and missions for the purpose of establishing strategic partnerships and delivering safety, injury, and fire prevention education programs, according to the following job performance requirements.

5.3.1 Explain the benefits to the organization of cooperating with allied organizations, given a specific problem or issue in the community, so that the purpose for establishing external agency relationships is clearly explained.

(A) Requisite Knowledge. Agency mission and goals and the types and functions of external agencies in the community.

(B) Requisite Skills. The ability to develop interpersonal relationships and to communicate orally and in writing.

5.4 Administration. This duty involves preparing a project or divisional budget, news releases, and policy changes, according to the following job performance requirements.

5.4.1 Develop a policy or procedure, given an assignment, so that the recommended policy or procedure identifies the problem and proposes a solution.

(A) **Requisite Knowledge.** Policies and procedures and problem identification.

(B) **Requisite Skills.** The ability to communicate in writing and to solve problems.

5.4.2 Develop a project or divisional budget, given schedules and guidelines concerning its preparation, so that capital, operating, and personnel costs are determined and justified.

(A) **Requisite Knowledge.** The supplies and equipment necessary for ongoing or new projects; repairs to existing facilities; new equipment, apparatus maintenance, and personnel costs; and appropriate budgeting system.

(B) **Requisite Skill.** The ability to allocate finances, to relate interpersonally, and to communicate orally and in writing.

5.4.3 Describe the process of purchasing, including soliciting and awarding bids, given established specifications, in order to ensure competitive bidding so that the needs of the organization are met within the applicable federal, state/provincial, and local laws and regulations.

(A) **Requisite Knowledge.** Purchasing laws, policies, and procedures.

(B) **Requisite Skills.** The ability to use evaluative methods and to communicate orally and in writing.

5.4.4 Prepare a news release, given an event or topic, so that the information is accurate and formatted correctly.

(A) **Requisite Knowledge.** Policies and procedures and the format used for news releases.

(B) **Requisite Skills.** The ability to communicate orally and in writing.

5.4.5 Prepare a concise report for transmittal to a supervisor, given fire department record(s) and a specific request for details such as trends, variances, or other related topics, so that the information required for the AHJ is accurate and documented.

(A) **Requisite Knowledge.** The data processing system.

(B) **Requisite Skills.** The ability to communicate in writing and to interpret data.

5.4.6 Develop a plan to accomplish change in the organization, given an agency's change of policy or procedures, so that effective change is implemented in a positive manner.

(A) **Requisite Knowledge.** Planning and implementing change.

(B) **Requisite Skills.** The ability to clearly communicate orally and in writing.

5.5 Inspection and Investigation. This duty involves conducting fire investigations to determine origin and preliminary cause, according to the following job performance requirements.

5.5.1 Determine the point of origin and preliminary cause of a fire, given a fire scene, photographs, diagrams, pertinent data, and/or sketches, to determine if arson is suspected so that law enforcement action is taken.

(A) **Requisite Knowledge.** Methods used by arsonists, common causes of fire, basic origin and cause determination, fire growth and development, and documentation of preliminary fire investigative procedures.

(B) **Requisite Skills.** The ability to communicate orally and in writing and to apply knowledge using deductive skills.

5.6 Emergency Service Delivery. This duty involves supervising multi-unit emergency operations, conducting pre-incident planning, and deploying assigned resources, according to the following job requirements.

5.6.1 Produce operational plans, given an emergency incident requiring multi-unit operations, the current edition of *NFPA 1600*, and AHJ-approved safety procedures, so that required resources and their assignments are obtained and plans are carried out in compliance with *NFPA 1600* and approved safety procedures resulting in the mitigation of the incident.

(A) **Requisite Knowledge.** Standard operating procedures; national, state/provincial, and local information resources available for the mitigation of emergency incidents; an incident management system; and a personnel accountability system.

(B) **Requisite Skills.** The ability to implement an incident management system, to communicate orally, to supervise and account for assigned personnel under emergency conditions, and to serve in command staff and unit supervision positions within the Incident Management System.

5.6.2 Develop and conduct a post-incident analysis, given multi-unit incident and post-incident analysis policies, procedures, and forms, so that all required critical elements are identified and communicated and the approved forms are completed and processed.

(A) **Requisite Knowledge.** Elements of a post-incident analysis, basic building construction, basic fire protection systems and features, basic water supply, basic fuel loading, fire growth and development, and departmental procedures relating to dispatch response, strategy tactics and operations, and customer service.

(B) **Requisite Skills.** The ability to write reports, to communicate orally, and to evaluate skills.

5.6.3 Prepare a written report, given incident reporting data from the jurisdiction, so that the major causes for service demands are identified for various planning areas within the service area of the organization.

(A) **Requisite Knowledge.** Analyzing data.

(B) **Requisite Skills.** The ability to write clearly and to interpret response data correctly to identify the reasons for service demands.

5.7 Health and Safety. This duty involves reviewing injury, accident, and health exposure reports, identifying unsafe work environments or behaviors, and taking approved action to prevent reoccurrence, according to the following job requirements.

5.7.1 Analyze a member's accident, injury, or health exposure history, given a case study, so that a report including action taken and recommendations made is prepared for a supervisor.

(A) **Requisite Knowledge.** The causes of unsafe acts, health exposures, or conditions that result in accidents, injuries, occupational illnesses, or deaths.

(B) **Requisite Skills.** The ability to communicate in writing and to interpret accidents, injuries, occupational illnesses, or death reports.

ProBoard Assessment Methodology Matrices for NFPA 1021, 2014 Edition

NFPA 1021 - Fire Officer I - 2014 Edition

INSTRUCTIONS: In the column titled 'Cognitive/Written Test' place the number of questions from the Test Bank that are used to evaluate the applicable JPR, RK, RS, or objective. In the column titled 'Manipulative/Skill Station' identify the skill sheets that are used to evaluate the applicable JPR, RS, or objective. When the Portfolio or Project method is used to evaluate a particular JPR, RK, RS, or objective, identify the applicable section in the appropriate column and provide the procedures to be used as outlined in the NBFSPQ Operational Procedures, COA-5. Evaluation methods that are not cognitive, manipulative, portfolio, or project based should be identified in the 'Other' column.

OBJECTIVE / JPR, RK, RS		COGNITIVE	MANIPULATIVE			
SECTION	ABBREVIATED TEXT	WRITTEN TEST	SKILLS STATION	PORTFOLIO	PROJECTS	OTHER
4.1	Meet the requirements of Fire Fighter II & Fire Instructor I					Chapter 1 (p 5)
4.2.1	Assign tasks or responsibilities					Chapter 4 (pp 66–68), Chapter 5 (pp 86–99), Chapter 7 (pp 129–130, 132–134), Chapter 16 (pp 327–328), Chapter 17 (pp 342–344)
4.2.1(A)	RK: Verbal communications during emergency situations					Chapter 4 (pp 66–68), Chapter 5 (pp 92–94), Chapter 7 (pp 130, 132–134), Chapter 16 (pp 327–328), Chapter 17 (pp 342–344)
4.2.1(B)	RS: Condense instructions for frequently assigned unit tasks					Chapter 4 (pp 66–68), Chapter 5 (pp 86, 89), Chapter 7 (pp 129–130, 132–134), Chapter 16 (pp 327–328), Chapter 17 (pp 342–344)
4.2.2	Assign tasks or responsibilities					Chapter 3 (pp 42–43), Chapter 4 (pp 62–63, 65–66), Chapter 7 (pp 129–130)
4.2.2(A)	RK: Verbal communications under nonemergency situations					Chapter 3 (pp 42–43), Chapter 4 (pp 62–63, 65–66), Chapter 7 (pp 129–130)
4.2.2(B)	RS: Instructions for frequently assigned unit tasks					Chapter 3 (pp 42–43), Chapter 4 (pp 62–63, 65–66), Chapter 7 (pp 129–130)
4.2.3	Direct unit members during a training evolution					Chapter 5 (p 95), Chapter 8 (pp 148, 150–151)
4.2.3(A)	RK: Verbal communication techniques					Chapter 5 (p 95), Chapter 8 (pp 148, 150–151)
4.2.3(B)	RS: Distribute issue-guided directions					Chapter 5 (p 95), Chapter 8 (pp 148, 150–151)
4.2.4	Recommend action for member-related problems					Chapter 9 (pp 166–169)
4.2.4(A)	RK: Signs and symptoms of member-related problems					Chapter 9 (pp 166–169)
4.2.4(B)	RS: Recommend a course of action for a member in need of assistance					Chapter 9 (pp 166–169)

Appendix B

OBJECTIVE / JPR, RK, RS		COGNITIVE	MANIPULATIVE			
SECTION	ABBREVIATED TEXT	WRITTEN TEST	SKILLS STATION	PORTFOLIO	PROJECTS	OTHER
4.2.5	Apply human resource policies and procedures					Chapter 3 (pp 54–55), Chapter 6 (pp 118–120), Chapter 9 (pp 166–169), Chapter 10 (pp 185–187, 189–190)
4.2.5(A)	RK: Human resource policies and procedures					Chapter 3 (pp 54–55), Chapter 6 (pp 118–120), Chapter 9 (pp 166–169), Chapter 10 (pp 185–187, 189–190)
4.2.5(B)	RS: Communicate verbally and in writing					Chapter 3 (pp 54–55), Chapter 6 (pp 118–120), Chapter 9 (pp 166–169), Chapter 10 (pp 185–187, 189–190)
4.2.6	Coordinate the completion of assigned tasks					Chapter 6 (pp 113–118, 120–121), Chapter 7 (pp 129–130)
4.2.6(A)	RK: Principles of supervision					Chapter 6 (pp 113–118, 120–121), Chapter 7 (pp 129–130)
4.2.6(B)	RS: Plan and to set priorities					Chapter 6 (pp 120–121), Chapter 7 (pp 129–130)
4.3.1	Initiate action on a community need					Chapter 11 (pp 207–213)
4.3.1(A)	RK: Community demographics					Chapter 11 (pp 204–206)
4.3.1(B)	RS: Familiarity with public relations					Chapter 11 (pp 207–213)
4.3.2	Initiate action to a citizen's concern					Chapter 12 (p 235)
4.3.2(A)	RK: Interpersonal relationships					Chapter 12 (p 235)
4.3.2(B)	RS: Familiarity with public relations					Chapter 12 (p 235)
4.3.3	Respond to a public inquiry					Chapter 11 (pp 207–212), Chapter 12 (p 235)
4.3.3(A)	RK: Written and verbal communication techniques					Chapter 11 (pp 207–212), Chapter 12 (p 235)
4.3.3(B)	RS: Relate interpersonally and to respond to public inquiries					Chapter 11 (pp 207–212), Chapter 12 (p 235)
4.4.1	Recommend changes to a departmental policy					Chapter 12 (pp 234–235), Chapter 19 (p 395)
4.4.1(A)	RK: Written and oral communication					Chapter 12 (pp 234–235), Chapter 19 (p 395)
4.4.1(B)	RS: Relate interpersonally					Chapter 12 (pp 234–235), Chapter 19 (p 395)
4.4.2	Execute routine unit-level administrative functions					Chapter 3 (pp 42–43)
4.4.2(A)	RK: Administrative policies and procedures and records management					Chapter 3 (pp 42–43)
4.4.2(B)	RS: Communicate verbally and in writing					Chapter 3 (pp 42–43)
4.4.3	Prepare a budget request					Chapter 14 (pp 272–273, 276–277, 279)
4.4.3(A)	RK: Policy and procedures					Chapter 14 (pp 272–273, 276–277, 279)
4.4.3(B)	RS: Communicate orally and in writing					Chapter 14 (pp 272–273, 276–277, 279)

OBJECTIVE / JPR, RK, RS		COGNITIVE	MANIPULATIVE			
SECTION	ABBREVIATED TEXT	WRITTEN TEST	SKILLS STATION	PORTFOLIO	PROJECTS	OTHER
4.4.4	Explain the purpose of each management component of the organization					Chapter 7 (pp 130, 132–134), Chapter 19 (p 395)
4.4.4(A)	RK: Organizational structure					Chapter 19 (p 395)
4.4.4(B)	RS: Ability to communicate verbally					Chapter 7 (pp 130, 132–134), Chapter 19 (p 395)
4.4.5	Explain the needs and benefits of collecting incident response data					Chapter 4 (pp 68–70)
4.4.5(A)	RK: Agency's records management system.					Chapter 4 (pp 68–70)
4.4.5(B)	RS: Ability to communicate					Chapter 4 (pp 66–68)
4.5.1	Describe the procedures of the AHJ for conducting fire inspections					Chapter 13 (pp 252–256, 258–264)
4.5.1(A)	RK: Inspection procedures					Chapter 13 (pp 252–256, 258–264)
4.5.1(B)	RS: Ability to communicate					Chapter 13 (pp 252–256, 258–264)
4.5.2	Identify construction, alarm, detection, and suppression features					Chapter 13 (pp 249–256)
4.5.2(A)	RK: Fire behavior; building construction; inspection and incident reports					Chapter 13 (pp 249–256)
4.5.2(B)	RS: Ability to use evaluative methods					Chapter 13 (pp 249–256)
4.5.3	Secure an incident scene					Chapter 18 (pp 366-368)
4.5.3(A)	RK: Types of evidence					Chapter 18 (pp 366-368)
4.5.3(B)	RS: Ability to establish perimeters					Chapter 18 (p 367)
4.6.1	Develop an initial action plan					Chapter 7 (pp 132–134), Chapter 15 (pp 293–294), Chapter 16 (pp 320–327, 330), Chapter 17 (pp 348–349)
4.6.1(A)	RK: Elements of a size-up					Chapter 15 (pp 293–294), Chapter 16 (p 320), Chapter 17 (pp 348–349)
4.6.1(B)	RS: Ability to analyze emergency scene conditions					Chapter 7 (pp 132–134), Chapter 15 (pp 293–294), Chapter 16 (pp 320–327, 330), Chapter 17 (pp 348–349)
4.6.2	Implement an action plan					Chapter 15 (pp 293–294, 296–297), Chapter 17 (pp 348–349)
4.6.2(A)	RK: Standard operating procedures					Chapter 17 (pp 348–349)
4.6.2(B)	RS: Ability to implement an incident management system					Chapter 15 (pp 293–294, 296–297)
4.6.3	Develop and conduct a post-incident analysis					Chapter 5 (p 99), Chapter 15 (pp 294–296)
4.6.3(A)	RK: Elements of a post-incident analysis					Chapter 5 (p 99), Chapter 15 (pp 294–296)
4.6.3(B)	RS: Ability to write reports					Chapter 5 (p 99), Chapter 15 (pp 294–296)
4.7.1	Apply safety regulations at the unit level					Chapter 5 (pp 95–99)

SECTION	OBJECTIVE / JPR, RK, RS ABBREVIATED TEXT	COGNITIVE WRITTEN TEST	MANIPULATIVE SKILLS STATION	PORTFOLIO	PROJECTS	OTHER
4.7.1(A)	RK: Most common causes of personal injury and accident to members					Chapter 5 (pp 87–89)
4.7.1(B)	RS: Ability to identify safety hazards					Chapter 5 (pp 95–99)
4.7.2	Conduct an initial accident investigation					Chapter 5 (p 99)
4.7.2(A)	RK: Procedures for conducting an accident investigation					Chapter 5 (p 99)
4.7.2(B)	RS: Ability to communicate					Chapter 5 (p 99)
4.7.3	Explain the benefits of being physically and medically capable					Chapter 5 (pp 86–99)
4.7.3(A)	RK: National death and injury statistics					Chapter 5 (pp 86–99)
4.7.3(B)	RS: Ability to communicate orally					Chapter 5 (pp 86–99)

NFPA 1021 - Fire Officer II - 2014 Edition

INSTRUCTIONS: In the column titled 'Cognitive/Written Test' place the number of questions from the Test Bank that are used to evaluate the applicable JPR, RK, RS, or objective. In the column titled 'Manipulative/Skill Station' identify the skill sheets that are used to evaluate the applicable JPR, RS, or objective. When the Portfolio or Project method is used to evaluate a particular JPR, RK, RS, or objective, identify the applicable section in the appropriate column and provide the procedures to be used as outlined in the NBFSPQ Operational Procedures, COA-5. Evaluation methods that are not cognitive, manipulative, portfolio, or project based should be identified in the 'Other' column.

SECTION	OBJECTIVE / JPR, RK, RS ABBREVIATED TEXT	COGNITIVE WRITTEN TEST	MANIPULATIVE SKILLS STATION	PORTFOLIO	PROJECTS	OTHER
5.1	Requirements of Fire Instructor I					Chapter 1 (p 16)
5.2.1	Initiate actions to maximize member performance					Chapter 6 (pp 121–122), Chapter 7 (pp 135–136), Chapter 9 (pp 171–177)
5.2.1(A)	RK: Human resource policies and procedures					Chapter 9 (pp 171–177)
5.2.1(B)	RS: Communicate orally and in writing					Chapter 7 (pp 135–136), Chapter 9 (pp 171–177)
5.2.2	Evaluate the job performance of assigned members					Chapter 9 (pp 171–174)
5.2.2(A)	RK: Human resource policies and procedures					Chapter 9 (pp 171–174)
5.2.2(B)	RS: Communicate orally and in writing					Chapter 9 (pp 171–174)
5.2.3	Create a professional development plan					Chapter 8 (pp 155–156)
5.2.3(A)	RK: Development of a professional development guide					Chapter 8 (pp 155–156)
5.2.3(B)	RS: Ability to communicate					Chapter 8 (pp 155–156)
5.3.1	Explain the benefits to the organization of cooperating					Chapter 11 (pp 213–217), Chapter 13 (pp 264–265), Chapter 14 (pp 280–283)
5.3.1(A)	RK: Agency mission and goals					Chapter 11 (pp 213–217), Chapter 13 (pp 264–265), Chapter 14 (pp 280–283)

OBJECTIVE / JPR, RK, RS		COGNITIVE	MANIPULATIVE			
SECTION	ABBREVIATED TEXT	WRITTEN TEST	SKILLS STATION	PORTFOLIO	PROJECTS	OTHER
5.3.1(B)	RS: Ability to develop interpersonal relationships					Chapter 11 (pp 213–217), Chapter 13 (pp 264–265), Chapter 14 (pp 280–283)
5.4.1	Develop a policy or procedure					Chapter 9 (pp 172–173)
5.4.1(A)	RK: Policies and procedures and problem identification					Chapter 9 (pp 172–173)
5.4.1(B)	RS: Communicate in writing and to solve problems					Chapter 9 (pp 172–173)
5.4.2	Prepare a project or divisional budget request					Chapter 14 (pp 274–275, 280–286)
5.4.2(A)	RK: Supplies and equipment necessary for ongoing or new projects					Chapter 14 (pp 274–275, 280–286)
5.4.2(B)	RS: Allocate finances					Chapter 14 (pp 274–275, 280–286)
5.4.3	Process of purchasing					Chapter 14 (pp 276–277, 279)
5.4.3(A)	RK: Purchasing laws, policies and procedures					Chapter 14 (pp 276–277, 279)
5.4.3(B)	RS: Ability to use evaluative methods and to communicate					Chapter 14 (pp 276–277, 279)
5.4.4	Prepare a news release					Chapter 4 (p 76), Chapter 11 (pp 215–216)
5.4.4(A)	RK: Policies and procedures and the format used for news releases					Chapter 4 (p 76), Chapter 11 (pp 215–216)
5.4.4(B)	RS: Communicate orally and in writing					Chapter 4 (p 76), Chapter 11 (pp 215–216)
5.4.5	Prepare a concise report for transmittal to a supervisor					Chapter 4 (p 75)
5.4.5(A)	RK: Data processing system					Chapter 4 (p 75)
5.4.5(B)	RS: Communicate in writing and to interpret data					Chapter 4 (p 75)
5.4.6	Develop a plan to accomplish change					Chapter 19 (p 395)
5.4.6(A)	RK: Planning and implementing change					Chapter 19 (p 395)
5.4.6(B)	RS: Ability to clearly communicate					Chapter 19 (p 395)
5.5.1	Determine the point of origin and preliminary cause of a fire					Chapter 18 (pp 368–375)
5.5.1(A)	RK: Methods used by arsonists					Chapter 18 (pp 375, 377–378)
5.5.1(B)	RS: Communicate orally and in writing and to apply knowledge using deductive skills					Chapter 18 (pp 368–375, 377–380)
5.6.1	Produce operational plans					Chapter 13 (pp 264–265), Chapter 15 (pp 293–294, 306–309), Chapter 17 (pp 352–354, 356)
5.6.1(A)	RK: Standard operating procedures					Chapter 13 (pp 264–265), Chapter 15 (pp 293–294, 306–309), Chapter 17 (p 356)
5.6.1(B)	RS: Implement an incident management system					Chapter 13 (pp 264–265), Chapter 15 (pp 293–294, 306–309), Chapter 17 (pp 352–354, 356)

OBJECTIVE / JPR, RK, RS		COGNITIVE	MANIPULATIVE			
SECTION	ABBREVIATED TEXT	WRITTEN TEST	SKILLS STATION	PORTFOLIO	PROJECTS	OTHER
5.6.2	Post-incident analysis					Chapter 15 (pp 294–296)
5.6.2(A)	RK: Elements of a post-incident analysis					Chapter 15 (pp 294–296)
5.6.2(B)	RS: Write reports					Chapter 15 (pp 294–296)
5.6.3	Prepare a written report					Chapter 4 (pp 73, 75)
5.6.3(A)	RK: Analyzing data					Chapter 4 (pp 73, 75)
5.6.3(B)	RS: Ability to write clearly and to interpret response data					Chapter 4 (pp 73, 75)
5.7.1	Analyze a member's accident, injury, or health exposure history					Chapter 5 (pp 104–105)
5.7.1(A)	RK: Causes of unsafe acts					Chapter 5 (pp 104–105)
5.7.1(B)	RS: Communicate in writing and to interpret accidents, injuries, occupational illnesses, or death reports					Chapter 5 (pp 104–105)

Principles of Fire and Emergency Service Administration (FESHE) Correlation Guide

Principles of Fire and Emergency Service Administration (FESHE) Course Outcomes	Corresponding Chapter(s)	Corresponding Page(s)
1. Acknowledge career development opportunities and strategies for success.	1, 2, 3, 8, 11	10, 26, 29–33, 35–36, 47–49, 152–156, 205–206, 212–217
2. Recognize the need for effective communication skills, both written and verbal.	4, 7, 8, 11, 12, 13, 14, 16, 18	62–63, 65–66, 68–73, 75, 132–134, 143–146, 150–152, 206–207, 209, 213–217, 230, 232–236, 243–244, 247, 256, 260, 265, 277, 279, 283–286, 326, 328–331, 370, 374–375, 378–380
3. Identify and explain the concepts of span and control, effective delegation, and division of labor.	1, 6, 7, 15	11–13, 112–113, 117–118, 120–122, 130–134, 295–299, 302–304, 306–309
4. Select and implement the appropriate disciplinary action based on an employee's conduct.	5, 6, 9, 10, 12, 19	88, 95, 99, 103–105, 114–115, 117–118, 122, 166–169, 171–177, 186–187, 189–190, 224–225, 230, 232–236, 387–389
5. Explain the history of management and supervision methods and procedures.	1, 5, 7, 10, 19	7–10, 113–120, 135–136, 184–185, 190–197, 386–388, 392–393
6. Discuss the various levels of leadership, roles, and responsibilities within the organization.	1, 3, 4, 5, 6, 7, 8, 9, 11, 13, 14, 15, 16, 17, 18, 19	11–14, 16, 46–47, 49–50, 68–71, 73, 75, 89–92, 94, 105, 114–118, 120–121, 128–130, 132–134, 143–148, 150–152, 164–165, 169, 171–177, 206–207, 213–217, 242–247, 251–252, 258–260, 264–265, 272–273, 275–276, 296–301, 306–309, 320–324, 326–330, 342–344, 347–350, 352–354, 368–374, 379–380, 388–389, 391–395
7. Describe the traits of effective versus ineffective management styles.	1, 3, 5, 6, 7, 9, 10, 11, 12, 14, 18	49–50, 53–55, 113–115, 117–118, 128–130, 132–133, 135–136, 165–168, 173–174, 177, 185–186, 191–197, 224–229, 232–236
8. Identify the importance of ethics as it relates to fire and emergency services.	1, 3, 5, 6, 7, 9, 10, 11, 12, 14, 18	14, 16, 50–51, 99, 103–105, 114, 122, 134–135, 167–168, 173–174, 177, 189–190, 195–197, 206, 213, 232, 235–236, 275–276, 379–380
9. Identify the roles of the National Incident Management System (NIMS) and Incident Management System (ICS).	4, 5, 15, 16, 17, 18, 19	66–68, 89–90, 92–94, 292–294, 296–299, 302–304, 306–309, 320–322, 324, 326, 331–333, 344–345, 350, 352, 356, 366–368, 391–392

Glossary

Accelerant An agent—often an ignitable liquid—used to initiate a fire or increase the rate of growth or spread of fire.

Accident An unplanned event that interrupts an activity and sometimes causes injury or damage; a chance occurrence arising from unknown causes; an unexpected happening due to carelessness, ignorance, and the like.

Accreditation A system whereby a certification organization determines that a school or program meets with the requirements of the fire service.

Actionable items Employee behavior that requires an immediate corrective action by the supervisor; dozens of lawsuits have shown that failing to act in the face of such behavior will create a liability and a loss for the department.

Active failures Unsafe acts committed by people who are in direct contact with the situation or system.

Administrative fire officer IAFC description of a person who has worked as a managing fire officer for 3 to 5 years, is certified at the NFPA Fire Officer III level, and has accomplished formal education equivalent to a bachelor's degree.

Adoption by reference Method of code adoption in which the specific edition of a model code is referred to within the adopting ordinance or regulation.

Adoption by transcription Method of code adoption in which the entire text of the code is published within the adopting ordinance or regulation.

Arbitration Resolution of a dispute by a mediator or a group rather than a court of law. Any civil matter may be settled in this way; some labor–management agreements include a binding arbitration clause.

Arson The crime of maliciously and intentionally, or recklessly, starting a fire or causing an explosion.

Artifacts The remains of the material first ignited, the ignition source, or other items or components in some way related to fire ignition, development, or spread. An artifact may also be an item on which fire patterns are present, in which case the preservation of the artifact is not focused on the item itself, but rather on the fire pattern that appears on the item.

Assessment centers A series of simulation exercises to identify a candidate's competency to perform the job that is offered in the promotional examination.

Assistant or division chief A midlevel chief who often has a functional area of responsibility, such as training, and answers directly to the fire chief.

Authority having jurisdiction An organization, office, or individual responsible for enforcing the requirements of a code or standard, or for approving equipment, materials, an installation, or a procedure.

Automatic sprinkler system A system of pipes with water under pressure that allows water to be discharged immediately when a sprinkler head operates.

Base The location at which the primary logistics functions are coordinated and administered. The incident command post may be co-located with the base. There is only one base per incident.

Base budget The level of funding required to maintain all services at the currently authorized levels, including adjustments for inflation, salary increases, and other predictable cost changes.

Battalion chief Usually the first level of fire chief; also called a district chief. These chiefs are often in charge of running calls and supervising multiple stations or districts within a city. A battalion chief is usually the officer in charge of a single-alarm working fire.

Bond A certificate of debt issued by a government or corporation; a bond guarantees payment of the original investment plus interest by a specified future date.

Brainstorming A method of shared problem solving in which all members of a group spontaneously contribute ideas.

Branch A supervisory level established in either the operations or logistics function to provide an appropriate span of control.

Branch director A supervisory position in charge of a number of divisions and/or groups. This position reports to a section chief or the incident commander.

Budget An itemized summary of estimated or intended expenditures for a given period, along with proposals for financing them.

Business Continuity Planning An ongoing process to ensure that the necessary steps are taken to identify the impact of potential losses and maintain viable recovery strategies, recovery plans, and continuity of services.

Catastrophic theory of reform An approach in which fire prevention codes or firefighting procedures are changed in reaction to a fire disaster.

Central tendency An evaluation error that occurs when a fire fighter is rated in the middle of the range for all dimensions of work performance.

Chain of command The superior–subordinate authority relationship that starts at the top of the organization hierarchy and extends to the lowest levels.

Char Carbonaceous material that has been burned and has a blackened appearance.

Chief's trumpet An obsolete amplification device that enabled a chief officer to give orders to fire fighters during an emergency; a precursor to the bullhorn and portable radio.

Chronological statement of events A detailed account of the fire company activities as related to an incident or accident.

Class specification A technical worksheet that quantifies the knowledge, skills, and abilities (KSAs) by frequency and importance for every classified job within the local civil service agency.

Coaching A method of directing, instructing, and training a person or group of people with the aim to achieve some goal or develop specific skills.

Collective bargaining Method whereby representatives of employees (unions) and employers determine the conditions of employment through direct negotiation, normally resulting in a written contract setting forth the wages, hours, and other conditions to be observed for a stipulated period (e.g., 3 years). This term also applies to union–management dealings during the terms of the agreement.

Command staff Positions that are established to assume responsibility for key activities in the incident management system that are not a part of the line organization; these include safety officer, public information officer, and liaison officer.

Community Emergency Response Team (CERT) A fire department training program to help citizens understand their responsibilities in preparing for disaster and increase their ability to safely help themselves, their families, and their neighbors in the first 72 hours of a catastrophe.

Company journal A log book at the fire station that creates an extemporaneous record of the emergency, routine activities, and special activities that occurred at the fire station. The company journal also records any fire fighter injuries, liability-creating events, and special visitors to the fire station.

Compensation and benefits Human resources system to identify and determine the pay, leave, and fringe benefits for each position in the organization.

Complaint Expression of grief, regret, pain, censure, or resentment; lamentation; accusation; or fault finding.

Conflict A state of opposition between two parties. A complaint is a manifestation of a conflict.

Consensus document A code or standard developed through agreement between people representing different organizations and interests. NFPA codes and standards are consensus documents.

Construction type The combination of materials used in the construction of a building or structure, based on the varying degrees of fire resistance and combustibility.

Contrast effect An evaluation error in which a fire fighter is rated on the basis of the performance of another fire fighter and not on the classified job standards.

Controlling Restraining, regulating, governing, counteracting, or overpowering.

Crew integrity A system in which fire fighters stay together as a team of two or more members.

Crew resource management (CRM) A behavioral modification training system developed by the aviation industry to reduce its accident rate.

"Data dump" question A promotional question that asks the candidate to write or describe all of the factors or issues covering a technical issue, such as suppression of a basement fire in a commercial property.

Decision making The process of identifying problems and opportunities and resolving them.

Defensive operation Conduct of suppression operations outside the fire structure; these operations feature the use of large-capacity fire streams placed between the fire and the exposures to prevent fire extension.

Demographics The characteristics of human populations and population segments, especially when used to identify consumer markets; generally includes age, race, sex, income, education, and family status.

Demonstrative evidence Tangible items that can be identified by witnesses, such as incendiary devices and fire scene debris.

Demotion A reduction in rank, with a corresponding reduction in pay.

Dimensions Attributes or qualities that can be described and measured during a promotional examination. On average, 5 to 15 dimensions are measured on a promotional examination. The six most commonly addressed are oral communication, written communication, problem analysis, judgment, organizational sensitivity, and planning/organizing.

Direct supervision A type of supervision in which the fire officer is required to observe the actions of a work crew directly; it is commonly employed during high-hazard activities.

Discipline A moral, mental, and physical state in which all ranks respond to the will of the leader. Also, the guidelines that a department sets for fire fighters to work within.

Diversity A characteristic of a fire workforce that reflects differences in terms of age, cultural background, race, religion, sex, and sexual orientation.

Division A supervisory level established to divide an incident into geographical areas of operations.

Division of labor The production process in which each worker repeats one step over and over, achieving greater efficiencies in the use of time and knowledge; also, the formal assignment of authority and responsibility to job holders.

Division supervisor A supervisory position in charge of a geographical operation at the tactical level.

Documentary evidence Evidence in written form, such as reports, records, photographs, sketches, and witness statements.

Education The process of imparting knowledge or skill through systematic instruction.

Electrical arc Luminous discharge of electricity from one object to another, typically leaving a blackening of objects in the immediate area.

Employee assistance program (EAP) An employee benefit that covers all or part of the cost for employees to receive counseling, referrals, and advice in dealing with stressful issues in their lives. These problems may include substance abuse, bereavement, marital problems, weight issues, or general wellness issues.

Enclosed structure A building that lacks windows or doors of sufficient number and size to provide for prompt ventilation and emergency evacuation.

Environmental noise A physical or sociological condition that interferes with the message in the communication process.

Equity theory Motivational theory in which people evaluate the outcomes they receive for their inputs and compare them with the outcomes others receive for their inputs.

Ethical behavior Decisions and behavior demonstrated by a fire officer that are consistent with the department's core values, mission statement, and value statements.

Evidence The documentary or oral statements and the material objects admissible as testimony in a court of law.

Expanded incident report narrative A report in which all company members submit a narrative on what they observed and which activities they performed during an incident.

Expectancy theory Motivational theory in which people act in a manner that they believe will lead to an outcome they value.

Expenditure The act of spending money for goods or services.

Failure analysis A logical, systematic examination of an item, component, assembly, or structure and its place and function within a system, conducted to identify and analyze the probability, causes, and consequences of potential and real failures.

Fair Labor Standards Act (FLSA) Federal legislation passed in 1938 that provides the minimum standards for both wages and overtime entitlement and spells out administrative procedures by which covered work time must be compensated. Public safety workers were added to FLSA coverage in 1986.

Finance/administration section A section of the incident management system that is responsible for the accounting and financial aspects of an incident, as well as any legal issues that may arise.

Fire analysis The process of determining the origin, cause, development, and responsibility for a fire or explosion, as well as the failure analysis of a fire or explosion.

Fire chief The highest-ranking officer in charge of a fire department; the individual assigned the responsibility for management and control of all matters and concerns pertaining to the fire service organization.

Fire Ground Location (FGL) program A two-person unit that responds to working incidents to track fire fighters operating in immediately dangerous to life and health (IDLH) environments and to monitor the medical status of members who rotate through the rehabilitation sector during an event.

Fire mark Historically, an identifying symbol on a building to let fire fighters know that the building was insured by a company that would pay them for extinguishing the fire.

Fire patterns Physical marks left on an object by the fire.

Fire prevention division or hazardous use permit A local government permit that is renewed annually after the fire prevention division performs a code compliance inspection. A permit is required if the process, storage, or occupancy activity creates a life-safety hazard. Restaurants with more than 50 seats, flammable liquid storage, and printing shops that use ammonia are examples of occupancies that may require a permit.

Fire scene reconstruction The process of re-creating the physical scene during fire scene analysis through the removal of debris and the replacement of contents or structural elements in their prefire positions.

Fire tax district A special service district created to finance the fire protection of a designated district.

Fiscal year A 12-month period during which an organization plans to use its funds. Local governments' fiscal years generally run from July 1 to June 30.

Floor plans Views of a building's interior. Rooms, hallways, cabinets, and the like are drawn in the correct relationship to each other.

Flow path The volume between an inlet and an exhaust that allows the movement of heat and smoke from a higher-pressure area within the fire area toward lower-pressure areas accessible via doors, windows, and other openings.

Followership The act or condition of following a leader; adherence. The characteristic that leaders can be effective only to the extent that followers are willing to accept their leadership.

Form of material What the ignited material is being used for; for example, the form of cotton material might be clothing or bales of cotton.

Formal communication An official fire department communication. Such a letter or report is presented on stationery with the fire department letterhead and generally is signed by a chief officer or headquarters staff member.

Formal written reprimand An official negative supervisory action at the lowest level of the progressive disciplinary process.

Frame of reference An evaluation error in which the fire fighter is evaluated on the basis of the fire officer's personal standards instead of the classified job description standards.

Fuel limited Fire in a compartment (or building) that has adequate air supply. Without intervention, all of the fuel will be consumed by the fire.

Fuel load The total quantity of all combustible products found within a room or space.

General orders Short-term directions, procedures, or orders signed by the fire chief and lasting for a period of days to 1 year or more.

Good faith bargaining A legal requirement of both the union and the employer arising out of Section 8(d) of the National Labor Relations Act. Enforced by the National Labor Relations Board, the parties are required to meet regularly to bargain collectively for wages, hours, and other conditions of employment.

Governmental Accounting Standards Board (GASB) An organization whose mission is to establish and improve the standards of state and local governmental accounting and financial reporting, thereby resulting in useful information for users of financial reports, and to guide and educate the public, including issuers, auditors, and users of those financial reports.

Grievance A dispute, claim, or complaint that any employee or group of employees may have in relation to the interpretation, application, and/or alleged violation of some provision of the labor agreement or personnel regulations.

Grievance procedure A formal structured process that is employed within an organization to resolve a grievance. In most cases, the grievance procedure is incorporated in the personnel rules or the labor agreement and specifies a series of steps that must be followed in a particular order.

Group A supervisory level established to divide the incident into functional areas of operation.

Group supervisor A supervisory position in charge of a functional operation at the tactical level. This position reports to a branch director, the operations section chief, or the incident commander.

Halo and horn effect An evaluation error in which the fire officer takes one aspect of a fire fighter's job task and applies it to all aspects of work performance.

Hazard Any arrangement of materials and heat sources that presents the potential for harm, such as personal injury or ignition of combustibles.

Health and safety officer The member of the fire department assigned and authorized by the fire chief as the manager of the safety and health program.

Health Insurance Portability and Accountability Act (HIPAA) Enacted in 1996, federal legislation that provides for criminal sanctions and civil penalties for releasing a patient's protected health information in a way not authorized by the patient.

Health, safety, and security Human resources activities intended to provide and promote a safe work environment.

Hierarchy of needs Maslow's description of human needs as a pyramid or ladder that starts with physiological needs and ends with self-actualization.

High-risk property Structure that has the potential for a catastrophic property or life loss in the event of a fire.

High-value property Structure that contains equipment, materials, or items that have a high replacement value.

Human resources development All activities to train and educate employees.

Human resources planning The process of having the right number of people in the right place at the right time who can accomplish a task efficiently and effectively.

Humanistic management A management strategy that emphasizes human need and attitude; motivation comes from within the employee and not from authoritarian control. It leads to Maslow's hierarchy of needs.

Hygiene factors Conditions external to the individual, such as pay and work conditions.

ICS general staff The group of incident managers composed of the operations section chief, planning section chief, logistics section chief, and finance/administration section chief.

Immediately dangerous to life and health (IDLH) Any condition that would do one or more of the following: (1) pose an immediate or delayed threat to life, (2) cause irreversible adverse health effects, or (3) interfere with an individual's ability to escape unaided from a hazardous environment.

Impasse A situation in which the parties in a dispute have reached a deadlock in negotiations; also described as the demarcation line between bargaining and negotiation. A declaration of an impasse in labor–management negotiations brings in a state or federal negotiator who will start a fact-finding process that will lead to a binding arbitration resolution.

In-basket exercise A promotional examination component in which the candidate deals with correspondence and related items that have accumulated in a fire officer's in-basket.

Incident action plan (IAP) The objectives reflecting the overall incident strategy, tactics, risk management, and member safety that are developed by the incident commander. Incident action plans are updated throughout the incident.

Incident Command System (ICS) A system that defines the roles and responsibilities to be assumed by personnel and the operating procedures to be used in the management and direction of emergency operations; also referred to as an Incident Management System (IMS).

Incident commander The person who is responsible for all decisions relating to the management of the incident and is in charge of the incident site.

Incident safety officer An individual appointed to respond to or assigned at an incident scene by the incident commander to perform the duties and responsibilities specified in NFPA 1521, *Standard for Fire Department Safety Officer*.

Incident safety plan The strategies and tactics developed by the incident safety officer based on the incident commander's incident action plan and the type of incident encountered.

Incident scene rehabilitation The tactical-level management unit that provides for medical evaluation, treatment, monitoring, fluid and food replenishment, mental rest, and relief from climatic conditions of an incident.

Informal communications Internal memos, e-mails, instant messages, and computer-aided dispatch/mobile data terminal messages. Informal reports have a short life and may not be archived as permanent records.

Initial rapid intervention crew (IRIC) Two members from the initial attack crew who are assigned for rapid deployment to rescue lost or trapped members.

Interrogatory A series of formal written questions sent to the opposing side of a legal argument. The opposition must provide written answers under oath.

Investigation A systematic inquiry or examination.

Involuntary transfer or detail A disciplinary action in which a fire fighter is transferred or assigned to a less desirable or different work location or assignment.

Job description A narrative summary of the scope of a job. It provides examples of the typical tasks.

Job instruction training A systematic four-step approach to training fire fighters in a basic job skill: (1) prepare the fire fighters to learn, (2) demonstrate how the job is done, (3) try them out by letting them do the job, and (4) gradually put them on their own.

Job-content/criterion-referenced validity A type of validity obtained through the use of a technical committee of job incumbents who certify that the knowledge being measured is required on the job and referenced to known standards.

Knowledge, skills, and abilities (KSAs) The traits required for every classified position within the municipality. KSAs are defined by a narrative job description and a technical class specification.

Latent conditions Inevitable "resident pathogens" within the system.

Leadership A complex process by which a person influences others to accomplish a mission, task, or objective and directs the organization in a way that makes it more cohesive and coherent.

Leading Guiding or directing in a course of action.

Legacy dwellings Single-family dwellings constructed before 1980.

Liaison officer The incident commander's representative or a point of contact for representatives from outside agencies.

Limited At the point, edge, or line beyond which something cannot or may not proceed; confined or restricted within certain limits.

Line-item budget A budget format in which expenditures are identified in a categorized line-by-line format.

Lobby control officer The fire officer who controls the entry and exit of both civilians and fire fighters in the lobby at a high-rise fire incident; this officer also oversees the use of the elevators, operates the local building communication system, and assists in the control of the heating, ventilating, and air-conditioning systems.

Logistics section Responsible for providing facilities, services, and materials for the incident. Includes the communications, medical, and food units within the service branch, as well as the supply, facilities, and ground support units within the support branch. The logistics section chief is part of the general staff.

Logistics section chief A supervisory position that is responsible for providing supplies, services, facilities, and materials during the incident. The person in this position reports directly to the incident commander.

Loudermill hearing A predisciplinary conference that occurs before a suspension, demotion, or involuntary termination is issued. The term "Loudermill" refers to a U.S. Supreme Court decision.

Managing fire officer The description from the IAFC *Officer Development Handbook* for the tasks and expectations for a Fire Officer II. In this role, the company officer is encouraged to acquire the appropriate levels of training, experience, self-development, and education to prepare for the Chief Fire Officer designation.

Masonry wall A wall that consists of brick, stone, concrete block, terra cotta, tile, adobe, precast, or cast-in-place concrete.

Mayday A situation where a fire fighter is experiencing a life-threatening emergency.

Measured Careful, restrained, calculated, and deliberate.

Mediation The intervention of a neutral third party in an industrial dispute. The object is to enable the two sides to reach a compromise solution to their differences, which the mediator usually does by seeing representatives of both sides separately and then together.

Mentoring A developmental relationship between a more experienced person (a mentor) and a less experienced person (a protégé).

Mini/max codes Codes developed and adopted at the state level for either mandatory or optional enforcement by local governments; these codes cannot be amended by local governments.

Mistake An error or fault resulting from defective judgment, deficient knowledge, or carelessness; a misconception or misunderstanding.

Mitigation Measures taken to limit or control the consequences, extent, or severity of an incident that cannot be reasonably prevented.

Model codes Codes generally developed through the consensus process with the use of technical committees developed by a code-making organization.

Modern dwellings Single-family dwellings constructed since 1980; they are typically larger structures with an open house geometry, lightweight construction materials, and exponentially increased fuel load.

Motivation factors An individual's internal desire for recognition, achievement, responsibility, and advancement.

National Fire Incident Reporting System (NFIRS) A nationwide database at the National Fire Data Center under the U.S. Fire Administration that collects fire-related data in an effort to provide information on the national fire problem.

National Incident Management System (NIMS) Provides a consistent nationwide template to enable federal, state, tribal, and local governments; the private sector; and nongovernmental organizations to work together to prepare for, prevent, respond to, recover from, and mitigate the effects of incidents, regardless of cause, size, location, or complexity, so as to reduce the loss of life, property, and harm to the environment.

National Response Framework (NRF) A comprehensive, national, all-hazards approach to domestic incident response that describes specific authorities and best practices for managing incidents; it builds upon the National Incident Management System, which provides a consistent template for managing incidents.

Naturalistic decision making The process by which commanders make decisions in their natural environment.

Negotiation Mutual discussion and arrangement of the terms of an agreement.

Occupancy type The purpose for which a building or a portion thereof is used or intended to be used.

Offensive operation An advance into the fire building by fire fighters with hose lines or other extinguishing agents to overpower the fire.

Ongoing compliance inspection Inspection of an existing occupancy to observe the housekeeping and confirm that the built-in fire protection features, such as fire exit doors and sprinkler systems, are in good working order.

Opened structure A building that has windows or doors of sufficient number and size to provide for prompt ventilation and emergency evacuation.

Operational period A term used in a written incident action plan identifying a period of time during a long-term incident that a specific incident action plan covers. For federally funded incidents, the operational period is 12 hours; local incidents may use operational periods of 8 hours.

Operations section Responsible for all tactical operations at the incident. In the national model, the operations section can be as large as 5 branches, 25 divisions/groups or units, or 125 single resources, task forces, or strike teams.

Operations section chief A supervisory position that is responsible for the management of all actions that are directly related to controlling the incident. This position reports directly to the incident commander.

Oral reprimand, warning, or admonishment The first level of negative discipline. Considered informal, this discipline action remains with the fire officer and is not part of the fire fighter's official record.

Ordinance A law established by an authorized subdivision of a state, such as a city, county, or town.

Organizing Putting resources together into an orderly, functional, structured whole.

Performance log An informal record maintained by the fire officer that lists fire fighter activities by date and includes a brief description; it is used to provide documentation for annual evaluations and special recognitions.

Performance management The process of setting performance standards and evaluating performance against those standards.

Personal bias An evaluation error that occurs when the evaluator's perspective skews the evaluation such that the classified job knowledge, skills, and abilities are not appropriately evaluated.

Personal study journal A personal notebook to aid in scheduling and tracking a candidate's promotional preparation progress.

Personnel accountability report (PAR) A systematic method of accounting for all personnel at an emergency incident.

Personnel accountability system A method of tracking the identity, assignment, and location of fire fighters operating at an incident scene.

Planning Developing a scheme, program, or method that is worked out beforehand to accomplish an objective.

Planning section Responsible for the collection, evaluation, dissemination, and use of information about the development of the incident and the status of resources. It includes the situation status, resource status, and documentation units as well as technical specialists. The planning section chief is part of the general staff.

Planning section chief A supervisory position that is responsible for the collection, evaluation, dissemination, and use of information relevant to the incident. This position reports directly to the incident commander.

Plot plan A representation of the exterior of a structure, identifying doors, utilities access, and any special considerations or hazards.

Point of origin The exact physical location where a heat source and a fuel come in contact with each other and a fire begins.

Policies Formal statements that provide guidelines for present and future actions. They often require personnel to make judgments.

Political action committee (PAC) An organization formed by corporations, unions, and other interest groups that solicits campaign contributions from private individuals and distributes these funds to political candidates.

Power The capacity of one party to influence another party.

Preincident plan A written document resulting from the gathering of general and detailed data to be used by responding personnel for determining the resources and actions necessary to mitigate anticipated emergencies at a specific facility.

Pretermination hearing An initial check to determine if there are reasonable grounds to believe that charges against an employee are true and support the proposed termination.

Problem A condition in which the desired situation is different from the current situation.

Professional development Skills and knowledge attained for both personal development and career advancement.

Progressive negative discipline A process for dealing with job-related behavior that does not meet expected and communicated performance standards. The level of discipline increases from mild to more severe punishments if the problem is not corrected.

Public information officer A command staff position that is responsible for gathering and releasing incident information to the news media and other appropriate agencies. This position reports directly to the incident commander.

Pyrolysis The destructive distillation of organic compounds in an oxygen-free environment that converts the organic matter into gases, liquids, and char.

Rapid intervention crew (RIC) A dedicated crew of four fire fighters who are assigned for rapid deployment to rescue lost or trapped members.

Recency An evaluation error in which the fire fighter is evaluated only on incidents that occurred over the past few weeks rather than on the entire evaluation period.

Recognition-primed decision making (RPD) The process by which commanders can recognize a plausible plan of action.

Recommendation report A decision document prepared by a fire officer for the senior staff. Its goal is to support a decision or an action.

Regulations Orders written by a governmental agency in accordance with the statute or ordinance authorizing the agency to create the regulation. Regulations are not laws but have the force of law.

Rehabilitation The process of providing rest, rehydration, nourishment, and medical evaluation to members who are involved in extended or extreme incident scene operations.

Reinforcement theory Motivational theory in which behavior is a function of its consequences.

Reliability The characteristic where a test measures what it is intended to measure on a consistent basis.

Restrictive duty A temporary work assignment during an administrative investigation that isolates the fire fighter from the public and usually is an administrative assignment away from the fire station.

Revenues The income of a government from all sources.

Right to work A worker cannot be compelled, as a condition of employment, to join or not to join or to pay dues to a labor union.

Risk assessment The process of identifying hazards, monitoring those hazards, determining the likelihood of their occurrence, and assessing the vulnerability of people, property, the environment, and the entity itself to those hazards.

Risk management Identification and analysis of exposure to hazards, selection of appropriate risk management techniques to handle exposures, implementation of chosen techniques, and monitoring of results, with respect to the health and safety of members.

Risk Watch A comprehensive NFPA school-based program focused on injury prevention.

Risk–benefit analysis A decision made by a responder based on a hazard and situation assessment that weighs the risks likely to be taken against the benefits to be gained for taking those risks.

Rules and regulations Directives developed by various government or government-authorized organizations to implement a law that has been passed by a government body.

Rules of Engagement Rules developed by the International Association of Fire Chiefs to promote safety for fire fighters and fire officers working at the task level of fire suppression operations.

Safety officer The person who is responsible for monitoring and assessing safety hazards and unsafe conditions; develops measures to ensure personnel safety.

Safety unit A member or members assigned to assist the incident safety officer; the tactical-level management unit that can be composed of the incident safety officer alone or with additional assistant safety officers assigned to assist in providing the level of safety supervision appropriate for the magnitude of the incident and the associated hazards.

Scientific management The breakdown of work tasks into constituent elements. The timing of each element is based on repeated stopwatch studies; the fixing of piece-rate compensation based on those studies; standardization of work tasks on detailed instruction cards; and generally, the systematic consolidation of the shop floor's brain work.

Service branch A major division within the logistics section of the ICS; it oversees the communications, medical, and food units.

Shop steward A union member appointed or elected to be the first line of labor representation at the workplace. The steward enforces the contract, collective agreement, or memorandum of understanding and represents the union members at that fire station or work location.

Situational awareness The process of evaluating the severity and consequences of an incident and communicating the results.

Size-up A systematic process of gathering and processing information to evaluate the situation and then translating that information into a plan to deal with the situation.

Social media Digital communications through which users create online communities to share information, ideas, personal messages, videos, pictures, and other content.

Source of ignition Devices or equipment that, because of their intended modes of use or operation, are capable of providing sufficient thermal energy to ignite flammable gas–air mixtures.

Span of control The maximum number of personnel or activities that can be effectively controlled by one individual (usually three to seven).

Special evaluation period A designated period of time when an employee is provided additional training to resolve a work performance/behavioral issue. The supervisor issues an evaluation at the end of the special evaluation period.

Spoils system Also known as the patronage system; the practice of making appointments to public office based on a personal relationship or affiliation rather than because of merit. The spoils system scandals of the New York City "Tweed Ring" and the Tammany Hall political machine (1865–1871) resulted in Congress passing the Pendleton Civil Service Reform Act of 1883.

Staffing The process of attracting, selecting, and maintaining an adequate supply of labor, as well as reducing the size of the labor force when required.

Staging A specific function in which resources are assembled in an area at or near the incident scene to await instructions or assignments.

Stairwell support group A group of fire fighters who move equipment and water supply hose lines up and down the stairwells at a high-rise fire incident. The stairwell support unit leader reports to the support branch director or the logistics section chief.

Standard operating guidelines (SOGs) Written organizational directives that identify a desired goal and describe the general path to accomplish the goal, including critical tasks or cautions.

Standard operating procedures (SOPs) Written organizational directives that establish or prescribe specific operational or administrative methods to be followed routinely for the performance of designated operations or actions.

Standard time–temperature curve A recording of fire temperature increase over time.

Standpipe system An arrangement of piping, valves, hose connections, and allied equipment installed in a building or structure, with the hose connections located in such a manner that water can be discharged in streams or spray patterns through attached hose and nozzles, for the purpose of extinguishing a fire, thereby protecting a building or structure and its contents in addition to protecting the occupants. This is accomplished by means of connections to water supply systems or by means of pumps, tanks, and other equipment necessary to provide an adequate supply of water to the hose connections.

Strategic level Command level that entails the overall direction and goals of the incident.

Strike A concerted act by a group of employees who withhold their labor for the purposes of effecting a change in wages, hours, or working conditions.

Strike team A specific combination of the same kind and type of resources, with common communications and a leader.

Strike team leader A supervisory position that is in charge of a group of similar resources.

Supervising fire officer The description from the IAFC *Officer Development Handbook* for the tasks and expectations for a Fire Officer I. In this role, the company officer is encouraged to acquire the appropriate levels of training, experience, self-development, and education to prepare for the Chief Fire Officer designation.

Supervisor's report A form that is required by most state worker's compensation agencies and that is completed by the immediate supervisor after an injury or property damage accident.

Supplemental budget Proposed increases in spending to provide additional services.

Support branch A major division within the logistics section of the ICS; it oversees the supply, facilities, and ground support units.

Suspension A negative disciplinary action that removes a fire fighter from the work location; he or she is generally not allowed to perform any fire department duties.

T-account A documentation system similar to an accounting balance sheet listing credits and debits, in which a single-sheet form is used to list the employee's assets on the left side and liabilities on the right side, so that the result resembles the letter "T."

Tactical level Command level in which objectives must be achieved to meet the strategic goals. The tactical-level supervisor or officer is responsible for completing assigned objectives.

Tactical worksheet A form that allows the incident commander to ensure all tactical issues are addressed and to diagram an incident with the location of resources on the diagram.

Task force Any combination of single resources assembled for a particular tactical need, with common communications and a leader.

Task force leader A supervisory position that is in charge of a group of dissimilar resources.

Task level Command level in which specific tasks are assigned to companies; these tasks are geared toward meeting tactical-level requirements.

Termination A situation in which the organization ends an individual's employment against his or her will.

Testimonial evidence Witnesses speaking under oath.

Theory X McGregor's description of the management assumption that people do not like to work and must be closely watched and controlled.

Theory Y McGregor's description of the management assumption that people like to work and need to be encouraged, not controlled.

Trailers Materials used to spread fire from one area of a structure to another.

Training The process of achieving proficiency through instruction and hands-on practice in the operation of equipment and systems that are expected to be used in the performance of assigned duties.

Transitional operation A situation in which an operation is changing or preparing to change.

Two-in/two-out rule A guideline created in response to OSHA Respiratory Regulation (29 CFR 1910.134), which requires a two-person team to operate within an environment that is immediately dangerous to life and health (IDLH) and a minimum of a two-person team to be available outside the IDLH atmosphere to remain capable of rapid rescue of the interior team.

Type of material What the ignited material is made of; for example, the type of material might be cotton.

Unconsciously competent The highest level of the Conscious Competence Learning Matrix developed by Dr. Thomas Gordon in the 1970s. At the Unconsciously Competent level, the skill becomes so practiced that it enters the unconscious parts of the brain—it becomes second nature.

Unfair labor practices Employer or union practices forbidden by the National Labor Relations Board or state/local laws, subject to court appeal. It often involves the employer's efforts to avoid bargaining in good faith.

Unit Either a geographical or a functional assignment.

Unity of command The management concept that a subordinate should have only one direct supervisor, and that a decision can be traced back through subordinates to the manager who originated it.

Use group A category in the building code classification system in which buildings and structures are grouped together by their use and by the characteristics of their occupants.

Ventilation limited Fire in a compartment (or building) that has inadequate air supply. It will flare up when air is introduced into the compartment.

Vigilant On the alert and watchful.

Work improvement plan A written document that is part of a special evaluation period. The plan identifies performance deficiencies and lists the improvements in performance or changes in behavior required to obtain a "satisfactory" evaluation.

Yellow dog contracts Pledges that employers required workers to sign indicating that they would not join a union as long as the company employed them. Such contracts were declared unenforceable by the Norris-LaGuardia Act of 1932.

Index

Note: Page numbers followed by *f*, or *t* indicate material in figures, or tables, respectively.

A

academic accreditation, 153–154, 157
accelerants, 375, 381
access pathways, 260–261, 262*f*
accident
    data analysis, 104–105
    definition of, 99, 107
    investigation, 99
    postincident analysis of, 99
    reports, 70–71, 71*f*, 105
accidental fire causes, 374
accommodation, 117–118
accountability
    fire fighter, 300–301
    system, personnel, 90, 90*f*, 107
accreditation, 153, 158
accuracy of communication, 65
acting out, 234
actionable items, 53, 57
active failures, 387, 397
active listening, 63, 63*f*, 65, 230, 232
activities, delegation of, 130
ADA. *See* Americans with Disabilities Act
ADEA. *See* Age Discrimination Employment Act
administrative fire officer
    definition of, 42, 57
    morning report to, 68
    vital tasks of, 42–45
adoption by reference, 248, 267
adoption by transcription, 248, 267
advance notice of substandard employee evaluation, 172–173
advance-notice procedure, 173
affection, needs for, 116
AFL. *See* American Federation of Labor
Age Discrimination Employment Act (ADEA), 122
aggression, 234
AHJ. *See* authority having jurisdiction
air management, 90–91
alarm box, 8
alternative disciplinary actions, 176
aluminum siding, safety precautions, 375
American Federation of Labor (AFL), 191
American Society for Testing and Materials (ASTM), 340
Americans with Disabilities Act (ADA), 51, 122
Anne Arundel County, Maryland fire fighter/paramedic law suit, 195
announcements, 73
annual evaluations
    conduct, 171–173
    process, steps for, 171
    six weeks before, 173

annual expenses for FFSL, 283–284
antitrust protection, 193
appearance for television interview, 216, 216*f*
application of training skills, 145
arbitration, 191, 198
arson, 375, 377–378, 381
arsonists, motives of, 377–378
artifacts, 367, 381
asbestos exposure, 350
asphyxiation, fire fighter deaths from, 89, 91
assembly occupancy, 254, 262
assertive statement process, 388–389
assessment centers
    definition of, 30, 37
    in-basket exercises, 30–31, 30*f*, 37, 120
assignments
    completion of, 120–121
    follow-up for, 120
Assistance to Firefighters Grant Program, 280
assistant chiefs, 11, 19
assistant incident safety officers, 94
ASTM. *See* American Society for Testing and Materials
attendance, reasons for reporting late to work, 166*f*
attention to problems, 225, 226*f*
audiovisual aids, 210
authority having jurisdiction (AHJ), 247, 267, 279, 364
authority, source of, 10–11
autocratic leadership style, 129, 129*f*, 130
automatic sprinkler systems, 249, 249*f*, 267
aviation industry, 328–329

B

bad news, receipt of, 226
balance, 118
balloon frame construction, 354
base, 356, 358
base budget, 272–273, 287
battalion chiefs (BC), 6, 11, 19, 43, 323
BC. *See* battalion chiefs
BCP. *See* business continuity planning
behavior
    accommodating, 117–118
    cultural factors and, 205
    ethical, 51, 52, 57
    integrity, 50–51
    leading to immediate negative discipline, 174
    neutral, 117
    unacceptable, correction of, 55, 55*f*, 174–177
benefits, 121, 123
bicentennial layoff, 276
bidding process, 279
bingo, 274–275
Bishop, Todd, 388–389
Blake, Robert, 117
blogs, 76

bloodborne pathogens, 147
bond, 282, 287
    referendums and capital projects, 283–284
brainstorming, 226–227, 238
branch, 307, 311
branch director, 307, 311
British fire service, 207
brush abatement, 212, 214
budget
    base, 272–273, 287
    cost recovery and reduction, 284–286
    cycle, 272–273
    definition of, 272, 287
    increases, 272–273
    process, navigating, 283–286
    proposal, developing, 283
    supplemental, 272–273, 287
building
    classification of, 252–256
    codes, 9–10
        development of, 10
        vs. fire code, 247–248
    collapses, 33
    deterioration of, 17
    pre-plan, example of, 245
built-in fire protection systems, 249–252, 261
burning structure fires, firefighter deaths in, 17
burns, fire fighter fatalities from, 89
business continuity planning (BCP)
    incident prevention, 264
    resource management and logistics, 264–265
    training, exercises, and evaluation, 265
business occupancy, 254, 254*f*, 262
business work location, fire station as, 55–56

C

call boxes, public, 8
CAN. *See* conditions–actions–needs
Canada, fire loss statistics, 9
Candidate Physical Ability Test (CPAT), 197
capital expenditures, 282
capital projects, 282–283
    bond referendums and, 283–284
car safety seats for children, 205
carbon dioxide extinguishing systems, 251
catastrophe planning, 211
catastrophic theory of reform, 247, 267
cats, rescuing from tree, 207
cause of fire
    classifications
        accidental fire, 374
        incendiary fire, 375
        natural fire, 374–375
        undetermined fire, 375
    common, 364, 365*t*
    determination of, 369–374. *See also* fire investigations

cause of fire (*Continued*)
   in vehicle fires, 372–373, 373*f*
   in wildland fire, 373–374
cellular telephones, 9
Center for Public Safety Excellence (CPSE)
   designations, 156
central processing unit, 72
central tendency, 173, 178
CERT. *See* Community Emergency
   Response Team
certification programs for fire fighters, 154
CFAI. *See* Commission on Fire Accreditation
   International
chain of command, fire department, 11, 11*f*, 19
char, 369, 369*f*, 381
CHEA. *See* Council for Higher Education
   Accreditation
check-off system, 260
chemical extinguishing systems, 251
Chief Fire Officer, 155, 156
chief's trumpet, 9, 9*f*, 19
child safety seats, 205
chronological statement of events, 71, 80
CIO. *See* Congress of Industrial Organizations
CISM. *See* critical incident stress management
citizen complaints, 235
Civil Rights Act of 1964, 51, 122
civil service system, 24
class specification, definition of, 26, 37
Clayton Act of 1914, 193
cleaning of contaminated clothing, 96
Cleveland Board of Education v. Loudermill, 176
Clinton, Bill, 195
closed shops, 192
clothing
   contaminated, cleaning of, 96
   protective, 96, 284
coaching
   definition of, 142, 158
   ensuring competence and confidence, 147
   mentoring and, 146, 389
   teamwork and, 389–391
code enforcement, 256
code of conduct, 55
coercive power, 130
cold zone, 350
Collapse of Burning Buildings (Dunn), 104
collective bargaining
   definition of, 184, 198
   for federal employees, 192
   legislative framework for, 190–192
   state legislation, 190–191
collisions, motor vehicle, 88
colonial fire fighters, 7
command
   establishing, 298
   functions of, 299–300

   levels of, 297
   mode, as option, 298–299
   responsibilities of, 297–298
   transfer of, 300
command presence, 47
command staff, 302, 311
   assignments, 344
commander, fire officer as, 47, 47*f*, 48*f*
Commission on Fire Accreditation International
   (CFAI), 113
communications, 8–9
   computer, 72
   cycle
      feedback, 63
      medium, 63
      message, 62–63
      receiver, 63
      sender, 63
   emergency, 66–68
   formal, 68, 81
   grapevine, 65
   importance of, 74
   informal, 68, 81
   news releases, 76
   order model for, 67
   reporting, 68–72. *See also* reports
   skills
      basic, 63–66
      CRM model and, 388–389, 396
   software, 72
   written, 72–75
Community Emergency Response Team (CERT),
   210–212, 211*f*, 219
   course schedule, 211
   maintaining involvement of, 211–212
community fire safety, fire officer role in, 242–243,
   242*f*, 243*f*
community, working in, 203–219
   media relations and, 213–217, 214*f*
   public education programs. *See* public
      education programs
   risk reduction, responding to public inquiries
      and, 206–207
   social media outreach, 217
   understanding, 204–206
company journal, 68, 69*f*, 80–81
company-level officer, 307
   as trainer, 46, 46*f*
compensation, 121, 123
competence
   ensuring, 147
   related to classified job description, 165
competitive grant proposals, 280
competitiveness, 168
complainants, 236
complaints, 238
   citizen, 235

   definition of, 224
   expectations, 236
   letter, example of, 236*f*
   viewpoint, 233–234
compromise, 118
computers, 75
conditions–actions–needs (CAN), 68
confidence, ensuring, 147
confidentiality of discrimination complaint, 55
confinement, 349
conflict, 225*f*, 238
   definition of, 224
   handling, 391
   managing, 230, 232–234
   resolution model, 230, 232
Congress of Industrial Organizations (CIO), 191
Conscious Competence Learning Matrix, 146
consensus document, 10, 19
consistency, 116
consolidation of services, 276
construction type, 252, 267
   fire resistive, 252–253
   heavy timber, 253
   limited combustible (ordinary), 253
   noncombustible, 253
   wood frame, 253–254
contracting of services, 276
contrast effect, 174, 178
controlling, 117
   as management function, 13, 19
   persons, 117
Corps of Vigiles, 7
correction report, writing, 260
cost recovery, 280–281, 284–286
   and reduction, 284–286
Council for Higher Education Accreditation
   (CHEA), 153
CPAT. *See* Candidate Physical Ability Test
CPSE. *See* Center for Public Safety Excellence
crew integrity, 89, 324, 335
crew resource management (CRM), 386–399
   communication skills, 388–389
   concepts, researching/validating, 386–387
   critical decision making, 392
   debriefing, 395
   definition of, 386, 397
   human error and, 387–388
   implementing change, 395
   managerial grid and, 117
   model of, 396
      communication skills, 388–389
      critical decision making, 392
      situational awareness, 392–394
      task allocation, 391–392
      teamwork, 389–391
   origins of, 386
   recommending change, 395

situational awareness, 392–394, 397
  task allocation in, 391–392
  teamwork, 389–391
crime concealment as arson motive, 378
criminally negligent manslaughter, after live fire training, 150
critical decision making, 392
critical incident stress management (CISM), 197
critical situations, 132–133
critique, incident management, 295–296, 296t
CRM. *See* crew resource management
cultural diversity in fire department, 17
cultural sensitivity, 205
customer service *vs.* satisfaction, 235–236, 236f

D

"daily" file, 121
data analysis for accident investigation, 104–105
"data dump" question, 31, 37
data storage systems, 247
databases
  for computer-aided dispatch system, 75
  description of, 75
dead smoke alarms, 214
deadlines for problem solution, 227
death, fire fighter
  analyzing, 101–102, 101f
  from asphyxiation, 89, 91
  from burns, 89
  and injury trends, 86–92
  nature of injury, 102f
  reducing
    from fire suppression operations, 89–92
    from motor vehicle collisions, 88
    from sudden cardiac arrest, 87–88
debriefing, 395
decision-based errors, 103
decision documents, 75–76
decision making
  by administrative fire officers, 44
  alternative solutions, idea sharing and, 226–227
  approaches, 129
  critical in CRM model, 392
  definition of, 17, 19
  evaluating results, 228–229
  general procedures, 225–229
  generating alternative solutions, 226–227
  high-quality, systematic approach to, 225
  skills, 224
decontamination procedures for infection control, 97
defensive fires, 321
defensive operation, 347, 348f, 358
defer scheduled expenditures, 275–276
dehydration, 94
delayed notification of fire, 366
delegation, 120, 121

deluge systems, 250
democratic leadership style, 129, 129f, 130
demographic analysis techniques, 205
demographics, 205, 219
demonstrative evidence, 367, 381
demotion, 176, 178
Department of Homeland Security (DHS), 195
Department of Labor, 122
deputy chief, 47
derogatory/racist terminology, 53
DETC. *See* Distance Education and Training Council
detention and correctional occupancy, 255, 263
deteriorating conditions, 323–324, 323f, 330
development, management concepts and, 113
DHS. *See* Department of Homeland Security
dimensions, 25, 37
direct mail fundraising, 274
direct supervision, 122, 123
directed questioning method, 65
directives, unpopular, 46–47, 47f
disabled built-in fire protection, 366
discipline
  definition of, 12, 19, 166, 178
  informal, 169
  negative, 164, 166, 174–177
  positive, 164, 166, 167–169
  predetermined policies, 176
  procedures, 169
  progressive nature of, 164
discrimination complaints, handling, 55
dispatch and alarm assignments, 229
dispatch center, 133–134
Distance Education and Training Council (DETC), 154
diversity, 51–53, 57
division chiefs, 11
division of labor, 12, 19
division supervisor, 306–307, 311
divisions, 306–307, 306f, 311
documentary evidence, 367, 381
documentation
  of accident investigation, 99
  of evaluation and discipline, 171
  investigation report, 379
  legal proceedings, 379–380
  performance log, 171
  preliminary investigation, 378–379
  T-account, 171
Dodd J. Miller Training Academy, 165
dominant personality, 117
dormitories, 255–256
Douglass, McDonnell, 386
driver training program, 88
drowning, fire fighter fatalities from, 102f
dry chemical extinguishing systems, 251
dry-pipe systems, 250

Dunn, Vincent, 47
Dupont, Gordon, 387
duty key, 350
duty uniforms, 96

E

EAP. *See* employee assistance program
ecological power, 130
education. *See also* training
  cultural change issues, 153
  definition of, 152, 158
  public. *See* public education programs
  training and, 10, 152
educational occupancy, 255, 262
EEOC. *See* Equal Employment Opportunity Commission
Effective Company Command (Page), 46
egress pathways, 260–261, 262f
Ehrlich, Robert, 195
electrical arc, 369, 382
electrical fire, 374
electrical systems, identifying in preincident plan, 245
electrical wiring, inspection of, 261
electronic communications, 78f
electronic version of personal library, 48
emergency communications, 66–68, 80
  key points for, 66–67
  order model, 67
  radio report, 67–68
emergency incidents
  functional components, 302
  injury prevention, 95–96
  leadership at, 129, 129f, 130–134
  media coverage of, 213
  simulations of, 31, 31f
  situational awareness, maintaining, 393–394
emergency lighting, 261, 262
emergency management, 264–265
emergency medical services (EMS), 185, 198
  calls to, 16, 17
  systems performance measurement, 196
emergency operation considerations, identifying in preincident plan, 245–246
emergency operations/response plan, 265
emergency services
  community information, 205
  training program, immediately learned skills for, 147–148
  workload changes, 17
Emergency Services Definition Act, 195
Emergency Support Functions (ESFs), 306, 306t
emergency traffic, mayday *vs.*, 331–333
empathy in conflict resolution, 235
employee assistance program (EAP), 177, 178
employee counseling and discipline steps, 164
employee evaluation, substandard, 172–173

employee health, safety, and security, as human resource management function, 121, 123
employee relations, 118
employee work improvement plan, 32, 32f
empowerment, 168
EMS. *See* emergency medical services
enclosed structure, 319, 335
English as second language, 206
environmental noise, 65–66, 80, 81
    types of, 66
Equal Employment Opportunity Act, 51
Equal Employment Opportunity Commission (EEOC), 51
equipment, fire fighting, 8, 91, 91f
    tampered/altered as indicator of incendiary fire, 366, 381
equity theory, 136, 138
error management model, 388, 388f, 395, 396
ESFs. *See* Emergency Support Functions
esteem, need for, 116–117
ethical behavior, 51, 52, 57
ethics, 14–16
evacuations, 133
evaluations, 164–166
    annual performance, 164, 171
    conducting annual evaluation, 171–173
    errors in, 173–174
    procedures, 169
    process for new fire fighter, 165
    providing feedback after incident/activity, 165, 168f
    purpose of, 164
    substandard, advance notice of, 172–173
    of training skills, 145–146
Everyone Goes Home® program, 86–87
evidence
    definition of, 367, 382
    physical for accident investigation, 99
    protection of, 367–368, 368f
exceptional violations, 103
excitement as arson motive, 378
excuses, conflict management and, 232
Executive Order #10988, 192
exit interview, 260
exit lighting, 262
exit signs, 261
expanded incident report narrative, 69, 81
expectancy theory, 136, 138
expectations, setting, 168
expenditure, 281–282
    capital, 282
    classification of, 281–282
    definition of, 272, 287
    operating, 281–282, 282f
    personnel, 281, 283
expert power, 130
expert witness testimony, 379

explanations, conflict management and, 232
exposures, 304, 349
external department issues, encountered by fire officers, 225
extinction, behavioral, 135, 136
extinguishing systems, special, 250–251
extinguishment of fire, 349
extremism as arson motive, 378

F

face-to-face conversation, 68
factory industrial occupancy, special concerns, 262–263
facts in size-up process, 345–346
failure analysis, 370, 382
Fair Labor Standards Act (FLSA), 122, 194, 198
Familia Publica, 7
Family and Medical Leave Act (FMLA), 122
family dwelling units, 255
fast-attack mode as command option, 298
FAST truck. *See* fire fighter assist search team truck
FBI. *See* Federal Bureau of Investigation
FDNY. *See* Fire Department of New York
fear *vs.* trust, 226
Federal Bureau of Investigation (FBI), 17
Federal Emergency Management Agency (FEMA), 17, 211
federal employees, collective bargaining for, 192
Federal Labor Relations Council, 192
Federal Labor–Management Conflicts, 193
federal legislation on collective bargaining, 190–192
feedback, 146
    in communication cycle, 63, 74, 80
    in evaluating problem solution, 228–229
    providing after incident/activity, 165, 168f
    receiving, 232
FEMA. *See* Federal Emergency Management Agency
FESHE. *See* Fire and Emergency Services Higher Education
FFSL. *See* Fire Fighter Safety and Location program
FGC. *See* fire ground commander
FGL. *See* Fire Ground Location
FGL SCBA interface, 284
finance/administration section, 304, 311
financial management, 275
fire
    alarm and detection systems, 251–252
    analysis, 370, 382
    behavior graph, 340
    cause of. *See* cause of fire
    development phases of, 246
    growth/development of, 365–366
    in industry, 305
Fire Analysis and Research Division, 319
Fire and Emergency Services Definition Act, 195
Fire and Emergency Services Higher Education (FESHE), 152–153

fire attack, 338–359
    supervision and, 342–344
    tactical safety considerations, 349–352
fire chief
    brainstorming alternative solutions and, 226–227
    definition of, 10, 19
    responsibilities of, 11
fire codes
    building *vs.*, 247–248
    classifying by building or occupancy, 252–256
    compliance inspections, 252
    development of, 7
    local, 248
    model, 248
    requirements, 247
        retroactive, 248–249
    reviewing, 258
    state, 248
fire companies
    inspections, 252
    multiple, supervision of, 352–354
    supervision of, 343
fire considerations, structure of, 354, 356
fire department
    administrative procedures, 26
    battalion chiefs, 11, 19
    building strong relationship with media, 214–215
    chain of command, 11, 11f, 19
    combination, 7, 13
    cultural diversity in, 17
    duties for, 17
    functional organization of, 12
    geographic organization of, 13
    mission, 26
        statement, 120
    organization of, 10–13, 12f
    organizational principles
        discipline, 12, 19
        division of labor, 12, 19
        span of control, 12, 20
        unity of command, 11–12, 20
    PIO, 213–215
    reductions, 276
    source of authority, 10–11
    staffing. *See* staffing
    supervision and motivation, 16–17
    volunteer, 204
    working with other organizations, 16
    workplace diversity in, 51–53
Fire Department of New York (FDNY), 276
    deaths in structure fires, 104
    dispatch and alarm assignments, 229
    Fire College, 154
fire extinguishers, portable, 261
fire fighter assist search team (FAST) truck, 90

Fire Fighter Safety and Location program (FFSL), 283
fire fighters. *See also* fire officers
   death and injury trends, 86–92
   expectations of, 52
   to fire officer, transition from, 45–46
   safety and deployment study, 113–114, 195–196
   staying together, 324, 326
   survivability inside structure fires, 319
fire ground commander (FGC), 292
fire-ground communications, monitoring, 327–328, 328f
Fire Ground Location (FGL)
   administrative office, 284
   annual personnel and operating expenditures, 283–284
   capital budget, 284
   program, 283
   uniforms and protective clothing, 284
   vehicle, 284, 285f
fire-ground skills on-the-job training, 148
fire history, review of, 258
fire insurance companies. *See* insurance companies
fire investigations
   documentation/reports, 378–380
   example of, 370
   finding point of origin, 368–369
   legal considerations, 366–368
   nature of, 368
   preliminary documentation, 378–379
   report, 379
   responsibilities of protect fire scene, 375
fire investigator
   request for, 365
   role of, 380
fire loss statistics, Canadian, 9
fire marks, 8, 10, 10f, 19
fire officers
   certification programs, 154
   challenges for, 16–17
   as commander, 47, 47f, 48f
   development, four-step method of, 143–146
   as follower, 128–129
   knowledge of neighborhood, 49
   level I
      administrative duties, 5
      candidates for, 5
      nonemergency *vs.* emergency duties, 5
      requirements of, 4–5
      roles and responsibilities of, 5
   level II, 4, 5
      requirements for, 16
      roles and responsibilities for, 16
   level III, 4
   level IV, 4

   as manager, 112–113
   managing/supervising, 42
   new, 46
   obligation to work with supervisor, 49–50
   poor performance, addressing, 295
   in preparation for live fire training, 150
   problem-solving scenarios, 49
   responsibilities of, 86, 143, 150, 165
      greater alarm, 307–308
   role, 128, 164
      in community fire safety, 242–243, 242f, 243f
      in incident management, 296–297
      as supervisor, grievance procedure, 186–190
      in workplace diversity, 51–53
   as spokesperson, 215–217
   as supervisor, 46–47
   supervisory activities, 184
   tasks for, 42–45
   theory X *vs.* Y for, 115
   as trainer, 47–49, 48f
   training responsibilities, 143–147
   transition from fire fighter to, 45–46
   working with supervisors, 49–50, 50f
fire patterns, 369, 369f, 382
fire prevention division, 261, 267
   coordinating activity with, 258
Fire Prevention Week, 209–210
fire protection district, 11
fire protection systems
   alarms, responding to, 17
   built-in, disabled, 366
   identifying in preincident plan, 245
   water-based, 249–250
fire pumps, 250, 251f
fire research structure, 340–342, 341f
fire resistive construction, 252–253
fire scene reconstruction, 367–368, 382
fire service
   history of, 7–8, 8f
   paying for, 10, 10f
   in United States, 5–10
fire service joint labor–management wellness–fitness task force, 197
fire spread, anticipating areas of, 246
fire sprinkler alarms, 251
fire station
   as business work location, 55–56
   as municipal work location *vs.* fire fighter home, 134–135
   safety, 96–97
   uniforms, 134, 134f
   workday, typical, 44
fire suppression operations, 89–92
fire suppression system, 257
fire tax district, 274, 287

FIrefighting REsources of Southern California Organized for Potential Emergencies (FIRESCOPE), 292
Firehouse (Halberstam), 128
FIREPAC, 195
fireplace ashes, 214
FIRESCOPE. *See* FIrefighting REsources of Southern California Organized for Potential Emergencies
fiscal year, 272, 273t, 287
flashover, 17
floor plans, 244, 267
flow path, 342, 358
FLSA. *See* Fair Labor Standards Act
FMLA. *See* Family and Medical Leave Act
foam systems, 251
focus during communication, 65
followership, 128–129, 138, 391, 397
food preparation activities, 97
forcible entry tool, 96
form of material, 370, 382
formal communications, 68, 72, 81
   announcements, 73
   general orders, 73
   legal correspondence, 73
   recommendation report, 73, 75
   SOGs, 73
   SOPs, 72–73
formal written reprimand, 174–175, 175f, 178
four-step system, 259
frame of reference, 173, 179
Franklin, Benjamin, 7, 8
fuel limited, 340, 341f, 358
fuel load, 246
function, fire department organization and, 12
fund-raising method, 274–275
Fundamentals of Fire Tactics (Layman), 345
fundraising for volunteer fire departments, 274
furloughs (unpaid leave), 194

G

gaming activities, to raise money, 274–275
Garcia v. San Antonio MTA, 194
gas meters, 280
GASB. *See* Governmental Accounting Standards Board
gasoline, amateur arsonists and, 366
general orders, 68, 73, 81
geography, fire department organization and, 13
"Get Your Message Out" media primer, 214
Gilbert v. Homar, 176
global positioning system (GPS), 283
gloves, 96
goal-setting theory, 136
goals for fire fighters, annual, 171–172
good faith bargaining, 191, 198

Governmental Accounting Standards Board (GASB), 281, 287
governments
　　fire department and, 10
　　as source of authority for fire department, 10–11
GPS. *See* global positioning system
Graham's Rules for Enhancing Firefighter Safety (GREFS), 104
grants, 280
grapevine, 65
Gravely, Jack W., 53
Great Chicago Fire of 1871, 7, 8*f*, 209, 210
Great Depression, 194
Great Strike Wave of 1933–1934, 191
greater alarm, 307–308
GREFS. *See* Graham's Rules for Enhancing Firefighter Safety
grid theory, 117
grievances, 186, 198–199, 230
　　form, sample, 190*f*
　　informal, 189
　　procedures, 186–190, 199, 232
group supervisor, 300, 311
groups, 306, 306*f*, 307, 311

H

habits to improve safety, 87
Halberstam, David, 128
halo and horn effect, 174, 179
Halon 1301, 251
handling conflict in teamwork, 391
handouts for fire station visits, 210
hands-on skill drills, 169
harassment complaints, handling, 55
Hawthorne effect, 114
hazardous conditions, investigation of, 17
hazardous materials awareness and operations, 148
hazardous occupancies, 261, 263
hazardous use permit, 261, 267
hazards
　　definition, 93, 107
　　identification signs, 261
　　mitigation, 105
　　special, identifying in preincident plan, 245
HAZWOPER, 148
health and safety officer, 99, 107
Health Insurance Portability and Accountability Act (HIPAA), 76, 81, 122
healthcare occupancies, 255, 263
heart attack, fire fighter fatalities and, 87–88
heat ignition, source and form of, 370
heating, ventilation, and air-conditioning (HVAC) system, 245
heavy timber construction, 253
helmet cams, 77

Helmreich's error management model, 388*f*
Herzberg, Frederick, 136
HFACS. *See* Human Factors Analysis and Classification System
hierarchy of needs, 115–117, 123
high-expansion foam system, 251
high-profile incidents, encountered by fire officers, 225
high-reliability organizations, 395
high-rise structural considerations, for fire attack, 356
high-risk property, 243, 267
high-value property, 243, 267
HIPAA. *See* Health Insurance Portability and Accountability Act
Homeland Security Act, 304
Homeland Security Presidential Directive 5 (HSPD-5), 148, 157, 293
"hook-and-ladder truck," 7
hostile workplace, 53–55
hot zone, 350
hotels, 255
housekeeping, around fire station, 96–97
HSPD-5. *See* Homeland Security Presidential Directive 5
human error, 387–388
Human Factors Analysis and Classification System (HFACS), 103–104
human resource division requirements, 173
human resource management, 118–122
　　development of, 112–113, 123
　　four borders of, 122
　　functions of, 118
　　utilization of, 120–121
human resources development, 120, 123
human resources planning, 118, 123
Human Side of Enterprise, The (McGregor), 114
humanistic management, 114–118
　　definition, 1, 123
　　hierarchy of needs and, 115–117, 115*f*, 123
　　managerial grid and, 117–118
　　theory X and theory Y, 114–115
HVAC system. *See* heating, ventilation, and air-conditioning system
hygiene factors, 136, 138

I

IAFC. *See* International Association of Fire Chiefs
IAFF. *See* International Association of Fire Fighters
IAP. *See* incident action plan
IC. *See* incident commander
ICS. *See* Incident Command System
ideas, sharing, 227
IDLH. *See* immediately dangerous to life and health
IFSAC. *See* International Fire Service Accreditation Congress

ignition
　　factor or cause, 370
　　source of, 370, 381
immediately dangerous to life and health (IDLH), 89, 107, 148, 283, 301, 343
immigrants in U.S. population, 205
impasse, 194, 199
in-basket exercises, 30–31, 30*f*, 37
in-house issues, encountered by fire officers, 225
incendiary fires
　　causes, 374, 375
　　indicators
　　　　accelerants and trailers, 375, 377
　　　　multiple points of origin, 377
incident action plan (IAP), 345, 358
　　assigning resources in, 353–354
　　components, 303, 311
　　definition, 86, 107
　　developing, 89, 348–349, 355
　　elements, 89
　　tactical priorities, 349
　　task assignment determination, 353
Incident Command System (ICS), 245, 352
　　building, 302–304
　　definition, 5, 16, 19, 93, 107, 293, 311
　　during emergency situations, 344
　　general staff, 302, 311
　　levels of command, 297
　　organizational chart, expanded, 303*f*
　　structure, 297*f*
incident commander (IC), 74, 294, 297, 311, 393, 394
incident management
　　developing one system, 293, 293*f*
　　fire officers role in, 296–297
　　origin, 292–293
　　postincident review
　　　　critique, 295–296, 296*t*
　　　　documentation, 296
　　　　follow-up, 296
　　　　information preparation, 294–295
　　strategic-level, 297–301
　　system, building
　　　　command staff, 302
　　　　general staff functions, 302–304
　　　　location designators, 304
　　task level, 308–309
incident safety officer, 92–95
　　assistant, 94
　　definition, 92, 107
　　and incident management, 93
　　knowledge requirements for, 93
　　qualifications, 93
　　tasks, 94
incident safety plan, 94, 107
incident scene rehabilitation, 94, 95*f*, 107

incident stabilization, as incident action plan priority, 348
incidents
    emergency. See emergency incidents
    large/complex, assistant incident safety officers for, 94
    nonfire, 17
    "nothing showing" calls, 298
    prevention, 264
    priorities for incident action plan, 348
    reports, 68
        sample, 70f
        for types of incidents, 69–70
    review
        conducting critique of, 295–296
        documentation, 296
        follow-up, 296
        preparing information for, 294–295
        questions for, 296t
indifference, 117
individual needs, 206
industrial occupancy, 255, 255f
industrial revolution, management concept and, 112–113
infection control program, 97–99
infectious disease exposure, 97–99, 98f
informal communications, 68, 72, 81
informal grievance, 189
informal reprimands, 169, 174
informal work performance reviews, 172, 173
information management, 72
information power, 130
information technology, reports using, 72
infrequent reports, 71–72
initial incident commander, 47, 298–300
initial rapid intervention crew (IRIC), 350, 358
initial report, 67
injury
    patterns, from training, 150
    prevention, 95–96
    reports, 70
inspection
    assembling tools and references for, 258–259
    conduct, 259–260
    preparing for, 258–259
    reports
        prior, review of, 258
        writing, 260
insurance companies, 10
    fire investigations and, 380
insurance investigator, role of, 380
integrity, 50–51
interior work team, 91
internal department issues, encountered by fire officers, 225

International Association of Fire Chiefs (IAFC), 113, 195, 318
    Fire and Emergency Services Definition Act an, 195
    fire progression, structural stability, and survivability comparison by, 322f
    Labor–Management Initiative, 196–197
    professional development, 155
International Association of Fire Fighters (IAFF), 113, 185, 185f, 198, 301
    charter of, 193
    growth as political group, 195
    Labor–Management Initiative, 196–197
    objectives of, 185, 195
    statistical reports, 86
International Fire Service Accreditation Congress (IFSAC), 155
interpersonal interaction exercise, 31–32, 31f, 32f
interrogatory, 73, 81
interviews
    for fire cause determination, 371–373
    with media, 215–217
investigations, 238
    of accident, 99
    definition, 232
    emotions/sensitivity and, 232–234
    of fires. See fire investigations
    follow up, 232
    mode, as command option, 298
    purpose, 232
    report, 233, 233f
    taking action, 232
involuntary transfer or detail, 176, 179
IRIC. See initial rapid intervention crew

J
job-content/criterion-referenced validity, 29, 37
job description
    competencies, evaluation during probationary period, 165
    definition, 26, 37
    sample, 166f
job instruction training, 143, 158

K
Kennedy, John F., 192
Klein, Gary, 392
knowledge, skills, and abilities (KSAs), 26, 29, 37

L
labor
    actions in fire service, 193, 193f
    contract, 184
    distinction from management, 186
    force reductions, as human resource management function, 120
    laws, 122, 192

Labor-Management Reporting and Disclosure Act, 191
labor unions
    bill of rights for, 191
    for fire fighters, 192–193
labor–management alliances, 195
    EMS systems performance measurement, 196
    Fire Fighter Safety and Deployment Study, 195–196
    Fire Service Joint Labor–Management Wellness–Fitness Task Force, 197
    IAFC/IAFF Labor–Management Initiative, 196–197
Labor–Management Initiative (LMI), IAFC/IAFF, 196–197
labor–management relations, 184, 186, 200
    conflicts, 193
    positive, 186
LAFD. See Los Angeles Fire Department
laissez-faire leadership style, 129
Landrum-Griffin Act of 1959, 191, 198
language immersion programs, 206
laptop computers, developing personal library, 48
latent conditions, 387–388, 397
laws, labor, 192
Layman, Lloyd, 345, 349
leaders, confused, indecisive and uncertain, 133
leadership
    authoritative, 344
    challenges, 134–135
    curiosity and, 390
    definition, 13, 19, 128, 138
    at emergency scenes, 129–134, 129f
        critical situations, 132–133
        dispatch center, 133–134
        methods of assigning tasks, 132
        other responding units, 134
    in routine situations, 130
    styles, 129
    triangle, 389, 389f
    in volunteer fire service, 135
Leadership in Energy & Environmental Design (LEED), 262
leading, 13, 19
leave-without-pay (LWOP), 175
LEED. See Leadership in Energy & Environmental Design
legacy dwellings, 341, 358
legacy single-family vs. modern dwellings, 340–341, 342f, 358
legal considerations for fire investigations
    evidence, 367
        protecting, 367–368
    scene, securing, 367
    searches, 366–367
legal correspondence, 73

legal proceedings, expert witness testimony, 379–380
legislative framework for collective bargaining, 190–192
legitimate power, 130
leniency, 173
LEPC. *See* Local Emergency Planning Council
lesson plan, 157
    components, 144
    purpose, 145, 152
letter, preparing for fire chief signature, 73
liaison officer, 302, 311
life safety, as incident action plan priority, 348, 349
Life Safety Code®, 248
life-threatening hazards, 260
lifting techniques, 97
limited, 322, 335
limited combustible (ordinary) construction, 253
line-item budget, 281, 287
listening
    active, 63, 63*f*, 65
    skills, 65
live fire training, 148–151
    fire officer preparation responsibilities, 150
    near-miss report, 151
    prohibited activities, 151
    student prerequisites, 148–150
lobby control officer, 356, 358
"local alarms," 249, 249*f*
local emergency plan, 353–354
Local Emergency Planning Council (LEPC), 244
local fire codes, 248
local government, 204
    sources, 274
locally developed programs, public education, 212–213
logistics, resource management and, 264–265
logistics section, 303, 311
logistics section chief, 303, 311
Los Angeles Fire Department (LAFD), 77, 196
Loudermill hearing, 176, 179
low-expansion foam system, 251
low-pressure warning devices, 91
low-rise multiple-family dwellings, structural considerations for fire attack, 354, 356
lower revenue options, 275–276
LWOP. *See* leave-without-pay

M

MABAS. *See* Mutual Aid Box Alarm System
management
    definition, 112
    distinction from labor, 186
    functions, 13
    humanistic. *See* humanistic management

    performance, 121, 123
    scientific, 113–114, 123
    toolbox, 14
managerial grid, 117–118
managing fire officer, 11, 16, 19, 29, 156
marking system, NFPA 704, 256
Maslow, Abraham H., 115
masonry walls, in heavy timber construction, 253, 267
material, first ignited, 370
mayday, 331, 335
    declaration, 330–331, 333
    vs. emergency traffic, 331, 333
    rescue, 133
Mayo, George Elton, 114
McGregor, Douglas, 114–115
MDC. *See* mobile data computer
measured risk, 323, 335
media relations, 213–217, 214*f*
    fire department PIO, 213–214
    fire officer as spokesperson, 215–217
    press releases, 215, 216*f*
mediation, 186, 199
Medic One paramedic ambulance service, 196
medical examination in preventing heart attacks, 102
medium in communication, 63, 80
meetings, in establishing supervisor/employee relationship, 185–186
memorandum of understanding (MOU), 184
mentoring
    definition, 146, 158
    rookie, 118
    teamwork and, 389
mercantile occupancy, 255, 255*f*, 263
merit-based pay systems, 121
message in communication cycle, 8, 62–63
Metropolitan Fire Department in 1865, 25
Michigan v. Clifford, 366, 367
Michigan v. Tyler, 366, 367
mid-year review, 172
mini/max codes, 247, 248, 267
mission statement, 120
mistakes, 224, 238
mitigation, 264, 267
mixed occupancy, 256, 264
mobile data computer (MDC), 279
model codes, 248, 267
modern *vs.* legacy single-family dwellings, 340–341, 342*f*, 358
monthly activity and training report, 68
morning report, 68, 80
motivation, 16–17, 135–136
motivation factors, 136, 138
motivation-hygiene theory, 136
motor vehicles

    operating costs, 281
    replacement fees, 281
MOU. *See* memorandum of understanding
Mouton, Jane, 117
multiple-choice written examination, 29–30
multiple-fatality fires, 247
Municipal City Fire and Rescue Department, 205
municipal work location *vs.* fire fighter home, 134–135
Murrah Federal Building bombing, 304
Mutual Aid Box Alarm System (MABAS), 276

N

National Aeronautics and Space Administration (NASA), 386, 396
National Board on Fire Service Professional Qualifications (NBFSPQ), 155
National Fallen Firefighters Foundation (NFFF)
    Everyone Goes Home® program, 86–87
    situational awareness, 93
National Fallen Firefighters Memorial Weekend, 102
National Fire Academy (NFA), size-up process, 346–347
National Fire Incident Reporting System (NFIRS), 17, 69, 81, 296, 364
National Fire Prevention Week, 210, 210*f*
National Fire Protection Association (NFPA)
    intentional fire statistics, 377
    marking system, 256, 256*f*
    NFPA 704 marking system, 256, 256*f*
    NFPA 1021 standard, 4
    statistical reports, 86, 87
National Fire Service Incident Management System Consortium, 293
National Incident Management System (NIMS), 67, 148, 293–294, 311
National Industrial Recovery Act (NIRA), 191
National Institute for Occupational Safety and Health (NIOSH), 75, 86, 91
National Institute of Standards and Technology (NIST), 196, 343*f*, 356
National Labor Relations Board (NLRB), 191
National Near-Miss Reporting System, 327
National Professional Development Model, 155, 155*f*
national public education programs, 209–212
National Response Framework (NRF), 293, 304, 306, 311
natural fire causes, 374–375
naturalistic decision making, 392, 397
NBFSPQ. *See* National Board on Fire Service Professional Qualifications
near-miss reports
    analyzing, 103–104
    national fire fighter, 87

Index

needs
  assessment, for training program, 151
  hierarchy of, 115–117, 115f, 123
negative reinforcement, 135
negotiation, 184, 199
neighborhood, knowledge of, 49, 49f
news releases, preparing, 76
Newsham, Richard, 8
NFFF. See National Fallen Firefighters Foundation
NFIRS. See National Fire Incident Reporting System
NFPA. See National Fire Protection Association
NIMS. See National Incident Management System
NIOSH. See National Institute for Occupational Safety and Health
NIRA. See National Industrial Recovery Act
NIST. See National Institute of Standards and Technology
Nixon, Richard, 192
NLRB. See National Labor Relations Board
"no surprises" rule, 44
noise interference in communication, 66
noncombustible construction, 253
nonemergency activities, rules for, 134
nonfire incidents, 17
nonhostile environment, 54
non–life-threatening code enforcement issues, 260
nontraditional revenue sources, cost recovery, 280–281
Norris-LaGuardia Act of 1932, 191, 198
"nothing showing" calls, 298
notification of fire, 366
NRF. See National Response Framework

O

objectives, establishing for training program, 151–152
objects, fatal contact with, 102
occupancy, classification, 252–256
occupancy type
  definition, 254, 268
  identifying special/unusual characteristics for, 246–247
occupant considerations, identifying during preincident planning, 244–245
occupant survival profile, determination of, 320–321
Occupational Safety and Health Administration (OSHA), 89, 147, 301
odor investigations, 17
"off the record" remarks, 217
offenses, predetermined disciplinary policies for, 176
offensive operation, 347, 358
Oklahoma City Murrah Federal Building bombing, 304
on-duty speech, 54, 55
on-duty wallet, 350

on-scene activity, 148
on-scene observations for size-up, 345
on-scene size-up, initial, 67
on-the-job digital images, 77
on-the-job training, 147–148
One Thousand Strikes of Government Employees (Ziskind), 194
ongoing compliance inspection, 258, 268
open shops, 192
opened structure, 319, 335
operating expenditures, 281–284, 282f
operating system, computer, 72
operational period, 89, 107
operations section, 303, 311
operations section chief, 303, 312
oral interview for promotional examination, 32, 32f
oral reprimand, warning or admonishment, 169, 179
orders
  model for emergency communications, 67
  need for, 116
  unpopular, 46–47, 47f
ordinance, 248, 268
organizational influences in near-miss report, 103–104
organizational structure, management concepts and, 113
organized labor, 184, 194, 196
organizing, as management function, 13, 20
OSHA. See Occupational Safety and Health Administration
overhaul, 349, 379
overtime pay, 195

P

PAC. See political action committee
Page, James O., 46, 48
paid time off (PTO), 277
PAR. See personnel accountability report
paraphrasing, 232
PASS. See personal alert safety system
PATCO. See Professional Air Traffic Controllers Association
PATCO Air Traffic Controllers' Strike of 1981, 193
pay, overtime, 195
pay systems, 121
Pendleton Civil Service Reform Act, 24
perception-based errors, 103
performance
  in context (whole-skill training), 151
  feedback, 165, 168f
performance log, 171, 172, 172f, 173, 179
performance management, human resource management function, 121, 123
personal alert safety system (PASS), 157, 283, 301
personal bias, 173, 179
personal competence, 389

personal journal, 48
personal protective equipment (PPE), 96, 350
personal study journal, 33, 35, 37
personal training library, development of, 48, 48f, 49f
personal wire cutters, 96
personnel accountability report (PAR), 90, 352, 358
personnel accountability system, 90, 90f, 107
personnel conflicts, 230
personnel expenditures, 281, 283–284
Peshtigo firestorm, 7
petty cash, 277
Phoenix Fire Department
  relations by objectives, 196
  training drill scenario, 92
physical elements, identifying for preincident plan, 244
physical fitness
  fire fighter, 88
  programs, in preventing heart attacks, 87, 88f
physiological needs, 115
PIO. See public information officer
plan B, 227–228
plan of operation in size-up process, 346
planning, as management function, 13, 20
planning section, 303, 312
planning section chief, 303, 312
plot plans, 244, 268
PO. See purchase orders
points of origin
  definition, 368, 382
  finding, 368–369
  fire growth/development and, 365–366, 380
  fire patterns and, 369, 369f
  multiple, 377
police department, 354
policy/policies
  changes in, 234
  definition, 14, 20
  implementation, 234–235
  recommendations, 234
political action committee (PAC), 195, 199
portable fire extinguishers, 261
portfolio management, 275
positive pressure ventilation (PPV), 353
positive reinforcement, 135
post-incident review, preparing information for, 294–296
Post-Katrina Emergency Management Reform Act, 293
postal workers strike of 1970, 193
postincident analysis, 99
power
  definition, 129, 138
  types of, 130
PPE. See personal protective equipment

PPV. *See* positive pressure ventilation
"practice like you play," 132
pre-incident information, for size-up process, 346–347
preaction sprinkler systems, 250
prearrival information, 345
preburn plan, for live fire training, 150
preconditions, to unsafe acts, 103
predisciplinary conference, 176–177
preincident plan, 107
    definition, 243, 268
    example, 243f
    goal, 247
    purpose, 243
    putting data to use, 247
    review, 258
    systematic approach, 244–247
    written report with, 247
preparation
    for being official spokesperson, 215–217
    for fire officer training, 143–145
presentation
    of fire fighter training, 145
    of report, 75–76
presentation software, 72
press releases, 215, 216f, 217
pretermination hearing, 176, 179
Principles of Scientific Management (Taylor), 113
privatization, 276
proactive outreach, 215
probabilities in size-up process, 346
probationary period, 165, 176
problem solutions
    alternative, generation of, 226–227
    changing plan, 228
    deadline for, 227
    evaluating results, 228–229
    implementing, 227
    plan B, 227–228
    selecting, 227
problem-solving process
    legitimate, 227
    people involved in, 227
    for training, 49
problems, 225f, 238
    definition, 224
    emergency incidents. *See* emergency incidents
    encountered by fire officers, 225
    identifying, in developing local public education programs, 212
procedures, for handling harassment and hostile workplace complaints, 54
Professional Air Traffic Controllers Association (PATCO), 193
professional development, 152, 158
    academic accreditation, 153–154
    education, cultural change issues in, 153

fire fighter certification programs, 154
IFSAC, 155
NBFSPQ, 155
plan, building, 155–156
training vs. education, 152–153
proficiency, of existing skill sets, 146
profit as arson motive, 377–378
progressive negative discipline, 174, 179
project control document, 227, 228f
promotion
    to chief officer, 46
    preparing for, 23–39
    sizing up opportunities for, 25–26
    test preparation committee, 26, 29
promotional examinations
    components, 29–32
        assessment centers. *See* assessment centers
        emergency incident simulations, 31, 31f
        interpersonal interaction, 31–32, 31f, 32f
        multiple-choice written examination, 29–30
        technical skills demonstration, 32
        writing/speaking exercise, 32, 32f
    developing process, 24
    dimensions, 25, 37
    eligibility lists
        banded, 25
        rank ordered, 25
    mastering content, 33
    oral interview for, 32, 32f
    origin, 24
    postexamination considerations, 25–26, 25f
    preparation, 26–29, 33
    role-playing exercises, 35–36
    variations in, 25
property conservation, as incident action plan priority, 348
property damage reports, 70
protective clothing, 96, 284
psychomotor skill levels, 147
PTO. *See* paid time off
public assembly occupancy, 261–262
public education programs, 207–213
    community information, 205
    delivery, 209
    goal, 209, 209f
    locally developed, 212–213
        designing, 213
        evaluating, 213
        identification, 212
        implementing, 213
        method selection, 212
    national and regional, 209–212
        CERT, 210–211
        Fire Prevention Week, 209–210
        Risk Watch, 210
    objectives, 209
public employees, collective bargaining rights for, 192

public information officer (PIO), 213–215, 302, 312, 344
public inquiries, responding to, 206–207
public safety education messages, 210
punishment, 135
purchase orders (PO), 277, 279
purchasing process, 276
    bidding process, 279
    petty cash, 277
    purchase orders, 278, 279
    requisitions, 279
pyrolysis, 374, 382

Q

questions
    "data dump," 31, 37
    for incident review, 296t
    for interviews in fire cause determination, 371
    supervisory on promotional examination, 29
    technical on promotional examination, 29

R

radio
    messages, 67, 80
    reports, 67–68, 298, 299
    system testing, during preincident planning, 244
RAND Corporation, 228
rapid assist team (RAT), 90
rapid intervention company operations (RICO), 90
rapid intervention crew (RIC), 90, 107, 301, 350, 352, 358
rapid intervention team (RIT), 6, 301
RAT. *See* rapid assist team
RBO. *See* relations by objectives
Reagan, Ronald, 193
real estate management, 275
"reasonable person" standard, 53
receiver in communication cycle, 63
recency, 173, 179
RECEO VS, 349
recognition-primed decision making (RPD), 392, 397
recommendation report, 73, 75, 81
record keeping, employment-related, 171
record of infectious disease exposure, 97
recordings of radio messages, 67
recruitment
    probationary period for, 165
    in volunteer fire departments, 6
red flag, 329
references for inspection, 258–259
referent power, 130
reflexive explanation, 232
regional public education programs, 209–212
regionalization of services, 276
regulations, 252, 268
rehabilitation, 94, 95f, 107

Index

reinforcement theory, 135–136, 138
relations by objectives (RBO), 196
reliability, 29, 37
reporting, 68–72. *See also* reports
reports
    infrequent, 70–71
    presenting, 75–76
    routine, 68, 69
    special, 70
    types of, 68–72
    using information technology, 72
    verbal, 68
    writing, 75
representative, meeting with, 259
reprimands, 169
request for proposal (RFP), 279
requisitions, 279
rescue, 349
residential fire deaths, injuries and, 207
residential occupancy, 255–256, 263
resource management, logistics and, 264–265
resources, assigning, 353–354
respect, 50
respiratory injuries, fire fighter deaths from, 102
responding to alarms, 148
responsibility in CRM model, 391
restrictive duty, 176, 179
retention in volunteer fire departments, 6
retirement, 25, 25*f*
retroactive code, 248–249
revenge as arson motive, 378
revenue
    definition, 272, 287
    from grants, 280
    local government sources, 274
    nontraditional sources, 280–281
    for volunteer fire departments, 274–275
reviews, postincident, 294–296
reward power, 130
RFP. *See* request for proposal
RIC. *See* rapid intervention crew
RICO. *See* rapid intervention company operations
right to work, 192, 192*t*, 199
risk assessment, 264, 268
risk management, 104, 107
risk-reduction programs, 206
Risk Watch, 210, 219
risk–benefit analysis, 91–92, 107
    in size-up process, 347–348
RIT. *See* rapid intervention team
ROE. *See* Rules of Engagement
role-playing exercises, preparing for, 35–36
Roman Empire, fire service in, 7
rookie officers, assigning mentors for, 118
rooming houses, 255
routine reports, 68, 69
routine situations, leadership in, 130

routine violations, 103
RPD. *See* recognition-primed decision making
Rules of Engagement (ROE)
    definition, 318, 335
    Rules, 320–333
    scope of problem, 319
rules/regulations, 14, 20

S

safety
    community information, 205
    enhancing fire fighter, 104
    fire fighter and deployment study, 113–114, 195–196
    fire station, 96–97
    hazards in fire station, 96
    need for, 115–116
    policies and procedures, 95
    vehicle, 90
safety culture, creation of, 387
safety officer, 302, 312
safety-related items, for fire fighters, 96
safety unit, 94, 107
salvage, 349, 379
savable lives, risk extended to protection, 323–324
savable property, risk extended to protection, 322–323
SCBA. *See* self-contained breathing apparatus
scene of fire
    initial on-scene size-up, 67
    safety considerations, 350
    securing for fire investigation, 367
scientific management, 113–114, 123
searches, legal aspects, 366–367
seat belts, 88
Seattle Medic One paramedic ambulance service, 196
security, need for, 115–116
self-actualization, 117
self-assessment of followers, 391
self-contained breathing apparatus (SCBA), 89–91, 89*f*, 96, 282, 331, 350
    air supply, 326, 326*f*, 327
    fit testing, 148
self-evaluation, before annual evaluation, 173
sender in communication cycle, 63, 80
service branch, 356, 358
Service Life Extension Program (SLEP), 275
severity, 173
sexual harassment, 53–54
Sherman Act, 193
shift report, beginning of, 42–43, 42*f*, 43*f*
shop steward, 184, 199
single-family dwellings, structural considerations for fire attack, 354
site considerations, identifying for preincident plan, 244

situation assessment in size-up process, 346
situational awareness, 91, 93, 257, 327, 335, 392–394, 397
size-up process, 344, 358
    elements of, 320
    initial, 347
    Layman's five-step, 345–346
    National Fire Academy, 346–347
    on-scene observations, 345
    ongoing, 347
    prearrival information, 345
skill-based errors, 103
skills
    communication, 63–66, 388–389
    fire-ground, on-the-job training, 148
    necessary to stay alive, 148
    new/revised, provision of, 146–147
skills-based pay systems, 121
SLEP. *See* Service Life Extension Program
SMACSS. *See* social media–assisted career suicide syndrome
smoke
    on-scene observations, 345
    volume/color, 347
social competence, 389
social media, 76, 81, 217, 219
    challenges, 217
    engaging community through, 77, 79
    helmet cams and on-the-job digital images, 77
    outreach goal of, 217
    policy, fire department, 76–77, 78*f*
social media–assisted career suicide syndrome (SMACSS), 217
social needs, 116
sociological environmental noise, 66
SOGs. *See* standard operating guidelines
SOPs. *See* standard operating procedures
sound behavioral model, 118
source of ignition, 370, 382
souvenirs for fire station visits, 210
span of control, 12, 20
Spanish as primary language, 206
speaking exercise in promotional examination, 32
special evaluation period, 176, 179
special occupancies, 246–247
"special properties" class, 263
speech, on-duty, 54, 55
spite as arson motive, 378
spoils system, 24, 37
spreadsheets, 72
sprinkler systems, 346
    activation, 250
    detention, 263
staffing
    fire department organization and, 13
    for fire department organizations, 7

staffing (*Continued*)
    as human resource management function, 118, 120, 123
    in shift report, 42–43
    strikes for, 194
Stafford Act, 293, 294*f*
staging, 308, 312
stairwell support group, 356, 358
standard operating guidelines (SOGs), 14, 73, 81
standard operating procedures (SOPs), 132, 344, 348
    definition, 14, 20, 72–73, 81
    during emergency situations, 132
    in reducing deaths from fire suppression operations, 89
standard time–temperature curve, 340, 358
standpipe systems, 250, 268
Stapleton, Leo D., 48
state fire codes, 248
state fire marshal, organizational patterns, 248
state labor laws, 192
station uniforms, 96
status, need for, 116–117
status quo, 118
Stop, Drop, and Roll, 209, 209*f*, 210
storage occupancy, 256, 263–264
strategic level, command, 297, 312
strategic-level incident management
    definition, 297
    establishing command, 298
    fire fighter accountability, 300–301
    functions of command, 299–300
    responsibilities of command, 297–298
    transfer of command, 300
strategies, 348
strike team leader, 308, 312
strike teams, 308–309, 309*f*, 312
strikes
    for better working conditions, 194
    definition, 193, 199
    historical examples, 193
    impact on public safety, 194
    negative impact, 194
    to preserve wages, 194
    for staffing, 194
    by state employees, 192
structure fires
    with no surviving occupants, 320*f*
    U.S. fire fighter *vs.* civilian fire fatalities, 319*f*
student prerequisites for live fire training, 148–150
study guides, 35
subobject code, 281
substandard employee evaluation, 172–173
sudden cardiac arrest
    fire fighter fatalities from, 101–102
    reducing deaths from, 87–88
supervising fire officer, 5, 11, 20, 29

supervision
    closeness of, 343
    direct, 122, 123
    fire department, 16–17
    of fire fighters, 164
    of fire officer training/preparation, 155
    of multiple companies, 352–354
    of single company, 342–343
    unsafe, 103, 104*t*
supervisors
    accident report for worker's compensation, 71*f*
    communicating with, 65
    and employee relationship, establishing, 185–186
    fire officer working with, 49–50, 50*f*
    fire officers as, 46–47, 186–190
supervisor's report, 70, 71*f*, 81
supervisor–subordinate relationship, 50
supplemental budget, 272–273, 287
support branch, 356, 358
survivability profiling, 319
suspension, 174–176, 179
Swiss cheese model, 387, 387*f*
systemic needs, 206

T
T-account, 171, 172, 172*f*, 173, 179
tactical level
    definition, 297, 312
    division/group/unit supervisor responsibilities, 307
tactical priorities, 353, 353*f*
    for incident action plan, 349
tactical safety considerations, 349–352
tactical worksheet, 299, 299*f*, 312
Taft-Hartley Labor Act of 1947, 191, 198
Tammany Hall, 24, 25
task assignment, determination of, 353, 353*f*
task force, 308, 312
task force leader, 308, 312
task-level of incident management, 308–309, 312
tasks
    allocation, in CRM model, 391–392
    assigning, 132
tax revenues, local, 274
Taylor, Frederick Winslow, 113
teamwork, 168, 389–391
technical class specification, 26
    worksheet, 27*f*–28*f*
technical skill demonstration in promotional examination, 32
telecommunication companies, renting land from volunteer fire departments, 274
television interviews, tips for, 217
termination, 174, 176, 179
testimonial evidence, 367, 381, 382
textbook study guides, 35

theory X, 115, 123
theory Y, 115, 123
thermal imaging device, 91, 91*f*
third-party inspectors, 257
13R automatic sprinkler systems, 354, 356
Thirty Years on the Line (Stapleton), 48
time efficiency, improvement of, 121
time studies (time and motion studies), 113
Title VII of Civil Rights Act, 51
tools for inspection, 258–259
tracking system, for occupants, 245
trailers, 375, 377, 382
trainer, fire officer as, 47–49, 48*f*
training, 140–158
    company-level, 143
    costs, 284
    definition, 142, 158
    vs. education, 10, 152
    for fire officers, 13, 392*f*
    injury patterns from, 150
    knowledge of neighborhood and, 49
    for managing fire officer, 156
    on-the-job, 147–148
    overview, 142–143
    programs
        delivery of, 152
        evaluating impact of, 153–154
        skills necessary to stay alive, 148
        specific, development of, 151–152
    report, monthly activity and, 68
    responsibilities of fire officers, 143–147
    use of problem-solving scenarios for, 49
transfer of command, 300
transitional operation, 348, 358
traumatic injuries, fire fighter fatalities from, 102
trouble alarm for fire sprinkler, 251
trumpet, chief's, 9, 9*f*, 19
trust, 50, 117
    fear *vs.*, 226
Tweed, William March "Boss," 25
207(k) exemption, 195
"two-in, two-out rule," 89–90, 301, 312
    overriding, 91
type of material, first ignition, 370, 382

U
U-shaped fire pattern, 369, 381
unconsciously competent, 143, 158
Underwriters Laboratory (UL), 320
undetermined fire causes, 374, 375
unfair labor practices, 191, 199
Uniformed Services Employment and Reemployment Rights Act (USERRA), 122
uniforms
    cleaning, 350
    FGL, 284
    formal *vs.* informal, 134, 134*f*

neatness of, 259
replacement costs, 281–282
United States
fire service in, 5–10
fire statistics, 8
units, 306, 306f, 307, 312
unity of command, 11–12, 20
unsafe acts, 103
unsafe conditions
immediate correction of, 165
reporting, 328–329
unsafe practices, reporting, 328–329
unusual occupancies, 256
urban areas, fire departments in, 204
U.S. Fire Administration (USFA), statistical reports, 86
U.S. fire fighter vs. civilian fire fatalities, 319f
U.S. postal workers strike of 1970, 193
use groups
classification, 256t
definition, 254, 268
inspection concerns for, 261–264
USERRA. See Uniformed Services Employment and Reemployment Rights Act
USFA. See U.S. Fire Administration

V

V-shaped fire pattern, 369, 369f, 381
vandalism as arson motive, 378
vehicle fires, determining cause of, 372–373, 373f
vehicle identification number (VIN), 373
vehicles. See motor vehicles
ventilation, 353
ventilation group, 307
ventilation limited, 340, 341f, 358
verbal presentations, 75, 80
verbal reports, 68
videos in reinforcing safety policies, 95
vigilant, 323, 335
VIN. See vehicle identification number
violations, 103
visit, arranging before inspection, 258
volunteer duty night, 45
volunteer fire departments, 6, 11, 204
leadership in, 135
periodic evaluations in, 164
probationary programs in, 165
revenue sources, 274–275
staffing for, 13
volunteer participation, phases of, 135

W

wages
minimum standards, 194, 198
strikes for, 194
Wagner-Connery Act of 1935, 191, 198
walk-throughs, 49
"walking the talk," 168, 168f
wallet, on-duty, 350
warm zone, 350
water-based fire protection systems, 249–250
water flow alarm, 251
water supply
identifying, in preincident plan, 245
need, calculation of, 150
websites, unofficial, 76
Wellness–Fitness Initiative (WFI) task force, 197
wet chemical extinguishing systems, 251
wet-pipe systems, 250
WFI task force. See Wellness–Fitness Initiative task force
Whittier Narrows earthquake in 1987, 210
wildland fire, cause of fire determination for, 373–374, 381
wildland–urban interface (WUI), 373
Wingspread Conference on Fire Service Administration, 152
witness
expert, testimony of, 379
interviews
for accident investigation, 99
for fire cause determination, 370–372
wood frame construction, 253–254
Worcester Polytechnic Institute (WPI), 113
word processors, 72
work environment
as municipal location vs. home, 134–135
safe, creating and maintaining, 95–97
work improvement plan, 172–173, 179
work performance reviews, informal, 172, 173
workday, fire station, typical, 44
workforce, reduction in, 276
working in community. See community, working in
workplace diversity
actionable items and, 53, 57
fire officer's role in, 51–53
workweek, average, 195
WPI. See Worcester Polytechnic Institute
written communication, 72–75
written reports
incident reports, 68–70
infrequent reports, 70–71
monthly activity and training report, 68
morning report to administrative fire officer, 68
routine reports, 68
written reprimand, 174–175, 175f
WUI. See wildland–urban interface

Y

yellow dog contracts, 191, 199
Yukl, Gary A., 128, 130

Z

zero-clearance chimneys, 356

2003 — 1 - I was born Dec 3, 2003 on a Wednesday. In south miami hospital. My parents brought me home to a small little apartment in Kendall.

2007 — 2 - My sister was born Jan 13, 2007. I was very excited to gain a baby sister. At that time we lived in Homestead.

2008 — 3 - My first day of kindergarten was fun. I was happy because all my friends where there. Ms. Cordiva was a very nice teacher.

2013 — 4 - In the summer of 2013 we went to Hawaii for family vacation. While in Hawaii I visited Pearl Harbor. I also was able to see a live volcano.

2013 — 5 - This was my first year of playing tackle football. I was awarded most Inproved player. Me and my team made it to the semi-finals

2014 — 6 - In basketball I was given MVP. I like basketball more than football.

My fi season was a diasaster we lost in the first playoff game.

7- These was the first time i been to rapids. We went on big thunder. we took air on one of the slides.

7- First day of 5th grade was cool.
2014 I meet a lot of new friends. I received honor roll the entire year.

2015 8- Going into middle school was challenging. Getting around is hard. I'm enjoying all of it.